Animal Genomics for Animal Health

Developments in Biologicals

Vol.132

This series"Developments in Biologicals" begins vith vol.102 and is the continuation of IABS symposia series "Progress in Immunobiological Standardization", Vol 1-5, "Immunobiological Standardization", Vols 1-22 and "Developments in Biological Standardization", Vols 23-101.

Organized and published by the International Association for Biologicals (IABS)

KARGER

Basel · Freiburg · Paris · London · New York · Bangalore · Bangkok · Shanghai · Singapore · Tokyo · Sydney

Animal Genomics for Animal Health

Paris, France
25-27 October 2007

Volume editors

Marie-Hélène Pinard
INRA/EADGENE

Cyril Gay
USDA-ARS

Paul-Pierre Pastoret
BBSRC/OIE

Betty Dodet
Dodet Bioscience

Technical editor

Marissa Vicari

Proceedings of an International Symposium organized by:
- Agricultural Research Services (ARS)
- Biotechnology and Biological Sciences Research Council (BBSRC)
- European Animal Disease Genomics Network of Excellence for Animal Health and Food Safety (EADGENE)
- Institut National de la Recherche Agronomique (INRA)
- International Association for Biologicals (IABS)
- World Organisation for Animal Health (OIE)

72 figures and 72 tables, 2008

KARGER

Basel · Freiburg · Paris · London · New York · Bangalore · Bangkok · Shanghai · Singapore · Tokyo · Sydney

••••••••••••••••••

Developments in Biologicals

Library of Congress Cataloging-in-Publication Data

International Symposium on Animal Genomics for Animal Health (2007 : Paris, France)
Animal genomics for animal health : Paris, France, 25-27 October 2007 / volume editors, Marie-Hélène Pinard ... [et al.] ; proceedings of an international symposium organized by Agricultural Research Services (ARS) ... [et al.].
p. ; cm. -- (Developments in biologicals, ISSN 1424-6074 ; v. 132)
This volume presents the proceedings of an international symposium held at the World Organization for Animal Health (OIE) Headquarters, Paris, France, 23-25 October 2007.
Includes bibliographical references.
ISBN 978-3-8055-8619-1 (soft cover : alk. paper)
1. Animal genetics--Congresses. I. Pinard, Marie-Hélène. II. United States. Agricultural Research Service. III. Title. IV. Series.
[DNLM: 1. Animals--Congresses. 2. Genomics--methods--Congresses. 3. Animal Diseases--genetics--Congresses. 4. Animal Diseases--prevention & control--Congresses. W1 DE997VKF v.132 2008 / QH 432 I61a 2008]

QH432.I575 2007
636.089'6042--dc22

2008028380

Bibliographic Indices. This publication is listed in bibliographic services, including Current Contents® and Index Medicus.

Drug dosage. The authors and the publisher have exerted every effort to ensure that drug selection and dosage set forth in this text are in accord with current recommendations and practice at the time of publication. However, in view of ongoing research, changes in government regulations, and the constant flow of information relating to drug therapy and drug reactions, the reader is urged to check the package insert for each drug for any change in indications and dosage and for added warnings and precautions. This is particularly important when the recommended agent is a new and/or infrequently employed drug.

Worldwide distribution by S. Karger AG, Basel (Switzerland).
Printed in France by IML, Saint-Matin-en-haut (France).
ISBN 978-3-8055-8619-1

The scientific integrity of the articles contained in this volume is the sole responsibility of the authors.

Steering Committee

Cyril Gay (Chairperson), USDA-ARS
Michel Lombard, IABS
Paul-Pierre Pastoret, BBSRC/OIE
Marie-Hélène Pinard, INRA/EADGENE
Bernard Vallat, OIE

Scientific Committee

Marie-Hélène Pinard-Van Der Laan, INRA, EADGENE, France (Chair)
Cyril Gay, USDA-ARS, USA (Vice Chair)
Stephen Bishop, Roslin Institute, UK
Abdenour Benmansour, INRA, France
Patrick Chardon, INRA, France
Hans Cheng, ARS-USDA, USA
Luiz Coutinho, University of Sao Paolo, Brazil
Lou Gasbarre, ARS-USDA, USA
Liz Glass, Roslin Institute, UK
Susan Lamont, Iowa State University, USA
Frederick C. Leung, University of Hong Kong, China
Bonnie Mallard, University of Guelph, Canada
Paul-Pierre Pastoret, BBSRC, UK; OIE, France
Mari Smits, ID-Lelystad/WUR, Netherlands
Gene Wijffels, CSIRO, Australia

Organisation of the Proceedings

Preface

Opening Remarks

Session I	Introduction
Session II	Review of Genomics and its Applications
Session III	Using Animal Genomics to Understand Animal Diseases
Session IV	The Search for Markers to Select for Desirable Animal Traits
Session V	Genomics in the Development of New Tools to Prevent and Control Animal Diseases
Session VI	Advancing Animal Genomics in Animal Health

Acknowledgements

The organizers of the International Symposium on Animal Genomics for Animal Health are grateful to the following organisations for their support:

Agricultural Research Services (ARS)
Biotechnology and Biological Sciences Research Council (BBSRC)
Cooperative and State Research, Education, and Extension Services (CSREES)
European Commission
European Animal Disease Genomics Network of Excellence for Animal Health and Food Safety (EADGENE)
Institut national de la recherche agronomique (INRA)
International Association for Biologicals (IABS)
World Organisation for animal Health (OIE)

Industry
Alpharma
Aviagen
AviTech
Cobb
France Hydrides
Genome Canada
Hendrix Genetics
International Federation for Animal Health (IFAH)
Intervet
Merial
Novus
Pfizer

Contents

Pinard M-H, Gay C, Pastoret P-P, Dodet B (eds): Animal Genomics for Animal Health. Dev Biol (Basel). Basel, Karger, 2008, vol 132, p XIII.

Preface

The era of the genome and the ensuing field of genomics are upon us. But can the animal health research community benefit from this new field of research that promises to revolutionize biomedical research? The input received from the 260 scientists from 28 countries that participated in the first international symposium on animal genomics dedicated to animal health was a resounding yes. Recent advances in genomics are now providing unprecedented opportunities for understanding the links between the genetic code and diseases. While it took 13 years to sequence the human genome, significant breakthroughs in sequencing technologies are providing the whole genome sequences of animals at record speed: the draft genome sequences of the chicken in 2005, dog in 2006, cattle and horse in 2007, and the cat and pig genomes in the queue in 2008 and 2009, respectively. Although it is abundantly clear that animal genome sequences alone are insufficient and that many of the genomic tools and infrastructure for animal health research need to be developed, we are now at a crossroad where we can begin to move animal health research to new dimensions and start to address some of the most difficult problems in animal health.

This volume presents the proceedings of this international symposium held at the World Organization for Animal Health (OIE) Headquarters, Paris, France, 23-25 October 2007. The primary aims of the conference were to identify key challenges for integrating animal genomics in animal health research and to bring together worldwide leaders in the fields of animal genomics and animal health to plan new directions that will fundamentally change the way we approach animal health research. Exceptional collaborations between these two communities will pave the way for addressing difficult problems in animal health by connecting those who are currently engaged in cutting edge genomics research with animal health scientists that are disease experts and understand the real world challenges facing animal health.

The proceedings include a review of the state of the art in animal genomics and animal health research, and the contribution of animal genomics in the construction of animal disease models. Research projects using animal genomics to understand animal diseases, disease susceptibility, desirable animal traits and the development of new tools to prevent and control animal diseases are presented herein. Importantly, the results of a round table discussion and a questionnaire given to the participants of this symposium provide four pillars to advance this field of research and key recommendations that will guide the way forward.

Cyril G. Gay
Paul-Pierre Pastoret
Marie-Hélène Pinard-van der Laan

Pinard M-H, Gay C, Pastoret P-P, Dodet B (eds): Animal Genomics for Animal Health. Dev Biol (Basel). Basel, Karger, 2008, vol 132, pp XV-XVI.

Opening Remarks

Dear Delegates,

It is a great pleasure to welcome you to this symposium on Animal Genomics for Animal Health.

The number of participating delegates in itself makes this symposium a success: over 250 delegates from 35 countries! I also welcome participants from industry, small and medium enterprises (animal breeding, animal health industry), associations, the European Commission and others.

During this conference, we will be discussing new societal demands and major questions, of both local and worldwide importance, that must be addressed.

In the late 1960s, the success of sanitary regulations, vaccines and antibiotics led many to believe that infectious diseases were no longer a threat to livestock. We now know that this was premature optimism and that infectious diseases are here to stay. In fact, the repertoire of microbes is so large that, through mutation, deletion, recombination, and the rapid exchange of genetic elements, they have the capability to continuously change their composition. Thus, under appropriate conditions, novel animal or human pathogens and new diseases may be generated. The globalization of all aspects of contemporary life, including farming practices, transport and trade, associated with a net trend towards global warming, provide a fertile environment for this to happen because accelerated changes in ecosystems continuously disturb the equilibrium between pathogens and their hosts.

Over the past decade, microbiology and infectious disease research has undergone a profound revolution through the application of genomics techniques. In less than a decade, genomics has provided the blueprints of thousands of pathogens, and more importantly an ability to explore the global diversity of living organisms.

The idea of genetic improvement for disease resistance of domestic animals has progressively been developed since 1970, based on population genetics and the first available genetic maps. In the past decade, this field was profoundly renewed by the genomic revolution and substantial progress is expected with the increasing availability of the genome sequence of several livestock species.

However, we must not forget that natural variation in disease resistance essentially refers to the evolutionary 'arms race' between pathogenic microbes and the hosts they target. We also have to bear in mind that significant genetic variation can involve everything from a single base change to millions of bases. This sets the framework for the formidable complexity researchers must embrace. The ability and power of genomics to explore the global diversity of living organisms now offers new research opportunities in this field. It is now possible to monitor changes in the pathogen genome and host immune response as they happen, with the ultimate goal of predicting pathogen evolution in response to host related factors.

As technology rapidly improves and the number of whole-genome sequences available for research multiplies, significant computational challenges inevitably arise.

The French National Institute for Agricultural Research (INRA) has a prime interest and a long tradition of research in the fields of animal genetics and animal health for the improvement of livestock performance, including better resistance to disease, and for animal product safety. It has naturally made significant investments in the development of basic research and new technologies. In this regard the INRA policy is translated on a day-to-day basis through the longstanding research activities of the Animal Genetic and Animal Health Divisions. In addition, special, large trans-disciplinary incentive actions were also launched in the past decade. Starting with the in-house AGENA programme in the 90's, these incentive actions are now continued with the GENANIMAL programmes funded at a national level through the National Research Agency (ANR). Special attention has also been given to support core disciplines mandatory for the development of genomics (mathematics, statistics, and bioinformatics) and to provide researchers with state of the art support through dedicated technological platforms (SIGENA, computing animal genomics; PICT, transcriptomics; GADIE, animal genome resources and genomics; AGMIAL, microbe genomics, etc).

At the European Union (EU) level, INRA has a sustained and proactive policy and was at the origin of the EadGene network of excellence, which is meant to overcome fragmentation and promote multidisciplinary research in the field of genomics of host-pathogen interactions in animals. Notably, EADGENE is coordinated by an INRA researcher, Marie-Helene Pinard Van der Laan, who is also chairing the Scientific Committee of the current symposium and very actively involved in its Steering Committee. INRA is also actively involved in the EU Global Animal Health Technology platform, as well as in the future EU funded ERANet network for animal health, of specific relevance for the topic of this symposium.

The Scientific Committee has created a very exciting programme, which encompasses every field of the application of genomics to animal health and especially to infectious diseases.

I would like to thank and to congratulate them on your behalf, especially Marie-Hélène Pinard-Van der Laan (INRA), Chair, and Cyril Gay (ARS-USDA), Vice Chair.

I am very happy that INRA colleagues have very significantly contributed to this international programme, in partnership with the ARS, BBSRC and OIE, and I welcome all the participants to France.

With hopes that your meeting will bring new ideas to the field and foster new collaborative research,

Dr. Marion Guillou
President, INRA, France

Pinard M-H, Gay C, Pastoret P-P, Dodet B (eds): Animal Genomics for Animal Health. Dev Biol (Basel). Basel, Karger, 2008, vol 132, p 1.

SESSION I

Introduction

Pinard M-H, Gay C, Pastoret P-P, Dodet B (eds): Animal Genomics for Animal Health. Dev Biol (Basel). Basel, Karger, 2008, vol 132, pp 3-11.

The Future Impact of Animal Genomics in Animal Health Research

M.-H. Pinard-Van Der Laan[1], C.G. Gay[2]

[1] Département de Génétique Animale, Institut National de Recherche Agronomique, Jouy-en-Josas, France

[2] United States Department of Agriculture, Agricultural Research Service, National Program Staff, Animal Production and Protection, Beltsville, Maryland, USA

Keywords: Animal genomics, animal genomes, genetic markers, functional genomics, bioinformatics

Abstract: The first International Symposium on Animal Genomics for Animal Health (AGAH) provided an excellent venue for scientists working in the field of genomics to interact with animal health experts. This paper provides an introduction to genome-enabled tools and highlights some of the research projects in the AGAH proceedings. A brief review of animal genomes, the next generation of genetic markers, and the versatility of genome-enabled technologies and their many applications are discussed.

Introduction

Animal genomics is a new field of research defined by rapidly evolving genome-enabled technologies derived from the recent breakthroughs in sequencing whole animal genomes. Although the field originated decades ago with the identification of genetic markers that led to the assembly of genetic and physical maps, the promise of animal genomics was not realized until the sequencing and annotation of the human genome was achieved in 2002. Since then, the explosion of new high-throughput technologies arising from the genomics field has been enabling the analysis of entire animal genomes, transcriptomes, and proteomes, providing links between the genetic code and the diversity of living organisms. Importantly, genome-enabled technologies are providing new opportunities for understanding disease mechanisms and novel approaches for improving the health and welfare of animals.

Genome-Enabled Research Tools

Animal genomes

After 13 years of hard work and dedication, a major milestone in biomedical research was reached in 2003, by a consortium of international scientific institutions, with the completion of the human genome project [1]. The goal of the project was to identify the estimated 25,000 genes in human DNA, determine the three billion nucleic acid base pairs that make up human DNA, store this information in databases, develop and improve tools for data analysis, transfer related technologies to the private sector, and address ethical and social issues that may arise from these technologies. The human genome project has since generated ground-breaking research tools such as the haplotype map of the human genome, consisting of the genome-wide detection and characterization of single nucleotide polymorphisms (SNPs) [2,3]. These genome-enabled technologies have clearly catalyzed the multibillion-dollar biotechnology industry [4] and are fostering the development of powerful tools in biomedical research [5-7]. However, these initial accomplishments have highlighted the need for the sequence of multiple animal species and the need for comparisons to fully explore the unknown world of genomic information [8].

The animal health research community is now in a position to capitalise on the infrastructure built by the human genome project with the sequencing of important animal species [9]: the chicken (*Gallus gallus*) in 2004 [10,11]; the dog (*Canis lupus familiaris*) in 2004 [12,13]; the cow (*Bos taurus*) in 2004 [14]; the cat (*Felis catus*) in 2006 [15]; and sequencing initiated for the pig (*Sus scrofa*) [16] and horse (*Equus caballus*) [17] genomes. Together, these projects provide the foundation for the field of animal genomics. A critical area will be the sequencing of genomes within animal species that demonstrate differences in disease susceptibility. Initial results of whole genome comparisons are already predicting that the field of comparative genomics will yield rich returns in biomedical research [8]. Successful examples are already given by the development of animal models to address human diseases and new tools to create comparative phenotypic models [18]. Likewise, the animal health research community now has access to powerful research tools to discover effective measures to prevent, diagnose, and treat diseases.

Bioinformatics and computational biology

The sequencing of animal genomes has generated an unprecedented amount of information, starting with the identification of billions of nucleotide base pairs, the identification of regulatory sequences that control gene expression, genetic markers, regulatory sequences in non-coding RNA, the transcriptome and proteome, and the three-dimensional structure of native molecules. Transforming this information into practical understanding requires intensive data-mining using sophisticated computational and bioinformatics tools. Highly intensive computations using supercomputers have greatly facilitated this process. Still needed are the development of public databases of the completely annotated genomic sequence data sets of many hosts, sequence-based high-throughput expression profiling technologies, and integrated bioinformatics tools to analyze and interpret genomic data. Integrating genetic, functional genomic, and bioinformatics data in a systems biology approach to complex diseases is developing in a promising way.

A new generation of genetic markers

Genetic markers can be simply defined as variations that are genetically controlled and inherited in a Mendelian way. Historically, there has been a continuous evolution of DNA markers developed and used thanks to technological progress, from the generation of restriction fragment length polymorphisms (RFLPs) and variable number tandem repeats (VNTRs) to PCR techniques that enable the development of microsatellite markers, all of which have been extensively used for genetic studies like quantitative trait loci (QTL) detection of disease resistance traits (see examples in these proceedings).

Another important area has been the discovery of "candidate genes," which represent a special group of genetic markers that can be used as markers of closely linked genes or studied as direct candidate genes. In disease resistance association studies, and immunogenetics studies in livestock, Major Histocompatibility Complex (MHC) genes have historically been the candidate genes of choice and several examples are presented in these proceedings. But with advanced genomics tools like pyrosequencing-based SNP typing, the identification of novel polymorphisms will progress rapidly.

With the availability of animal genome sequences, a new era of genome-enabled technologies is foretelling the feasibility of rapidly identifying genetic markers using emerging high-throughput novel technologies. Good examples are high density SNP panels. This new generation of genetic markers is paving the way for conducting fully automated whole genome scans to quickly genotype a large number of animals for a significant number of SNPs. These genetic markers are now the tools of choice to further dissect complex traits like disease resistance. High throughput technologies using dense DNA chips to type tens of thousands of SNPs are also becoming routinely available. The impact of this research will not be achieved until the "annotation" of the animal genomes is completed, which will require identifying all the genes, their location and function. Large international annotation initiatives are needed to identify functional elements within the genome sequences that can potentially lead to the identification of SNPs responsible for the observed variations in disease resistance.

The discovery of other markers is likely to present as yet unexplored sources of genetic variation. For example, copy number variations (CNVs) are genomic structural variations observed within and between breeds that may have a possible effect on health traits [19]. Although, at the beginning of this paper, we defined genetic markers as being inherited in a Mendelian pattern, another area to consider will be the growing evidence that a significant part of the genetic variation in observed phenotypes may be due to polymorphisms in imprinted genes [20].

Genetic markers are essentially tools with two purposes: the first is identifying and mapping QTL/genes of traits of interest and the second is application in breeding (see below). Protocols to identify QTL of disease resistance in livestock species are often complex and costly and may be hampered by the lack of reliable and repeatable disease phenotypes in well structured populations. Study designs may combine dedicated experimental populations (custom bred) or commercial breeds challenged with controlled pathogens or epidemic situations. Recent successful examples include the identification of QTL for infectious pancreatic necrosis (IPN) in full-sib commercial salmon families using microsatellites [21] and markers of tick resistance in bovine F2 populations following natural infestation [22]. In order to improve the accuracy of QTL mapping, alternative designs are being used, such as advanced intercross lines (AIL). A good example is the use of an SNP panel in chicken AIL leading to the identification of genes and possible pathways for resistance to *Salmonella* colonization in the chicken [23].

Applications in Animal Health

Selection of animals with desirable health traits

What are the conditions and prerequisites to using genetic markers, as described previously, to improve disease resistance? Firstly, enough genetic variability evidenced in appropriate experiments for the disease trait of interest is required to ensure possible improvement. Secondly, as direct selection for the disease trait may be feasible in experimental conditions but not practical in commercial breeding situations, identified markers of closely linked genes having a significant effect on disease resistance phenotypes are needed. If these markers are polymorphic in the breed of interest, then marker assisted selection (MAS) strategies may be implemented. If not, alternative introgression strategies may be considered if proved to be worthwhile. As described previously, a shift can be observed from using microsatellite markers to SNPs. Promising examples are given from breeding companies, for instance by the Aviagen Genomics Initiative Project, which has implemented experimental SNP selection for early mortality in broilers using a two-step selection strategy [24].

Often, QTL regions identified in the research phase are too large to be operational. Several strategies to reduce the interval, such as fine mapping or positional cloning are proving to be effective in achieving this goal. Adding enough families and applying new generation markers may turn QTL into potential tools for MAS, as shown in fish breeding for IPN [21]. All these "from QTL to the gene" approaches are twofold: they produce reliable genetic tools to be used in selection, and identify the gene(s) and further elucidate the pathways responsible for disease resistance mechanisms (see below).

With the development of genome-wide SNP panels and the technological possibility of high-throughput genotyping, a new alternative to MAS has been conceived: genome wide selection (GWS) [25]. Although methodological developments are still needed to identify the best practice for using this approach in all cases, as demonstrated by Raadsma et al. for predicting merit in complex diseases such as mastitis in dairy cattle [26], GWS may become a powerful tool of choice in the near future. GWS using high density SNP panels opens the way to unraveling the myriad of genes and underlying pathways, especially if large collaborative projects are taking place [27].

Before implementing complex genetic strategies, the cost/benefit ratio of this approach needs to be assessed to evaluate the economic importance of the disease of interest and determine whether genetic improvement is the best alternative or merely an interesting complement to therapeutic methods. Social and ethical considerations should also be carefully investigated. From a breeder's point of view, integration of health traits into breeding programmes is very much the desired outcome of using genome-enabled technologies. However, other classical traits, such as production traits, need to be considered. The next step will be to identify the best genetic strategies to improve health and production traits in either crossbred and/or purebred animals and carry out several validation steps to confirm the success of the desired outcomes [28].

It is foreseen that the field of breeding animals for desired traits will have to take into account a number of parameters before it can start implementing genetic strategies that include disease resistance for infectious diseases. These parameters may have different levels of importance, depending on the animal species, the pathogen, the disease trait and its genetic control, and the environment in which the animal is raised. Since the genetics influencing host–pathogen interactions are still poorly understood, the cumulative and interactive effects of the environment and the evolution and ecology of the pathogen will need to be integrated into a genetic epidemiology approach to assess the long-term benefits and risks of selection [29]. An example of integration of genomics and modelization is given by a combined approach for control of the *Salmonella* carrier state in poultry [30]. Also, there is growing evidence that gene-by-gene interactions (epistasis) play a major role in the control of many traits including disease traits [20]. Marek's disease is one example where the interacting loci have been shown not to be located in the QTL regions associated with disease resistance [31], showing the difficulty of using specific genes in selection. We are still in the infancy of knowing how to use in practice the complex underlying source of genetic variability and more generally to which extent data obtained in given breeds is repeatable in other breeds. This emphasizes the critical necessity for collaborative research in animal genomics, to gather a tremendous amount of complex information from research organizations in the public and private sectors [27], and the willingness to share health data sets to improve knowledge and yield subsequent tools in the near future [28].

Using animal genomics to understand animal diseases

The sequencing of animal genomes is also providing sophisticated strategies for understanding mechanisms of disease and the biological complexities of host-pathogen interactions [32]. In particular, new approaches for understanding these interactions at the molecular level are now available through high-throughput, large scale expression analyses that are rapidly revolutionizing the analysis of whole genome responses of pathogens and their hosts [33]. In the near future, novel tools to control diseases will be available through a better understanding of disease processes, the mechanisms through which pathogens evade host immunity and the genetic basis of host–pathogen interactions.

Advanced genetic studies are now being developed towards integrating QTL data with expression data [34]. These 'genetical genomics' approaches look at genomic loci called expression quantitative loci (eQTL) from microarray studies, allied with whole genome SNP arrays (WGA) and a candidate gene approach. For example, using poultry inbred lines with defined phenotypes for susceptibility or resistance to *Salmonella*, Peter Kaiser et al. [35] has applied the "genetical genomics" approach to identify SNPs associated with over 400 genes linked to innate immunity. The segregation of 6,000 WGA SNPs across inbred lines yielded some 900 informative SNPs that can be used for a cross between any two lines. The function of candidate gene SNPs can now easily be tested between the inbred lines; this will contribute to our understanding of early immune responses to infection and eventually link phenotype, genotype, and eQTL.

An example of integrating molecular genetics and functional genomics to increase our knowledge of mechanisms of diseases is the identification of candidate genes flanking QTL previously shown to be associated with coccidiosis disease resistance in poultry [36]. Genome-wide searches for markers that control resistance to coccidiosis led to the identification of two major QTL associated with parasite reduction [37,38]. Although these two QTL markers provided important information on the genetics of resistance to coccidiosis, the region in which they are located is sufficiently large to contain hundreds, if not thousands of candidate genes. A functional genomics approach to identify candidate genes flanking the QTL potentially controlling coccidiosis disease resistance consisted of using an avian intestinal intraepithelial lymphocyte cDNA array [39] constructed from a library of genes expressed by intestinal intraepithelial lymphocytes of *Eimeria*-infected chickens to identify candidate genes affecting gut immune responses to *E. maxima* infection. Another approach is combining structural and functional data by using differentially expressed genes as first candidate genes for positional cloning as an efficient strategy to map functional genes [31] or by using comparative gene ontology analyses of differentially expressed genes [40].

Global profiling of host and pathogen gene expression presents an attractive approach to identifying the novel genes involved in disease processes since genes are usually only transcribed when and where their function is required. An example is given by the refined study of the dialogue between porcine cells and pseudorabies virus at the molecular level, which has helped decipher the host response evasion strategies employed by this virus [41]. Determining the conditions under which a given gene is expressed allows inferences to be made about its function. Thus, host genomic analyses will lead to a better understanding of the response to pathogenesis, the development of diagnostic gene expression profiles, and the characterization of genetic polymorphisms associated with diseases.

Discovering a new generation of tools to control animal diseases

As discussed above, genome-enabled technologies will facilitate the selection of animals with production and health traits that target specific production systems. One of the outcomes will be the development of commercial diagnostic tests to select animals with desired traits in outbred commercial herds. Although in its infancy, animal genomics also has the potential of yielding new and improved medical therapeutics for animals that will lower production costs and improve animal welfare by fighting devastating infectious diseases that affect animals worldwide. Understanding the molecular mechanisms of disease is likely to yield logical targets for drugs and biotherapeutics, and perhaps one day transgenic animals that are resistant to infectious diseases [42]. Another important application will be the evaluation of genetic influences on individual animal responses to vaccination and the selection of good responders to vaccination [32,43,44].

Preserving animal genetic diversity

Finally, an important consideration will be the use of genetic markers to assess and manage domestic and wildlife animal genetic diversity [29,45]. Interestingly, the sad circumstances leading to the extinction of endangered animal species worldwide, and the loss of genetic biodiversity in domesticated livestock and poultry, could be impacted by better management of biodiversity based on new genetic markers and novel methods for genomic selection. Indeed, contrary to former genetic selection methods, which relied directly on the trait itself or using a few markers, that were accompanied with an evident risk of losing genetic variability, genome-wide markers allow for the possibility of monitoring and preventing the loss of important non-selected components of animal genomes.

Conclusion

Genome-enabled technologies are driving fundamental changes in the approaches used to understand mechanisms of disease, the selection of animals for health traits, and the discovery of tools to control and mitigate animal diseases. Gene discovery and genetic variations can now be used in genotyping analyses. New research strategies employing high-throughput gene expression analysis are providing novel platforms for more comprehensive understanding of host–pathogen interactions. In particular, functional genomics is rapidly revolutionizing the analysis of whole genome responses of hosts and pathogens, which will ultimately lead to a better understanding of disease processes and the mechanisms through which pathogens evade host immunity; identification of the genetic basis of host–pathogen interactions; and discovery of novel vaccines, drugs, and biotherapeutics.

Acknowledgments

The authors are grateful to the Scientific Committee of the International Symposium on Animal Genomics for Animal Health (AGAH) for their support.

References

1 Collins FS, Green ED, Guttmacher AE, Guyer MS: A vision for the future of genomics research. Nature 2003;422:835-847.

2 International HapMap Consortium: A haplotype map of the human genome. Nature 2005;437:1299-1320.

3 The International HapMap Consortium: A second generation human haplotype map of over 3.1 million SNPs. Nature 2007; 449:851-861.

4 Wang E, Panelli MC, Marincola FM: The impact of genomics on the biotechnology industry. Expert Opin Biol Ther 2001;1:749-751.

5 Guttmacher AE, Collins FS: Realizing the promise of genomics in biomedical research. JAMA 2005; 294:1399-1402.

6 Trevino V, Falciani F, Barrera-Saldaña HA: DNA microarrays: a powerful genomic tool for biomedical and clinical research. Mol Med 2007;13(9-10):527-541.

7 Heller MJ: DNA microarray technology: devices, systems, and applications. Annu Rev Biomed Eng. 2002;4:129-153.

8 Buza TJ, McCarthy FM, Burgess SC. Experimental-confirmation and functional-annotation of predicted proteins in the chicken genome. BMC Genomics. 2007;8:425.

9 Eukaryotic Genome Sequencing Projects:Genome Project. National Center for Biotechnology Information (NCBI). http://www.ncbi.nlm.nih.gov/genomes/leuks.cgi. Last accessed on May 25, 2008.

10 International Chicken Genome Consortium: Sequence and comparative analysis of the chicken genome provide unique perspectives on vertebrate evolution. Nature 2004;32:695-716.

11 International Chicken Polymorphism Map Consortium: A genetic variation map for chicken with 2.8 million single-nucleotide polymorphisms. Nature 2004;432:717-722.

12 Lindblad-Toh K, Wade CM, Mikkelsen TS, Karlsson EK, Jaffe DB, Kamal M, et al: Genome sequence, comparative analysis and haplotype structure of the domestic dog. Nature 2005;438:803-819.

13 Kirkness EF, Bafna V, Halpern AL, Levy S, Remington K, Rusch DB, et al: The dog genome: survey sequencing and comparative analysis. Science, 2003;301:1898-1903.

14 Everts-van der Wind A, Larkin DM, Green CA, Elliott JS, Olmstead CA, Chiu R, et al: A high-resolution whole-genome cattle-human comparative map reveals details of mammalian chromosome evolution. Proc Natl Acad Sci USA 2005;102:18526-18531.

15 Pontius JU, Mullikin JC, Smith DR, Agencourt Sequencing Team, Lindblad-Toh K, Gnerre S, et al: Initial sequence and comparative analysis of the cat genome. Genome Res 2007;17:1675-1689.

16 Wernersson R, Schierup MH, Jørgensen SG, GorodkinJ, Panitz F, Staerfeldt HH, et al: Pigs in sequence space: a 0.66X coverage pig genome survey based on shotgun sequencing. BMC Genomics, 2005;6:70.

17 Swinburne JE, Boursnell M, Hill G, Pettitt L, Allen T, Chowdharry B, et al: Single linkage group per chromosome genetic linkage map for the horse, based on two three-generation, full-sibling, crossbred horse reference families. Genomics 2006;87:1-29.

18 Schook LB, Kuzmuk K, Adam S, Rund L, Chen K, Rogatcheva R, et al: DNA-based animal models of human disease: from genotype to phenotype. Dev Biol 2008;132:15-25.

19 Liu GE, Van Tassell CP, Sonstegard TS, Li RW, Alexander LJ, Keele JW, et al: Detection of germline and somatic copy number variations in cattle. Dev Biol 2008;132:231-237.

20 Georges M: Mapping, fine mapping, and molecular dissection of quantitative trait loci in domestic animals. Annu Rev Genomics Hum Genet 2007;8:131-162.

21 Houston RD, Haley CS, Guy DR, Hamilton A, Tinch A, Taggart JB, et al: Detection and confirmation of a major QTL affecting resistance to infectious pancreatic necrosis (IPN) in Atlantic salmon (*Salmo Salar*). Dev Biol 2008;132:199-204.

22 Regitano LCA, Ibelli AMG, Gasparin G, Miyata M, Azevedo ALS, Coutinho LL, et al: On the search for markers of tick resistance in bovines. Dev Biol 2008;132:225-230.

23 Hasenstein JR, Hassen AT, Dekkers JCM, Lamont SJ: High resolution, advanced intercross mapping of host resistance to Salmonella colonization. Dev Biol 2008;132:213-218.

24 Long N, Gianola D, Rosa GJM, Weigel KA, Avendano S: Machine learning classification procedure for selecting SNP's in genomic selection: Application to early mortality in broilers. Dev Biol 2008;132:373-376.

25 Meuwissen THE, Hayes BJ, Goddard ME: Prediction of total genetic value using genome wide dense marker maps. Genetics 2001;157:1819-1829.

26 Raadsma HW, Moser G, Crump RE, Cavanagh JJ, Khatkar M, Sölkner J, et al: Predicting genetic merit for mastitis and fertility in dairy cattle using genome wide selection and high density SNP screens. Dev Biol 2008;132:219-223

27 Goddard ME: The use of high density genotyping in animal health. Dev Biol 2008;132:383-389.

28 Mellencamp A, Galina-Pantoja L, Gladney CD, Torremorell M: Improving pig health through genomics: a view from the industry. Dev Biol 2008;132:35-41.

29 Gibson JP, Bishop SC: Use of molecular markers to enhance resistance of livestock to disease: a global approach. Rev sci tech Off int Epiz 2005;24(1):343-353.

30 Beaumont C, Lecerf F, Protais J, Prevost K, Pitel F, Burlot T, et al: An integrated approach of genetic resistance to Salmonella carrier state in fowls: from genetics to genomics and modelling. Dev Biol 2008;132:353-357.

31 Cheng H, Niikura M, Kim T, Mao W, MacLea KS, Hunt H, et al: Using integrative genomics to elucidate genetic resistance to Marek's disease in chickens. Dev Biol 2008;132:365-372.

32 Gay CG, Zuerner R, Bannantine JP, Lillehoj HS, Zhu JJ, Green R, Pastoret PP: Genomics and vaccine development. Rev sci tech Off int Epiz 2007; 26 (1): 49-67.

33 Smits MA, Rebel JMJ, te Pas MFW, Hulst MM: Exploiting genomics to improve animal health. Dev Biol 2008;132:27-33

34 de Koning DJ, Carlborg O, Haley CS: The genetic dissection of immune response using gene-expression studies and genome mapping. Vet Immunol Immunopathol 2005;15;105:343-352.

35 Kaiser P, Howell J, Fife M, Sadeyen J-R, Salmon N, Rothwell L, et al: Integrated immunogenomics in the chicken: Deciphering the immune response to identify disease resistance genes. Dev Biol 2008; 132:57-66.

36 Lillehoj HS, Hong Y, Kim C: Quantitative genetic and functional genomics approaches to investigating parasite disease resistance and protective immune mechanisms in avian coccidiosis. Dev Biol. 2008; 132:67-75.

37 Zhu JJ, Lillehoj HS, Allen PC, van Tassell CP, Sonstegard TS, Cheng H, et al: Mapping quantitative trait loci associated with resistance to coccidiosis and growth. Poul Sci 2003;82:9-16.

38 Kim ES, Hong YH, Min W, Lillehoj HS: Fine-mapping of coccidia-resistant quantitative trait loci in chickens. Poul Sci, 2006; 85:2028-2030.

39 Kim CH, Lillehoj HS, Hong YH, Keeler CL Jr.: Comparison of transcriptional changes associated with *E. acervulina* and *E. maxima* infections using cDNA microarray technology. Dev Biol 2008; 132: 121-130.

40 Tuggle CK, Wang YF, Couture OP, Qu L, Uthe JJ, Kuhar D, et al: Computational integration of structural and functional genomics data across species to develop information on porcine inflammatory gene regulatory pathway. Dev Biol 2008;132:105-113.

41 Flori F, Rogel-Gaillard C, Mariani V, Lemonnier G, Cochet M, Hugot K, et al: A combined transcriptomic approach to analyse the dialogue between pseudorabies virus and porcine cells. Dev Biol 2008; 132: 99-104.

42 Doran TJ, Wise TG, Schafer DS, Lowenthal JW: The use of RNAi and transgenics to develop viral disease resistant livestock. Dev Biol 2008;132:377-382.

43 Glass EJ: Genetic variation and responses to vaccines. Rev sci tech Off int Epiz 2004;5(2):197-208.

44 O'Neill RG, Woolliams JA, Glass EJ, Williams JL, Fitzpatrick JL: Quantitative evaluation of genetic and environmental parameters determining antibody response induced by vaccination against bovine respiratory syncytial virus. Vaccine 2006;24 (18): 4007-4040.

45 Simianer H: Report of the International Society of Animal Genetics (ISAG) and FAO advisory group on animal genetic diversity, May 2007.

Pinard M-H, Gay C, Pastoret P-P, Dodet B (eds): Animal Genomics for Animal Health. Dev Biol (Basel). Basel, Karger, 2008, vol 132, p 13.

SESSION II

Rewiew of Genomics and its Application

Pinard M-H, Gay C, Pastoret P-P, Dodet B (eds): Animal Genomics for Animal Health. Dev Biol (Basel). Basel, Karger, 2008, vol 132, pp 15-25.

DNA-based Animal Models of Human Disease: from Genotype to Phenotype

L.B. Schook[1,2], K. Kuzmuk[1], S. Adam[3], L. Rund[1], K. Chen[1], M. Rogatcheva[1], M. Mazur[1], C. Pollock[1], C. Counter[3]

[1] Laboratory of Comparative Genomics, Department of Animal Sciences, University of Illinois, Urbana, USA

[2] Regenerative Biology and Tissue Engineering Theme, Institute for Genomic Biology, University of Illinois, Urbana, USA

[3] Department of Pharmacology and Cancer Biology, Duke Medical Center, Durham, NC, USA

Key words: pigs, animal models, splice variants, SNPs, tumorigenesis, atherosclerosis, AT

Abstract: Biomedical research utilizes animal models to elucidate human disease processes at the cellular and molecular level and for the development of new therapies. Traditionally, mammalian models have been limited to the mouse, primarily because of well characterized genetic lines and the ability to manipulate the genome to directly test hypotheses regarding causal mutations and disease phenotypes. The emerging availability of genome sequences of other mammals (bovine, canine, equine, feline, and porcine) now permits utilization of the mammal in which the phenotype best approximates the human condition. Equally important is the use of somatic cell nuclear cloning (SCNT) coupled with targeted germline manipulation to create animals to resolve the molecular mechanisms of the disease state. Our efforts have focused on the pig, which has emerged as an important biomedical mammalian model due to its closer physiology to humans. The utility of porcine genetically-defined tumour, cardiovascular and neurological disease models is described.

INTRODUCTION

Public health programmes and advances in clinical medicine have increased the human lifespan from 47.3 years in 1900 to 76.9 years in 2000 [1]. Technological and medical advances have also improved the quality of life and provided prevention strategies and treatments. However, the incidence of chronic diseases, due to complex genetic and environmental interactions, has also dramatically increased in the past century. In fact, the majority of the top ten leading causes of death in the United States in 2000 (heart disease, cancer, stroke, chronic lower respiratory disease, diabetes, Alzheimer's disease, and renal disease) represent diseases shown to result from such interactions [2]. Resolving complex human diseases (that reflect lifestyle influences and genetic predisposition) is difficult (e.g., expenses associated with human clinical experiments and ethical issues) and thus appropriate biomedical models must be developed and validated. In the past, researchers have used two approaches to study human diseases. One strategy fully characterizes a human clinical disease and uses the most appropriate animal model based on criteria such as anatomical and/or physiological characteristics. The recent advent of techniques in genomics, transgenesis and cloning provide a new opportunity to create domesticated animals (pigs, cows, chickens, dogs) as model organisms.

The incidence of chronic diseases due to complex genetic and environmental interactions has dramatically increased [3]. Atherosclerosis, the primary cause of coronary artery disease and stroke, is a disorder with multiple genetic and environmental contributions. Successful discovery of these genetic factors requires using complementary approaches with appropriate animal models, large-scale human genetic studies, and functional experiments [4]. Although many advances have been made in understanding the genetics of atherosclerosis by using rodent and transgenic mouse models, an animal model that will permit studies on phenotypic changes not evident in the mouse is necessary. The link between the APOE gene, serum cholesterol levels, and the development of atherosclerosis has been well established in humans. Extensive association studies with disease risk have been performed for alleles of *APOE* [5]. These studies determined that the ε4 allele is associated with an increased risk for cardiovascular disease. We are investigating the potential of the pig as a model for atherosclerosis by utilizing sequence information to compare human *APOE* to pig *APOE*.

The transition from basic to clinical cancer research is hampered by the lack of a genetically malleable, large animal model. In particular, preclinical imaging studies as well as studies on hyperthermia, radiation, or photodynamic therapy of tumours are limited by this lack of suitable large animal models. Mice are ideal models for genetically inducing tumours but the organs and tumours are too small for the medical devices. For example, it is virtually impossible to do magnetic resonance tomography on mice due to their small tumour size and the energy of clinical accelerators. High-resolution intensity modulated treatment in rodents is further hampered by the same problems and devices used for hyperthermia treatment cannot be scaled down to be useful for studies in rodents. While companion animals (particularly dogs and cats) with spontaneous tumours are ideal for imaging studies and the above described therapies, it is very difficult and expensive to accrue enough animals with the desired tumour types for clinical trials. Moreover, tumours in these animals are genetically undefined and in some cases too rare to be studied [6]. In vivo cancer studies requiring large tumours would therefore be greatly aided by a genetically defined large mammal model of cancer. The pig represents such an animal. Like humans, the incidence of cancer in pigs is rare. The parallels in cancer biology between humans and pigs are conserved at the molecular level. For example, in both species, telomerase is suppressed in numerous tissues and reactivated during cancer. Tumours can also reach sizes seen in humans, which is particularly relevant for many preclinical applications. Additionally, pigs have the added advantage of being physiologically similar to humans in a number of aspects compared to other commonly used animal models in cancer biology [7]. For example, the size, and anatomy of the swine reflect that of humans in many respects, and as such the pig is often the primary biomedical model for a number of diseases, surgical research, and organ transplantation [3]. The genomic sequence homology between humans and pigs is also very high [3] and the porcine PXR protein that regulates p450 cytochrome CYP3A, which metabolizes almost half of prescription drugs in humans, is more similar to humans than, for example, the murine PRX [8]. We are utilizing the pig as a clinically-relevant, large animal model of cancer through the disruption of regulatory pathways using genetic engineering techniques.

Ataxia-telangiectasia (A-T) is a genetic disorder causing cerebellar degeneration, immune deficiency, cancer predisposition, chromosomal instability and radiation sensitivity. Among the mutations responsible for A-T, 85% represent truncating mutations that result in the production of shorter, highly unstable forms of ATM (AT-mutated) protein leading to a null ATM phenotype. Several ATM-deficient mice have been created, however none reflect the extent of neurological degeneration observed in humans. In an attempt to identify an alternative animal model, we have characterized the porcine ortholog of ATM (*pATM*) to support a pig model for AT.

Materials and Methods

RNA extraction, cDNA synthesis and sequencing

Total RNA was extracted from frozen porcine tissues using the TRIzol® reagent (Invitrogen), treated with DNase I, and purified. Samples were reverse transcribed (+RT samples) with Oligo(dT)12-18 and random primers and Superscript II reverse transcriptase. To determine the porcine *ATM* cDNA sequence, 0.5 kb to 1.8 kb overlapping cDNA fragments were amplified and sequenced. The porcine genomic sequence (AY587061) was used as a reference to design primers located in exons of the ATM gene. The 5' RACE analysis was performed on total RNA extracted using the FirstChoice RLM-RACE kit. Splicing of the 5' UTR was analysed by amplifying fragments from exon 1 to exon 8 with the forward 5' GATTCCACTTGGTTGTCCAGTC 3' and reverse Race ex8R primers. PCR fragments were purified using a QIAquick PCR purification kit and bi-directionally sequenced with PCR primers. Alternatively, if multiple amplicons were observed, the amplified fragments were cloned into the pCR®2.1-TOPO® vector system and then sequenced using M13 primers. DNA sequencing was performed with the ABI Prism® BigDye™ Terminator Cycle Sequencing Kit and analysed on an ABI 3730 automated DNA sequencer.

Quantitative RT-PCR of 5' UTR ATM transcript variants

For the identification and quantification of 5' UTR variants, quantitative RT-PCR analyses were performed on cDNA samples from porcine tissues using TaqMan 5' nuclease assays. Primers and probes were designed with Primer Express software to anneal with splice variant-specific sequences at exon-exon junctions. To provide more accurate splice variant discrimination, TaqMan® Probes with a nonfluorescent quencher and a minor groove binder at the 3′ end were used. In order to estimate the total amount of ATM transcripts, a probe that recognized a sequence common to all splice variants was used. The TaqMan® assay was performed using a 96-well plate format with a 20 µl reaction mix and the ABI PRISM 7900 Sequence Detection System instrument and software. The copy numbers of splice variants (SV) per reaction was determined for each tissue and the absolute copy number was normalized for the relative amount of 18s rRNA and background subtracted using the (-RT) control reactions. To quantify the porcine 18s rRNA, the TaqMan reagent Starter kit was used according to the manufacture's protocol, except that the probe-primer mix was diluted 10 fold. An absolute quantitation assay was used to estimate the copy number of each SV by interpolating their quantity from a standard curve. Standard curves were developed using serial dilutions (10^1 to 10^7 copies) of each SV standard. A standard curve was constructed for each SV using threshold cycle data plotted against the log copy numbers for the standard dilution sets.

Retroviral Vector Construction

The retroviral vectors encoding a single transgene, pBabepuro-cyclinD1•HA, pBabehygro-CDK4^{R24C}•HA, pBabeblasto-c-Myc^{T58A}, pBabeneo-p53^{DD}, pBabepuro-flag•H-Ras^{G12V} and pBabepuro-flag•hTERT, have been previously described [9,10]. The retroviral vectors encoding two transgenes were constructed based on these plasmids. pBabe-hTERT+p53^{DD} was created by subcloning the hTERT and PCR cloning the p53^{DD} transgenes into the *EcoRI-Sal*I and *Hind*III-*Cla*I sites of pBabepuro, respectively, thereby replacing the puromycin selection marker. pBabe-cyclinD1+CDK4^{R24C} was created by PCR cloning cyclin D1 and CDK4^{R24C} into the same *Eco*RI-*Sal*I and *Hind*III-*Cla*I sites. pBabe-c-Myc^{T58A}+H-Ras^{G12V} was created by PCR cloning the H-Ras^{G12V} transgene into the *Hind*III-*Cla*I sites of pBabeblasto-c-Myc^{T58A}, replacing the blastocidin selection marker.

Porcine transformed cell lines and tumour growth

At one day of age, ears were notched under sterile conditions for identification purposes and the samples were used to establish a primary culture of fibroblasts. As soon as these cultures were established, they were serially infected every 48 hrs with one of each of the retroviruses derived from the plasmids using standard approaches. RNA was isolated from the described cell lines 48 hrs after the last infection and reverse transcribed. Resultant cDNA was used to verify the absence or presence of expression of specific transgenes by PCR amplification with one primer specific to the transgene and another specific to a transcribed region of the pBabe plasmids. Transformed cells (10^3) per 3cm plate (in triplicate) were suspended in soft agar as described and colonies >30 cells scored after four weeks. Transformed cells ($1x10^7$) were mixed with Matrigel and injected somatic (SC) cell into the flanks of SCID/beige mice per cell line, after which tumour volumes were determined at regular intervals as described or $1x10^8$ of cells were mixed with a similar number of isogenic control normal cells and injected orthotopically behind each ear of the donor pig treated daily with cyclosporine (20 mg/kg), azothioprine (2mg/kg) and prednisolone (4 mg/kg).

Animals and diets

All procedures involving animals were approved by the Institutional Animal Care and Use Committee of the University of Illinois. A single Duroc gilt was used as the genomic template for producing cloned animals by nuclear transfer. Four clones, 14 months of age, were housed at the University of Illinois ISRL. Pigs were randomly assigned to one of two treatment groups: a control diet (CD) (n = 2) or a high-fat, high-cholesterol (HFHC) diet (n = 2). Control pigs were fed 2.0 kg of a control diet once daily in the morning and had free access to water. The second group of pigs were fed 1.5 kg of the HFHC atherogenic diet consisting of the control diet supplemented with 2.0% cholesterol, 17% hydrogenated coconut oil, 2.3% corn oil, and 0.7% sodium cholate once daily in the morning and had free access to water. The cholesterol diet ratios for metabolizable energy, calcium, phosphorus and the principal amino acids lysine, methionine, threonine, and tryptophan were made equivalent to the control diet.

Lipid measures

Blood samples were collected via the jugular vein from pigs fasted overnight at day 0 and every two weeks until the end of the study. Plasma was sent to Mayo Medical Laboratories for lipoprotein analysis (Rochester, MN). The lipoprotein metabolism profile reported total cholesterol and total triglycerides, and quantification of cholesterol and triglycerides in VLDL, HDL cholesterol, and LDL cholesterol.

Results

Porcine atherosclerosis model

A novel porcine model of atherosclerosis was created by utilizing genomic sequence information to characterize exon 4 of porcine *APOE*, establishing a nutritional regimen that induces the formation of atheromas, and characterizing the initial stage of atherosclerosis formation. The advantages of working with this animal model are that genetic variation is controlled for by using cloned animals, lipoprotein metabolism in swine is similar to that in humans, and the cardiovascular system, specifically coronary circulation, susceptibility to cardiovascular disease, and atherosclerotic lesion morphology is similar to humans. Unrelated pigs from ten breeds were genetically screened to identify animals that had the SNP associated with increased susceptibility associated with the *APOE* ε4 genotype. A Duroc female was selected for cloning by nuclear transfer. After reaching 14 months of age, the clones were moved to the experimental diet (Table 1). The animals received the same daily nutritional energy and were weighed to establish daily growth rates. Blood was collected weekly and was processed to establish plasma cholesterol, triglyceride and lipoprotein cholesterol levels in each animal. As shown in Table 2, after 4 weeks the clones fed the high fat, high cholesterol (HFHC) diet both had elevated total cholesterol, VLDL cholesterol, LDL and HDL cholesterol compared to the two control clones (CD). Thus, this diet was able to induce a rapid rise in plasma levels. In addition, these animals were shown following necropsy and histopathology to have developed atherosclerotic plaques [11].

Table 1: Diet compositions.

Composition of diets	Control diet	High fat/cholesterol diet
	Weight, g/100g	Weight, g/100g
Corn	85.94	50.89
Soybean meal	10.81	22.82
Limestone	1.07	1.4
Calcium phosphate	1.61	2.11
Premix minerals	0.35	0.45
Premix vitamins	0.2	0.26
Tylan-40	0.03	0.03
DL-methionine	0	0.03
Coconut oil	0	17
Corn oil	0	2.3
Cholesterol	0	2
Sodium cholate	0	0.7
Energy	3.304 kcal/g	4.239 kcal/g
Daily intake	2 kg/day	1.5 kg/day

Table 2: Plasma cholesterol and triglyceride levels and lipoprotein cholesterol levels in control and high fat, high cholesterol fed swine.

Serum value (mg/dL)	Week 0			
	HFHC1	HFHC2	CD1	CD2
Total cholesterol	73	84	79	82
VLDL cholesterol	3	1	2	3
LDL cholesterol	40	53	43	46
HDL cholesterol	30	30	34	33
Total triglyceride	33	35	41	43
VLDL triglyceride	27	20	33	34
LDL triglyceride	2	4	4	5
	Week 4			
	HFHC1	HFHC2	CD1	CD2
Total cholesterol	439	381	61	63
VLDL cholesterol	32	29	6	7
LDL cholesterol	337	300	30	30
HDL cholesterol	70	52	25	26
Total triglyceride	21	30	24	26
VLDL triglyceride	9	19	17	19
LDL triglyceride	9	9	4	4

Generation of porcine cells harbouring genetic changes commonly found in human cancers

To establish a genetically malleable porcine model of cancer, normal porcine cells were genetically engineered in culture to recapitulate changes commonly found in human cancers, and when returned back to the host (isogenic) swine formed tumours [12]. Primary skin cultures, confirmed to be vimentin-positive via western blot and displayed a fibroblast morphology were established from ear notch biopsies. Fibroblasts were chosen primarily because of the ease with which they can be isolated, the existence of established cultured conditions for these cells, and the fact that human fibroblasts can be converted to a tumourigenic state by disrupting the aforementioned p53, Rb, c-Myc, Ras and telomerase pathways. These early passage porcine fibroblasts were sequentially and rapidly infected over the course of six days with three retroviruses co-expressing pairs of available human and murine versions of the cDNAs for *hTERT* and $p53^{DD}$, *cyclin D1* and $CDK4^{R24C}$, and lastly, $c\text{-}Myc^{T58A}$ and $H\text{-}Ras^{G12V}$. RNA was isolated from the resulting polyclonal population, termed 6510T, and subjected to RT-PCR with one primer anchored in the transgene and another in an accompanying vector-specific untranslated region to specifically amplify the transgenic mRNAs. Using this approach, all six transgenes were verified*to be expressed in the pig fibroblast cells (Table 3).

Table 3: Genetic conversion of normal porcine fibroblasts to a tumourigenic state by disrupting pathways commonly altered in human cancers.

Genetic element (Δ, deletion of single gene from transfection)	Porcine fibroblast phenotype		
	ΔTransgene expression (RT-PCR)	Colony number	In vivo tumour (10^6)
Vector	No	0	No
$\Delta p53^{DD}$	No	900	Yes
ΔCyclin D1	No	0	NO
$\Delta CDK4^{R24C}$	No	0	Yes (small)
ΔMyc^{T58A}	No	400	No
ΔRas^{G12V}	No	0	No
ΔhTERT	No	1,000	Yes
$p53^{DD}$, Cyclin D1, $CDK4^{R24C}$, Myc^{T58A}, Ras^{G12V}, hTERT	Yes, all expressed	4,000	Yes

Confirmation of appropriate transgene expression by RT-PCR of primary porcine fibroblast cultures. Established cultures were infected with retroviruses encoding hTERT, p53DD, cyclin D1, CDK4R24C, c-MycT58A, and H-RasG12V, or combinations of five of these six cDNAs (cell lines ΔhTERT, Δp53DD, ΔcyclinD1, ΔCDK4R24C, ΔMycT58A and ΔRasG12V, which lack the indicated transgene), or as a negative control, empty vector(s).

Engineered genetic changes endowed porcine cells with the ability to grow in an anchorage-independent fashion in soft agar, one of the most rigorous in vitro assays for the transformed growth characteristic of cultured cancer cells. Transformed porcine fibroblasts, or as a negative control, cells serially infected with vectors lacking these transgenes, were seeded in soft agar and three weeks later colony growth was assessed. Only cells expressing all six transgenes were capable of robust transformed cell growth (Table 3). To address the requirements of transformation of porcine cells, all combinations of five of the six transgenes were introduced into primary normal cells by sequential infection with two retroviruses encoding four different transgenes, followed by infection with a retrovirus encoding only one transgene. The resultant six polyclonal populations (denoted by indicating the missing transgene) were confirmed by RT-PCR to express the desired combinations of five transgenes. When these populations were assayed for growth in soft agar, loss of any one transgene was found to reduce the number of colonies. As with cells from the first animal, expression of all six transgenes was again found to be required for robust transformed cell growth in soft agar. To determine whether expression of the described six transgenes was sufficient to induce tumor growth in vivo, the transformed cells were injected SC into immunocompromised mice, and tumour growth was monitored over time. These cells rapidly formed tumours at the site of injection, reaching maximum tumour burden within ten days in all animals, whereas vector control cells failed to grow in this fashion[12]. These cells maintained transgene expression in vivo, as assessed by RT-PCR, and formed vascularized tumours characterized by poorly differentiated cells with large nuclei and multiple nucleoli. To address requirements of tumorigenic growth, porcine cells expressing five of the six transgenes were injected into immunocompromised mice. These cells did not form tumours in the absence of cyclin D1, c-Myc^{T58A}, or H-Ras^{G12V}, but could still form tumours when either p53^{DD}, hTERT, or CDK4 proteins was omitted. Thus, multiple transgenes are also needed for tumour growth in vivo, although interestingly, not as many as required for robust transformed growth in vitro. Whether expression of these six proteins would drive other types of porcine cells (embryonic kidney, porcine testes and kidney epithelial cells) to a tumourigenic state was also tested. Transformed cell lines were injected into immunocompromised mice and tumours formed. Furthermore, when the porcine cell lines transformed with six transgenes were injected into immunocompromised isogenic animals, solid tumors formed at the site of injection. In conclusion, inactivation of the tumour suppressors *p53* and *Rb* concomitant with the activation of *c-Myc*, *Ras*, and telomerase is reproducibly tumourigenic in multiple porcine cell types, even when the cells are isolated from an outbred population [12]. The ability to genetically engineer porcine cells and induce tumorigenic growth in vivo provides a large animal model for therapeutic testing.

Ataxia-telangiectasia (A-T)

This genetic disorder causes cerebellar degeneration, immune deficiency, cancer predisposition, chromosomal instability and radiation sensitivity. Among the mutations responsible for A-T, 85% represent truncating mutations that result in the production of shorter, highly unstable forms of ATM (AT-mutated) protein leading to a null ATM phenotype. Several ATM-deficient mice have been created however none reflect the extent of neurological degeneration observed in humans. In an attempt to identify an alternative animal model, we have characterized the porcine ortholog of *ATM* (*pATM*) [13]. When compared to the human *ATM* (*hATM*), the *pATM* showed a high level of homology in the coding region, particularly in the regions coding for functional domains, and had extensive alternative splicing of the 5' UTR, characteristic for the human *ATM* mRNA. Overall, the porcine ATM protein consists of 3,058 aa and is highly homologous to the hATM protein (88% identity and 93% positive). Among all the non-primate mammalian species, the human protein has the highest homology with pATM, with identity greater than 92% in the FAT and PIK functional domains.

Six different 5' UTRs resulting from alternative splicing of the first three exons were identified. The porcine 5' UTRs varied in size, had multiple ATG codons and different secondary structures, supporting the possibility of complex transcriptional regulation. In contrast with *hATM*, alternative splicing of the 5' region involved not only the 5' UTR but also the coding region. Three of the six transcripts demonstrated splicing of the first coding exon, altering the translation start, and giving rise to a putative protein lacking the N-terminus substrate binding domain (82-89 aa) involved in activation of human p53 and BRCA1 pathways. Real-time PCR analysis revealed variable expression levels of total *ATM* transcripts in individual tissues (Fig. 1). Although each splice variant was ubiquitously expressed among the tissues, differences in the relative abundances of specific 5' UTRs were observed. The longest splice variant represented no less than 60% of the total *ATM* transcript in each tissue examined.

Discussion

In the future, we will continue to investigate the effects of complex gene and environmental interactions on complex phenotypes. This paper shows how genomics can be used to develop animal models of human disease based on a single nucleotide polymorphism (atherosclerosis), disrupting regulatory pathways (tumourigenesis) and differences in splice variants (AT). With regard to atherosclerosis, additional studies will be performed with longer feeding trials to further characterize the developmental process of atherosclerosis and establish a timeline of progression. Once this is accomplished, intervention procedures can be performed and the most appropriate treatment for a particular stage of atherosclerosis can be determined. In addition, as our understanding of the genetics of atherosclerosis increases, genetic diagnosis will become increasingly important. One of the greatest potential benefits of genetic studies is testing for susceptibility. Once the genes contributing to common forms of the disease have been identified, along with the particular mutations involved, DNA-based tests will contribute to our ability to assess risk.

We also demonstrated that porcine cells can be genetically engineered to perturb pathways commonly altered in human cancers and become tumourigenic and when returned back to the host animal, provide the first method of genetically inducing tumours in a large mammal. The pig has a number of clear advantages that make it ideal for preclinical imaging studies, as well as studies on hyperthermia, radiation, or photodynamic therapy of tumours. First, the changes required to make porcine cells tumourigenic are more similar to those needed in human compared to murine cells, and hence tumourigenesis in pigs may be more similar to the process in humans. Second, like companion animal models of cancer, the resultant tumours in pigs can be grown to large sizes, ideal for a number of preclinical applications. Third, in this study the porcine tumours were genetically engineered, and hence it is possible to tailor-make tumours of a defined background, unlike with companion animal models. Fourth, as different cell types injected at different anatomical sites all yielded tumours, it should be possible to generate many different types of tumours essentially anywhere in the body. Fifth, the entire procedure is quite simple and nearly as rapid as existing mouse xenograft models. The ability to genetically generate experimentally malleable tumours of a size and genetic makeup akin to those found in humans rapidly, and at will in the pig, can now be exploited with relative ease as a relevant model for preclinical studies of human cancers.

Finally, an extremely important role of alternative splicing is expanding protein diversity. Approximately 75% of human genes are alternatively spliced [14] and the species-specific alternative splicing observed in human and pig *ATM* represents the most common type of splicing in mammalian species. The alternative splicing events that affect the coding sequence change protein functions and thus provide proteomic diversity. The multiple splice variants of porcine *ATM* gene that were identified in this study could possibly be translated into various proteins with yet unknown functions. Overall, the extensive alternative splicing of the porcine *ATM* gene resembles complex alternative splicing (AS) in humans and could impact differences observed between mice and humans with regards to the onset of A-T.

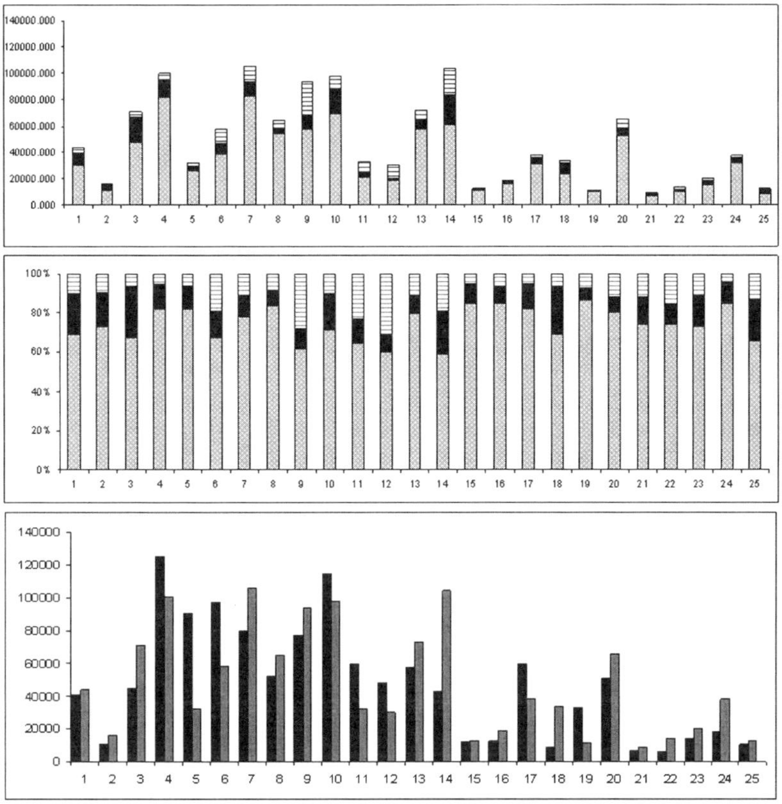

Fig. 1:Tissue expression of porcine ATM splice variants (SV1, SV2 and SV3).
RT-PCR was used to quantify cDNA from different porcine tissues for SV1, SV2, and SV3 transcripts. The TaqMan assay for quantitative RT-PCR was used to determine expression levels. Expression was normalized by using the level of 18s rRNA. Grey bars are for SV1 transcripts, black – for SV2 transcripts, and bars with stripes are for SV3 transcripts. Analysed tissues are: 1, skin; 2, fibroblasts; 3, spinal cord; 4, brain; 5, adrenal gland; 6, mandibular lymph node; 7, mesenteric lymph node; 8, spleen; 9, ovary; 10, uterus; 11, testis; 12, fat; 13, bone morrow; 14, heart; 15, muscle; 16, kidney; 17, thymus; 18, thyroid; 19, lung; 20, trachea; 21, small intestine; 22, colon; 23, stomach; 24, liver; and 25, bladder. **Upper Panel**: expression profile is given as copy numbers for each SV per 25 ng of amplified cDNA. **Middle Panel**: percentage (%) of each SV from the total amount of SV1, SV2 and SV3 ATM transcripts. **Lower Panel**: total amount of ATM transcripts. Black bars represent the copy number of transcripts obtained by TaqMan assay using the exon6-exon7 junction as a probe. Dotted bars represent the copy number calculated as the sum of transcripts detected with SV1, SV2 and SV3 TaqMan assays in the same tissue.

Conclusion

Comparative genomics can provide insights into the development of "ideal" animal models for addressing human diseases. In addition, tools (targeted gene manipulation, and NT cloning) are emerging to create comparative phenotypic models, however, more phenotypic metrics are required to recapitulate the human condition. The pig genome x environment (G x E) nutritional model of human cardiovascular disease has been validated and supports intervention strategies. Porcine drug metabolism is more relevant to human FDA studies than murine [8]. Targeting porcine pathways associated with human cancer provides a genetically malleable tumour model for pre-clinical applications. Both sequence diversity and splice variants contribute to the complex neurological disease phenotypes. Recombineering supports the development of specific allelic and SV lines for the study of developmental genes.

Acknowledgements

This work was supported in part by USDA/NRI-CSREES grant AG2001-35205-11698, USDA-ARS Cooperative Agreement AG58-5438-2-313, DOD Training Grant W81XWH-06-1-0437 and the AT Children's Project.

References

1 National Vital Statistics Reports. 2002. Estimated life expectancy at birth in years, by race and sex: Death-registration States, 1900-28, and United States, 1929-2000.

2 National Vital Statistics Reports. 2000. Deaths: Leading Causes for 2000.

3 Swanson KS, Mazur MJ, Vashisht K, Rund LA, Beever JE, Counter CM, Schook LB: Genomics and clinical medicine: Rationale for creating and effectively evaluating animal models. Exp Biol Med 2004;229:866-875.

4 Lusis AJ, Fogelman AM, Fonarow GC: Genetic basis of atherosclerosis: Part i: New genes and pathways. Circulation 2004;110:1868-1873.

5 de Knijff P, van den Maagdenberg AM, Frants RR, Havekes LM: Genetic heterogeneity of apolipoprotein e and its influence on plasma lipid and lipoprotein levels. Hum Mutat 1994;4: 178-194.

6 Dewhirst MD, Thrall D, Macewen E: Spontaneous Pet Animal Cancers, In Teicher BA (ed.): Tumor Models in Cancer Research, Totowa, NJ, Humana Press, 2000, pp 565-589.

7 Tumbleson M, Schook LB (ed.): Advances in Swine in Biomedical Research. New York, NY, Kluwer Academic, 1996.

8 Pollock CB, Rogatcheva MB, Schook LB: Comparative genomics of xenobiotics metabolism: a porcine-human PXR gene comparison. Mamm Genome 2007;18:210-219.

9 Hahn WC, Dessain SK, Brooks MW, King JE, Elenbaas B, Sabatini DM, DeCaprio JA, Weinberg RA: Enumeration of the simian virus 40 early region elements necessary for human cell transformation. Mol Cell Biol 2002;22:2111-2123.

10 Yeh E, Cunningham M, Arnold H, Chasse D, Monteith T, Ivaldi G, Hahn WC, Stukenberg PT, Shenolikar S, Uchida T, Counter CM, Nevins JR, Means AR, Sears R: A signalling pathway controlling c-Myc degradation that impacts oncogenic transformation of human cells. Nat Cell Biol 2004;6:308-318.

11 Mazur M, Pettigrew J, Zachary J, Schook LB: Dietary cholesterol induced atherosclerosis in genetically pre-disposed cloned pigs. Under review.

12 Adam SJ, Rund LA, Kuzmuk KN, Zachary JF, Schook LB, Counter CM: Genetic induction of tumorigenesis in swine. Oncogene 2007;26:1038-1045.

13 Rogatcheva MB, Fritz KL, Rund LA, Pollock CB, Beever JE, Counter CM, Schook LB: Characterization of the porcine ATM gene: Towards the generation of a novel non-murine animal model for Ataxia-Telangiectasia. Gene doi:10.1016/j.gene.2007;405:27-35.

14 Johnson JM, Castle J, Garrett-Engele P, Kan Z, Loerch PM, Armour CD, Santos R, Schadt EE, Stoughton, R, Shoemaker, DD: Genome-wide survey of human alternative pre-mRNA splicing with exon junction microarrays. Science 2003;302:2141-2144.

Dr. Lawrence B. Schook, PhD, Institute for Genomic Biology, University of Illinois, Room 2109, 1206 W. Gregory Drive, Urbana, IL 61801, USA.

E-mail: schook@uiuc.edu

Pinard M-H, Gay C, Pastoret P-P, Dodet B (eds): Animal Genomics for Animal Health. Dev Biol (Basel). Basel, Karger, 2008, vol 132, pp 27-33.

Exploiting Genomics to Improve Animal Health

M.A. Smits, J.M.J. Rebel, M.F.W. te Pas, M.M. Hulst

Animal Sciences Group, Wageningen UR, Lelystad, The Netherlands

Key words: genomic selection, host-pathogen interaction, nutrition, systems biology

Abstract: The research area of animal genomics is moving now from its sequencing era into an integrative functional genomics era. The fast growing sequence information of animal genomes provides exiting opportunities for improving animal health traits by genomics-assisted breeding approaches. In addition, data from functional genomics studies offers deeper insight into the biological mechanisms that underlie animal health phenotypes. Understanding host-pathogen relationships, for example, promises to forward the integration of health genetics into breeding programmes and the development of new tools and strategies for the diagnosis, prognosis, treatment and prevention of infectious diseases. Similarly, increased knowledge on nutrient-gene interactions provides information on the effects of nutrients on biological processes. This knowledge may be used to redefine and optimize the nutritional needs of healthy animals. In this paper, prospects, challenges, and requirements of animal genomics research for improving animal health will be presented.

LIVESTOCK GENOME SEQUENCING

In 2004 the first draft of the genome sequence of a farm animal species, *Gallus gallus*, was deposited into free public databases. The latest version of the chicken genome assembly was released in May 2006. Its genome comprises approximately 1.05 billion base pairs and 20.00-23.000 estimated genes. In addition, more than 3.2 million chicken SNPs are currently known. The sequencing of the bovine genome initiated in 2003 and the third version of the bovine genome assembly (Btau_3.1) was released in August 2006. The bovine genome is roughly the same size (3.25 billion base pairs) as that of the human. Currently, over 2.2 million cattle SNPs are included in public databases. For pig, no genome sequence is available yet, but the first draft is expected early 2009. Shotgun sequencing of its genome is in progress now to achieve a draft 4x sequence depth across the genome through a minimal tilepath BAC-by-BAC approach. The pig chromosomes 7 and 14 are sequenced to a depth of 6-8x. The pig sequence data is being released through Pre-Ensemble. Since many entirely sequenced genomes are available now, comparative genomic approaches are increasingly used to accelerate our knowledge gain on the structure and the annotation of farm animal genomes, particularly for the identification of regulatory genomic elements.

Technological Revolution of DNA Sequencing

A recent milestone in the genome sequencing domain was the establishment of the complete genome sequences of two human individuals [1] in a relatively short period of time and at limited costs. This is demonstrable for the rapid progress in DNA sequencing technologies during the last years. In addition, new ultra high throughput genome sequencing technologies (Roche/454, Illumina/Solexa, ABI/SOLiD, Heliscope/Helicos) have recently been developed that change radically the ease, speed, and costs at which sequence data can be generated. Using massively parallel sequencing approaches, these technologies increase the speed and capacity of sequencing by 100 to 100,000 fold from current methods and can generate more than a billion bases of data in a single run. Since the lengths of their current sequence reads is still rather limited, the new technologies are mainly used for re-sequencing and comprehensive surveys of sequence variation (SNP) within populations. In the coming years we can expect even further technological developments that will lead to the sequencing of a complete genome in less than a day for only $1,000. With this information, the genetic potential of individual animals can be documented and, theoretically, the complete genetic potential for performance traits can be assessed. New computational techniques will be required to exploit such huge datasets.

Exploiting Genome Sequence Information

One of the first applications of animal genomic research is the possibility to genotype individual animals for thousands of SNPs in a short period of time and at a cheap rate. This is a very important prerequisite for the development of genome-based selection procedures. "Genomic selection" aims to predict breeding values for very large numbers of consecutive genome-fragments across the genome in animals without performance records [2]. Such approaches accelerate the genetic selection process and reduce the phenotyping costs in progeny. In addition, more precise genetic improvement can be obtained since the selection of more than one trait can be monitored at the same time. Several breeding industries already started with the implementation of genomic selection procedures using 3000 – 50.000 SNPs. The success rate of these approaches are dependent on the availability and accessibility of genetic resource populations that are deeply phenotyped by standardized systems, especially for traits that are difficult to measure, such as disease resistance, well-being, and responsiveness to microbes and vaccines.

Based on the steadily increasing knowledge on the relationship between phenotype and genomic variation, a number of other potential applications can be envisaged for the future [3]. These include the prediction of the "genetic value" of individual animals, the design of precision mating systems, and the sorting of animals into optimal management systems. The latter may result in the application of preventive healthcare programmes that match the animal's genotype.

Critical Requirements for Fully Exploiting Genome Information

There are a number of critical requirements for a better exploitation of genome sequence information. One important requirement is to get insight into the extent of structural variation (other than SNPs) in animal genomes and its impact on animal performance traits. Recent studies in humans, mice and the chimpanzee have indicated that their genomes contain an abundance of DNA segments with structural (copy number) variation ranging in size from 1 Kb to several Mb [4,5]. The published data suggest that the greatest source of genetic diversity lies not in the millions of SNPs, but rather in these variable segments. Since many of these segments overlap known annotated genes, it is likely that they play a major role in determining quantitative phenotypic characteristics. In order to fully understand the genetic variation underlying phenotypic variations in animal genomes, it is important to characterize animal genomes for both SNP and variable segments on a genome-wide scale.

Another area, which is not well developed yet, is the study of heritable epigenetic information. It is known that epigenetic mechanisms mediate the adaptation of the genome to specific environmental conditions. This is accomplished by DNA modifications which are superimposed on the primary DNA sequence, i.e. DNA methylation of CpG islands and histone modification [6]. Improved knowledge on epigenetic mechanisms, together with gene interactions and epistatic effects, will be necessary to fully exploit genomic information for improving animal health.

Functional Genomics and Animal Health

Functional genomic studies fill the gap between genotype and phenotype and aim to provide insight into the dynamic molecular relationships in biological systems. The area deals with transcription, translation, protein interaction, protein activity, and biosynthesis of metabolites. The latter two research fields are not well developed yet for livestock species. In contrast, mRNA expression profiling has proven to be a powerful tool in studying the biology of all kinds of animal health traits. Whole-genome DNA microarrays, necessary for these large scale gene expression analyses, are (commercially) available now for all major farm animal species. The current research in this field has a strong focus on host-pathogen interactions, since the quality and quantity of these interactions determine important health traits. These studies aim to elucidate the principles of pathogen recognition and host responses, to identify the genes and signalling pathways involved in these processes, and to define the differences between susceptible and resistant hosts [7]. Genetic variation in resistance to infectious diseases has been observed in all major livestock species. It is hoped that this variability is reflected by differences in gene expression profiles and in the utility of specific signalling pathways. The current studies provide long lists of genes that are differentially expressed between the conditions under study, e.g. infected *versus* non-infected and susceptible *versus* resistant. They have identified new genes and pathways involved in processes such as pathogen recognition, signalling, defence, tissue repair, apoptosis, etc. Striking differences in host responses have been observed between breeding lines, cell types, infection models, and pathogen strains. The challenge will be to identify common mechanisms of host resistance.

Nutrigenomics and Animal Health

It is known that environmental factors influence the expression of genes that contribute to health traits. Besides pathogens, diet is one of the most potent factors that alters the environment of cells and tissues. This insight has led to the development of the relative new nutrigenomics research field which examines nutrient-gene interactions on a genome-wide scale. For humans this research field is focussing on the identification of genes which influence diet-related disease and on the development of dietary intervention strategies to recover normal homeostasis in diseased individuals. For animals the primary research focus is on nutrition-based health, improvement of product quality, and increasing fertility.

Up to now, dietary fat is probably the most intensively studied macronutrient in this respect, since the amount and type of fat has major effects on growth and health. These effects are caused by the regulation of key metabolic genes by transcription factors like PPAR and SREBP that are (in)activated through their direct interaction with specific fatty acids [8]. SREBP for example, activates in mice the expression of more than 30 genes involved in lipid and fatty acid metabolism. SREBP is also involved in bovine mammary lipid synthesis and is inactivated by *trans-10*, *cis-12* conjugated linolic acid that can be produced during rumen biohydrogenation.

Since all nutrients enter the body via the gut, the gastrointestinal tract is another major focal point of nutrigenomics studies. Here, a variety of interactions occur between diet and microbes (commensals and pathogens), diet and host cells, microbes and host cells, and between the various microbial species. Various nutrients and bioactive feed ingredients have an effect on these interactions and influence intestinal health by modulation of the host (immune) response, by inhibiting the growth of pathogen, or by modulating their pathogenicity. Our own nutrigenomics research, for example, provided evidence that linseed in piglet diet affects the intestinal expression of several genes, including genes involved in the formation of mucus and other compounds of the innate defence system [9]. In another study in pigs, we demonstrated that a probiotic *Lactobacillus* strain can induce the intestinal expression of a gene coding for a bactericidal protein (manuscript in preparation). Finally, Rebel et al [10] demonstrated in chicken that maternal diets have an effect on the intestine of the offspring, not only at the level of gene expression, but also on the proliferation of immune cells, and on the development of an intestinal infectious disease.

Potential Applications of Functional Genomics Information

As already indicated, there exists a major gap in our understanding of the relationship between genome structure and phenotypic trait. We need to fill up this gap first before functional genomics information can be exploited extensively. Nevertheless, it is expected that the integration of large scale expression data with genome-wide DNA mapping data will facilitate the identification of biological and positional relevant gene targets for breeding industries [11]. Increased functional genomic knowledge will allow for the selection of animals with improved disease resistance, improved responses to vaccines, or with a better adaptability to housing systems. The pharmaceutical industries already benefit from knowledge obtained by functional genomic approaches through the development of new generation vaccines that include components that activate innate immune mechanisms and route the adaptive immune response in specific directions.

Another promising application of functional genomic studies is the identification of gene expression "signatures" or biomarkers exhibiting the physiological "status" of a biological sample. In the future such biomarkers may be applied to measure the effects of treatments or may be used as a predictor for the "endpoint" of the ongoing physiological process. It can be envisaged that such biomarkers, probably in combination with new sensor technologies, will be developed into tools for the monitoring of animal health and the prognosis and diagnosis of early or unknown infections. Molecular signatures may also be used for the development of new strategies to enhance disease resistance by modulation of (the awareness of the) immune system by the use of environmental stimuli such as nutrients, bioactive compounds, pre- and probiotics, receptor ligands, avoidance of stress, etc. In addition, the identification of molecular signatures that are correlated with an increased protection level in animals would also be extremely beneficial to the development of improved vaccines. Finally, as already indicated, results from nutrigenomics studies may be developed into new strategies for the induction of nutrition-based health, improved fertility, and higher product quality.

Critical Requirements for Exploiting Functional Genomic Information

The currently available DNA microarrays are based on initial genome annotation efforts. However, genome annotations and pathway descriptions are steadily improving and this new knowledge should be taken into account for the next generation of DNA microarrays. In addition, they should include elements that allow for the measurement of the expression of regulatory RNA's (e.g. miRNA), protein binding (ChIP-on-chip), and DNA modification. Furthermore, tools and resources for the high throughput measurement of protein profiles in biological samples of livestock animals should be improved.

To extract the maximum value from the increasing number of functional genomics datasets, methods are needed to evaluate, integrate and cross-validate related datasets derived from different laboratories. Likely candidates for integration are datasets derived from various infection models, datasets derived from different hosts infected with the same pathogen, or datasets derived from a single host infected with different pathogens. To this end, reliable statistical models for performing meta-analyses of independently generated datasets are required. A further prerequisite for this, is the availability of the relevant datasets in publicly accessible databases in a form that enables the full analysis and interpretation of the results, e.g. MIAME compliant datasets for transcriptional studies.

Although functional genomic technologies have already delivered a wealth of information, the gain in our understanding of the biological processes underlying animal traits so far is limited. It is increasingly recognized, however, that data generated from different biological levels (DNA sequence, transcription, translation, pathways, networks of pathways, metabolic synthesis) and different time scales do not exist on their own, but are connected to each other; they all form part of the same biological system that is organized in multiple levels that influence each other in a dynamic way. In the "systems biology" approach these different types of data are integrated with each other. This requires the development of a standardized vocabulary or ontology for all the biological levels up to the phenotypic level. Systems biology is an emerging interdisciplinary research field combining biology, mathematics and computational science that aims at building models for dynamic molecular interactions at the cellular and sub-cellular level. It will enable researchers to predict the outcome of a biological system in response to a given external factor. As a first step in the systems biology approach, we have integrated microarray data of *Salmonella* infected chicken with pathway information present in public databases. We identified networks of pathways that are induced by the pathogen in a susceptible chicken line. Surprisingly, the basic expression levels of these pathways, under non-challenge conditions, seemed to be higher in a more resistant chicken line compared to the susceptible line [12].

Conclusion

Animal science has arrived at the threshold of the genomic revolution. It is in a position to make most effective use of genomic comparisons across many species. Improved knowledge on the structure and variation of animal genomes and a better insight into the biology of traits provides exiting opportunities to improve animal health issues. This field of research will benefit from close collaborations between interdisciplinary scientific communities that share resources, technologies, knowledge, and that integrate their joint datasets. With the development of systems biology approaches we are entering the era of a predictive theoretical biology for farm animal traits [13].

Acknowledgement

The FP6 EU Network of Excellence EADGENE is acknowledged for financial support.

References

1 Levy S, Sutton G, Ng PC, Feuk L, Halpern AL, Walenz BP, et al: The diploid genome sequence of an individual human. PLoS Biol 2007;5(10):e254.

2 Meuwissen THE, Hayes BJ, Goddard ME: Prediction of total genetic value using genome-wide dense marker maps. Genetics 2001;157:1819-1829.

3 Green RD, Burfening PC, Cockett NE, Kappes S, et al: Blueprint for USDA efforts in agricultural animal genomics 2008-2017. 2006; http://www.csrees.usda.gov/newsroom/news/2007news/blueprint.html

4 Redon R, Ishikawa S, Fitch KR, Feuk L, Perry GH, Andrews TD, et al: Global variation in copy number in the human genome. Nature 2006;444:444-454.

5 Korbel JO, Urban AE, Affourtit JP, Godwin B, Grubert F, et al: Paired-end mapping reveals extensive structural variation in the human genome. Science 2007;318:420-426.

6 Higgs DR, Vernimmen D, Hughes J, Gibbons R: Using genomics to study how chromatin influences gene expression. Annu Rev Genomics Hum Genet 2007;8:299-325.

7 van Hemert S, Hoekman AJW, Smits MA, Rebel JMJ: Early host gene expression responses to a Salmonella infection in the intestine of chickens with different genetic background examined with cDNA and oligonucleotide microarrays. Comp Biochem Physiol D 2006;1:292-299.

8 Salter AM, Tarling EJ: Regulation of gene transcription by fatty acids. Animal 2007;1:9:1314-1320.

9 Jansman AJM, Niewold TA, Hulst MM: Inclusion of linseed and linseed expeller meal in piglet diets affects intestinal gene expression profiles. Livestock Science 2007;108(1-3):23-25.

10 Rebel JMJ, Van Hemert S, Hoekman AJW, Balk FRM, Stockhofe-Zurwieden N, Bakker D, Smits MA: Maternal diet influences gene expression in intestine of offspring in chicken (*Gallus gallus*). Comp Biochem Physiol A 2006;145:502-508.

11 de Koning DJ, Carlborg Ö, Haley CS: The genetic dissection of immune response using gene-expression studies and genome mapping. Vet Immunol Immunopathol 2005;105:343-352.

12 te Pas MFW, van Hemert S, Hulsegge B, Hoekman AJE, Pool HM, Rebel JMJ, Smits MA: A pathway analysis tool for analyzing microarray data of species with low physiological information. Submitted

13 Quackenbush J: From omes to biology. Animal Genetics 2006;37(suppl. 1):48-56.

Dr. Mari A. Smits, PO Box 65, 8200 AB Lelystad, The Netherlands.
E-mail: mari.smits@wur.nl

Pinard M-H, Gay C, Pastoret P-P, Dodet B (eds): Animal Genomics for Animal Health. Dev Biol (Basel). Basel, Karger, 2008, vol 132, pp 35-41.

Improving Pig Health through Genomics: a View from the Industry

M.A. Mellencamp[1], L. Galina-Pantoja[2], C.D. Gladney[3], M. Torremorell[2]

[1] Ralco Nutrition, Inc., Marshall, MN, USA
[2] Genus Plc, Hendersonville, TN, USA
[3] Genus Plc, De Forest, WI, USA

Key words: genomics, pig health, F18+ E. coli, PRRSV

Abstract: Health is one of the most important contributors to animal welfare, productivity and profitability in pig production today. For the past 30 years, pig breeders have focused on genetic improvement of lean growth, feed efficiency, meat quality and reproduction. However, in recent years, selection objectives have been broadened to include livability, robustness and disease resistance. A DNA marker for selection of resistance to F18+ E. coli has been available for several years. This marker decreases mortality and improves growth on farms experiencing post-weaning scours and/or oedema disease. However, for most diseases affecting intensive production systems such as porcine reproductive and respiratory syndrome (PRRS), porcine circovirus type 2-associated diseases (PCVAD), *Haemophilus parasuis*, and swine influenza virus, resistance is a complex and polygenic trait. Selection for improved resistance to these diseases will be incremental and require use of multiple markers in complex breeding schemes. Novel technologies such as pig gene microarrays, single nucleotide polymorphism (SNP) panels and advanced bioinformatics are being used to identify new health candidate genes for these economically important diseases. Lagging behind, however, is availability of large DNA datasets from pedigreed populations with accurately measured health phenotypes that are needed to identify associations between SNPs and health traits. Increased focus on datasets with health traits will be the key to finding useable discoveries with new genomics technologies. Currently, the industry uses dozens of SNP markers to increase the accuracy of selection for complex breeding objectives, including disease resistance. As the pig genome is sequenced and barriers to genotyping thousand of markers are eliminated, genomic selection for health traits will receive increasing attention from commercial breeders.

INTRODUCTION

Traditionally, the pig breeding industry has focused its efforts on genetic improvement of heritable traits with high economic value, including lean growth, feed efficiency, carcass, meat quality and reproductive efficiency. This was done using extensive phenotypic databases with generations of pedigrees and sophisticated programs to improve the accuracy of estimated breeding values (EBVs). Trait measurement and selection of pure lines has taken place in well managed, nucleus herds, without the common health challenges present on commercial farms.

Disease is a key driver of profitability, productivity and welfare in commercial pig production. Despite health management schemes that include the use of antimicrobial agents, vaccines, diagnostic tests, age segregation, and rigorous biosecurity plans, swine diseases continue to have a negative impact on production. Porcine reproductive and respiratory syndrome virus (PRRSV), classical swine fever, and foot and mouth disease continue to have a huge economic impact on pig production worldwide. Swine influenza virus has resurfaced as an important swine pathogen, with new reassortant virus variants emerging in North American herds. Costs associated with poor pig health include death loss, reduced reproductive productivity, decreased growth and feed intake, increased days to market, inefficient building use, variation of market weight pigs and costly management interventions. Clearly, new strategies are needed to complement the traditional methods of disease control.

Currently, two major strategies are used for selection for disease resistance. The first is selection for resistance to a specific disease. This manner of selection may not improve resistance to diseases other than the target disease. Resistance to oedema disease caused by F18+ *E. coli* is a primary example of selection for resistance to a specific disease. A genetic test for resistance has been available for several years.

The second approach is to improve immune responsiveness to the many disease agents encountered by animals in the commercial environment. While not implemented on a large scale by commercial breeders, this approach has been studied for several years. It has several advantages, including that the disease targets do not have to be identified and it may improve resistance to subclinical infections that play a role in reducing performance on commercial farms. Research has shown that various immunological traits have additive genetic variation [1], and selection using panels of immune traits has been correlated with disease resistance and improved performance in some cases [2]. However, testing can be very expensive and is available in specialized laboratories only.

Marker Assisted Selection for Disease Resistance

Molecular genetics has been integrated into traditional breeding for several years. The candidate gene approach identifies DNA markers in genes of interest that are thought to contribute to variation in the trait under consideration. The initial breakthroughs were associated with single genes with major effects. In 1991, "Halothane" gene testing became available to eliminate the recessive mutant halothane allele, which was associated with pale, soft, exudative pork [3]. Other tests soon became available, including the RN (Rendement Napole) "acid meat" gene test [4] and a DNA marker test for increased litter size [5].

One of the most attractive targets for using DNA markers for genetic improvement is in the area of health and disease. These traits are difficult and expensive to measure and have low heritability. The first commercial DNA marker test for disease resistance in pigs was for resistance to oedema disease caused by F18+ *E. coli*. A DNA marker test for a mutation in the FUT-1 gene on SSC6 [6] became available in 2003. Resistance to disease is clear cut. Studies of weaned piglets that were challenged with F18+ E. coli showed a reduction in death loss from 22% to less than 1% in pigs with two copies of the resistance allele (Fig. 1). In surviving pigs, resistant pigs demonstrated a 30% improvement in average daily gain (0.57 pounds/day) compared with susceptible pigs (0.4 pounds/day) (Fig. 2) [7].

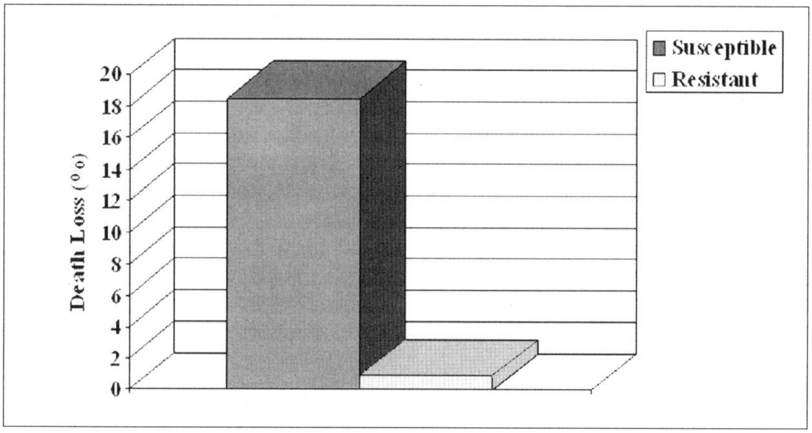

Fig. 1: Mortality of F18+ *E. coli* resistant and susceptible pigs following oral challenge.

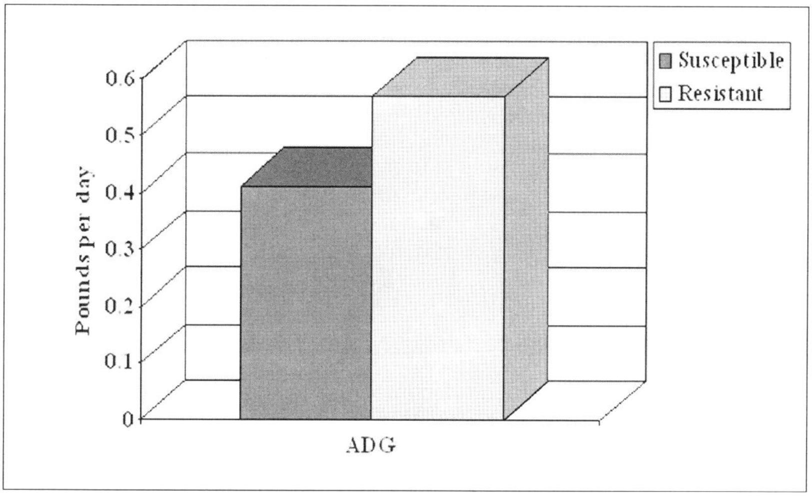

Fig. 2: Average daily gain of surviving F18+ *E. coli* resistant and susceptible pigs following oral challenge.

Resistance to another *E. coli* disease, pre-weaning scours caused by F4+ (K88) *E. coli*, has been known since the 1970's [8]. Genetic studies mapped this recessive trait to SSC13 in the mid-1980's. Since then, several research teams have embarked on using positional cloning to identify likely candidate genes. Several viable candidate genes have been proposed in the past few years.

Resistance to other pig diseases, including PRRS, *Haemophilus parasuis* and Salmonellosis, is complex and associated with multiple genes. Susceptibility is weakly heritable. "All-or-none" resistance may not be possible and the implementation of DNA markers may result in incremental reduction in susceptibility.

Pig Improvement Company's (PIC) genomics research program has focused on several pig diseases in the last few years, including PRRS. PRRS has been estimated to cost the North American swine industry approximately $560 million dollars annually [9]. Infection is not easily controlled. In this study, associations of 60 SNPs with litter traits of sows infected with PRRSV were studied. Candidate genes targeted known virus replication and receptor genes, macrophage genes and other innate immunity genes. Datasets from two farms were used for the association studies. Both farms were naïve for PRRSV before the outbreak and a programme of intentional exposure assured that all sows and gilts were infected. Several reproductive traits were evaluated, including number born alive, total number born, mummies, stillbirths, abortions and farrowing interval. Results showed that several SNPs were associated with reproductive traits. Figure 3 shows associations of four markers with a single trait: number of pigs born alive during the PRRSV outbreak. Sows with the "11" genotype for marker B showed nearly 2 additional pigs born alive compared with sows with the 22 genotype. A greater effect was observed for sows with the 22 genotype for marker C [10].

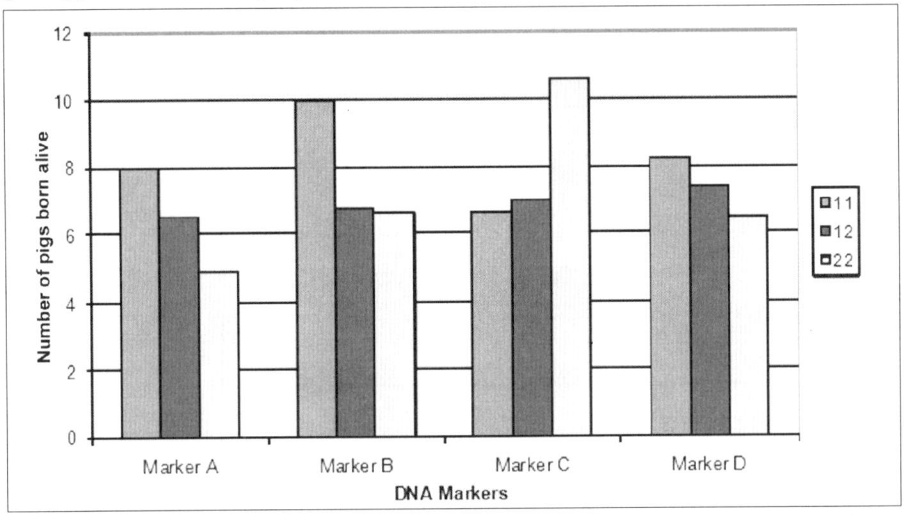

Fig. 3: Effects of four DNA markers on the number of piglets born alive from sows infected with PRRSV.

One of the challenges of implementation of DNA markers from studies such as the one described above is how to select the "best" DNA marker or combination of markers. In the PRRSV study (as with other disease-related studies), markers with the greatest effects for some traits were not always associated with improvement in others. Some markers showed consistent results across datasets while others did not. Datasets with multiple traits describing the disease phenotype were particularly challenging. Finding a single trait that best represented resistance to PRRSV in these datasets was impossible. Marker interactions were of concern, as were possible negative effects on performance, reproduction or meat quality traits.

Validation and Implementation of DNA Markers in Commercial Pig Breeding Programs

Once DNA marker associations are discovered, markers must be subjected to an intense validation scheme to determine their impact in multiple pig lines and across different traits in the breeding program. At PIC, DNA markers are routinely validated by testing against a DNA panel from several hundred pigs from several diverse lines with performance, mortality, carcass, meat quality, and reproduction traits. Such testing provides a broad picture of the effects of markers before implementation. Once fully validated, markers are implemented in line development. A balanced approach to the use of disease resistance markers is the key to successful implementation. DNA marker profiles are only one component of the breeding program. Other components include performance in the genetic nucleus and in commercial situations, carcass evaluation and meat quality traits (Fig. 4). Simultaneous selection of health markers along with other important traits means that it may be years before the customer sees the results. Intense selection pressure on health traits decreases pressure on other traits. Breeding goals have evolved over the past 30 years and now routinely include selection for robustness, including disease resistance and mortality, and behaviour (Fig. 5).

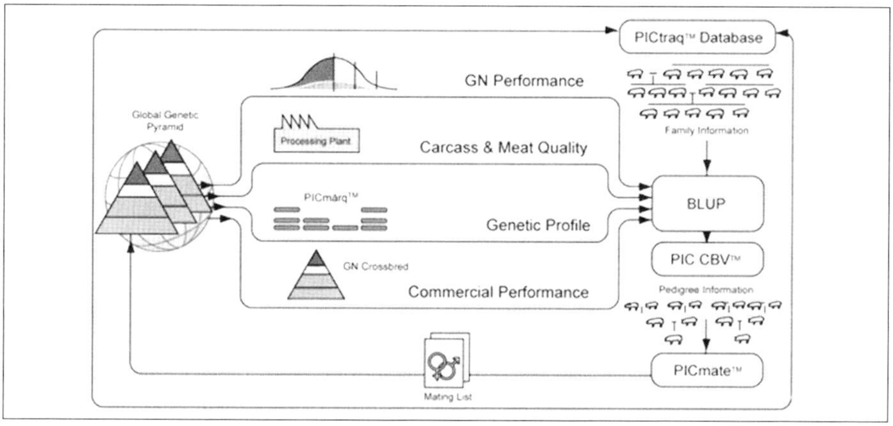

Fig. 4: Schematic of PIC breeding tools.

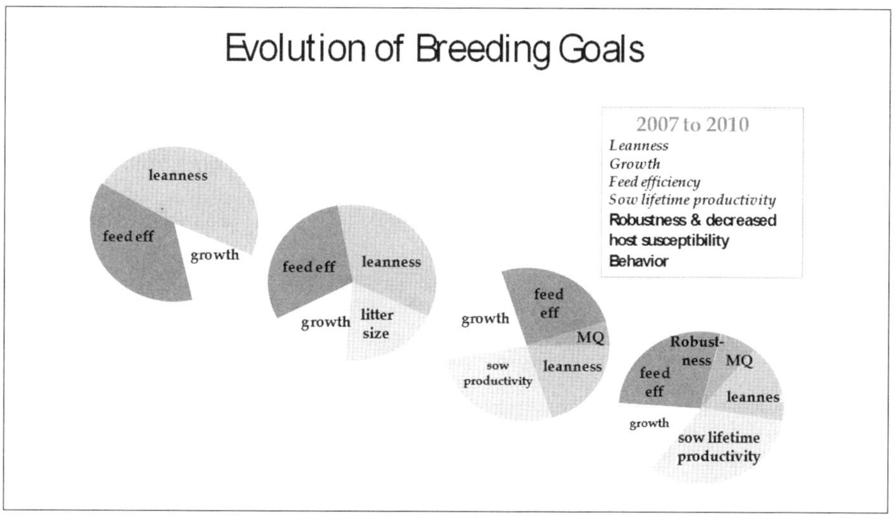

Fig. 5: Evolution of breeding goals from 1980-2010 (projected).

Challenges and Opportunities

One of the most challenging aspects of selection for disease resistance is the creation of DNA datasets for marker discovery and validation. Disease traits are difficult and expensive to measure. It is difficult to agree on a single or few traits that represent resistance to a particular disease, and these traits may not be applicable to the next disease. Experimental disease challenge studies present attractive populations because they have many disease-associated traits which are measured in detail. However, these studies often lack the numbers needed for adequate statistical power and do not have pedigrees. One solution to generating large populations for validation of disease resistance markers is to measure traits in commercial cross-bred populations that are raised under specific health and management environments. Health-related traits can be measured in these populations. Currently, the focus is on traits associated with robustness: preweaning survival, nursery mortality, grow-finish mortality, culls, and lightweight pigs.

Conclusion

Breeding for improved resistance to disease has clear benefits for the industry: reducing performance losses associated with acute and subclinical disease, decreasing costs associated with controlling and treating disease, improving pig welfare by reducing disease expression in commercial herds, decreasing reliance on antimicrobial agents, addressing issues of antibiotic resistance and improved food safety and security. Selection for disease resistance has had some success in the past few years, but new genomics technologies will allow for more rapid progress, provided resources are available. Integration of many new disease resistance markers into current selection tools has been accomplished, and this will result in genetic improvements moving into the marketplace. The use of markers to improve selection for disease and robustness traits will continue. The development of DNA datasets for health trait discovery and validation will be important for continued success.

Acknowledgements

The authors would like to thank the Genus R&D and Quantitative Sciences groups, including Dr. Nader Deeb, Dr. Matt Cleveland, Becky Geiger, Dr. Alan Mileham and Saad Sebbana for their invaluable contributions to these studies.

References

1 Edfors-Lilja I, Wattrang E, Marklund L, Moller M, Andersson-Eklund L, Andersson L, Fossum C: Mapping quantitative trait loci for immune capacity in the pig. J Immunol 1998;161:829-835.

2 Mallard BA, Wilkie BN, Kennedy BW, Quinton M: Use of estimated breeding values in a selection index to breed yorkshire pigs for high and low immune and innate resistance factors. Anim Biotechnol 1992;3:257-280.

3 Fujii J, Otsu K, Zorzoto F, DeLeon S, Khanna VK, Weiler JE, O'Brien PJ, McLennan DH: Identification of a mutation in porcine ryanodine receptor associated with malignant hyperthermia. Science 1991;253:448-451.

4 MacLenna DH, Duff C, Zozato F, Fujii J, Phillips M, Korneluk RG, Frodis W, Britt BA, Worton RG: Ryanodine receptor gene is a candidate for predisposition to malignant hyperthermia. Nature 1990;343:559-561.

5 Rothschild MF, Messer LA, Vincent A: Molecular approaches to improve pig fertility. J Reprod Fertil Suppl 1997;52:227-236.

6 Vogeli P, Bertschinger HU, Stamm M, Stricker C, Hagger C, Fries R, Rapacz J, Stranzinger G: Genes specifying receptors for F18 fimbriated *Escherichia coli*, causing oedema disease and postweaning diarrhoea in pigs, map to chromosome 6. Anim Genet 1996;27:321-328.

7 Mellencamp MA, Balint-Kurti P, Galina L: Genetic resistance to diseases in swine. Proc Ann Meet Am Assoc Swine Vet 2003;1:261.

8 Sellwood R: *Escherichia coli* diarrhoea in pigs with or without the K88 receptor. Vet Rec 1979;105:228-230.

9 Cho JG, Dee SA: Porcine reproductive and respiratory syndrome virus. Theriogenology 2006;66:655-662.

10 Galina-Pantoja L, Torremorell M, Deeb N, Geiger B, Gladney C, Mellencamp M: DNA markers associated with reproductive traits during PRRSV infection. Proc 19th Cong Int Pig Vet Soc 2006.

Dr. Martha A. Mellencamp, Director of Immunology, Ralco Nutrition, Inc., 1600 Hahn Road, P.O. Box 1083, Marshall, MN 56258 USA.

E-mail: marnie.mellencamp@ralconutrition.com

Pinard M-H, Gay C, Pastoret P-P, Dodet B (eds): Animal Genomics for Animal Health. Dev Biol (Basel). Basel, Karger, 2008, vol 132, pp 43-53.

Proteomic Dissection of Viral Pathogenesis

H.C. Liu[1], J. Hicks[1], D. Yoo[2]

[1] Department of Animal Science, North Carolina State University, Raleigh, USA
[2] Department of Pathobiology, University of Illinois at Urbana-Champaign, Urbana, USA

Key words: proteomics, viral pathogenesis, vaccine development

Abstract: Many factors play into the complexity of viral pathogenesis. Understanding viral pathogenesis is key to developing vaccines and treatments for viral diseases. One emerging area of research is proteomics, which is the study of the protein complement and functions of the genome. Many different proteomic approaches have been utilized by researchers worldwide to further elucidate viral pathogenesis. For example, a high throughput MALDI-MS approach was recently employed to study the antigenicity of the influenza virus. Another study utilized MALDI-TOF MS and liquid chromatography MS/MS of proteins present in lipid droplets of hepatoma cell lines to identify proteins involved in the pathogenesis of hepatitis C virus that contribute to its carcinogenic properties. In conjugation with MS, yeast two-hybrid systems have also been shown to be useful in identifying potential host receptors of various viruses as well as revealing the interaction of viral proteins with other host proteins and viral proteins. In this review, the focus is on various proteomic approaches to dissecting the mechanisms of viral pathogenesis.

INTRODUCTION

In the last decade proteomics has become an integral branch of molecular biology. Proteomics can generally be defined as the study of the protein complement of the genome of an organism [1]. Some of the most popular proteomics methods in use today, such as the yeast two-hybrid system and co-immunoprecipitation, are designed for identifying and characterizing specific protein-protein interactions. Other techniques, such as two-dimensional electrophoresis, have been developed for studying patterns and changes in global protein expression. Recently, various mass spectrometry techniques, in combination with well established structural analysis techniques, such as X-ray crystallography and nuclear magnetic resonance, have been applied towards identifying the amino acids that comprise a protein and how they come together to form the structure of a protein.

PROTEOMIC METHODS

The two-hybrid system

The yeast two-hybrid system [2] has become popular as a relatively quick and simple method for screening for protein-protein interactions. A yeast two-hybrid screen is often one of the first steps taken in characterizing a protein and its function(s). The principle of the technique relies upon reconstitution of a severed transcription factor's ability to activate expression of a reporter gene. The yeast protein GAL4 transcription factor consists of two distinct and independent functional domains, and its function is to regulate the expression of galactolytic enzymes. One domain of GAL4 specifically binds a unique DNA sequence near the promoter of galactolytic enzyme genes. The second domain interacts with RNA polymerase, leading to the activation of gene expression. In order for gene expression to occur, the DNA binding domain and the activation domain must be in contact. The GAL4 protein is manipulated for use in the yeast two-hybrid system by fusing a "bait" protein to the DNA-binding domain and fusing "prey" protein(s) to the activation domain. Only when the bait and prey interact will the two domains be brought together and reconstitute a functional GAL4 molecule, leading to the expression of a reporter gene, of which β-galactosidase is the most commonly used [reviewed by 3]. Positive interactions between bait and prey are visualized by growing the yeast in the presence of the chromogenic substrate X-gal, which turns blue upon hydrolysis by the β-galactosidase. This makes it relatively simple to screen for positive interactions, which is one of the reasons why yeast two-hybrids screens are so popular today.

Co-immunoprecipitation

This technique is frequently used for discovery of protein-protein interactions [4,5] and can confirm interactions identified by two-hybrid analysis. In co-immunoprecipitation, an antibody directed against a specific protein within a complex of interacting proteins is added to a cell lysate. The antibody binds its target to form an immune complex, which is then captured on a solid support cross-linked to Protein A or Protein G, which bind to the Fc region of the antibody. Capture of the immune complex serves to "precipitate" the target protein and any other proteins associated with it from solution. Proteins that are not precipitated are washed away from the solid support. The bound proteins are then eluted from the support by boiling and analyzed by SDS polyacrylamide gel electrophoresis. The identity of the proteins in the complex can then be determined by Western blot analysis or by mass spectrometry.

Two-dimensional polyacrylamide gel electrophoresis (2D-PAGE)

2D-PAGE is another popular proteomic technique and is a rather robust means for characterizing global protein expression. In 2D-PAGE proteins are separated in two dimensions. In the first dimension, proteins are separated by their isoelectric points. The protein sample is applied to the end of a polyacrylamide gel rod or strip containing a pH gradient and an electric current is applied. The proteins migrate through the gel until they enter the region where the pH equals their isoelectric point. Here they assume a net charge of zero and cease to migrate any further in the gel. After the proteins have been separated by their isoelectric points, they are separated by their mass. This separation is similar to standard SDS-PAGE gel electrophoresis. The gel rod containing the proteins separated in the first dimension is placed onto a polyacrylamide gel containing the anionic detergent sodium dodecyl sulfate (SDS). SDS denatures and binds the protein, causing it to assume an equal charge to mass ratio. Because the proteins are denatured, their migration through the gel is not affected by their secondary structure. In the second dimension, proteins with a lower mass migrate faster than proteins with a higher mass. Finally, to visualize the proteins, the gel is stained. This reveals a very clear picture of the proteins in the sample, as they have been separated by their isoelectric point and mass [reviewed by 6]. Two-dimensional electrophoresis is not only useful for visualizing global protein expression but is also useful for characterizing protein modifications. For example, if a protein is phosphorylated, it can appear as several spots in the gel due to different stages of phosphorylation. This information can help to better understand how a particular protein is synthesized and modified and perhaps give insight into its function.

Mass spectrometry methods

Mass spectrometry analyzes the characteristics of a substance based on its mass and charge. These characteristics can be used to identify an unknown compound. In regard to proteomics, various mass spectrometry devices have been developed to study various aspects of proteins. In the most general sense mass spectrometry is used to determine the molecular weight of a protein by separating its molecular ions based on their mass to charge (m/z) ratio. In order to analyze a protein, it must first be ionized by an ionizing source, such as an atom beam or a laser light. These ions are carried into a mass analyzer and are separated based on their mass-to-charge ratio. This produces a mass spectrum, a representation of the mass to charge ratios of all of the ions of the sample, which can then be used to identify the sample. Mass spectrometry is a vital component of proteomic analysis because it allows for the identification of proteins. Many mass spectrometry methods are currently in use in proteomics. These methods vary in their degree of fragmentation, their resolution, how much sample is required, etc. Which method is chosen depends upon the type of information that is sought.

In electron ionization, electrons are accelerated from a heated filament to form an ion beam, and the sample is desorbed into this beam. The electrons then transfer energy to the sample which ionizes it. This method has sub-picomole sensitivity, and there are several databases available for identifying the sample. However, this method is limited in the mass range that can be analyzed, and it leads to extensive fragmentation of the sample, which can result in no observable molecular ions.

Another type of mass spectrometry is fast atom bombardment [reviewed by 7]. With this method the sample is mixed with a liquid matrix, often glycerol, and placed on the tip of a solid probe. Then an atom beam, either Cs or Xe, is directed at the sample causing it to ionize. The matrix absorbs some of the energy which assists with the ionization of the sample and helps limit fragmentation of the sample. However, as this method is only accurate for masses up to 7 kDa, it is limited to the analysis of small proteins.

A very popular type of mass spectrometry used in proteomic studies is matrix-assisted laser desorption/ionization-mass spectrometry (MALDI-MS) [reviewed by 7]. In MALDI-MS the sample is dissolved into a solid matrix. A laser beam is then directed onto the matrix/sample, and the energy vaporizes and ionizes the sample. MALDI-MS is popular because it is a soft ionization form of mass spectrometry, meaning that little to no fragmentation occurs. Consequently a fully intact protein can be ionized, making it easier to determine a protein's molecular weight because there will only be one peak on the spectrum. MALDI-MS is also useful because it works for large molecules and is very tolerant of salt, which is often needed to keep a protein in solution.

Electrospray ionization (ESI) is the softest ionization method of mass spectrometry. With this technique even noncovalent interactions can be preserved [reviewed by 8]. In ESI the sample is sprayed from the tip of a probe, and the resulting droplets enter a strong electric field where a dry and/or heated gas is applied, which causes evaporation of solvent from the droplet surface. Eventually the repulsion of like charges overcomes the force of the surface tension, which causes ions to leave the droplet and enter the mass analyzer. Because of its soft ionization approach, extensive information about a protein can be obtained.

Understanding the structure of a protein is key for understanding how it functions. To determine the structure of a protein there are two major methods currently in use, X-ray crystallography and nuclear magnetic resonance (NMR). Though they function in different manners both methods can give high resolution structures of a protein.

X-ray crystallography

In this method an X-ray beam is directed onto a crystal of the protein, which diffracts the beam into a diffraction pattern, which is then mathematically analyzed to determine the protein's structure [reviewed by 9]. A well ordered crystal is needed for X-ray diffraction to work. There are several ways to produce a crystal; the most commonly used method is the hanging-drop method. In this method a drop of the protein solution is placed on a glass plate which is then sealed upside down over a concentrated precipitant solution. The precipitant solution draws water out of the protein droplet, causing the protein to come out of solution and to form crystals. Ultimately, the diffraction data is used to produce an electron density map of the amino acid sequence of the protein, which can then be fitted onto the map to determine the structure of the protein. The quality of this structure, or its resolution, is measured in angstroms (Å). For a good quality structure, the resolution should be 3 Å or less. At 3 Å, amino acids can be differentiated from one another and a resolution of 1.5 Å will give a very detailed protein structure.

Nuclear magnetic resonance (NMR)

NMR is another popular method for determining protein structure. This method is based on the principal that nuclei with odd mass numbers, such as H1, have a half integral spin and thus assume either a lower energy state α or a higher energy state β [reviewed by 9]. An external magnetic field is applied, causing the nuclei to resonate between the two fields. When all the nuclei act together they cause a measurable coherent effect between the two energy states that is influenced by neighbouring atoms. This information can then be used to determine how the different atoms that make up the protein interact with each other, which in turn can be used to determine the structure of the protein.

Examples of Proteomic Applications in Virology

Herpes simplex virus (HSV)

HSV is a large enveloped DNA virus and a member of the alpha herpesvirus group. Herpes simplex virus type 1 (HSV-1) causes oral cold sores and herpes simplex virus type 2 (HSV-2) causes genital herpes. Various proteomic techniques have been employed for dissecting the mechanisms of HSV pathogenesis. One of the earliest proteomic techniques to be used to study HSV was two-dimensional gel electrophoresis. In 1981, Haarr and Marsden [10] utilized 2D gel electrophoresis to study polypeptides and glycoprotein processing of HSV-1. They discovered that 230 polypeptides, both of host and viral origin, were induced by the virus, demonstrating the wide and diverse effects that viral infections have. This study also revealed that the viral glycoproteins gB, gC, and gD contain sialic acid, and that the synthesis of these proteins takes several steps.

The yeast two-hybrid system is an excellent method for determining what host proteins a virus overtakes to replicate itself and so can give great insight into viral pathogenesis. This method is often used in conjugation with other methods, such as co-immunoprecipitation to confirm the interaction(s). These approaches have been utilized to study many different HSV gene products. One study employed co-immunoprecipitation to identify interacting partners of ICP8, a single-stranded DNA (ssDNA) binding protein [11]. It was shown that ICP8 associates with over fifty other proteins, including both host and viral proteins. ICP8 co-immunoprecipitated with proteins involved with DNA repair and recombination and chromatin remodelling and various transcription factors. These discoveries provided new insights into the role of ICP8 and host factors in HSV replication.

If the function of a protein is unknown, identifying the proteins' interactive partners can suggest its putative function. For example, the protein encoded by the HSV-1 UL56 gene is a type II membrane protein of unknown function. Using UL56 as a bait in a yeast two-hybrid screen, followed by confirmation with a glutathione S transferase (GST) pull-down assay, determined that UL56 interacts with KIF1A, a member of the kinesin-3 family, a group of motor proteins involved in the transport of various proteins [12]. This led to the hypothesis that UL56 acts as a receptor for neuron motor proteins and may be involved in vesicular trafficking in infected cells [12].

In conjunction with mass spectrometry, co-immunoprecipitation was also used to identify proteins that interact with HSV protein ICP27, an immediate-early gene [13]. Because ICP27 is an immediate-early gene, it plays a role in viral late gene transcription, polyadenylation, and RNA export. Immediate-early genes generally act to regulate the expression of early and late genes. This study found that ICP27 interacts with various cellular translation initiation factors, such as poly A binding protein (PABP), eukaryotic initiation factor 3 (eIF3), and eukaryotic initiation factor 4G (eIF4G). These observations confirmed the importance of ICP27 in initiating gene expression, thus providing insight into which host proteins are manipulated by immediate-early genes to replicate the virus.

As mentioned earlier, X-ray crystallography is a technique in which a crystallized form of a protein is used to determine the structure of the protein. The structure of the core of the HSV protein VP16, which makes up part of a transcriptional regulatory complex, was determined in this manner [14]. The structure of the core of VP16 was determined at 2.1Å resolution, which provided decent resolution of the protein's structure. Several unique characteristics of this protein were revealed by its crystal structure. The core has a seat-like structure, which is a novel structure not seen before. It was also determined that the C-terminus has a more disordered structure and associates with other proteins of the transcriptional regulatory complex, and that the N-terminus, which is involved in virion assembly and DNA binding, is more ordered. X-ray crystallography illuminated novel characteristics of VP16, which provided insight as to how VP16 interacts with other factors to transcribe viral genes. Understanding how HSV genes are transcribed is a key factor in understanding its pathogenesis, and understanding this process can also be very helpful in developing new vaccines and drugs to prevent and treat infection.

Human cytomegalovirus (HCMV)

HCMV is a beta-herpesvirus and a leading cause of congenital defects. In healthy adults HCMV is usually asymptomatic, but it can cause disease in immunocompromised individuals. Several proteomic techniques have been employed to better understand HCMV pathogenesis.

Two-dimensional gel electrophoresis, in conjugation with mass spectrometry, has been used to elucidate the protein composition of the HCMV virion [15]. A diverse group of proteins was identified in HCMV virions, including 71 HCMV proteins and 70 host proteins. These approaches also revealed 12 HCMV proteins not previously identified in the virion and another 12 proteins encoded by novel open reading frames (ORFs). By determining the global protein composition of the HCMV virion, a number of novel proteins were found to be associated with virion, providing additional insight into the entry and replication mechanisms of this virus.

Yeast two-hybrid screens have been performed on several different HCMV proteins to better understand how their interactions with both host and viral proteins lead to disease. The HCMV protein pp71 is important in the activation of the immediate-early promoter of HCMV. Using pp71 as bait in a yeast two-hybrid screen, it was determined that pp71 interacts with the host protein human Daxx (hDaxx) [16]. When pp71 and hDaxx were co-transfected into host cells, hDaxx enhanced the ability of pp71 to activate the immediate-early promoter. This observation revealed a specific viral-host interaction that enhances the pathogenesis of HCMV and demonstrated how a viral protein can manipulate the host for its own benefit.

The yeast two-hybrid system can also reveal which domain of a protein interacts with its partner protein. For example, HCMV pUL80a and pUL80.5 are scaffold proteins involved in virion assembly. By mutating different domains and screening the mutants by yeast two-hybrid, it was determined that a conserved amino-terminal domain is required for both intramolecular and intermolecular interactions [17].

Marek's disease virus (MDV)

MDV is an oncogenic herpesvirus that is the causative agent of Marek's disease, a lymphoproliferative disease in chickens. Susceptibility to MDV infection is dependent on many factors that are both genetic and environmental. An infectious clone of MDV was found to have lost its oncogenicity, and it was discovered that loss of oncogenicity was due to an insertion of a long terminal repeat sequence upstream of the MDV *SORF2* gene [18]. A yeast two-hybrid screen of SORF2 revealed an interaction between SORF2 and chicken growth hormone [19], which was confirmed by a co-immunoprecipitation assay. Further analysis of growth hormone suggested that it contributes to the resistance of infection. This proteomic approach demonstrates how host factors can contribute to the degree of viral pathogenesis.

Herpesvirus infections consist of a lytic and a latent phase. To better understand herpesvirus pathogenesis it is critical to understand the switch from a lytic to a latent infection and vice versa. Similar to other herpesviruses, the lytic/latent switch in MDV is not well understood. To better understand this switch, global protein expression of lytically infected cells was determined by fractionating an infected cell lysate by strong cation exchange chromatography and then subjecting the fractions to microcapillary reversed-phase liquid chromatography-tandem mass spectrometry [20]. This approach revealed that viral proteins associated with DNA replication and capsid formation, i.e. proteins that are likely involved in a productive infection, were frequently identified proteins in lytically infected cells. Comparing these proteins to those expressed during latent infection will likely give good insight into the switch between lytic and latent infection.

Influenza virus

Influenza virus is an RNA virus with a segmented genome. Proteomic approaches have been very useful in furthering our understanding of influenza infection and discerning what makes certain flu strains so much more virulent than others, such as the strain that led to the pandemic of 1918.

The influenza protein NS1 is the only nonstructural protein of the flu virus, but its function is unknown. To develop a clearer idea of the function(s) of NS1, a yeast two-hybrid screen was performed to identify host proteins that specifically interact with this viral protein [21]. The screen revealed an interaction between NS1 and the human homolog of the Drosophila Staufen protein (hStaufen), which was confirmed by co-immunoprecipitation. It was suggested that this interaction could function to localize viral mRNAs to specific regions within the infected cell [21]. By determining that NS1 interacts with hStaufen, an idea concerning NS1's role in an influenza infection was developed, increasing our understanding of how influenza proteins are produced.

The flu pandemic of 1918 caused millions of deaths worldwide. Understanding what made this particular influenza strain so virulent may provide insight towards preventing another pandemic. To assist with this, two studies undertook the task of determining the structure of the haemagglutinin (HA) protein of the 1918 flu strain [22,23]. These studies found that the human influenza HA protein possesses properties similar to avian influenza HA protein. For example, the human HA protein has an avian-like receptor-binding site [22,23] but has properties enabling it to bind human receptors. The fact that this deadly human influenza has many avian influenza characteristics but binds human receptors helps to explain why this particular flu strain was so deadly.

Severe acute respiratory syndrome-associated coronavirus (SARS-CoV)

Severe acute respiratory syndrome (SARS) is caused by SARS-CoV. This virus causes a severe pneumonia that has resulted in hundreds of deaths worldwide [24]. Because SARS-CoV is a newly emerged virus and has only recently been identified, little is known about it. The various proteomic techniques available have provided for a better understanding of SARS pathogenesis.

A recent study identified an interaction between the SARS-CoV M (membrane) protein and host proteins that is likely a contributing factor to its pathogenesis [25]. The M protein of SARS-CoV shares some homology with the M proteins of other coronaviruses and is likely involved in virion assembly by interacting with the N (nucleocapsid) protein [26]. SARS-CoV M protein was also shown to interact with the host protein IKKβ kinase (IKKβ) via a co-immunoprecipiation assay [25]. This interaction was shown, via luciferase reporter assays, to suppress the expression of tumor necrosis factor alpha (TNFα). M protein was also shown to induce nuclear factor kappa B (NF-κB) activation, which suppresses cox-2, an important inflammatory protein, thereby contributing to the severe development of disease seen with SARS-CoV infection [25]. Identification of the interacting partners of SARS-CoV M protein has thus given insight into how this virus causes such severe disease.

Another study examined global interactions of SARS-CoV proteins using a yeast two-hybrid matrix analysis and identified 65 interactions [27]. It was shown that nsp8 (nonstructural protein 8) interacted not only with itself but with a wide range of replicase proteins required for SARS-CoV replication. The study indicated that nsp8 is probably a component of the SARS-CoV replication complex.

Nuclear magnetic resonance has been employed to better understand the lifecycle of the SARS virus. Chang et al. [28] utilized NMR to determine the structure of the SARS-CoV nucleocapsid protein (N protein) dimer. Interestingly they found that SARS-CoV N protein has a structure similar to the porcine respiratory and reproductive syndrome virus (PRRSV) N protein. NMR analysis of SARS-CoV N protein revealed that its dimer consists of five α helices and one β-hairpin. As with the PRSSV N protein, the interface of the SARS-CoV N dimer consists of a four-stranded β-sheet superimposed by two long α helices [25]. This observation suggests that SARS-CoV N protein function may be similar to that of the PRRSV N protein, and that the SARS nucleocapsid may assemble like that for the PRRSV nucleocapsid.

Porcine reproductive and respiratory syndrome virus

This virus causes respiratory problems in young pigs and late term abortions in pregnant sows. PRRSV is a small RNA virus of ~15 kDa and is comprised of nine open reading frames. As stated earlier, the N protein of PRRSV interacts with itself through noncovalent interactions. The N protein monomer is transported through the Golgi apparatus by a currently known mechanism, where it is then linked by disulfide bonds to form the N protein dimer, which ultimately forms the viral nucleocapsid [29]. N protein is also localized to the nucleus and nucleolus by an unknown mechanism. Confocal microscopy has shown that N protein colocalizes with the small nucleolar RNA (snoRNA)-associated protein fibrillarin [30]. A specific interaction between these two proteins was demonstrated using a mammalian two-hybrid assay where both N protein and fibillarin genes were co-transfected and also by a glutathione S transferase (GST) pull-down assay. By using a series of deletion mutants, it was determined that amino acids 30 to 37 of the N-protein are critical for interaction with fibrillarin. Conversely, it was found that the first 80 amino acids of fibrillarin contain the site of interaction with the N protein. It was also determined that a region spanning amino acids 41 to 47 of N protein binds full length genomic PRRSV RNA. Interestingly, this region overlaps the nuclear localization signal, suggesting that RNA binding to the N protein may be involved in its transport to the nucleus [30]. Finally, it was discovered that the interaction of N protein and fibrillarin is dependent upon RNA binding, suggesting that N protein may function to regulate the host cell via RNA processing and ribosome biogenesis.

Conclusion

Proteomics is a broad field of study that will become more popular as we enter the post-genome era. Proteomics has greatly aided our understanding of viral pathogenesis. From understanding how proteins are structured to how they interact, from looking at a single protein to global protein expression, proteomics will continue to enable us to dissect and understand the complexity of viral pathogenesis.

Acknowledgments

This research was supported in part by a grant from the National Research Initiative of the USDA Cooperative State Research, Education and Extension Service (CSREES)(Award number#NRI 05-35604-15420) awarded to H.C. Liu.

References

1 Wasinger VC, Cordwell SJ, Cerpa-Poljak A, Yan JX, Gooley AA, Wilkins MR, Duncan MW, Harris R, Williams KL, Humphery-Smith I: Progress with gene-product mapping of the Mollicutes: Mycoplasma genitalium. Electrophoresis 1995;16:1090-1094.

2 Fields S, Song O: A novel genetic system to detect protein-protein interactions. Nature 1989;340:245-246.

3 Miller J, Stagliar I: Using the yeast two-hybrid system to identify interacting proteins. Met Mol Biol 2004;261:47-262.

4 Phizicky EM, Fields S: Protein-Protein Interactions: Methods for detection and analysis. Microbiological Reviews 1995;59:94-123.

5 Adams PD: Identification of associated proteins by coimmunoprecipitation, in Golemis E (ed): Protein-Protein Interactions - A Molecular Cloning Manual. Cold Spring Harbor Laboratory Press 2002; pp 59-74.

6 Wittmann-Liebold B, Graack HR, Pohl T: Two-dimensional gel electrophoresis as tool for proteomics studies in combination with protein identification by mass spectrometry. Proteomics 2006;6: 4688-4703.

7 Zaikin VG, Halket JM: Derivatization in mass spectrometry-8. Soft ionization mass spectrometry of small molecules. Eur J Mass Sepctrom 2006;12:79-115.

8 Smith RD, Tang, Shen Y: Ultra-sensitive and quantitative characterization of proteomes. Mol Biosyst 2006;2:221-230.

9 Lacapere JJ, Pebay-Peyroula E, Neumann JM, Etchebest C: Determining membrane protein structures: still a challenge! Trends Biochem Sci 2007;32:259-270.

10 Harrar L, Marsen HS: Two-dimensional gel analysis of HSV type 1-induced polypeptides and glycoprotein processing. J Gen Virol 1989;52:77-92.

11 Taylor TJ, Knipe DM: Proteomics of herpes simplex virus replication compartments: association of cellular DNA replication, repair, recombination, and chromatin remodeling proteins with ICP8. J Virol 2004;78:5856-5866.

12 Koshizuka T, Kawaguchi Y, Nishiyama Y: Herpes simplex virus type 2 membrane protein UL56 associates with the kinesin motor protein KIF1A. J Gen Virol 2005;86:527-533.

13 Fontaine-Rodriguez EC, Taylor TJ, Olesky M, Knipe DM: Proteomics of herpes simplex virus infected cell protein 27: association with translation initiation factors. Virology 2004;330:487-492.

14 Liu Y, Gong W, Huang CC, Herr W, Cheng X: Crystal structure of the conserved core of the herpes simplex virus transcriptional regulatory protein VP16. Genes Dev 1999;13:1692-1703.

15 Varnum SM, Streblow DN, Monroe ME, Smith P, Auberry KJ, Pasa-Tolic L, Wang D, Camp DG 2nd, Rodland K, Wiley S, Britt W, Shenk T, Smith RD, Nelson JA: Identification of proteins in human cytomegalovirus (HCMV) particles: the HCMV proteome. J Virol 2004;78:10960-10966.

16 Hofmann H, Sindre H, Stamminger T: Functional interaction between the pp71 protein of human cytomegalovirus and the PML-interacting protein human Daxx. J Virol 2002;76:5769-5783.

17 Loveland AN, Nguyen NL, Brignole EJ, Gibson W: The amino-conserved domain of human cytomegalovirus UL80a proteins is required for key interactions during early stages of capsid formation and virus production. Virol 2007;81:620-628.

18 Witter RL, Li D, Jones D, Lee LF, Kung KJ: Retroviral insertional mutagenesis of a herpesvirus: a Marek's disease virus mutant attenuated for oncogenicity but not for immunosuppression or in vivo replication. Avian Dis 1997;41: 407-421.

19 Liu HC, Kung HJ, Fulton JE, Morgan RW, Cheng HH: Growth hormone interacts with the Marek's disease virus SORF2 protein and is associated with disease resistance on chicken. Proc Natl Acad Sci 2001;98:9203-9208.

20 Liu HCS, Soderblom EJ, Goshe MB: A mass spectrometry-based proteomic approach to study Marek's disease virus gene expression. J Virol Methods 2006;135: 66-75.

21 Falcon AM, Fortes P, Marion RM, Beloso A, Ortin J: Interaction of influenza virus NS1 protein and the human homologue of Staufen in vivo and in vitro. Nucleic Acids Res 1999;27:2241-2247.

22 Stevens J, Corper AL, Basler CF, Taubenberger JK, Palese P, Wilson IA: Structure of the uncleaved human H1 hemagglutinin from the extinct 1918 influenza virus. Science 2004;303:1866-1870.

23 Gamblin SJ, Haire LF, Russell RJ, Stevens DJ, Xiao B, Ha Y, Vasisht N, Steinhauer DA, Daniels RS, Elliot A, Wiley DC, Skehel JJ; The structure and receptor binding properties of the 1918 influenza hemagglutinin. Science 2004;303:1838-1842.

24 Kumar P, Gunalan V, Liu B, Chow VT, Druce J, Birch C, Catton M, Fielding BC, Tan YJ, Lal SK: The nonstructural protein 8 (nsp8) of the SARS coronavirus interacts with its ORF6 accessory protein. Virology 2007;366:293-303.

25 Fang X, Gao J, Zheng H, Li B, Kong L, Zhang Y, Wang W, Zeng Y, Ye L: The membrane protein of SARS-CoV suppresses NF-kappaB activation. J Med Virol 2007;79:1431-1439.

26 Fang X, Ye L, Timani KA, Li S, Zen Y, Zhao M, Zheng H, Wu Z: Peptide domain involved in the interaction between membrane protein and nucleocapsid protein of SARS-associated coronavirus. J Biochem Mol Biol 2005;38:381-385.

27 von Brunn A, Teepe C, Simpson JC, Pepperkok R, Friedel CC, Zimmer R, Roberts R, Baric R, Haas J: Analysis of intraviral protein-protein interactions of the SARS coronavirus ORFeome. PLoS ONE 2007;2:e459.

28 Chang CK, Sue SC, Yu TH, Hsieh CM, Tsai CK, Chiang YC, Lee SJ, Hsiao HH, Wu WJ, Chang CF, Huang TH: The dimer interface of the SARS coronavirus nucleocapsid protein adapts a porcine respiratory and reproductive syndrome virus-like structure. FEBS Lett 2005; 579:5663-5668.

29 Wootton SK, Yoo D: Homo-oligomerization of the porcine reproductive and respiratory syndrome virus nucleocapsid protein and the role of disulfide linkages. J Virol 2003;77:4546-4557.

30 Yoo D, Wootton SK, Li G, Song C, Rowland RR: Colocalization and interaction of the porcine arterivirus nucleocapsid protein with the small nucleolar RNA-associated protein fibrillarin. J. Virol 2003;77:12173-12183.

Dr. Hsiao-Ching Liu, 232D Polk Hall, Campus Box7621, Dept. Animal Science, North Carolina State University, Raleigh, NC27695, USA.

E-mail: hc_liu@ncsu.edu

Pinard M-H, Gay C, Pastoret P-P, Dodet B (eds): Animal Genomics for Animal Health. Dev Biol (Basel). Basel, Karger, 2008, vol 132, p 55.

SESSION III

Using Animal Genomics to Understand Animal Diseases

Pinard M-H, Gay C, Pastoret P-P, Dodet B (eds): Animal Genomics for Animal Health. Dev Biol (Basel). Basel, Karger, 2008, vol 132, pp 57-66.

Integrated Immunogenomics in the Chicken: Deciphering the Immune Response to Identify Disease Resistance Genes

P. Kaiser[1], J. Howell[1], M. Fife[1], J.-R. Sadeyen[1], N. Salmon[1], L. Rothwell[1], J. Young[1], P. van Diemen[1], M. Stevens[1], T.-Y. Poh[1,2], M. Jones[1,3], P. Barrow[1,3], P. Wigley[1,4], H. He[5], C. Swaggerty[5], M. Kogut[5], J. Smith[6], D. Burt[6]

1 Institute for Animal Health, Compton, Berkshire, UK
2 Duke-NUS Graduate Medical School, Singapore
3 School of Veterinary Medicine and Science, University of Nottingham, Sutton Bonington, UK
4 Department of Veterinary Pathology, University of Liverpool, Leahurst, UK
5 USDA-ARS, Southern Plains Agricultural Research Center, College Station, TX, USA
6 Roslin Institute, Roslin, Midlothian, UK

Key words: SNP, eQTL, innate immunity, TLR, cytokine, chemokine

Abstract: Resistance to infection takes place at many levels, and involves both non-specific and specific immune mechanisms. The chicken has a different repertoire of immune genes, molecules, cells and organs compared to mammals. To understand the role of any disease resistance gene(s), it is therefore important to understand these different repertoires, and the bird's response to a particular pathogen. Our studies focus on the innate immune response, as responses of macrophages from inbred lines of chickens, and heterophils from commercial birds, correlate with resistance or susceptibility to *Salmonella* infection with a variety of *Salmonella* serovars and infection models. To map disease resistance genes, we are using a combination of expression quantitative trait loci (eQTLs) from microarray studies, allied with whole genome SNP arrays (WGA) and a candidate gene approach. There are over 500 human genes with the Gene Ontology term "innate immunity." We have identified over 400 of these genes in the chicken genome, and are actively identifying informative SNPs in them. The segregation of 6,000 WGA SNPs across all of our inbred lines was also assessed, which should yield approximately 900 informative SNPs for a cross between any two lines. The initial focus of these studies is on mapping resistance genes in our inbred lines, but the studies will be extended to commercial flocks.

INTRODUCTION

The availability of genome sequences, not only for pathogens, but also now for the chicken host [1] represents a major shift in our ability to understand host-pathogen interactions. In previous work, we have shown that chickens differ greatly in their susceptibility to a wide range of diseases and vaccination response [2,3]. Our focus is now on understanding the role of cytokines and chemokines in the avian immune response [4], including their potential to act as vaccine adjuvants, and to use the new resources available to map disease resistance genes (e.g. see [5,6]).

The immune system Th1-Th2 paradigm is well established in mammals. Th1 responses control inflammatory reactions to viruses and other intracellular pathogens, and Th2 responses to helminthic worm infections, extracellular pathogens and allergens. Until recently, there was limited information as to whether this paradigm applied to non-mammalian species, such as birds. Cytokines control immune responses and drive them towards Th1 or Th2 responses, and the recent cloning of the first non-mammalian Th2-controlling cytokines [7] means that this question of fundamental importance to understanding the evolution of the immune response can now be addressed. Using these reagents, we demonstrated, for the first time, that this paradigm also extends to chickens [8], in that infections with intracellular pathogens result in a Th1-dominated immune response and infections with extracellular pathogens in a Th2-dominated response. Further, analysis of chicken expression sequence tags (ESTs) and genome sequences has allowed the chicken's full repertoire of cytokines and chemokines and other genes of the avian immune system to be catalogued [1,4]. The results correlate with certain differences in the biology of the chicken compared to biomedical species. For example, chickens lack functional eosinophils, and they also lack IgE, IL-5 and the eotaxins, all molecules crucial for the function of eosinophils in mammals. Chickens also lack lymph nodes, and knockout mouse models have shown that the development of secondary lymphoid organs, such as lymph nodes, is dependent on expression of the lymphotoxin genes. The chicken lacks the genes for lymphotoxins and their receptors.

In recent years, we have played a major role in developing new tools and resources for the analysis of complex traits, such as disease resistance in poultry. This started with the creation of genetic linkage maps of the chicken and recently maps based on single nucleotide polymorphisms or SNPs [6]. These tools have been used to map hundreds of QTL for a wide range of traits (e.g. [5], and summarised in the genetic variation database). The chicken EST project funded by the Biotechnology and Biological Sciences Research Council (BBSRC) played a major part not only in gene discovery but also in annotation of the genome [1]. We have used these ESTs to build high-density gene expression microarrays [9] and recently in collaboration with Affymetrix and others, gene chips have been developed for all known chicken genes (http://www.affymetrix.com/support/technical/byproduct.affx?product=chicken).

Demonstration of the crucial importance of the innate immune response is one of the major advances in immunology of the past few years. Rather than just being an ancestral simple immune response, it is now obvious that the innate immune response both provides an important initial response to pathogens that can limit or even prevent infection, and crucially determines the course of an adaptive immune response, and hence immunological memory (i.e. the ability to respond to future infection with the same pathogen - the basis of vaccination).

There are polymorphisms in the genes controlling innate responses in mammals, which affect function (e.g. in the *TLR4* gene between He/J and He/N mice). Previously we have concentrated on the chicken chemokine gene families and their receptors [4], which are central to controlling the recruitment of immune cells of differing types to sites of infection, the Toll-like receptors ([10] and Burt et al, unpublished data), which are fundamental to the initial recognition of pathogens, and on the defensins [11], whose products are directly harmful to bacteria. Investigating these genes it rapidly became apparent that all of these have been subject to very rapid and distinct evolution since mammals and chickens diverged, both in the sequences of the individual genes, with polymorphisms between the different inbred lines for some genes and in the duplications and chromosomal movements which have taken place in these gene families. This is probably due to strong selective pressures on the genes, arising from the need to develop better immune responses, or to prevent pathogens evading or manipulating the immune response by mimicry of these genes.

There is also evidence that different lines of chickens differ in their innate immune responses to pathogen challenge (particularly the enteric pathogen *Salmonella*), and that this correlates with resistance to infection with those pathogens [12-14]. It therefore seems reasonable to suppose that this differential response is under genetic control, and that similar mechanisms may in part explain the different resistance profiles of the Institute for Animal Health (IAH) inbred lines (see Table 1) of chickens to all the pathogens so far tested (viral, bacterial and protozoan parasites) [2,3].

The identification of innate immune resistance mechanisms also has strategic industrial relevance. The UK poultry industry faces numerous challenges in order to remain sustainable. These include the imminent move to more extensive rearing systems; the withdrawal of prophylactic and many therapeutic antibiotics, and other drugs such as anti-coccidials; and resistance and residue problems with anti-helminthics. These challenges will all have an impact on poultry health. It is important that poultry breeders are able to select for genetic improvement in performance when birds are reared in such environments, and one obvious phenotype would be "robustness." Improved innate resistance would contribute to robustness and would be a general phenomenon, as by its very nature the innate immune response does not differentiate between different pathogens, but rather recognises and responds to pathogen-associated molecular patterns that are conserved across pathogen classes.

The recent availability of the chicken genome marks a quantum shift in our ability to understand the general biology of this important model and commercial organism, and its immune response to infection in particular. Before the genome era, genes involved in immune function have not always been easy to identify, as they are under heavy selective pressure due to the ongoing host-pathogen "arms-race" and thus have limited amino acid identity to their mammalian orthologues. However, many of the genes related to innate immunity (including those encoding defensins, chemokines and their receptors, pro-inflammatory cytokines, the type I interferons (IFNs), natural killer (NK) cell receptors and Toll-like receptors) have already been cloned and sequenced, and the availability of the chicken genome sequence and some comparative breed data from the Beijing Genomics Institute [6] means that additional genes can readily be identified in the chicken by comparison with their mammalian orthologues.

Table 1: Disease resistance of IAH inbred lines of chickens.

	Protozoan parasites[a]							Enteric bacteria[b]						Viruses[c]		
Line[d]	Ea	Ep	En	Ema	Emi	Eb	Et	St	Sg	Se	Sp	St col	C col	IBV	IBDV	MDV
BrL	M	S	S	M	S	M	M							R	S	M
Wl	S	R	R	R	R	R	S	R	R	R	R			M	M	M
15I	S	S	S	S	S	S	R	S	S	S	S			S	R	M
7_2	S	S	R	M	S	S	R	S	S	S	S			S	R	S
6_1	S	S	R	S	S	S	R	R	R	R	R	R	R	M	R	R
C	R	R	R	R	S	S	S	S	S	S	S			R	R	S
N	M	R	R	M	S	R	R	R	R	R	R	S	S	M	S	R
0	R	S	M	M	S	R	R	S	S	S	S				R	R
P															R	S
Sykes RIR															R	S

R = resistant; M = intermediate; S = susceptible.
[a]Ea = *Eimeria acervulina*; Ep = E. *praecox*; En = E. *necatrix*; Ema = E. *maxima*; Emi = E. *mitis*; Eb = E. *brunetti*; Et = E. *tenella*.
[b]St = *Salmonella* Typhimurium; Sg = S. Gallinarum; Se = S. Enteritidis; Sp = S. Pullorum; St col = S. Typhimurium colonisation; C col = *Campylobacter jejuni* colonisation.
[c]IBV = infectious bronchitis virus; IBDV = infectious bursal disease virus; MDV = Marek's disease virus.
[d]BrL = brown leghorn; Wl = Wellcome line; RIR = Rhode Island red.

Materials and Methods

Genomic analysis and identification of innate immune function genes

All genome analysis for identification of chicken innate immune genes was based on Ensembl release 46.2d which is assembly WASHUC2 at http://genome.wustl.edu/genome.cgi?GENOME=Gallus%20gallus. Where gene orthologues were not already identified as such in Ensembl, it and the UCSC genome browser were used extensively for examination of syntenic context of newly identified genes. Chicken genes were considered to have conserved synteny if flanked by at least three recognisable orthologues of the same genes as the orthologous human genes, with the same arrangement and relative orientations.

SNP identification

Primers were designed for PCR amplification and sequencing of portions (800 to 1,000 base pairs) of innate immune candidate genes in a number of birds of each line of interest (lines 6_1 and N). PCR was performed by standard methods and sequencing carried out on a Beckman Coulter CEQ 8000 sequencer. The area of each gene to be sequenced was generally selected based on the presence of a number of SNPs already shown in Ensembl sequence. This denoted that there was likely to be a good level of variation in that area and thus an increased possibility of an informative SNP between our lines. These selected areas were mainly intronic or promoter areas as exons usually contain a much lower level of genetic variation. Sequences were compared between lines using CodonCode Aligner software. Informative SNPs were selected as ones that were different between the two lines and showed no heterozygosity within either line. If no informative SNP was identified in a selected area of a candidate gene, another area of that gene was selected for further sequencing and possible informative SNP detection.

Whole genome analysis using the Illumina beadstation

Four oligonucleotide pools (OPAs) made available by our collaborators (Aviagen and USDA) were screened for SNPs that are fully informative between lines 61 and N for QTL mapping studies, which are polymorphic between lines yet monomorphic within a line. Genomic DNA was extracted from the blood of 10 birds from each of the chicken lines. Whole genome genotyping was performed on 250 ng of DNA for each of the four OPAs (custom 1536 SNP panels, 96-sample Sentrix array matrix) using the Illumina Golden Gate genotyping platform. Automatic clustering of the samples and genotype calling was done with the Illumina BeadStudio software (version 3.0.27) and each locus confirmed independently by two investigators. BeadStudio was also used to assess each assay for Genetrain score, heterozygosity, cluster separation and line-specific homozygosity. This software integrates a genotype call score to give weighted genotype calls. Only assays with a high call rate and reproducibility were selected.

Results

A search of the human Gene Ontology database with the term "innate immunity" identified approximately 510 genes known to have a role in innate immune responses. Of these, we have identified around 480 in the chicken genome. In part, the difference in numbers can be explained by the fact that in general, although chickens have the same multigene families of innate immune response genes as mammals (e.g. TLRs, defensins, proinflammatory chemokines, etc.), there are fewer members of these families in chickens than in mammals.

To date, we have concentrated on identifying SNPs between lines 61 and N in a study on disease resistance (specifically resistance to *Salmonella* and *Campylobacter* gut colonisation), focusing on some of the better characterised "headline" genes (see Table 2), 99 in all. Altogether, 71 informative SNPs (i.e. SNPs that segregate absolutely between the two lines) in 36 genes have been identified. There are also 20 other genes in which SNPs were identified, but these were heterozygous within one or both lines, and thus not usable in mapping studies. Strikingly, we have identified very few SNPs (an informative SNP in TLR 1/6/10 gene 1, and uninformative SNPs in the TLR2 type2, TLR3 and TLR5 genes) in the TLR genes, having sequenced several kilobases of these genes, including promoter regions, exons and introns. In contrast, we readily find SNPs in genes encoding signalling molecules downstream of the TLRs. Of the 71 SNPs, 49, covering all 36 genes, will be incorporated into the final SNP panel for line 6_1 and line N comparisons.

Table 2: Candidate gene approach – innate immune gene panel. Rs = receptors.

TLR pathway:	Interleukins:	Chemokines:	Chemokine Rs:	Defensins:	Other AMPs:
<u>*Tollip*</u>	*IL1B*	*XCL1*	*XCR1*	*AvBD1*	<u>*LEAP2*</u>
IRAK-4	<u>*IL2*</u>	*CCLi1*	*CCRa*	*AvBD2*	
Mal/TIRAP	*IL3*	***CCLi2***	*CCRb*	*AvBD3*	Other PRRs:
MyD88	***IL4***	*CCLi3*	***CXCR1***	<u>*AvBD4*</u>	*NOD1*
TRAF6	***IL5***	*CCLi4*		*AvBD5*	*MDA5*
MAP3K7IP1	***IL6***	*CCLi5*	TNFSF:	*AvBD6*	
MAP3K7IP2	<u>*IL7*</u>	<u>*CCLi6*</u>	*TNFSF4 (OX40L)*	***AvBD7***	IL Rs:
MAP3K7	<u>*IL9*</u>	<u>*CCLi7*</u>	***TNFSF5 (CD40L)***	*AvBD8*	***IL1R1***
TLR1/6/10 gene 1	***IL10***	*CCLi8*	*TNFSF6 (FASL)*	*AvBD9*	***IL1RL2***
TLR1/6/10 gene 2	***IL12A***	<u>*CCLi9*</u>	***TNFSF8 (CD30L)***	*AvBD10*	***GP130***
TLR2 type 1	*IL12B*	*CCLi10*	***TNFSF10 (TRAIL)***	*AvBD11*	
<u>*TLR2 type 2*</u>	***IL13***	***CXCLi1***	*TNFSF11 (RANKL)*	*AvBD12*	Other genes:
<u>*TLR3*</u>	***IL15***	<u>*CXCLi2*</u>	<u>*TNFSF13B (BAFF)*</u>	***AvBD13***	*Caspase 1*
TLR4	<u>*IL16*</u>	***CXCLi3***	***TNFSF15 (VEGI)***		*MIF*
<u>*TLR5*</u>	<u>*IL17A*</u>		<u>*TRAIL-L*</u>		*NRAMP1*
TLR7	***IL17B***				
TLR15	*IL17D*		CSFs:		
TLR21	<u>*IL17F*</u>		*GCSF*		
IKKA	<u>*IL18*</u>		***GMCSF***		
IKKB	***IL19***				
IKBA	<u>*IL21*</u>		TGFs:		
NFKB1	***IL22***		*TGFB2*		
NFKB2	***IL26***		*TGFB3*		

Genes in bold have informative SNPs. Genes underlined had SNPs, but they were heterozygous within lines 6^1 and/or N.

Whole genome analysis

Analysis of this data identified a panel of approximately 900 markers that will be used to identify informative transmissions between lines 6_1 and N (Fig. 1). Furthermore, it was noted during the genotyping of these samples that a number of SNP assays failed for an entire line (Fig. 2). This was thought to result from differential amplification of alleles that contain divergent probe sequence between the lines. Subsequent sequencing of the regions flanking the assay locus proved this to be the case, resulting in identification of line-specific variation at this site (Fig. 3). An additional set of SNP assays for these "allelic dropout" regions has been incorporated into the final SNP panel for line 6_1 and line N comparisons.

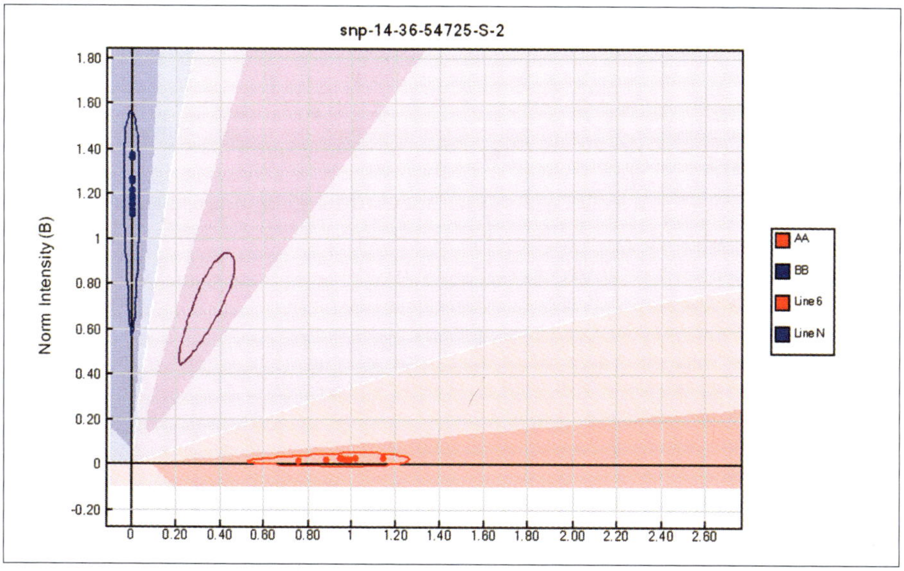

Fig. 1: Example of one of the "fixed informative" line 61/line N SNPs identified in the screening of a chicken 3072 SNP panel (Cartesian Plot).
Line 61 is a "GG" genotype whereas line N is "AA" for this locus. These genotypes carry a high call score, indicating this assay has a good call rate and accuracy.

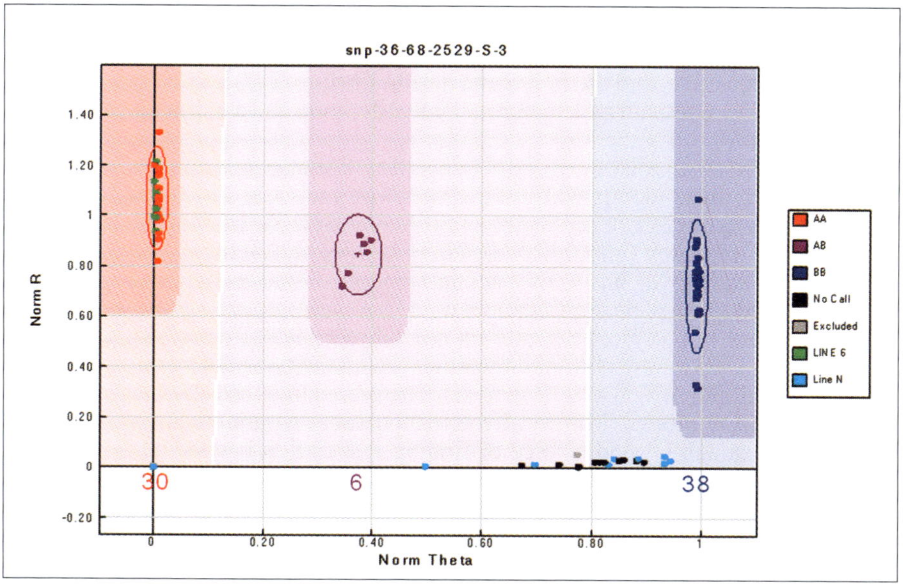

Fig. 2: Allelic dropout – line 6_1 calls as an "AA" genotype whereas line N fails to amplify for this assay, as a result of line-specific SNPs in the probe sequence (Polar Plot). Note that other birds tested are a mix of heterozygote and homozygote at this locus, indicating that the assay is not failing.

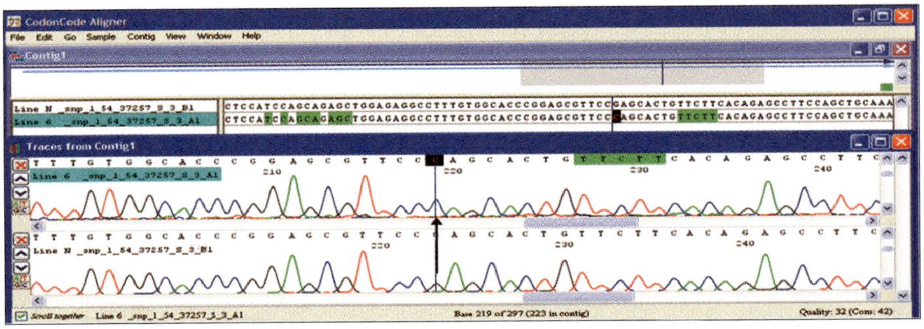

Fig. 3: Allele-specific differences confirmed by DNA sequencing. line 61 and line N show a single nucleotide polymorphism change within the probe sequence for this assay (as indicated by the arrow). This SNP will now act as the informative SNP at this locus for the Illumina assay.

Discussion

The paucity of SNPs in the TLR genes in lines 6_1 and N is both surprising, as SNPs in TLR genes have been described for other chicken lines, and interesting. It suggests that in these inbred lines, at least, any genetic control is presumably at the level of the genes encoding signalling molecules downstream of the TLRs (or potentially at the level of the genes that encode the molecules that control TLR expression). This hypothesis is supported by the fact that numerous "read-out" genes of the induced innate response (e.g. pro-inflammatory cytokines and chemokines) are differentially expressed between our two lines. It is highly unlikely that there are multiple SNPs in each of these "read-out" genes [which have completely different chromosomal locations and are in linkage disequilibrium (LD)], and the signalling molecules therefore remain a focus for SNP identification and candidates as disease resistance genes.

Our next step is to finalise the line 6_1-N-specific OPA set, combining the candidate gene SNPs with those identified from the 6k WGA analyses, and adding in SNPs from a further 18k WGA run across the same genomic DNAs by Martien Groenen, Wageningen Universiteit (unpublished). This line-specific WGA will be put across both archived and novel line 6_1/N backcross populations, to hopefully identify markers with genome-wide significance.

These studies initially concentrate on our inbred lines of defined phenotype. They are being used to test a defined hypothesis with the candidate gene approach, but with the addition of WGA SNP analysis. The inbred lines are all of layer origin, and have large LD blocks (Martien Groenen, personal communication), so there should be a reasonably good chance of finding association between markers and the phenotype being measured. The function of candidate gene SNPs can easily be tested between the inbred lines. Then any association in more outbred, commercial flocks will be assessed.

Microarrays have been run for both lines following infection with both *Salmonella* and *Campylobacter*, using the "Roslin" chicken whole genome microarray, and analysis to identify potential eQTLs is ongoing.

Our overall approach is primarily to understand immune responses to infection in the chicken, an animal that is not just a mouse with feathers. Birds have a different repertoire of immune organs, cells, molecules and genes than mammals, and therefore we need to understand the bird's immune response to disease so as to understand which genes might be important in that response. The aim is to link phenotype, genotype and eQTLs, using a candidate gene approach, whole genome scans with SNP panels and microarrays. Of course, this approach can be used for any disease-relevant cross, not just the one currently under study. We have also characterised all of the IAH inbred lines for their vaccine responsiveness, at least to killed vaccines, as part of an EU project, and therefore could use a similar approach to investigate the genetics of vaccine responses in birds.

Acknowledgements

The authors acknowledge funding from the BBSRC, DEFRA-LINK, the EU (EADGENE and SABRE), Cobb-Vantress, Aviagen and USDA. We would also like to thank our partners in SABRE, in particular Rima Zoorob (CNRS, Villejuif, France), Dave Burt and Jacqueline Smith (Roslin Institute, Scotland), Catherine Beaumont (INRA, Tours, France), Alain Vignal (INRA, Toulouse, France) and Annemarie Rebel (ASG Lelystad and Wageningen University, The Netherlands).

References

1 International Chicken Genome Sequencing Consortium: Sequence and comparative analysis of the chicken genome provide unique perspectives on vertebrate evolution. Nature 2004;432:695-716.

2 Bumstead N, Millard BM, Barrow P, Cook JKA: Genetic basis of disease resistance in chickens, in Owen JB, Axford RFE (eds): Breeding for Disease Resistance in Farm Animals. Wallingford, UK, CABI, 1991, pp 10-23.

3 Bumstead N: Genetic resistance to avian viruses. Rev Sci Tech 1998;17:249-255.

4 Kaiser P, Poh TY, Rothwell L, Avery S, Balu S, Pathania US, Hughes S, Goodchild M, Morrell S, Watson M, Bumstead N, Kaufman J, Young JR: A genomic analysis of chicken cytokines and chemokines. J Interferon Cytokine Res 2005;25:467-484.

5 De Koning DJ, Haley CS, Windsor D, Hocking PM, Griffin H, Morris A, Vincent J, Burt DW: Segregation of QTL for production traits in commercial meat-type chickens. Genetic Res 2004;83:211-220.

6 International Chicken Polymorphism Map Consortium: A genetic variation map for chicken with 2.8 million single nucleotide polymorphisms. Nature 2004;432:717-722.

7 Avery S, Rothwell L, Degen WDJ, Schijns VECJ, Young J, Kaufman J, Kaiser P: Characterization of the first non-mammalian T2 cytokine gene cluster: the cluster contains functional single-copy genes for IL-3, IL-4, IL-13 and GM-CSF, a gene for IL-5 which appears to be a pseudogene, and a gene encoding another cytokine-like transcript, KK34. J Interferon Cytokine Res 2004;24:600-610.

8 Degen WDJ, van Daal N, Rothwell L, Kaiser P, Schijns VECJ: Th1/Th2 polarization by viral and helminth infection in birds. Vet Micro 2005;105:163-167.

9 Burnside J, Neiman P, Tang J, Basom R, Talbot R, Aronszajn M, Burt D, Delrow J: Development of a cDNA array for chicken gene expression analysis. BMC Genomics 2005;6:13.

10 Kogut MH, Iqbal M, He H, Philbin V, Kaiser P, Smith A: Expression and function of Toll-like receptors in chicken heterophils. Devel Comp Immunol 2005;29: 791-807.

11 Lynn DJ, Higgs R, Lloyd AT, O'Farrelly C, Herve-Grepinet V, Nys Y, Brinkman FS, Yu PL, Soulier A, Kaiser P, Zhang G, Lehrer RI: Avian beta-defensin nomenclature: a community proposed update. Immunol Letts 2007;110:86-89.

12 Ferro PJ, Swaggerty CL, Kaiser P, Pevzner IY, Kogut MH: Heterophils isolated from chickens resistant to intraintestinal *Salmonella enteritidis* infection express higher levels of pro-inflammatory cytokine mRNA following infection than heterophils from susceptible chickens. Epidemiol Infection 2004;132:1029-1037.

13 Swaggerty CL, Kogut MH, Ferro PJ, Rothwell L, Pevzner IY, Kaiser P: Differential cytokine mRNA expression in heterophils isolated from *Salmonella*-resistant and -susceptible chickens. Immunology 2004;113:139-148.

14 Swaggerty CL, Ferro PJ, Pevzner IY, Kogut MH: Heterophils are associated with resistance to systemic *Salmonella enteritidis* infections in genetically distinct chicken lines. FEMS Immunol Med Micro 2005;43:149-154.

Dr. Pete Kaiser, Avian Genomics Group, Institute for Animal Health, Compton, Berkshire RG20 7NN, UK.
E-mail: pete.kaiser@bbsrc.ac.uk

Pinard M-H, Gay C, Pastoret P-P, Dodet B (eds): Animal Genomics for Animal Health. Dev Biol (Basel). Basel, Karger, 2008, vol 132, pp 67-75.

Quantitative Genetic and Functional Genomics Approaches to Investigating Parasite Disease Resistance and Protective Immune Mechanisms in Avian Coccidiosis

H.S. Lillehoj, Y. Hong, C. Kim

Animal Parasitic Diseases Laboratory, Animal and Natural Resources Institute, Beltsville Agricultural Research Center, Beltsville, MD, USA

Key words: chicken, Eimeria, QTL, microarray, disease

Abstract: A major challenge for the commercial poultry production is controlling enteric disease-causing pathogens in the absence of prophylactic drugs. Although traditional genetic approaches have contributed to enhanced efficiency of poultry production and increased the ability of the industry to provide safe, high quality, and low cost meat products to consumers, economic losses due to enteric diseases remains a significant problem. Therefore, there is a need to develop alternative control strategies against poultry enteric pathogens of economic importance. Recent advances in molecular genetics and functional genomics are now allowing for rapid progress in understanding the molecular mechanisms of disease resistance against major infectious agents of poultry. It is anticipated that identifying the nature of host-pathogen interactions will facilitate the development of novel vaccines and therapeutics. In this paper, recent progress in identifying genes that influence resistance to avian coccidiosis, the most economically important disease of poultry, will be discussed. This work provides an excellent example of how the integration of new molecular genetic and functional genomics tools increases our knowledge of the mechanism of genetic resistance to complex diseases.

INTRODUCTION

Intestinal mucosal surfaces are defended against enteric pathogens by the gut-associated lymphoid tissues. The primary effecter cells localized here are intraepithelial lymphocytes (IELs) of the outer mucosal epithelial layer that recognize and destroy pathogens that breach the intestinal epithelium [1]. IELs play an important role in the host response to avian coccidiosis, an intestinal disease caused by a ubiquitous intestinal protozoan belonging to the genus *Eimeria* that seriously impairs the growth and feed utilization of infected chickens [2]. Chicken intestinal IELs and their cytokines generated during a primary immune response against *Eimeria* parasites are principally responsible for protective immunity against subsequent secondary infections [1,3]. However, detailed investigation of host-pathogen interactions in avian coccidiosis at the cellular and molecular levels has been hampered for a variety of reasons, mainly the lack of DNA sequence homology between mammalian and avian immune-regulated genes.

Over the past few decades, in-feed antibiotics and active vaccination have been the two main strategies used to control avian coccidiosis. However, increasing public concerns are prompting more strict government regulations on the use of growth-promoting drugs in animal production, and the ability of current vaccines to protect against emerging hypervirulent strains of many poultry pathogens is becoming a major issue. Therefore, there is a timely need to develop alternative control strategies against avian diseases of economic importance. Genetic selection of disease resistant chickens offers a powerful method of disease control, particularly in combination with other state-of-the-art protocols, such as those utilizing new molecular genetic and functional genomics tools. Although tremendous improvements in commercial chicken production traits (e.g. growth rate and feed conversion efficiency) have been accomplished using classical genetic breeding techniques, selection of commercial poultry stocks with improved disease resistance using similar techniques has been relatively unsuccessful. However, with the advent of quantitative trait loci (QTL) mapping strategies, DNA markers that are associated with growth performance and disease resistance can now be identified, in particular genotypes, and this information subsequently used in marker-assisted selection of breeding stocks. For example, recent studies have reported QTL mapping of body weight gains and resistance to Marek's disease in chickens [4,5].

A consensus chicken genomic linkage map was published in 2000 based on three resource mapping populations (broiler, layer, and Chinese Silkie) comprising 1,889 loci [6]. The physical map and draft sequence of the chicken genome were completed at the Washington University Genome Sequencing Center and Beijing Genome Institute on these three domestic chicken varieties to identify and evaluate chicken genome variation. For the comprehensive mapping of disease resistance traits, many more microsatellites are needed to obtain complete genome coverage, due to the relatively low abundance of microsatellites on many microchromosomes [4]. As an alternative, single nucleotide polymorphisms (SNPs) represent the most common form of DNA variation in the genome. More than 3 million candidate chicken SNPs are available in the public database (http://www.ncbi.nlm.nih.gov/SNP). SNPs are single base changes or nucleotide variations between individuals (usually biallelic variations) that can occur in gene promoters, exons or introns or between genes (intergenic). SNPs within coding sequences are categorized as either synonymous (do not result in an amino acid change) or non-synonymous (results in an amino acid change). Non-synonymous SNPs are of interest due to their potential effect on the phenotype of an individual. The optimal approach for SNP discovery and validation is to perform a comparative analysis of the DNA sequences from different individuals to identify nucleotide variations in a gene. Because of their high density in genomes and ease of detection and statistical analysis, SNPs will undoubtedly find future use in several areas of poultry science such as genetic linkage map construction and QTL mapping.

Our initial attempt at genome-wide searches for markers that control resistance to coccidiosis led to the identification of a QTL associated with reduction of parasite growth [7]. Using 119 microsatellite markers in 314 F2 birds segregating for coccidiosis disease susceptibility, which covered 80% of the genome with an average interval of 25 cM, we identified the LEI0101 marker, located at 259 cM on chromosome 1, as being significantly associated with disease resistance to *E. maxima* (LOD = 3.46) (Fig. 1). Subsequent multipoint linkage analysis studies using eight additional microsatellite markers linked to LEI0101 identified LEI0071 located at 242 cM on chromosome 1 with an even higher association with disease resistance (LOD = 3.74) [8]. Although these two QTL markers have provided important information on the genetics of resistance to coccidiosis, the region in which they are located is sufficiently large to contain hundreds, if not thousands, of candidate genes. Therefore, in the current study, a functional genomics approach was undertaken to identify candidate genes flanking the LEI0071 and LEI0101 QTLs potentially controlling disease resistance. As a supplemental approach, our avian intestinal intraepithelial lymphocyte cDNA array (AVIEA), constructed from a library of genes expressed by intestinal intraepithelial lymphocytes (IELs) of *Eimeria*-infected chickens [9], was also used to identify candidate genes affecting gut immune responses to *E. maxima* infection. The results of these studies are described below.

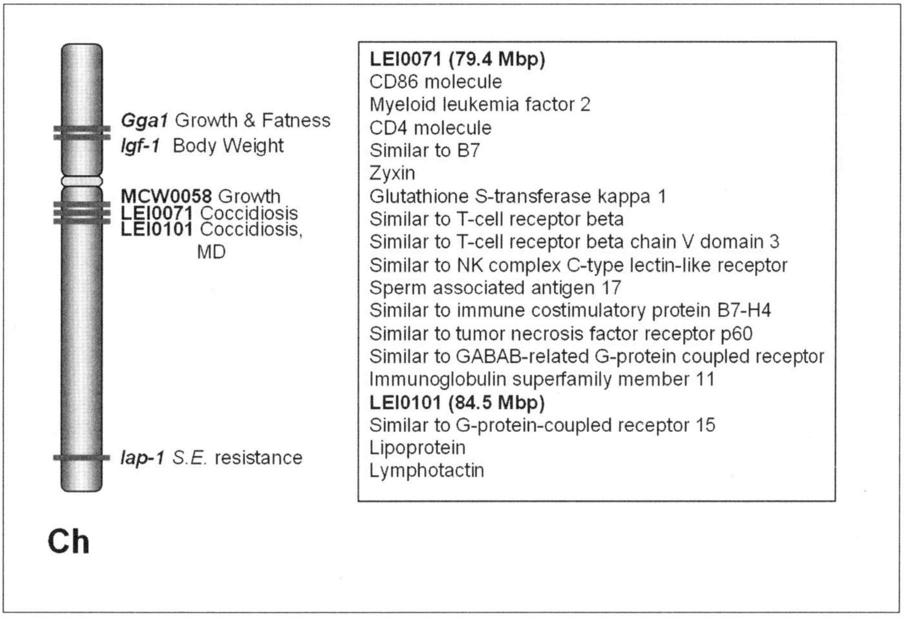

Fig. 1: Map of coccidiosis-resistance candidate genes and microsatellite markers on chromosome 1. Immune-related genes located near the LEI0071 and LEI0101 QTLs are listed in the box.

Materials and Methods

Identification of DNA markers associated with coccidiosis resistance in commercial broiler chickens

In the initial QTL mapping study which we reported in 2003 [7], three commercial broiler breeding lines (Lines 1, 2 and 3) were separately challenged with the *RB1B* highly virulent Marek's disease (MD) virus, *E. maxima* oocysts, or sheep red blood cells (SRBC) as a positive control, to evaluate their immune responses and disease resistance. These three lines showed wide variations in immune responses to the different inoculations. Line 1 exhibited the least susceptibility to coccidiosis and MD whereas Line 3 was identified as the most susceptible to infections and gave the lowest antibody titres to SRBC. Utilizing Lines 1 and 3, F^1 and F^2 generation animals were produced to study the development of resistance to coccidiosis. The F^2 offspring were challenged with a pre-determined optimal dose of *E. maxima* oocysts and five disease-resistance-associated parameters were measured (body weight gain, faecal oocyst shedding, and plasma levels of NO^{2}- + NO^{3}-, carotenoid, and interferon-γ). Genotypic data collected from the F^1 and F^2 chickens was analyzed with CRIMAP version 2.4 to test the agreement between the linkage map of these families and the chicken consensus map. From these studies, the LEI0071 and LEI0101 markers were identified.

Fine mapping studies were carried out to identify additional QTLs near the LEI0101 marker using eight additional microsatellite markers selected based on linkage map, physical map, and marker informativeness [6]. The data was adjusted for hatch, sex, and interaction between hatch and sex. Sequential oligogenic linkage analysis (SOLAR) was used to perform QTL linkage analysis and the threshold of significance was based on the guidelines suggested for genome scans by Lander and Kruglyak [10], i.e. a log of the odds (LOD) score (A statistical estimate of whether two loci are likely to lie near each other on a chromosome and are therefore likely to be inherited together) of 2.7 for single point linkage analysis and 3.6 for multipoint linkage analysis. Identification of SNPs linked to coccidiosis resistance was investigated in several candidate genes located near the LEI0071 and LEI0101 QTLs. Several candidate genes related to host immunity and located near these QTLs are shown in Figure 1.

Construction of an avian intestinal IEL cDNA array (AVIEA) and global gene expression analysis

The AVIEA contained expressed sequence tagged (EST) cDNAs that were selected from our previously reported 14,409 EST clones from a chicken intestinal IEL library [9]. Plasmid DNAs were prepared using the REAL Prep 96 kit (Qiagen, Valencia, CA) and cDNA inserts were amplified using PCR SuperMix (Invitrogen, Carlsbad, CA). Purified PCR products were spotted onto Telechem SuperAmine slides (Telechem International, Sunnyvale, CA) using an OmniGrid Accent spotter (GeneMachines, San Carlos, CA) as duplicated 24 subarrays, each as a 21 x 21 grid. A total of 9,668 ESTs were spotted in duplicate. Of these, 6,654 elements (69%) were identified as known genes and 3,014 (31%) were unique singletons of unknown identity, as determined by Blast analysis. A subset of 5,872 elements containing at least one replicate spot that provided high quality data was used for the microarray analysis. All sequence data files were obtained from NCBI (http://www.ncbi.nlm.nih.gov). IEL cDNA elements used to create the AVIEA were mapped to the chicken genome reference assembly (version 2.1) and reference RNA sequences (formatted database for Blast, May 2006) using NCBI Blast (version 2.2.13). Entrez gene data and Homologene data (October 2006) from NCBI were used to gather gene information (identification number, symbol, and name). Chicken Entrez gene identities were mapped to Homologene IDs for human or mouse, which were used for analyses of Gene Ontology and pathway classification from the PANTHER (http://www.pantherdb.org) and DAVID databases (http://david.abcc.ncifcrf.gov). Python (version 2.2.4) and SQL scripts were used to process and manipulate Blast results and data processing, respectively. Data for genes from the chicken genome were obtained from Ensembl database release 42 (http://www.ensembl.org).

Experimental coccidiosis was achieved by oral inoculation of White Leghorn SPAFAS chickens (Charles River Laboratories, Storrs, CT) with sporulated oocysts of the wild-type 41A *E. maxima* strain [3]. For primary infection, three groups of chickens were inoculated with 1.0×10^4 oocysts/chicken at 3 weeks of age. For secondary infection, three groups received the same primary inoculation followed by oral infection at 6 weeks of age with 2.0×10^4 oocysts. Two remaining groups served as uninfected controls. IELs were isolated from the intestinal jejunum (the region from the Meckel's diverticulum to the ileum) from uninfected control and inoculated animals at 1, 2, and 3 days post-primary or post-secondary infection. Total RNA was prepared using TRIzol (Invitrogen) and the RNeasy Mini RNA Purification Kit (Qiagen), and aminoallyl-labeled RNA was prepared using the Amino Allyl Message Amp II aRNA Amplification Kit according to the manufacturer's protocol (Ambion, Austin, TX). Briefly, first strand cDNA was prepared by reverse transcription from 2.0 μg of RNA using a modified oligo-dT primer containing a T7 RNA polymerase binding site at the 5' end, followed by second strand cDNA synthesis. The double-stranded cDNA was transcribed to aminoallyl-labeled RNA using T7 RNA polymerase with aminoallyl-UTP. Two 20 µg aliquots of each aminoallyl-RNA sample were fluorescently labeled with AlexaFluor 555 or AlexaFluor 647 (Invitrogen) according to the manufacturer's instructions and labeled RNAs were column purified using the RNA Amplification Kit (Ambion). For microarray hybridization, a circular loop design was employed and two-colour microarray hybridizations were performed using HybIt hybridization buffer (Telechem) in Mica hybridization chambers (GeneMachines) at 50°C overnight. After hybridization, the slides were scanned with an ArrayWoRx scanner (Applied Precision, Issaquah, WA) using appropriate filters. Spot and background intensities were acquired using SoftWoRx tracker (Applied Precision) and data analysis was performed using GeneSpring version 7.3 (Silicon Genetics, Redwood City, CA). For each spot, the mean spot intensity was obtained by subtraction of the mean background intensity from the mean foreground spot intensity and normalized to the slide mean background-subtracted spot intensity. Within the primary and secondary infection groups, normalized spot intensities generated using RNAs from days 1, 2, and 3 post-infection were compared with the corresponding intensities obtained using day 0 uninfected control RNAs. Elements exhibiting greater than 2.0-fold alterations in signal intensity during at least one of the three time points were analyzed by ANOVA and considered statistically significant at $P < 0.05$.

Results

Identification of DNA markers influencing coccidiosis resistance in broiler chickens

We previously identified the LEI0101 QTL on chromosome 1 as significantly associated with resistance to coccidiosis (i.e. reduced oocyst shedding) with an LOD score of 3.14 (Fig. 1) [7]. The heritability (hq^2) at this locus was 0.39. Interestingly, a QTL associated with viremia in chickens challenged with MD virus also was mapped near LEI0101 [9]. In addition, 3 potential loci affecting chicken growth were identified, MCW0058 (LOD = 2.20, $hq^2 = 0.37$) and MCW0020 (LOD = 1.84, $hq^2 = 0.30$) on chromosome 1, and ADL0142 (LOD = 2.43, $hq^2 = 0.26$) on chromosome 6. The MCW0058 marker was previously reported to be significantly associated with chicken growth [4]. The MCW0058 and LEI0101 markers are 20 cM apart. Subsequently, fine mapping analysis using 8 additional microsatellite markers in F^1 and F^2 animals (N = 314) revealed a new QTL (LEI0071) linked to LEI0101. Using single locus analysis of association with oocyst shedding, the LOD scores at 5 marker loci between 240 cM and 265 cM were less than 0.5 based on the data from 4 families classified as informative. In contrast, single point linkage analysis of marker LEI0071 at 242 cM on chromosome 1 gave an LOD score of 2.45, and the marker LEI0101 at 259 cM was also confirmed as significantly associated with disease resistance (LOD = 2.90). Multipoint linkage analysis with 1.0 cM increments based on the four selected families revealed a LOD score of 3.74 at 254 cM. The maximum LOD point was located between the LEI0071 and LEI0101 markers. The F^2 offspring of the largest family were divided into three groups based on their genotype of alleles 1 and 3 (1/1, 1/3, 3/3) with the marker genotype 1/3. The difference between genotypes 1/1 and 3/3 of this locus was significant, based on the multiple comparison test ($P < 0.05$). Thus, allele 3 was potentially linked to the high oocyst shedding QTL allele.

Several potential disease resistance genes linked to LEI0071 and LEI0101 were identified (Fig. 1). Due to their known roles in immunity, the CD4 and zyxin genes were selected as candidates that could potentially influence resistance to coccidiosis for SNP analysis. Of the SNPs from these genes that were analyzed in association with the five phenotypic parameters of disease resistance mentioned above, only plasma carotenoid level was significantly linked to zyxin. In this regard, plasma carotenoid concentrations were previously shown to be a more sensitive measurement of avian coccidiosis than body weight gain [11]. None of the CD4 SNPs were associated with disease resistance.

Local gene expression profiling in broiler chickens infected with *E. maxima*

In an alternative attempt to identify chicken genes conferring protective immunity to coccidiosis, our previously described AVIEA microarray, which was constructed from genes expressed in intestinal IELs following *Eimeria* infection, was used. Initially, chicken intestinal IEL RNA expression levels at days 1, 2, and 3 following primary or secondary *E. maxima* infections were determined, and these values were compared with those of the corresponding RNAs from uninfected control animals. Using an arbitrary cut-off of > 2.0 alteration in gene expression, 406 (6.9%) of the gene transcripts exhibited increased or decreased levels following primary infection. Of these, 74 were significantly changed (51 increased, 23 decreased) ($P < 0.05$). Similarly, 717 (12.2%) of the mRNAs were altered more than 2.0-fold during secondary infection, of which 308 were significantly modulated (62 increased, 246 decreased). It is interesting to note that while the majority of genes with altered expression levels during primary infection were upregulated, more genes were downregulated following secondary inoculation. However, only 3 genes were commonly modulated during both primary and secondary infections (2 unidentified ESTs [GenBank accession numbers CD734414 and CD734999] and 1 mapped gene [etoposide-induced 2.4 mRNA].

Functional analyses of the genes whose expression was altered during the course of primary or secondary *E. maxima* infections were performed using the PANTHER and DAVID databases. Biological process analysis showed that the largest groupings of modulated genes were related to protein and nucleic acid metabolism, signal transduction, and intracellular protein transport. Similarly, molecular function analysis of these genes revealed that the most prominent functions were related to nucleic acid binding, regulatory molecules, transcription factors, and transferase activities. Finally, the analysis was specifically focused on immune-related pathways; it was observed that many of the modulated genes were related to peroxisome proliferator-activated receptor (PPAR) signaling, apoptosis, interleukin signaling, Toll-like receptor signaling, antigen processing, the defence response, and the ubiquitin/proteosome system. The genes involved in interleukin signaling were especially interesting in this regard since our prior studies have documented an important role for these cytokines in protective immunity to coccidiosis [1]. We observed that at least 7 genes relevant to interleukin signaling (*IL1R, FOXP2, MAPK6, JAK, STAT3, STAT4*, and *STAT5B*) were either up- or downregulated during primary or secondary *E. maxima* infections.

Next, the microarray analysis was specifically focused on three of the candidate genes linked to the LEI0071 and LEI0101 QTLs identified above, specifically zyxin, glutathione S-transferase kappa 1 (*GSTK1*), and lymphotactin. As shown in Figure 2, most changes in gene expression levels were observed at 2 or 3 days post-primary and post-secondary infection. Transcripts for zyxin were significantly upregulated at 3 days post-secondary infection, whereas those for *GSTK1* were upregulated at 2 days post-primary infection but downregulated at days 2 and 3 post-secondary infection.

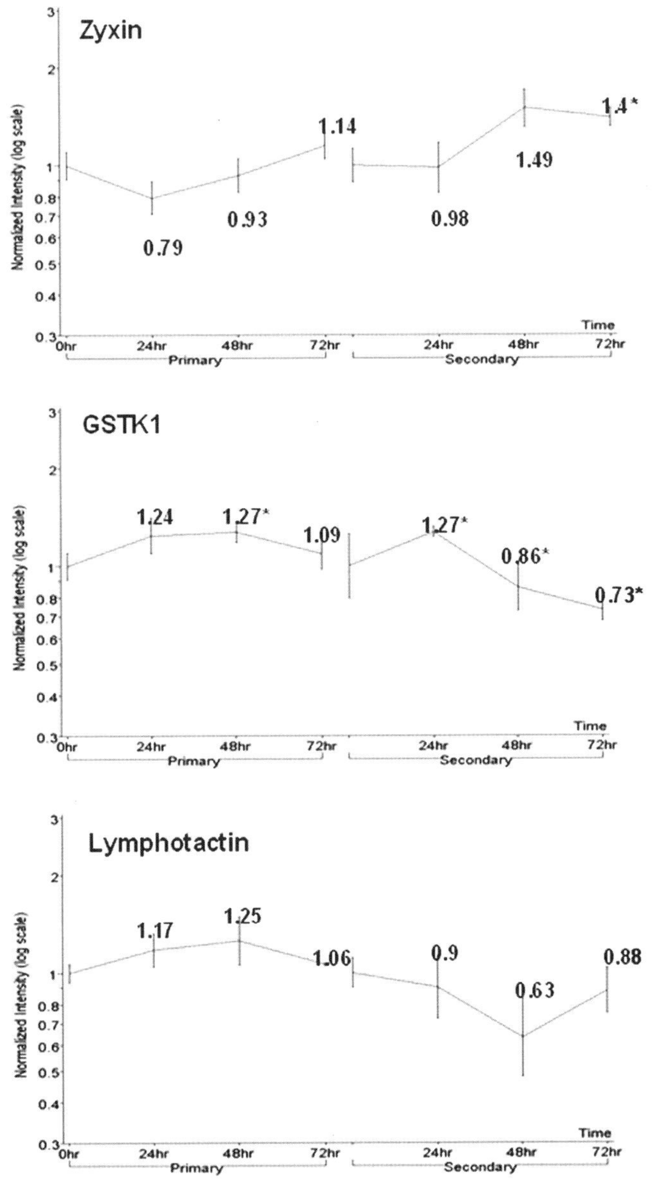

Fig. 2: Avian intestinal cDNA microarray data analysis for expression of the zyxin, GSTK1, and lymphotactin genes in *E. maxima* infected chickens. The numbers indicate up-regulation (> 1.0) or down-regulation (< 1.0) of gene expression. * $P < 0.05$ compared with uninfected controls (0 hr).

Discussion

The results of the current study confirm and extend our previous studies that identified two QTLs on chromosome 1 (LEI0071 and LEI0101) associated with resistance to avian coccidiosis. By SNP analysis, zyxin (a component of focal adhesions that regulates actin filament assembly), was most closely associated with disease resistance as assessed by plasma carotinoid levels. Because *Eimeria* protozoa invasion of intestinal epithelia leads to widespread tissue destruction, it is tempting to hypothesize that part of the host's physiological response in the gut to infection involves tightening of the epithelial barrier, which may be accomplished, in part, by strengthening the interaction between epithelial cells and the extracellular matrix through focal adhesions.

Additionally, by microarray analysis it was observed that the expression of at least 7 genes relevant to interleukin signaling was altered during *E. maxima* infection (*IL1R, FOXP2, MAPK6, JAK, STAT3, STAT4*, and *STAT5B*). We speculate that the increased MAPK6 and decreased *STAT3* observed following primary *E. maxima* infection indicate innate immune activation at an early phase of pathogen challenge. Furthermore, augmentation of *JAK* and *STAT5B* expression, and repression of *IL1R* and *MAPK6* expression at day 1 post-secondary infection may signify rapid switching from an early innate response to a later adaptive immune response. The current study also indicates that other signaling pathways involved in driving innate and adaptive immune responses were activated during coccidiosis. PPARs are cell surface receptors that act as transcription factors and belong to the ligand-activated nuclear receptor superfamily [12]. In addition to mediating anti-inflammatory effects, PPARs also are involved in lipid oxidation and homeostasis, and adipocyte differentiation.

In summary, this report provides the first comprehensive analysis of chicken intestinal IEL gene expression following infection with *Eimeria*. The results clearly illustrate the utility of the AVIEA as a tool for elucidating the genetic mechanisms used by immune cells in local immune responses. Furthermore, the integration of modern molecular genetics and functional genomics tools has provided new information for identifying genes controlling complex traits such as disease resistance. With rapidly developing technologies in functional genomics and computational biology, it is anticipated that with the integration of these new findings will enable the formulation of new paradigms to control coccidiosis. Furthermore, understanding what types of genes are involved in controlling the host response to pathogens will enable the development of novel vaccination and other strategies.

Acknowledgments

This work was supported by funds from the National Research Initiative of the USDA Cooperative State Research, Education and Extension Service, grants # 2002-35204-12368 and # 2004-35204-14798.

References

1 Lillehoj HS, Min WG, Dalloul RA: Recent progress on the cytokine regulation of intestinal immune responses to *Eimeria*. Poult Sci 2004;83:611-623.

2 Dalloul RA, Lillehoj HS: Poultry coccidiosis: recent progress on control measures and vaccination strategies. Expert Rev Vaccines 2006;5:143-163.

3 Hong Y, Lillehoj HS, Lillehoj EP, Lee SH: Changes in immune-related gene expression and intestinal lymphocyte subpopulations following *Eimeria maxima* infection of chickens. Vet Immunol Immunopathol 2006;114:259-272.

4 Van Kaam JB, Arendonk JA, Groenen M, Bovenhuis H, Vereijken AL, Crooijmans RP, van der Poel JJ, Veenendaal A: Whole genome scan for quantitative trait loci affecting body weight in chickens using a three-generation design. Livest Prod Sci 1998;54:133-150.

5 Yonash N, Bacon LD, Witter RL, Cheng HH: High resolution mapping and identification of new quantitative trait loci (QTL) affecting susceptibility to Marek's disease. Anim Genet 1999;30:126-135.

6 Groenen MA, Cheng HH, Bumstead H, Benkel BF, Briles WE, Burke T, Burt DW, Crittenden LB, Dodgson J, Hillel J, Lamont S, de Leon LP, Soller M, Takahashi H, Vignal A: A consensus linkage map of the chicken genome. Genome Res 2000;10:137-147.

7 Zhu JJ, Lillehoj HS, Allen PC, van Tassell CP, Sonstegard TS, Cheng H, Pollock D, Sadjadi M, Min W, Emara MG: Mapping quantitative trait loci associated with resistance to coccidiosis and growth. Poult Sci 2003;82:9-16.

8 Kim ES, Hong YH, Min W, Lillehoj HS: Fine-mapping of coccidia-resistant quantitative trait loci in chickens. Poult Sci 2006;85:2028-2030.

9 Min W, Lillehoj HS, Ashwell CM, van Tassell C, Matukumalli LK, Han J, Lillehoj EP. EST analysis of *Eimeria*-activated intestinal intraepithelial lymphocytes in chickens. Mol Biotech 2005;30:143-150.

10 Lander E, Kruglyak L: Genetic dissection of complex traits: guidelines for interpreting and reporting linkage results. Nat Genet 1995;11:241-247.

11 Conway DP, McKenzie ME, Dayton AD: Relationship of coccidial lesion scores and weight gain in infections of Eimeria. Poult Sci 1990;69:694-700.

12 Rizzo G, Fiorucci S: PPARs and other nuclear receptors in inflammation. Curr Opin Pharmacol 2006;6:421-427.

Dr. Hyun S. Lillehoj, Animal Parasitic Diseases Laboratory, United States Department of Agriculture, BARC-East, Building 1040, Beltsville, MD 20705 - USA.

E-mail: Hyun.Lillehoj@ARS.USDA.GOV

Pinard M-H, Gay C, Pastoret P-P, Dodet B (eds): Animal Genomics for Animal Health. Dev Biol (Basel). Basel, Karger, 2008, vol 132, pp 77-88.

Identification of Genes Involved with Tick Infestation in *Bos taurus* and *Bos indicus*

K. Kongsuwan[1], E.K. Piper[2], N.H. Bagnall[1], K. Ryan[3], P. Moolhuijzen[3], M. Bellgard[3], A. Lew[4], L. Jackson[4], N.N. Jonsson[2]

Funding body-Cooperative Research Centre for Beef Genetic Technologies.

1 CSIRO Livestock Industries, Queensland Bioscience Precinct, St Lucia, Australia
2 School of Veterinary Science, University of Queensland, St Lucia, Australia
3 Centre for Comparative Genomics, Murdoch University, Perth, Australia
4 Department of Primary Industries and Fisheries, Moorooka, Australia

Key words: cattle tick, microarray, *Bos taurus*, *Bos indicus*, tick resistance

Abstract: Tick resistant cattle could provide a potentially sustainable and environmentally sound method of controlling cattle ticks. Advances in genomics and the availability of the bovine genome sequence open up opportunities to identify useful and selectable genes controlling cattle tick resistance. Using quantitative real-time PCR and the Affymetrix bovine array platform, differences in gene expression of skin biopsies from tick resistant *Bos indicus* (Brahman) and tick susceptible *Bos taurus* (Holstein-Friesian) cattle following tick challenge were examined. We identified 138 significant differentially-expressed genes, including several immunological/host defence genes, extracellular matrix proteins, and transcription factors as well as genes involved in lipid metabolism. Three key pathways, represented by genes differentially expressed in resistant Brahmans, were identified; the development of the cell-mediated immune response, structural integrity of the dermis and intracellular Ca^{2+} levels. Ca^{2+}, which is implicated in host responses to microbial stimuli, may be required for the enhancement or fine-tuning of transcriptional activation of Ca^{2+}-dependant host defence signalling pathways.

Introduction

Approximately 60% of Australia's beef exports originate from Queensland, primarily from cattle tick (*Rhipicephalus* (*Boophilus*) *microplus*) infested areas. It is well documented that *Bos indicus* cattle are naturally more resistant to *R. microplus* than *Bos taurus* breeds, although considerable variation occurs within breeds [1]. Several studies have demonstrated that the resistance status of both *Bos taurus* and

Bos indicus herds can be improved by selection [2]. The use of cattle that have an inherently higher tick resistance status makes sound economic sense and avoids many of the problems associated with chemical control of ticks, including environmental contamination and the rising incidence of acaricide resistance in tick populations.

This study was carried out to confirm methodologies for further planned studies, using cattle in which the host resistance to infestation was already stabilized and that were heavily infested with all stages of *R. microplus* at the time of the study. It was anticipated that using Holstein (*B. taurus*) and Brahman (*B. indicus*) cattle under these conditions would provide the most divergent results possible, although overlap was expected between breed and resistance status unless there was wide divergence among the small number of animals in each breed group. Biologically meaningful and statistically significant results from this study would justify proceeding with further studies using similar methodologies. The ultimate aim of this work and the secondary aim of this study in particular were to identify genes which are differentially expressed in the skin of tick resistant and tick susceptible cattle following infestation with *R. microplus*. Although much work was undertaken on the dynamics of naturally acquired host resistance in the 1970s and 80s, the biological mechanisms employed by the resistant host to prevent heavy infestation remain elusive. Recent techniques for global transcription profiling provide an opportunity to explore the biological pathways activated during infestation, the results of which present the basis for further study in this area.

Materials and Methods

Cattle and sampling of skin tissues

The trial was undertaken from March to June 2006 at the University of Queensland's Pinjarra Hills Campus. Six *Bos taurus* (Holstein-Friesian) and six *Bos indicus* (Brahman) heifers with previous exposure to *R. microplus* were randomly chosen, and each artificially infested with 10,000 *R. microplus* larvae per week, for five weeks. In addition, these cattle were exposed to ticks under natural conditions in tick infested pastures. Standard tick counts (counts of ticks sized between 4.5 and 8 mm) were undertaken for 11 weeks during and following the infestation period to rank animals according to their relative tick resistance. Several 8 mm skin punch biopsies were taken from the neck 24 h after the 4th infestation (Fig. 1A). Biopsies were taken at the site of a tick bite (lesion), determined by visual examination for the presence of an attached tick.

Larvae used to infest cattle 7-14 days after hatching were maintained at 28°C and approximately 95% relative humidity until application to the cattle.

Quantitative real time PCR

Quantitative real time PCR (qRT-PCR) using Sybr Green fluorescent dye and the ABI Prism 7900 Sequence Detection System (PE Applied Biosystems, Foster City, CA) was used to assess mRNA abundance. A 1/10 dilution of cDNA derived from 2 µg of total RNA and four technical replicates were performed per sample for each gene. GeNorm [3] analysis determined acidic ribosomal protein PO (RPLPO) to be the least variable (among the 4 genes tested) and hence most suitable internal control. Data is expressed as the normalized mean expression values with standard deviation and was analysed by analysis of variance (ANOVA). Differences between expression values were considered significant if probability values of $P < 0.05$ were obtained.

GeneChip hybridization

Genes expressed in Brahman and Holstein-Friesian cattle were analysed by Affymetrix GeneChip Bovine Array containing 24,128 probe sets representing ~23,000 bovine transcripts and 19,000 UniGene clusters. Target preparation and microarray processing procedures were performed as described in the Affymetrix GeneChip Expression Analysis manual. Scanning was performed with an Agilent Microarray Scanner (Agilent Technologies, Palo Alto, CA).

Microarray Data analysis

Two different but commonly used normalization algorithms were used to process microarray data; RMA (Robust Multi-chip Analysis), and MAS5 (Microarray Analysis Suite). Within each algorithm, gene lists were filtered using a minimum 2-fold expression change and ANOVA with a false discovery rate of 0.01%. For MAS5, gene lists were also filtered on P (present) calls in 50% of the gene chips. In addition, raw intensity values were manually scrutinised to minimize false positives. The final gene list (138 in total) consists of overlapping genes from the RMA and MAS5 analyses.

Results and Discussion

Bos indicus cattle are generally more resistant to infestation with the tick *R. microplus* than are *Bos taurus* breeds [4]. This resistance is predominantly manifested by non-attachment of larvae during the first 24 h of infestation [5]. Initial tick attachment and creation of a feeding lesion are important to the ticks ability to feed and develop from larva to engorged adult. In order to identify differentially expressed genes at the tick attachment site during infestation, the present study investigated gene expression within skin samples of *Bos indicus* and *Bos taurus* cattle collected 24 h after artificial tick challenge.

Overall, cattle tick counts for Brahman are lower than that of the Holstein-Friesian cattle (Fig 1B), supporting previous work demonstrating the heightened ability of the *Bos indicus* breed to resist heavy infestation [1]. Brahman cattle showed a higher degree of resistance from the first challenge and this was maintained throughout the trial. In contrast, Holstein-Friesian cattle showed relatively higher susceptibility from the first challenge and tick numbers increased over the course of the trial (Fig. 1B).

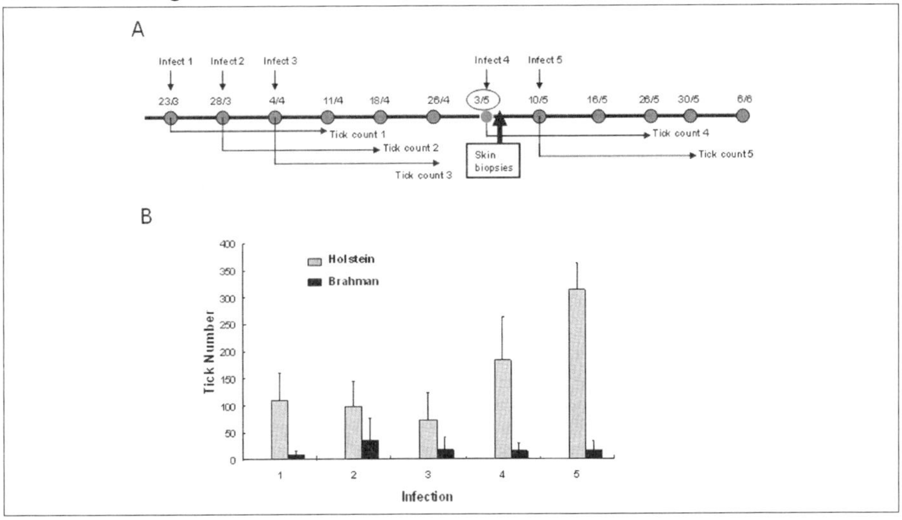

Fig. 1: Tick infestation protocol and tick counts.

(A) The animals were managed together at the University of Queensland's Pinjarra Hills Droughtmaster Unit, in a pasture naturally infested with larvae of *R. microplus*. Animals were artificially infested with 10,000 tick larvae per week for five weeks (on 23/3, 28/3, 4/4, 3/5 and 10/5). As ticks engorged between 19-24 days after larval infestation, tick counts of pre-engorged females were performed 20-22 days following infestation (on 11/4, 18/4, 26/4, 26/5 and 30/5, respectively). Tick resistance was assessed by counting the number of engorging ticks, 4.5-8.0 mm in size, on one side of the animal. Skin biopsies were collected 24 h following the 4th infestation (on 4/4/07).

(B) Tick count comparison of Brahman and Holstein heifers.

qRT-PCR using selected genes

Cattle ticks feed by ingesting host blood that pools at the site of tick attachment [6]. Forty-nine genes (Table 1) were selected from various pathways, including cytokines, chemokines and components of the toll-like receptor (TLR) signalling pathway, skin structural proteins and candidate genes that were previously shown to be upregulated in highly tick resistant Hereford Shorthorn (*Bos taurus*) cattle [7]. Fifteen of the genes tested were significantly upregulated at Holstein-Friesian bite sites. The significantly differentially expressed genes included factors involved in response initiation, chemokines and pro-inflammatory genes; *NFKBP50, CD1A, CXCL10/IP10, CCL26, CCL3L1, TLR2, TLR5, CD14, CCR1, FTH1, BCL2, P4HB, MYD88* and *TRAF6* (Fig. 2). Taken together, the qRT-PCR results of selected candidate genes suggest that TLR signalling and pro-inflammation pathways of exposed Holstein-Friesians were highly activated (in response to tick substances) at bite sites compared to Brahmans, which may be advantageous to the tick. It might be possible that blunting the immune response and maintaining homeostasis while avoiding inflammation contributes to tick resistance, as observed in Brahmans.

Table 1: List of candidate genes whose expression was examined by qRT-PCR comparing tick bite sites in Holstein and Brahman cattle. Oligonucleotide primers for the genes and the internal controls (*RPLPO, GAPDH*) were designed from GenBank (cDNA) sequence information. Most primers were as described previously [7,14,15], however for ease of reference all primers used in this study are shown below.

Gene name	Abbreviation	AccessionNo.	Forward primer	Reverse primer	Size (bp)
Acidic ribosomal protein P0	*RPLP0*	BT021080	5' caaccctgaagtgcttgacat 3'	5' aggcagatggatcagcca 3'	220
Glyceraldehyde 3 phosphate dehydrogenase	*GAPDH*	BC102589	5' cctggagaaacctgccaagt 3'	5' gccaaattcattgtcgtacca 3'	226
Beta actin	*bACTIN*	NM173979	5' agcaagcaggagtacgatgagt 3'	5' atccaaccgactgctgtca 3'	246
18S ribosomal RNA	*18S rRNA*	DQ066896	5' gtaacccgttgaaccccatt 3'	5' ccatccaatcggtagtagcg 3'	151
Interleukin 1 beta	*IL1B*	M35589	5' aaatgaaccgagaagtggtgtt 3'	5' ttccatattcctcttggggtaga 3'	185
Interleukin 6	*IL6*	X57317	5' ctgggttcaatcaggcgat 3'	5' cagcaggtcagtgtttgtgg 3'	205
Interleukin 8	*IL8*	AF232704	5' ctgtgtgaagctgcagttct 3'	5' atggaaacgaggtctgccta 3'	216
Interleukin 10	*IL10*	NM_174088	5' cttgtcggaaatgatccagt 3'	5' tctcttggagctcactgaag 3'	208
Tumour necrosis factor alpha	*TNFa*	AF011926	5' ctggttcagacactcaggtcct 3'	5' gaggtaaagcccgtcagca 3'	183
b-defensin	*b-defensin*	NM_175703	5' tcttcctggtcctgtctgct 3'	5' ccgaacaggtgccaatctgt 3'	130
Nuclear factor kappa B p50	*NFkBP50*	TC244499	5' ctggaagcacgaatgacaga 3'	5' gctgtaaacatgagccgtacc 3'	207
Nuclear factor kappa B p65	*NFkBP65*	TC235598	5' aacaaccccttccaagttcc 3'	5' ctcccagagttccgattcac 3'	203
CD1a	*CD1A*	btcn49853	5' tgtgccacgtctcaggatt 3'	5' tgtcctggtctcctagactgc 3'	204
Chemokine (C-C motif) receptor 1	*CCR1*	AY575855	5' ctgctggtgatgattgtctg 3'	5' tgctctgctcacacttacgg 3'	191
Chemokine (C-C motif) ligand 2	*CCL2*	NM_174006	5' ccctcctgtgcctgctact 3'	5' tcaaggctttggagtttggt 3'	284
Chemokine (C-C motif) ligand 3 L1	*CCL3L1*	NM_174511	5' ggtcttctcggcaccattt 3'	5' ccaggtcggtgatgtattcc 3'	209
Chemokine (C-X-C motif) ligand 10	*CXCL10 / IP10*	XM_866821	5' agtggaagcccctgcagtaaa 3'	5' agtcccagccttgctactgaca 3'	176

Chemokine (C-C motif) ligand 26	*CCL26*	AF124601	5' cctggaagcaggtgcatagt 3'	5' caggagtcttggagcgaaaa 3'	191
Chemokine (C-C motif) ligand 28	*CCL28*	XM_615382	5' tgtgacttggctgctgtcat 3'	5' cgatgtgcccctttactgtt 3'	173
Toll-like receptor 1	*TLR1*	AY634628	5' ctgcccatatgccaagagtt 3'	5' aaaccaactggaggatcgtg 3'	421
Toll-like receptor 2	*TLR2*	NM_174197	5' gcttcctgtgacttcctgtcc 3'	5' ccgaaagcacaaagatggtt 3'	504
Toll-like receptor 3	*TLR3*	AY124007	5' gaggcaggtgtccttgaact 3'	5' gctgaatttctggacccaag 3'	329
Toll-like receptor 4	*TLR4*	AB056444	5' aggcagccataacttctcca 3'	5' ggttgagtaggggcatttga 3'	410
Toll-like receptor 5	*TLR5*	BE753283	5' aacgctttgctcaaacacct 3'	5' accctctgatggactgatgc 3'	301
Toll-like receptor 6	*TLR6*	BE753283	5' aacgctttgctcaaacacct 3'	5' taccctctgatggactgatgc 3'	681
Toll-like receptor 7	*TLR7*	BI683325	5' gctgtgtggtttgtctggtg 3'	5' ttctctctcgggtcttccaaa 3'	395
Toll-like receptor 8	*TLR8*	NC_007331	5' ccacctggaagagagtgagg 3'	5' tcatccattagcctctgcaa 3'	201
Toll-like receptor 9	*TLR9*	AJ509825	5' ctggaggagctgaacctgag 3'	5' tggttgtaggacagcagcag 3'	320
Toll-like receptor 10	*TLR10*	NW_381204	5' tctgaatgggtgaagtatga 3'	5' aatggcaccattcagtctgg 3'	190
Myeloid differentiation primary response gene (88) F2/R2	*MYD88*	NM_002468	5' cgcggcttacttgtcct 3'	5' ctgccagtcgtccaggag 3'	200
CD14	*CD14*	D84509	5' ggtgctacccgatgtgtctg 3'	5' aagggatttccgtccagagt 3'	191
TNF alpha receptor associated factor 6	*TRAF6*	CB221913	5' cggtgactctctccagctct 3'	5' tggacatttgtgacctgcat 3'	194
Ferritin heavy polypeptide 1	*FTH1*	NM_174062	5' gaggagagggaacatgctg 3'	5' tcacacagatggggatcat 3'	212
Peroxiredoxin 4	*PER4*	NM_174433	5' cgaagacaaggaggacttgg 3'	5' tgcctgaaccaaacgtagt 3'	213
Duox-1	*DUOX1*	CF118082	5' agtttgagtggctggctgac 3'	5' aacagactccggttcagcac 3'	169
Proapoptotic marker	*P53*	X81704	5' ctttgaggtgcgtgtttgtg 3'	5' gtggtttcttctttggctgtg 3'	161
Antiapoptotic marker	*BCL2*	BC103230	5' tgctgctgtttctgcctaca 3'	5' gcacttttgcatgggtcaa 3'	143
Nuclear transcription factor	*CMYC*	BC113343	5' gaaaaagcccccaaggtagt 3'	5' ggtcaatttaggcgcaagag 3'	166
CD44 antigen	*CD44*	CF763972	5' ccctcctgacgatgagatga 3'	5' cccgtggtgtggtagaaatg 3'	172
Keratin 6A	*KRT6A*	CF765461	5' cagaacctggagcctttgtt 3'	5' tgctgtgcgtttgttgattt 3'	165
Keratin 6L	*KRT6L*	CF764540	5' gcacttctctccccactcc 3'	5' cacccctcccatcctgtt 3'	159
Keratin, type II cytoskeletal 6C (Cytokeratin 6C)	*CK6C*	XM867210	5' taggcgtgagtggaggaagt 3'	5' ggaggaggaggtggtggt 3'	155
Major allergen BDA20	*BDA20*	CF762806	5' tgactgtgaatccctctccat 3'	5' ggccttcagtcatttttgtga 3'	212
Procollagen-proline, 2-oxoglutarate 4-dioxygenase	*P4HB*	CF763975	5' gacattgccagcgtgagag 3'	5' cccgtcttcctttgctttt 3'	171
Collagen, type I alpha1	*COL1A1*	CF117427	5' tggtgacaagggtgagacag 3'	5' gggagaccattgagtccatc 3'	201
Collagen, type I alpha2	*COL1A2*	CF117427	5'ggtcgaagtggagagacagg 3'	5' agggttcacccacagatccag 3'	198
Collagen, type III alpha1	*COL3A1*	AB258724	5' atggggtcaaatgaaggtga 3'	5' gtccgcacgaaattcttgat 3'	201
Platelet-derived growth factor receptor-like	*PDGFRL*	CF764395	5' ccacaaggatctcccagagt 3'	5' ggctttccgtcagttcattt 3'	165

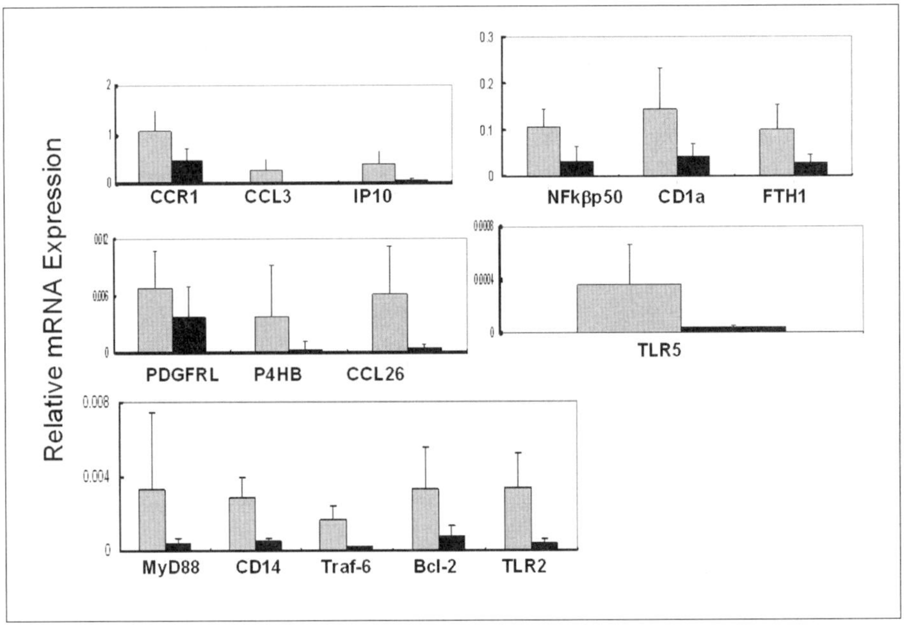

Fig. 2: Gene expression analysed by qRT-PCR.
Expression of 15 genes (a total of 49 tested) which were significantly upregulated ($P < 0.05$) in Holstein skin bite sites compared to Brahman bite sites. The x-axis shows the genes and the y-axis represents normalized gene expression levels, with vertical bars representing standard deviation. Results were normalized using the endogenous control gene RPLPO. Black bars represent Holstein and grey bars, Brahman.

Differential gene expression analysis using microarray

RNA samples from bovine neck skin samples were subjected to global transcription profiling using the Affymetrix GeneChip Bovine Array. Four animals from Brahman and four from Holstein-Friesian were chosen, as they provided a representative spread of tick counts observed within each breed. There were two technical replicates per animal. Breed comparisons were conducted using two different and commonly used normalization methods, RMA and MAS5. Using the RMA normalized data, 183 genes (78 downregulated and 105 upregulated in the Brahman samples) were differentially expressed. Using the MAS5-normalized data, 309 genes (160 downregulated and 148 upregulated in the Brahman samples) were differentially expressed. Overlap of RMA and MAS5 genelists generated a list of 138 genes differentially expressed between Brahman and Holstein samples (75 downregulated and 63 upregulated in relation to Brahman). Putative identities could be determined for 94 of the 138 genes, based on sequence similarity by BLAST searches (cut-off E value of 0.0001 was applied), while 44 had no significant similarity to sequences in the human RefSeq. Gene Ontology annotation was carried out using GOTERM (http://www.geneontology.org). Analysis of specific GO terms for biological processes assigned to the differentially expressed genes revealed various functional gene categories, including enzymes, transporters, structural proteins, signal transducers, and immune genes (Table 2).

Table 2: Transcripts significantly differentially expressed in Brahman and Holstein skin following tick infestation. Only transcripts with putative identity are shown. FC = Fold Change.

Genes upregulated in Brahmans			
HuRef	**Affy ID**	**FC**	**Gene**
Immune/Host Defence			
	Bt.22867.2.A1_at	70.63	MHC class II
NM_006419	Bt.28088.1.S1_at	10.64	Chemokine C-X-C motif ligand 13
NM_001935	Bt.8282.1.S1_at	4.479	Dipeptidyl-peptidase 4
NM_005194	Bt.1747.1.S1_at	2.735	CCAAT/enhancer binding protein (C/EBP), beta
NM_001025101	Bt.16377.1.S1_at	2.392	Myelin basic protein
NM_000587	Bt.18557.1.S1_at	2.131	Complement component 7
NM_014333	Bt.3806.1.A1_at	2.021	Immunoglobulin superfamily, member 4
Lipid metabolism/leukotriene/Inflammation response			
NM_001140	Bt.8917.1.S1_at	8.726	Arachidonate 15-lipoxygenase
NM_000896	Bt.27036.1.S1_at	4.794	Cytochrome P450
NM_001442	Bt.97.1.S1_at	3.122	Fatty acid binding protein 4, adipocyte
NM_001279	Bt.20167.1.S1_at	2.909	Cell death-inducing DFFA-like effector A
Protein modification			
NM_153498	Bt.17491.1.A1_at	3.766	Calcium/calmodulin-dependent protein kinase
NM_018960	Bt.20919.2.A1_at	3.382	Glycine N-methyltransferase
NM_006983	Bt.7179.1.S1_a_at	2.426	Matrix metallopeptidase 23B
Signal transduction and transcription regulation			
NM_004675	Bt.12283.1.A1_at	6.641	Rho-related GTP-binding protein RhoI
NM_052854	Bt.2236.1.S1_at	3.784	CAMP responsive element binding protein 3-like 1
NM_032682	Bt.28682.1.A1_at	3.74	Forkhead box P1
NM_007249	Bt.8505.1.A1_at	3.389	Kruppel-like factor 12
NM_012234	Bt.20494.1.S1_at	3.172	RING1 and YY1 binding protein
NM_199512	Bt.7788.1.S1_at	2.548	Steroid sensitive gene 1
NM_000125	Bt.24926.1.S1_at	2.417	Oestrogen receptor 1
NM_022439	Bt.4060.1.S1_at	2.183	MAL, T-cell differentiation protein
Protein transport and response to stimuli			
NM_006569	Bt.22210.1.S1_at	3.728	Cell growth regulator with EF-hand domain 1
NM_198901	Bt.15882.3.S1_at	2.765	Sorcin
NM_007035	Bt.5391.1.S1_at	2.7	Keratocan
NM_030959	Bt.7955.2.S1_at	2.569	Olfactory receptor, family 12, subfamily D
NM_002025	Bt.7816.1.S1_at	2.556	AF4/FMR2 family, member 2
NM_002901	Bt.23305.1.S1_at	2.313	Reticulocalbin 1, EF-hand calcium binding domain
Cytoskeleton organisation and cell matrix			
NM_147147	Bt.29555.1.A1_at	4.74	Blood vessel epicardial substance
NM_022144	Bt.21407.1.A1_at	3.565	Tenomodulin
NM_020962	Bt.27089.1.S1_at	2.845	Likely ortholog of mouse neighbour of PUNC E11
NM_006944	Bt.508.1.A1_at	2.654	Secreted phosphoprotein 2, 24KDA
NM_002593	Bt.7955.1.S1_at	2.342	Procollagen c-endopeptidase enhancer
NM_015719	Bt.11570.1.A1_at	2.262	Collagen, type V, alpha 3
Miscellaneous			
NM_030915	Bt.17942.1.S1_at	4.305	Hypothetical protein DKFZP566J091
NM_015393	Bt.24157.1.A1_at	3.639	Hypothetical protein DKFZP564O0823
	Bt.1701.2.A1_a_at	2.762	Coatomer protein complex, subunit zeta 2 (COPZ2)
	Bt.22629.1.A1_at	2.577	Putative protein C21orf62
	Bt.28472.1.S1_at	2.467	Olfactomedin-like 3 (OLFML3)
	Bt.3115.1.A1_at	2.429	Aspartylglucosaminidase (AGA)
	Bt.15685.1.A1_at	2.329	Hypothetical protein FLJ20605

Genes upregulated in Brahmans HuRef	Affy ID	FC	Gene
Immune/host defence			
NM_005091	Bt.9980.1.S1_at	2.932	Peptidoglycan recognition protein 1
NM_002123	Bt.29870.1.S1_at	2.242	Major histocompatibility complex, class II, DQ beta 1
Lipid metabolism/leukotriene/inflammation response			
NM_012212	Bt.14010.1.S1_at	3.355	Leukotriene B4
NM_181900	Bt.22144.2.S1_at	2.857	Start domain containing 5
NM_020139	Bt.1792.1.S1_at	2.487	3-hydroxybutyrate dehydrogenase, type 2
NM_001018056	Bt.5484.1.S1_at	2.32	Very low density lipoprotein receptor
NM_018441	Bt.15781.1.A1_at	2.159	Peroxisomal trans-2-enoyl-coa reductase
NM_021151	Bt.4599.1.S1_at	2.109	Carnitine o-octanoyltransferase
Protein modification			
NM_016123	Bt.9250.2.S1_at	2.994	Interleukin-1 receptor-associated kinase 4
NM_014038	Bt.5495.1.S1_at	2.958	Basic leucine zipper and W2 domains 2
NM_001762	Bt.6236.1.S1_at	2.07	Chaperonin containing TCP1,subunit 6A
NM_004582	Bt.20263.1.S1_at	2.04	RAB geranylgeranyltransferase, beta subunit
Signal transduction and transcription regulation			
NM_006023	Bt.3256.1.S1_at	5.464	Chromosome 10 open reading frame 7
NM_130387	Bt.18153.1.A1_at	4.545	Ankyrin repeat and socs box containing 14
NM_004815	Bt.18409.1.A1_at	3.846	RHO GTPase activating protein 29
NM_014720	Bt.25077.1.S1_at	3.831	STE20-like kinase (yeast)
NM_018433	Bt.12249.2.S1_at	3.23	Jumonji domain containing 1A
NM_002160	Bt.11061.1.S1_at	2.785	Tenascin C
NM_206963	Bt.24933.1.S1_at	2.247	Retinoic acid receptor responder
NM_018023	Bt.8251.1.A1_at	2.202	Yeats domain containing 2
Protein transport and response to stimuli			
NM_152565	Bt.11010.1.S1_at	3.759	Hypothetical protein FLJ38708
NM_014043	Bt.19768.1.A1_at	2.544	Chromatin modifying protein 2B
Cytoskeleton organisation and cell matrix			
NM_014751	Bt.24376.1.S1_at	4.524	Metastasis suppressor 1
	Bt.3611.1.S1_at	3.254	Keratin 5B
NM_000495	Bt.16495.1.A1_at	2.544	Collagen, type IV, alpha 5
NM_173086	Bt.7.1.S1_at	2.352	Keratin 6E
Blood coagulation			
NM_007351	Bt.28532.1.S1_at	2.159	Multimerin 1
NM_000129	Bt.19845.1.S1_at	2.1	Coagulation factor XIII
Melanin biosynthesis			
NM_000550	Bt.221.1.S1_at	6.211	Tyrosinase-related protein 1
NM_000372	Bt.9028.1.S1_at	4.566	Tyrosinase (oculocutaneous albinism IA)
NM_006928	Bt.5091.1.S1_at	2.415	Silver homolog (mouse)
Metabolic process			
NM_005651	Bt.24081.1.A1_at	34.965	Tryptophan 2,3-dioxygenase
NM_001359	Bt.13390.2.S1_a_at	5.813	2,4-dienoyl COA reductase 1, mitochondrial
NM178135	Bt.14669.2.A1_at	4.485	Short-chain dehydrogenase/reductase 9
NM_005896	Bt.13324.1.S1_a_at	2.32	Isocitrate dehydrogenase 1 (NADP+), soluble
NM_018394	Bt.24834.1.A1_at	2.004	Abhydrolase domain containing 10

Miscellaneous			
NM_020783	Bt.22143.1.A1_at	132.625	Synaptotagmin IV
	Bt.20372.1.S1_at	17.301	Paraoxonase 3
NM_001007242	Bt.2278.1.S1_at	10.638	Glycoprotein 2 (zymogen granule membrane)
NM_014915	Bt.24642.1.A1_at	8.264	Ankyrin repeat domain 26
	Bt.22599.1.A1_at	8.064	L-threonine dehydrogenase
	Bt.968.1.S1_at	6.06	Histidine ammonia-lyase
	Bt.28518.1.S1_s_at	4.291	Trypsin inhibitor
NM_006919	Bt.9693.1.S1_at	2.785	Serpin peptidase inhibitor
	Bt.160.2.S1_x_at	2.59	Desmocollin 1
	Bt.2966.1.A1_at	2.5	Cytokinesis 11
NM_138787	Bt.21720.1.S1_at	2.457	Hepis
	Bt.5005.1.S1_at	2.45	Alpha 2 type I collagen preproprotein
NM_004684	Bt.5153.1.S1_at	2.352	Sparc-like 1 (MAST9,HEVIN)
NM_005730	Bt.8606.1.S1_a_at	2.352	CTD (carboxy-terminal domain, RNA polymerase II)
NM_003675	Bt.17154.1.S1_at	2.309	PRP18 pre-mRNA processing factor 18 homolog (yeast)
NM_002712	Bt.1867.2.S1_at	2.304	Protein phosphatase 1, regulatory subunit 7
NM_030666	Bt.27909.1.S1_at	2.298	Serpin peptidase inhibitor
	Bt.4952.2.S1_a_at	2.298	Ribosomal protein L22
	Bt.24749.1.S1_at	2.237	Solute carrier family 35, member E3
NM_032796	Bt.8054.2.S1_a_at	2.004	Hypothetical protein FLJ14495

qRT-PCR validation of microarray results

Expression patterns of the top 20 genes in the microarray experiment were validated using qRT-PCR. qRT-PCR results confirmed microarray results for all tested genes except for one of the transcripts, Bt.18426.1.A1_s_at, whose function is unknown (Table 3). Discrepancies in observed microarray and qRT-PCR results may stem from probe and array effects and different hybridization kinetics of the probe sets and primers for each gene. Genes with low expression have been shown to correlate poorly between microarray and qRT-PCR data [8].

Differential pathway activation between breeds

Among the differentially expressed genes between Brahman and Holstein-Friesian skin samples (Table 2), the most notable categories were (1) immune pathways involving MHC class II, antibodies, and complement; (2) extracellular matrix proteins, especially genes implicated in skin (dermis) barrier function; and (3) genes encoding Ca^{2+} binding proteins which were upregulated in Brahmans.

It is not surprising that immune/host defence signalling genes accounted for a significant percentage of genes upregulated in Brahmans (seven out of 37 genes with putative immune functions). These upregulated genes are associated with cell-mediated responses or derived from dendritic cells and components of the complement system (component 7), MHC class II and the B lymphocyte attracting chemokine CXCL13 (Table 2). The basic leucine zipper transcription factor CCAAT/enhancer binding protein-beta (*C/EBPB*) which is also upregulated, is an important regulator for cellular differentiation and/or proliferation, and plays a role in the modulation of the adaptive immune response.

Table 3: Validation of microarray results by qRT-PCR of the top 7 genes up regulated in Holstein and top 13 genes up regulated in Brahman

AFFY_ID	Gene	Fold Change	
Holstein		**Microarray**	**qRT-PCR**
Bt.22143.1.A1_at	Synaptotagmin IV	132.625	68.26
Bt.24081.1.A1_at	Tryptophan 2,3-dioxygenase	34.965	7
Bt.20372.1.S1_at	Paraoxonase 3	17.301	16.2
Bt.24642.1.A1_at	Ankyrin repeat domain 26	8.264	6.91
Bt.22599.1.A1_at	L-threonine dehydrogenase	8.064	10.16
Bt.221.1.S1_at	Tyrosinase-related protein 1	6.211	5.18
Bt.13390.2.S1_a_at	2,4-dienoyl COA reductase 1, mitochondrial	5.813	5.3
Brahman			
Bt.8282.1.S1_at	CD26, adenosine deaminase complexing protein 2	4.479	5.05
Bt.29555.1.A1_at	Blood vessel epicardial substance	4.74	4.93
Bt.27036.1.S1_at	Cytochrome P450, family 4, subfamily F, polypeptide 3	4.794	4.36
Bt.9123.1.A1_at	Unknown function	4.924	5.6
Bt.18426.1.A1_s_at	Unknown function	5.091	0.014
Bt.25676.1.A1_at	Unknown function	5.1	5.2
Bt.17235.1.A1_at	Unknown function	6.322	7.96
Bt.12283.1.A1_at	Rho-rrelated GTP-binding protein RhoI	6.641	6.88
Bt.8917.1.S1_at	Arachidonate 15-lipoxygenase	8.726	7.6
Bt.23980.1.A1_at	Unknown function	9.944	6.74
Bt.28088.1.S1_at	Chemokine (C-X-C motif) ligand 13	10.64	14.8
Bt.25393.1.A1_at	Unknown function	17.75	7.08
Bt.7682.1.S1_at	Unknown function	112.5	20.28

Arachidonate 15-lipoxygenase and cytochrome P450 were both upregulated in Brahmans. These two enzymes, known to be involved in the production of lipid mediators from arachidonic acid (AA), and the lipid mediators play a role in the regulation of pro-inflammatory cytokines and chemokine production. The upregulation of MHC class II genes is consistent with previous studies that suggest an association of MHC class II genes (DRB genes) with tick resistance phenotypes in cattle [9].

Physical factors controlling tick attachment and feeding are functions of skin structure and include skin integrity, pH, and temperature. Consistent with prominent structural remodelling in the dermis, many deregulated transcripts were related to extracellular matrix composition and turnover as well as protein and lipid trafficking. Abundant skin structural proteins, such as keratin, various collagens and a number of extracellular matrix genes, were upregulated in both breeds. Collagen gene *COL5* was expressed at a higher level in Brahman, while Keratin 5 and 6 and *COL4* genes had higher expression in Holstein animals. Notably, procollagen C-endopeptidase enhancer 2 (*PCOLCE2*), the protein of which is involved in the post-translational modification of collagen, was upregulated in Brahmans. PCOLCE2 binds to type I procollagen and potentiates its cleavage by procollagen C-proteinases (BMP-1). In addition to collagens, other extracellular matrix and cell adhesion proteins such as peptidoglycan recognition protein 1, PUNC homologue, tenascin, and tenomodulin, also showed differences in expression between breeds. These extracellular matrix proteins most likely assist in forming a protective barrier to produce an environment inhospitable for tick attachment and feeding. Several genes encoding enzymes involved in the synthesis and binding of fatty acids, such as peroxisomal trans-2 enoyl CoA reductase, keratocan and fatty acid binding protein (FABP-4), were also upregulated in Brahman cattle. These genes are known to be implicated in cellular trafficking and storage of lipids. Alternatively, they may function in neutralising immunosuppressive factors that facilitate tick attachment and formation of the feeding lesion.

To successively feed to repletion, ticks secrete a cocktail of bioactive enzymes such as prostaglandins and anti-coagulation factors [10] that ensure the host blood remains in a fluid state. It is plausible to envisage the host countering parasite attack through promotion of vasoconstriction, platelet aggregation factors and an increase in blood coagulation mediators. Notably, Kruppel-like factor gene (*KLF2*) and blood vessel/epicardial substance (*BVES*) were upregulated in Brahmans while coagulation factor XIII was highly expressed in Holsteins. KLF-2 is a novel transcription factor involved in multiple biological processes including maintaining T cells in the quiescent state, stabilizing blood vessel walls through endothelial cell function, and regulating pro-inflammatory activation of monocytes through inhibiting NF-κB activation [11].

Numerous recent studies provide evidence that Ca^{2+} plays a pivotal role in many aspects of cellular physiology, including activating the cell surveillance system against attempted microbial invasion. Ca^{2+} acts as an early second messenger in transcriptional activation of an array of defence-related genes. For example, Ca^{2+} is required for parasite-mediated mitogen-activated protein kinase (MAPK) activation and IL-12 synthesis during Toxoplasma gondii infection [12]. In Brahmans, the upregulation of several Ca^{2+} binding proteins such as Ca^{2+} dependent protein kinase, sorcin, cell growth regulator with EF hand domain, and reticulocalbin (Table 2), may reflect an increase in Ca^{2+} levels and transport, especially from cytoplasm to membranes, thus leading to an enhanced host signalling response. Alternatively, these proteins may aid in sequestering Ca^{2+} away from the ticks as calcium and cyclic AMP are known to stimulate tick salivary gland secretion [13].

The major limitation to this current study is the absence of a comparison of skin samples from pre-tick infestation (0 h) and post-treatment (24 h after tick challenge) within breeds makes it difficult to rule out the possibility that naturally existing differences between Brahman and Holstein-Friesian cattle have contributed to the observed differential gene expression (Table 2). To confirm our findings, we plan to analyse gene expression profiles in skin cell lines derived from both breeds before and after exposure to tick antigen extracts. Second, in this analysis, gene expression changes in blood following tick infestation were not included although consequential immune responses against ticks may be derived from circulating immune cells. For example, macrophages, dendritic cells and other antigen presenting cells resident in the skin may signal to peripheral blood cells and direct an appropriate response to infestation via the production of specific cytokines or other biological mediators. In the future, we plan to investigate gene expression changes in blood samples.

Conclusion

This study has provided evidence that distinct and different pathways are activated in the skin of tick resistant Brahman cattle compared with tick susceptible Holstein-Friesian cattle during infestation with *R. microplus*. Further functional characterisation of these genes, to demonstrate their role in tick host resistance/susceptibility, is the next logical step. The results presented here will be validated in an independent herd of cattle, with the analysis of pre- and post-infestation samples, to confirm whether the differences observed in this study apply in other breeds of cattle and whether they are associated with tick resistance. For example, if the overt innate immune response observed in Holstein-Friesians is confirmed to promote susceptibility to ticks, then control of innate immunity may offer an attractive target for intervention. Alternatively, improving skin biomechanic properties by stimulating the production of the appropriate type of collagens could provide a tick resistance microenvironment. Beyond affecting livestock health, an understanding of tick resistance mechanisms also impacts on human health due to potential reduced acaricide use and reduced introduction of chemical residues into the food chain.

References

1 Utech KB, Wharton RH, Kerr JD: Resistance to *Boophilus microplus* in different breeds of cattle. Aus J Agric Res 1978;29:885-895.

2 Frisch JE, O'Neill CJ, Kelly MJ: Using genetics to control cattle parasites-the Rockhampton experience. Int J Parasitol 2000;30:253-264.

3 Vandesompele J, De Preter K, Pattyn F, Poppe B, Van Roy N, De Paepe A, et al: Accurate normalization of real-time quantitative RT-PCR data by geometric averaging of multiple internal control genes. *Genome Biology* 2002;3:1-11.

4 Seifert GW: Variations between and within breeds of cattle in resistance to field infestations of the cattle tick (*Boophilus microplus*). Aus J Agric Res 1971;22:159-168.

5 Wagland BM: Host resistance to cattle tick (*Boophilus microplus*) in Brahman (*Bos indicus*) Cattle. III Growth on previously unexposed animals. Aust J Agric Res 1978;29:401-409.

6 Mulenga A, Blandon M, Khumthong R: The molecular basis of the *Amblyomma americanum* tick attachment phase. Exp Appl Acarol 2007;41:267-287.

7 Wang YH, Reverter A, Kemp D, McWilliam SM, Ingham A, Davis C, et al: Gene expression profiling of Hereford Shorthorn cattle following challenge with *Boophilus microplus* tick larvae. Aust J Experi Agric 2007;47:1397-1407.

8 Czechowski T, Bari RP, Stitt M, Scheible WR, Udvardi MK: Real-time RT-PCR profiling of over 1400 Arabidopsis transcription factors: unprecedented sensitivity reveals novel root and shoot-specific genes. Plant J 2004;38:366-379.

9 Martinez ML, Machado MA, Nascimento CS, Silva MVGB, Teodoro RL et al: Association of BoLA-DRB3.2 alleles with tick (*Boophilus microplus*) resistance in cattle. Genet Mol Res 2006;5:513-524.

10 Gordon JR, Allen JR: Factor V and VII anticoagulent activities in the salivary glands of feeding *Dermacentor andersoni* ticks. J Parasitol 1991;77:167-170.

11 Ahmad N, Lingrel JB: Kruppel-like factor 2 transcriptional regulation involves heterogeneous nuclear ribonucleoproteins and acetyltransferases. Biochemistry 2005;44:6276-6285.

12 Masek KS, Fiore J, Leitges M, Yan SF, Freedman BD, Hunter CA: Host cell Ca^{2+} and protein kinase C regulate innate recognition of *Toxoplasma gondii*. J Cell Sci 2006;119:4565-4573.

13 Maritz-Olivier C, Louw AI, Neitz AW: Similar mechanisms regulate protein exocytosis from the salivary glands of ixodid and argasid ticks. J Insect Physiol 2005;51:1390-1396.

14 Strandberg Y, Gray C, Vuocolo T, Donaldson L, Broadway M, Tellam R: Lipopolysaccharide and lipoteichoic acid induce different innate immune responses in bovine mammary epithelial cells. Cytokines 2005;31:72-86

15 Menzies M, Ingham A: Identification and expression of Toll-like receptors 1-10 in selected bovine and ovine tissues. Vet Immunol and Immunopathol 2006;109:23-30.

Dr. K Kongsuwan, CSIRO Livestock Industries, Queensland Bioscience Precinct, 306 Carmody Road, University of Queensland St Lucia 4072 Australia.

E-mail: Kritaya.kongsuwan@csiro.au

Pinard M-H, Gay C, Pastoret P-P, Dodet B (eds): Animal Genomics for Animal Health. Dev Biol (Basel). Basel, Karger, 2008, vol 132, pp 89-91.

Genomics Approaches to Study the Biology Underlying Resistance to Trypanosomiasis - Some Unexpected Lessons

S.J. Kemp[1,3], O. Hanotte[3], M. Agaba[3], H.A. Noyes[1], J. Gibson[3,5], A. Archibald[4], C. Rennie[1], H. Hulme[2], A. Brass[2]

1 School of Biological Sciences, Liverpool, UK
2 The University of Manchester, Manchester, UK
3 International Livestock Research Institute, Nairobi, Kenya
4 Roslin Institute, Roslin, UK
5 Currently, Institute for Genetics and Bioinformatics, C.J. Hawkins Homestead, University of Armidale, Australia

Key words: trypanosomes, *T. congolense*, trypanosomosis, trypanosomiasis, expression analysis, networks, quantitative trait loci

Abstract: An international multidisciplinary consortium is conducting a programme of research on the host response to trypanosome infection. This builds upon quantitative trait loci (QTL) mapping which identified genome regions influencing susceptibility to pathology following *T. congolense* infection in both cattle and mice. The approach uses expression analysis to examine the response of both susceptible and resistant strains and a series of novel informatics tools to identify pathways which are activated as a result of challenge, and those which are differentially used by resistant and susceptible strains. Of particular interest are those pathways which simultaneously satisfy both criteria, i.e. are significantly differentially activated and contain genes within QTL regions. However, it is important to stress that it is not required that the genes within the QTL region are differentially expressed themselves.

Tsetse fly transmitted trypanosmiasis is of enormous economic importance; it is responsible for the exclusion of livestock from vast tracts of West and Central Africa. Closely related species of trypanosomes cause sleeping sickness in humans, which is of great, and increasing, public health significance. Resistance to infection has been described in several breeds of cattle, especially the N'Dama, and in mouse models. Following infection with *Trypanosoma congolense*, both resistant and susceptible cattle breeds become infected and show initial anaemia, but the resistant genotype quickly recovers and continues to grow while the susceptible animals show increasingly severe anaemia, cachexia and eventually death. The total cost to Africa has been estimated to exceed $1 billion per annum [1].

Quantitative trait loci (QTL) affecting resistance in both the mouse and cow have been mapped [2-4] and we are following this with an expression analysis. In the study of host resistance and susceptibility, one very important challenge in the analysis of microarray data (or indeed, any phenotype data) is distinguishing relevant from irrelevant differences, and differences that are consequences of pathology from those which are causes of pathology. There are many clear differences in the biology of different animal strains, and there are many differences between healthy and sick animals. The challenge is to use objective unbiased methods to analyse expression data in order to extract the causative changes. We have developed some novel methodologies for this [5].

The comparison of murine and bovine systems has proven to be highly informative. In particular, the use of congenic mouse lines in which fragments of resistant genome surrounding the QTL have been placed on a susceptible background has proven to be a powerful means of studying the action of QTL, and several trypanotolerance QTL co-localize to QTL involved in a range of other traits. There are a number of cases in which genes of susceptible origin show an altered pattern of expression as a result of the presence of a distant gene of resistant origin. For example, a structural polymorphism within Daxx within the QTL was correlated with altered expression of p53, which is regulated by Daxx and is on a different chromosome.

We have identified the genes that are most significantly differentially expressed over an infection time course and examined the pathways in which they are involved. This has proven to be a highly informative approach and there are strong indications that survival or death is a result of differential use of few very generic pathways. In all cases, there are candidate genes within QTL involved in these pathways, many of which are not differentially expressed themselves. There is strong crosstalk between pathways which are normally considered as separate entities. This is mediated by a small number of key molecules which are highly connected and represent critical points through which a wide range of responses may be modulated.

The power of such systems-based approaches to array data is that they can generate findings of relevance well beyond the system being investigated. The results of our trypanosomiasis studies illustrate this, having relevance for human survival following sepsis and many other forms of stress, and suggesting common mechanisms of adaptation to disease resistance across mammalian species.

References

1 Kristjanson PM, Swallow BM, Rowlands GJ, Kruska RL, de Leeuw PN: Measuring the benefits of controlling African animal trypanosomis and the returns to research. Agric Syst 1999;59:79-98.

2 Kemp S J, Iraqi F, Darvasi A, Soller M, Teale AJ: Localisation of genes controlling resistance to trypanosomiasis in mice. Nat Genet 1997;16:194-196.

3 Clapcott SJ: Localisation of genes controlling resistance to trypanosomiasis in mice (*Mus musculus* L.). PhD Thesis, School of Biological Sciences (University of Liverpool, Liverpool), 1998.

4 Hanotte O, Ronin Y, Agaba M, Nilsson P, Gelhaus A, Horstmann R, Sugimoto Y, Kemp S, Gibson J, Korol A, Soller M, Teale A: Mapping of QTL controlling resistance to trypanosomosis in African indigenous cattle. Proc Natl Acad Sci USA 2003;100(13):7443-7448.

5 Fisher P, Hedeler C, Wolstencroft K, Hulme H, Noyes H, Kemp S, Stevens R, Brass A: A systematic strategy for large-scale analysis of genotype phenotype correlations: identification of candidate genes involved in African trypanosomiasis. Nucl Acids Res 2007;35:5625-5633.

Pr. Steve J. Kemp, School of Biological Sciences, Biosciences Building, Liverpool L69 7ZB, UK. E-mail: S.J.Kemp@liverpool.ac.uk

Pinard M-H, Gay C, Pastoret P-P, Dodet B (eds): Animal Genomics for Animal Health. Dev Biol (Basel). Basel, Karger, 2008, vol 132, pp 93-98.

Bovine Viral Diarrhoea Virus Infection Alters Global Transcription Profiles in Bovine Endothelial Cells

J.D. Neill[1], J.F. Ridpath[1], A. Lange[2], R.L. Zuerner[3]

1 Virus and Prion Diseases of Livestock Research Unit, National Animal Disease Center, USDA/ARS, Ames, Iowa, USA
2 Institute of Virology, University of Veterinary Medicine, Hannover, Germany
3 Bacterial Diseases of Livestock Research Unit, National Animal Disease Center, USDA/ARS, Ames, Iowa, USA

Key words: serial analysis, gene expression, inflammation, cell culture contamination, endothelial activation

Abstract: Bovine viral diarrhoea viruses (BVDV) are significant pathogens of cattle worldwide. These viruses exist in both non-cytopathic and cytopathic biotypes. Non-cytopathic BVDV can establish persistent lifelong infections in cattle and are a frequent contaminant of biological reagents such as cell cultures and foetal bovine serum. We identified commercially available bovine aortic endothelial cells (BAECs) contaminated with BVDV. In this study, to determine if BVDV alters endothelial gene transcription patterns, serial analysis of gene expression (SAGE) was used to compare gene expression profiles from uninfected and BVDV contaminated BAEC. SAGE is an open ended, quantitative method for characterizing global patterns of transcription. Comparison of expression profiles of BVDV-contaminated and noninfected cells revealed significant increases in the transcription of many genes including P-selectin, tryptophan tRNA synthetase and prostaglandin D2 synthase. These changes were validated by real-time PCR. Additionally, real-time PCR demonstrated that the response to LPS and dsRNA by contaminated cells, as well as cells acutely infected with noncytopathic BVDV, is altered. The altered response may be through the high level of expression of *A20* and inhibition of activation of NF-κB. BAECs are commonly used as a model to study endothelial cell function in many different systems. As shown here, transcriptional and probable protein changes resulting from BVDV infection significantly alter cellular responses and may have a profound impact on experimental outcome. Transcriptomic analysis provided the initial clues leading to the characterization of this altered function.

INTRODUCTION

Bovine viral diarrhoea viruses (BVDVs) are major pathogens in cattle worldwide and cause millions of dollars in losses to the cattle industry in the United States (US) annually. Disease caused by BVDV can range from clinically unapparent to a severe acute hemorrhagic syndrome. Acute BVDV infections can also be associated with secondary infections by opportunistic pathogens that are more severe than the secondary pathogen causes alone [1,2]. Additionally, BVDV can establish persistent infections in cattle if a dam is infected in the first 125 days of pregnancy. The calf can be born persistently infected and immunologically tolerant to the infecting virus strain and spread virus for the rest of its life. These viruses are also found as contaminants of biologic reagents, foetal bovine serum (FBS) and cell cultures [3-5]. The contaminated cell cultures are not limited to those of bovine origin. Finding FBS free of BVDV and anti-BVDV antibodies is a major problem for research and diagnostic laboratories and biologics companies. It is unknown what effect contaminating BVDV in cell cultures has on the normal functions of the cells and how it affects the outcome of experiments utilising these cells.

The objective of this study was to examine cultured bovine aortic endothelial cells (BAECs) that were found contaminated with BVDV. Serial analysis of gene expression (SAGE) was used to compare gene expression in BAECs and in BVDV contaminated BAECs to determine the effect of BVDV on normal endothelial cell function. Altered expression of specific transcripts was validated with real time PCR. The data obtained from this transcriptomics study indicated that BVDV contamination does indeed alter normal functions of the cells and has a profound effect on data obtained using these cells.

MATERIALS AND METHODS

Low passage primary BAECs were obtained from a US company that supplies endothelial cells for research purposes. The cells were maintained in bovine endothelial cell medium containing horse serum that was purchased from the company of origin. The cells were cultured as recommended by the supplier. BAECs were tested for the presence of BVDV by reverse-transcriptase PCR as previously described [6]. This screening detects any BVDV strain that may be present and determines the genotype of the virus. The non-cytopathic BVDV type 2 strain 296nc was used to compare acutely infected cells with those that were chronically infected. BVDV-free BAECs (obtained from the same company at an earlier date) were infected with 296nc at an MOI of 2 and were processed at 24 hours post-infection for downstream analysis.

SAGE was done with total cellular RNA isolated from BVDV-free BAECs and from BVDV-contaminated BAECs using Trizol Reagent according to manufacturer's specifications (Invitrogen, Inc., CA). The two libraries were constructed as described previously [7,8]. The libraries were annotated as described [9]. Changes in gene expression as indicated by SAGE were validated by real-time PCR [9] using Superscript Platinum SYBR green rtPCR reagent (Invitrogen). Primer sets that amplified sequences from transcripts encoding housekeeping, apoptotic, protein translation accessory proteins and transcriptional activators were used (Table 1; primer sequences available upon request).

The standard test for determining identity and number of cells in an endothelial cell preparation is binding and uptake of acetylated low-density lipoprotein (aLDL) [10]. Fluorescently-labelled aLDL (Invitrogen) was used to determine if BVDV infection had an impact on LDL uptake. Briefly, 60% confluent BAECs in four well Lab-tek chamber slides (Nalge Nunc International, Inc., Rochester, NY) were washed with PBS and covered in medium containing 10 µg of aLDL/ml and incubated at 37 C in 5% CO_2. Medium was placed on the cells for 2, 3 or 4 hours. The cells were washed with PBS and removed from the slides by brief trypsinization. The cells were washed in ice-cold PBS and examined by flow cytometry. Total cells with LDL uptake and fluorescence intensity were measured.

To examine the effect of BVDV infection on endothelial function, noninfected, BVDV contaminated and 296nc-infected BAECs were treated with 1 µg/ml LPS (*E. coli* O111:B4; Sigma-Aldrich, Inc., St. Louis, MO) for 6 hours or 1 µg/ml double-stranded RNA (polyI:C; Sigma-Aldrich) for 18 hours. At the end of the incubation, the cells were washed with PBS and total cellular RNA was isolated. Real-time PCR (Table 1) was used to examine the effect of treatment on expression of specific genes of BVDV infected cells.

Table 1: Comparison of SAGE and real-time PCR.

Transcript	Tags noninfected	Tags contaminated	real-time PCR[1]
b-actin	67	101	1
b2-microglobulin	54	126	2.1
GAPDH	79	44	0.45
TRAM	15	16	1.5
Trp tRNA synthetase	10	66	5.2
Prostaglandin D2 synthase	4	61	9.45
BAD	2	0	1.4
BclXl	11	8	2.2
JunB	3	4	0.5
JunD	4	4	1
c-myc	7	4	0.5
c-Jun	3	3	1.2
c-Fos	0	2	2.8

[1]Fold change of level in contaminated cells compared to noninfected control

RESULTS

Routine testing of the low passage BAECs by reverse-transcriptase PCR demonstrated that the cells were contaminated with BVDV. Sequencing of the PCR product revealed that the virus belonged to the BVDV2 genus. Immunohistochemical staining of the cells showed that >90% of the cells were infected (data not shown). The BAEC showed no overt signs of viral infection but grew at a rate slightly slower than that of the noninfected BAECs based on the lower split ratio of the contaminated cells (1:10 vs. 1:6).

SAGE was done from total mRNA isolated from the two cell populations. The BAECs and BVDV contaminated BAEC libraries were sequenced to 65,000 and 80,000 tags, respectively. Comparison of the two libraries revealed that many of the transcripts examined showed little or no change. However, several transcripts with apparent changes in expression levels were identified (Table 1). Validation of altered gene expression levels was done using reverse-transcription real-time PCR. This analysis revealed that the changes indicated by SAGE were a reflection of what was taking place in these cells (Table 1).

The uptake of aLDL by BAECs, BVDV-contaminated BAECs and 296nc acutely infected BAECs was tested. Flow cytometry of cells from each group revealed a significant increase in the uptake of aLDL by the contaminated and 296nc-infected cells (Table 2). It is not clear what caused this, although an examination of the SAGE data indicated that there was increased transcription of genes involved in LDL uptake and processing (data not shown). These changes have yet to be validated.

The effect of BVDV infection on important functions of the BAECs, such as endothelial activation, was evaluated by treatment of the cells with either LPS or dsRNA. Real-time PCR was used to examine the increased expression of three indicator genes, *A20*, E-selectin and P-selectin, which are known to be induced following treatment with LPS or dsRNA. The noninfected BAECs showed increases in all three indicator genes with treatment of both LPS and dsRNA (Table 3). The contaminated BAECs showed higher levels of expression of the three transcripts when treated with LPS. The 296nc-infected BAECs showed an even higher level of expression of *A20* but significantly lower expression of P- and E-selectin. The three BAEC populations treated with dsRNA showed roughly equivalent levels of increased expression of *A20*. Interestingly, the P-selectin expression varied between cell populations. There were very low increases in the noninfected cells and the 296nc-infected cells but the contaminated cells showed a high level of expression. E-selectin in the dsRNA-treated cells showed similar large increases in expression levels in the noninfected and contaminated cells but only a low level increase in the 296nc-infected cells.

Table 2: Uptake of acetylated LDL by BAECs.

	BAEC population		
Length of incubation	**Noninfected[1]**	**Contaminated**	**296nc**
0 hr	22	33	155
1 hr	242	248	297
2 hr	794	852	1205
3 hr	1091	1312	1479
4 hr	1373	1587	1824

[1]Mean of fluorescence units in two experiments.

Table 3: Real-time PCR of LPS and dsRNA upregulated genes in BAECs.

Transcript	**Noninfected**	**Contaminated**	**296nc1**
	LPS		
A20	14.52	23.9	65.2
E-selectin	99.6	194.6	33.7
P-selectin	78.1	176.8	12.4
	dsRNA		
A20	10.6	10.5	8.6
E-selectin	86.5	63.6	7.9
P-selectin	2.6	65.4	4.7

[1]acute infection
[2]average fold increase over untreated control cells

Discussion

The contamination of biologics with BVDV has great importance because of potential downstream repercussions. Biologics companies pay a great deal of attention to BVDV contamination in their procurement of foetal bovine serum (FBS) as well as in their cell stocks that are used in vaccine production. Diagnostic laboratories are also greatly concerned with obtaining BVDV-free FBS for the accuracy of the diagnostic tests. However, little attention is paid to BVDV contamination of bovine cell cultures used for routine research purposes. Noncytopathic BVDV contamination of cell cultures is not readily apparent and most users of these cells are unaware of the presence of the virus. Exceptions to this would be laboratories that conduct BVDV research. Routine PCR testing of the BAECs described here revealed BVDV contamination. The contaminating BVDV was a noncytopathic BVDV2 strain. Testing of second lot of BAECs obtained from the same company revealed that it was contaminated with a noncytopathic BVDV1 strain (data not shown). BAECs are used in many different laboratories as a surrogate model for human endothelial cells. Little work to date has focused on any effect that contaminating BVDV may have on cell function and thus, the accuracy of the results obtained using these cells.

SAGE analysis revealed that there were indeed differences in gene expression between noninfected and BVDV contaminated BAECs (Table 1). Some of the changes that were observed could potentially affect the response of the cells to different stimuli. Real time PCR showed good agreement with SAGE results (Table 1). Initial experiments using fluorescently labelled aLDL supported this with significantly increased uptake in BVDV-infected BAECs (Table 2). Follow-up experiments in which the BAECs were exposed to LPS and dsRNA, stimuli that elicit strong responses by the cells, revealed that the BVDV-infected BAECs responded differently from those that were not infected. The cells infected acutely with 296nc showed the greatest effect overall with diminished increases in the three transcripts tested when stimulated with both LPS and dsRNA. The only exception to this was the higher level of *A20* transcripts in the LPS-treated 296nc-infected cells. This was an interesting result. *A20* was shown to block endothelial activation by LPS through toll-like receptor 4 (TLR4) by preventing the activation of NF-κB [11,12]. The response to bacterial LPS is mediated primarily through toll-like receptor 4. The decline in E- and P-selectin expression in the 296nc infected cells could be explained by decreased NF–κB activation. Animals that are acutely infected with BVDV often have secondary infections that are more severe than with the pathogen alone. The lack of adequate TLR-4 signalling in acutely infected cattle may explain, at least in part, the enhanced secondary infections.

The work described here confirms that a thorough transcriptomic analysis can reveal specific gene expression changes that may have a profound effect on how cells respond to specific stimuli. These results have provided insights into the altered responses of BAECs and information concerning the disease processes in BVDV-infected cattle.

Acknowledgements

The authors would like to thank Ms. Tracy Porter and Mr. William Boatwright for excellent technical assistance. Thanks are also given to Dr. David Alt and Ms. Karen Halloum for their assistance in the sequencing of the SAGE libraries and to Mr. Bruce Pesch for his assistance in the flow cytometric analysis.

REFERENCES

1 Potgieter LN: Bovine respiratory tract disease caused by bovine viral diarrhea virus. Vet Clin North Am Food Anim Pract 1997;13(3):471-481.

2 Fulton RW, Purdy CW, Confer AW, Saliki JT, Loan RW, Briggs RE, Burge LJ: Bovine viral diarrhea viral infections in feeder calves with respiratory disease: interactions with Pasteurella spp., parainfluenza-3 virus, and bovine respiratory syncytial virus. Can J Vet Res 2000;64(3):151-159.

3 Bolin SR, Ridpath JF: Prevalence of bovine viral diarrhea virus genotypes and antibody against those viral genotypes in fetal bovine serum. J Vet Diagn Invest 1998;10(2):135-139.

4 Bolin SR, Ridpath JF, Black J, Macy M, Roblin R: Survey of cell lines in the American Type Culture Collection for bovine viral diarrhea virus. J Virol Methods 1994; 48(2-3): 211-221.

5 Vilcek S: Identification of pestiviruses contaminating cell lines and fetal calf sera. Acta Virol 2001;45(2):81-86.

6 Ridpath JF, Bolin SR: Differentiation of types 1a, 1b and 2 bovine viral diarrhoea virus (BVDV) by PCR. Mol Cell Probes 1998;12(2):101-106.

7 Neill JD, Ridpath JF: Gene expression changes in BVDV2-infected MDBK cells. Biologicals 2003;31(2):97-102.

8 Velculescu VE, Zhang L, Vogelstein B, Kinzler KW: Serial analysis of gene expression. Science 1995;270(5235):484-487.

9 Neill JD, Ridpath JF, Liebler-Tenorio E: Global gene expression profiling of Bovine immature B cells using serial analysis of gene expression. Anim Biotechnol 2006;17(1):21-31.

10 Voyta JC, Via DP, Butterfield CE, Zetter BR: Identification and isolation of endothelial cells based on their increased uptake of acetylated-low density lipoprotein. J Cell Biol 1984;99(6):2034-2040.

11 Cooper JT, Stroka DM, Brostjan C, Palmetshofer A, Bach FH, Ferran C: A20 blocks endothelial cell activation through a NF-kappaB-dependent mechanism. J Biol Chem 1996;271(30):18068-18073.

12 Ferran C, Stroka DM, Badrichani AZ, Cooper JT, Wrighton CJ, Soares M, Grey ST, Bach FH: A20 inhibits NF-kappaB activation in endothelial cells without sensitizing to tumor necrosis factor-mediated apoptosis. Blood 1998;91(7):2249-2258.

Dr. John D. Neill, Virus and Prion Diseases of Livestock Research Unit (VPDRLU), National Animal Disease Center, USDA/ARS, 2300 Dayton Avenue, Ames, IA 50010 USA.

E-mail: john.neill@ars.usda.gov

Pinard M-H, Gay C, Pastoret P-P, Dodet B (eds): Animal Genomics for Animal Health. Dev Biol (Basel). Basel, Karger, 2008, vol 132, pp 99-104.

A Combined Transcriptomic Approach to Analyse the Dialogue between Pseudorabies Virus and Porcine Cells

L. Flori[1], C. Rogel-Gaillard[1], V. Mariani[1,2], G. Lemonnier[1], M. Cochet[3], K. Hugot[1], P. Chardon[1], S. Robin[4], F. Lefèvre[3]

[1] INRA, DGA, UMR 314, Laboratoire de Radiobiologie et d'Etude du Génome, Jouy-en-Josas, France ; CEA, DSV, IRCM, SREIT, Laboratoire de Radiobiologie et d'Etude du Génome, Jouy-en-Josas, France

[2] Centro Ricerche e Studi Agroalimentari, Parco Tecnologico Padano, Lodi, Italy

[3] INRA, DSA, UR 892, Unité de Virologie et Immunologie Moléculaires, Jouy-en-Josas, France

[4] AgroParisTech-ENGREF-INRA, UMR 518, Unité de Mathématiques et Informatique Appliquées, Paris, France

Key words: pig, pseudorabies virus, PK15, transcriptome, immune response, SLA

Abstract: The pseudorabies virus (PrV), a porcine Alphaherpesvirus, is a good model for the study of virus-host cell dialog. As PrV has a strong tropism for mucous epithelial cells, we chose to follow in vitro the PrV time course-infection of porcine PK15 cells. The viral and cellular transcriptome modifications were simultaneously analysed using a combined SLA/PrV cDNA microarray, the porcine Qiagen-NRSP8 oligonucleotides microarray and real time quantitative PCR. A high increase in viral gene expression was found from 4 h post-infection (PI), concomitantly to the first viral progeny and most viral genes were differentially expressed 12 h PI. No early global cellular shutoff was observed but many cellular genes were downregulated between 8 and 12 h PI, when *UL*41 transcripts encoding the virion shutoff protein, were first detected. Several genes involved in the MHC class I mediated antigenic pathway were downregulated including *SLA-Ia, TAP1, TAP2, PSMB8* and *PSMB9* genes. These results suggested that PrV prevents the viral antigen presentation by epithelial cells to cytotoxic T lymphocytes by decreasing transcription levels of SLA Ia mediated antigenic pathway genes. Other genes involved in the immune response, the apoptosis pathway, nucleic acid metabolism and cytoskeleton also appeared to be regulated during PrV infection. The combined approach will help to decipher host response evasion strategies developed by PrV and to study early cellular modifications.

Introduction

Pseudorabies virus (PrV) is a well-known pig pathogen responsible for the Aujeszky's disease, which has a considerable economical impact worldwide in this species. PrV is a herpesvirus belonging to the *Alphaherpesvirinae* subfamily as the human pathogen HSV-1 and is a pertinent model for the in vitro study of the biology of this subfamily of viruses and their interactions with host cells. The PrV genome (about 140 Kb) contains 70 genes encoding structural and non-structural proteins, the expression of which is coordinated during the lytic cycle [1]. PrV has a strong tropism for epithelial cells of the oronasal respiratory tract, which are the first cells interacting with virions. It can easily multiply in these cells and is sometimes responsible for respiratory signs [1].

The knowledge of cellular and viral genes that are up- or downregulated during viral infection is essential to better understand the physiopathology of infection and to identify molecules involved in host resistance/susceptibility mechanisms. Over the past few years, the gene expression profiles of host cells or pathogens have been widely studied and DNA microarrays have proven to be very efficient tools for this type of experiment [2,3]. At the outset of this experiment, transcriptome analyses of host cells infected by PrV had only been carried out in heterologous models [4-6] and the PrV transcriptome had not yet been studied by microarray experiment. In this context, we aimed at simultaneously following both PrV and porcine PK15 epithelial cell line transcriptomes during a time course of infection with a microarray experiment to determine which host metabolic pathways are modulated during PrV infection in epithelial cells in relation with PrV gene expression, and the specific effects of PrV infection on genes located in the pig major histocompatibility complex (MHC) or swine leucocyte antigen (SLA) region. This region, which is located on submetacentric chromosome 7 (SSC7p1.1–1q1.1), is the first region of the pig genome that has been entirely sequenced and annotated. It contains several genes encoding molecules involved in SLA antigen presentation pathways and other genes encoding immune proteins. SLA class I molecules, which are expressed on almost all nucleated cells, present non-self peptides including peptides derived from viral antigens. These cells are continuously under surveillance by cytotoxic T lymphocytes (CTLs), which play a critical role in the defence against viral infection. Herpesviruses have evolved strategies to interfere with the MHC class I antigenic pathway. PrV infection decreases the expression of MHC class I molecules on the cell surface at least via the inhibition of a transporter associated with antigen processing (TAP) activity [7,8].

To simultaneously follow transcriptional modifications in cells and the virus, a DNA/cDNA microarray (SLA/PrV microarray) targeting the genes of the SLA region, additional genes encoding other important immunological molecules and all the viral genes was first constructed. The Qiagen-NRSP8 commercial array containing 13,297 oligonucleotides representative of 8,541 different porcine genes was also used [9]. Both microarrays are complementary in function: the SLA/PrV microarray to investigate variation in the expression of specific genes, the Qiagen-NRSP8 microarray to more widely explore the physiological pathways.

MATERIALS AND METHODS

PK15 cells were infected with purified virions from the wild type NIA3 strain of Pseudorabies virus at a multiplicity of infection (MOI) of 20 pfu/ml or mock-infected by incubation in virus- free culture medium. For transcriptomic analysis, the same pool of cells was used to perform four time-course replicate experiments on infected and mock-infected cells. Cells were harvested for RNA extraction just after infection (I) or mock-infection (MI) and 1, 2, 4, 8 and 12 hours after I or MI. Total RNA was extracted with the Trizol-chloroform method (Invitrogen, France), dissolved in 40µl DEPC treated water and quantified (Nanodrop, Nyxor Biotech). For quantitative real-time RT-PCR (qRT-PCR) RNA samples were treated with DNase I and cleaned up to remove any DNA contamination (DNase set and cleanup, Qiagen, France). The RNA samples were diluted at 1µg/µl final concentration and stored at - 80°C.

Two microarrays were used in this experiment. The SLA/PrV microarray, composed of 1,789 probes, is a dedicated DNA/cDNA array. Four distinct probe sets are spotted on these slides: 420 probes mapping to a segment of chromosome 7 which included the extended-SLA region and represented 272 unique sequences, 73 probes specific of genes encoding molecules involved in immunity and mapping outside the SLA region, 80 PrV probes specific to the 70 viral genes and 1,170 probes randomly chosen for data normalisation from porcine cDNA AGENAE library. This microarray was constructed in the CRB-GADIE (Centre de Ressources Biologiques-Génomique des Animaux Domestiques et d'Intérêt Economique, LREG, INRA, Jouy-en Josas, France). The second microarray is a commercial generic microarray spotted on slides (Qiagen-NRSP8) which contains 13,297 oligonucleotides (70-mers) specific to 8,541 porcine genes [9].

Five µg of each RNA sample was reverse-transcribed and labelled with Cy3 and Cy5 (reagents: Pronto kit, Corning and Amersham, Biosciences, France). Labelled targets were quantified (Nanodrop, Nyxor Biotech, France), evaporated and the pellets were resuspended in hybridisation buffer (Pronto Kit, Corning, France) at a final concentration of 2 pmoles/µl.

All the time points (T0, T1, T2, T4, T8 and T12 h post-infection or PI) were used for SLA/PrV hybridisations and only four time points for Qiagen-NRSP8 hybridisations (T1, T2, T4 and T8 h PI). Twelve and eight different conditions were considered for the SLA/PrV and Qiagen-NRSP8 hybridisations, respectively. A balanced loop design with two independent loops was used, each loop containing two replicates of PrV infection and MI. Twenty and 40 pmoles of labelled cDNA for each were used for the SLA and Qiagen-NRSP8 slides, respectively (Pronto Kit, Corning, France).

SLA and Qiagen-NRSP8 slides were scanned on the Scanarray scanner (Perkin Elmer, France) and on the Microarrayreader scanner (Virtek, France), respectively. Signals were quantified with the Imagene version 5.1 software (Biodiscovery, USA). All the results were stored in the BioArray Software Environment (BASE) of Sigenae (Système d'Information du projet d'Analyse des Génomes des Animaux d'Elevage).

The normalisation and statistical analysis steps were performed with scripts written with the R software. Raw data were log2 transformed, normalized by lowess (f=0.3) and by subtraction of the median of each block spotted on the slide. Normalised data was analysed with a linear model. Selected contrasts were tested with the Student t test: I conditions against MI conditions at each time point. The false discovery rate (FDR) was calculated after the Student t test (R software, varmixt package). Differentially expressed genes with a FDR = 0.05 were selected.

Hierarchical clustering analysis (HCL, Eisen) was performed to analyse viral and cellular genes that were differentially expressed during infection (Euclidian distance, average linkage) using TMeV software. The "k-means" method was used to identify groups of genes with similar expression kinetics (TMeV software). The "Gene Ontology" annotations of differentially expressed cellular genes were obtained with the Panther classification system based on human orthologous genes (http://www.pantherdb.org/). The Ingenuity Pathways Analysis (IPA) program allowed for determination of the significant networks and the top functions associated with the differentially expressed genes at each time point (IPA 5.0, Ingenuity Systems Inc., USA).

RNA corresponding to mock-infected T0 and infected T0, T2, T4, T8, T12 triplicate samples were used for real-time PCR analysis. Triplicate reactions were set up in a 20 µl volume using 10 ng cDNA, PCR primers at a final concentration of 300 nM and SYBR Green PCR Master Mix (Applied Biosystems, USA). An ABI PRISM 7900 HT sequence detection system was used for monitoring the level of SYBR green fluorescence.

RESULTS

The expression of viral genes was detected between 2 and 12 h PI and increased over time. Most of the genes seemed to be expressed at 8 and 12 h PI. The number of differentially expressed viral genes between infected and mock-infected cells increased drastically between 2 and 8 h and reached a plateau after 8 h (Table 1).

Nineteen viral genes were not differentially expressed between infected and mock-infected cells at any time. The first differentially expressed genes observed 1 h PI were *US1* and *UL29* encoding two non-structural proteins (RSP40 and ICP8) and *UL49.5* encoding the virion envelope glycoprotein, gN. *UL41* transcripts encoding the virion host shutoff protein (vhs) were first detected as differentially expressed at 8 h PI.

With both microarrays, SLA/PrV and Qiagen-NRSP8, the number of differentially expressed cellular genes was observed to increase especially between 4 and 12 h PI (Table 1). No early global cellular shutoff was observed. A lot of cellular genes were found downregulated from 4 h PI and this number increased until 12 h PI.

Many biological processes and cellular functions were found to be modulated during PrV infection, using the GO classification and the IPA software. In particular, PrV infection modifies the expression of genes involved in MHC class I antigenic presentation pathways. Indeed, several genes involved in this pathway (*SLA-Ia, TAP1, TAP2* and *PSMB8*) were downregulated. Many genes involved in interferon-mediated immunity (*IRF1, IRF2, IRF3, IRF5, IRF9, IFNAR2, IFI6, IFNA6* and *IFI30*) were modulated during PrV infection. Genes involved in the apoptosis pathway (several genes belonging to the BCl2 family, *FAIM2*, several caspases and HSP), in nucleic acid metabolism (several histones and histone deacetylases) and cytoskeleton signalling (*ACTG1, ACTC1, ACTRT2, ACTN4, ACTL6B, MYO1D* and *MYO5A*) were also modulated during PrV infection.

It was confirmed by qRT-PCR that SLA class I genes were downregulated during infection at 8 h PI and *TAP1* and *TAP2* at 8 and 4 h PI, respectively. An early downregulation of *PSMB8* and *PSMB9* was detected before 2 h PI.

Table 1: Number of viral and cellular probes differentially expressed at each time point.

Probe set	Microarray	Probe subset	T0	T1	T2	T4	T8	T12
PrV	SLA/PrV	Probes homologous to a single transcript	0	3	7	18	26	30
		Total	0	3	13	34	52	58
Pig	SLA/PrV	up-regulated SLA/Immune probes	0	0	0	1	4	3
		down regulated SLA/Immune probes	0	0	0	11	18	13
		up-regulated random probes	1	0	2	4	15	11
		down-regulated random probes	0	0	0	20	70	80
		Total	1	0	2	36	107	107
	Qiagen-NRSP8	up-regulated probes	ND*	27	134	1262	3184	ND*
		down-regulated probes	ND*	4	47	1494	3509	ND*
		Total	ND*	31	181	2756	6693	ND*

*ND: not determined

Discussion

This study is the first report on PrV transcriptome expression over the time course of infection. It showed that virus and host cells can be analysed with a unique microarray combining viral and host probe sets and confirmed the feasibility of this kind of approach, which is highly relevant for establishing a direct link between the pathogen and cellular gene expression. Moreover, the Qiagen-NRSP8 microarray allowed for wider exploration of the porcine transcriptome.

A global PrV gene transcription picture during the lytic cycle was obtained with 90% of infected cells. The increase in cellular transcript levels and in the number of differentially expressed cellular genes detected at 4 h PI correlated with viral growth and coincided with the beginning of the release of extracellular progeny. These results suggested that the transcriptional machinery is fully active at 8 h PI thus permitting the massive production of viral progeny. The temporary ordered gene transcription, which is classically described, can be observed. However the immediate early gene (*IE180*) encoding a transactivator of early gene expression was detected as differentially expressed only at 4 h PI when the transcript level reaches its peak, suggesting that a low level of *IE180* transcript is sufficient to induce the transcription of other genes.

Even if a global cellular shutoff was not observed, the results revealed that the cellular activity of the vhs protein, usually described as responsible for this shutoff [1], seems to be attributed to the newly synthesised protein and not to the vhs protein present in the virion tegument at the moment of infection.

PrV develops strategies of host response evasion likely targeting several biological pathways including the MHC class I presentation pathway. Several genes belonging to this pathway (*SLA-Ia, TAP1, TAP2, PSMB8* and *PSMB9*) were downregulated in this experiment, suggesting that PrV developed complementary strategies to evade this pathway: turning off the peptide pump with inhibition of TAP activity [8,10] and transcription alteration of key players. Interestingly, the viral gene *UL49.5* encoding a TAP inhibitor protein was differentially expressed as soon as 2 h PI, indicating that viral evasion through TAP inhibition may start earlier than the cellular gene shutoff. In addition, several other genes playing a role in anti-viral response were regulated during PrV infection, such as genes belonging to the IFN signalling pathway.

The transcriptome analysis confirmed that many other biological processes and function were modulated during PrV infection in PK15 cells as previously observed in other studies. Genes belonging to the apoptosis pathway were modulated during PrV infection. Viral infection tends to generate pro-apoptotic signals to limit viral replication but viruses and in particular herpesviruses produce molecules acting as inhibitors of apoptosis [1]. *US3*, which plays an anti-apoptotic role [11], was upregulated from 8 h PI suggesting a possible late anti-apoptotic role. Genes belonging to nucleic acid metabolism were found differentially expressed from very early time points, with the repression of many genes encoding histones and nucleosome components and the modulation of histone deacetylases. US3 could inhibit histone acetylation during infection [12]. PrV also seemed to be involved in regulation of the expression of genes belonging to cytoskeleton signaling. US3 can also induce cellular cytoskeleton modifications with an increase of PrV spread [13].

Conclusion

The results presented here show that gene expression of PrV and porcine cells can be analysed together with microarrays, providing a chronology of PrV gene transcription never described before and a direct temporal link between viral and host gene expression.

Acknowledgements

The authors wish to thank Dr. René L'Haridon, Dr. Isabelle Oswald (LPT, INRA, Toulouse, France), Dr. Xiao Xiang Hu (State Key Laboratory for Agrobiotechnology, China Agricultural University, Beijing, China), Dr. Christine Renard (LREG, INRA, Jouy en Josas, France), Ms. Céline Urien, Mr. Jerome Lecardonnel and Ms. Céline Ducroix-Crepy (LREG, INRA, Jouy en Josas, France), Dr. Sophie Pollet, Dr. Claudia Bevilacqua and Ms. Emmanuelle Zalachas (PICT Plateform, INRA, Jouy en Josas, France), Dr. Sylvie Chevillard (LTE, CEA, Fontenay aux Roses, France), and Mr. Philippe Bardou (SIGENAE, INRA, Toulouse, France). This work was supported by Genanimal funding (PRA 2004-B01, Agence Nationale de la Recherche, 2004-2007).

References

1 Pomeranz LE, Reynolds AE, Hengartner CJ: Molecular biology of pseudorabies virus: impact on neurovirology and veterinary medicine. Microbiol Mol Biol Rev 2005;69(3):462-500.

2 Jenner RG, Young RA: Insights into host responses against pathogens from transcriptional profiling. Nat Rev Microbiol 2005;3(4):281-294.

3 Jansen A, Yu J: Differential gene expression of pathogens inside infected hosts. Curr Opin Microbiol 2006;9(2):138-142.

4 Ray N, Enquist LW: Transcriptional response of a common permissive cell type to infection by two diverse alphaherpesviruses. J Virol 2004;78(7):3489-3501.

5 Brukman A, Enquist LW: Suppression of the interferon-mediated innate immune response by pseudorabies virus. J Virol 2006;80(13):6345-6356.

6 Blanchard Y, Le MN, Le CM, Blanchard P, Leger J, Jestin A: Cellular gene expression survey of PseudoRabies Virus (PRV) infected Human Embryonic Kidney cells (HEK-293). Vet Res 2006;37(5):705-723.

7 Mellencamp MW, O'Brien PC, Stevenson JR: Pseudorabies virus-induced suppression of major histocompatibility complex class I antigen expression. J Virol 1991;65(6):3365-3368.

8 Ambagala AP, Hinkley S, Srikumaran S: An early pseudorabies virus protein down-regulates porcine MHC class I expression by inhibition of transporter associated with antigen processing (TAP). J Immunol 2000;164(1):93-99.

9 Zhao SH, Recknor J, Lunney JK, Nettleton D, Kuhar D, Orley S, Tuggle CK: Validation of a first-generation long-oligonucleotide microarray for transcriptional profiling in the pig. Genomics 2005;86(5):618-625.

10 Ambagala AP, Gopinath RS, Srikumaran S: Inhibition of TAP by pseudorabies virus is independent of its vhs activity. Virus Res 2003;96(1-2):37-48.

11 Geenen K, Favoreel HW, Olsen L, Enquist LW, Nauwynck HJ: The pseudorabies virus US3 protein kinase possesses anti-apoptotic activity that protects cells from apoptosis during infection and after treatment with sorbitol or staurosporine. Virology 2005;331(1):144-150.

12 Poon AP, Gu H, Roizman B: ICP0 and the US3 protein kinase of herpes simplex virus 1 independently block histone deacetylation to enable gene expression. Proc Natl Acad Sci USA 2006;103(26):9993-9998.

13 Favoreel HW, Van MG, Adriaensen D, Nauwynck HJ: Cytoskeletal rearrangements and cell extensions induced by the US3 kinase of an alphaherpesvirus are associated with enhanced spread. Proc Natl Acad Sci USA 2005;102(25):8990-8995.

Dr. Laurence Flori, Laboratoire de Radiobiologie et d'Etude du Génome, UMR INRA-CEA 314, Domaine de Vilvert, 78350 Jouy en Josas, France.
E-mail: laurence.flori@jouy.inra.fr

Pinard M-H, Gay C, Pastoret P-P, Dodet B (eds): Animal Genomics for Animal Health. Dev Biol (Basel). Basel, Karger, 2008, vol 132, pp 105-113.

Computational Integration of Structural and Functional Genomics Data across Species to Develop Information on the Porcine Inflammatory Gene Regulatory Pathway

C. K. Tuggle[1], Y. F. Wang[1], O. P. Couture[1], L. Qu[1], J. J. Uthe[2], D. Kuhar[3], J. K. Lunney[3], D. Nettleton[4], J. C. Dekkers[1], S. M. D. Bearson[2]

[1] Department of Animal Science and Center for Integrated Animal Genomics, Iowa State University, Ames, IA, USA
[2] National Animal Disease Center, USDA-ARS, Ames, IA, USA
[3] Animal Parasitic Diseases Laboratory, USDA-ARS-BARC, Beltsville, MD, USA
[4] Department of Statistics, Iowa State University, Ames, IA, USA

Key Words: pig, *Salmonella*, transcript profiling, microarray, qPCR, NF-κB, gene regulation, innate immune response

Abstract: We are investigating the porcine gut immune response to infection through gene expression profiling. Porcine Affymetrix GeneChip data was obtained from RNA prepared from mesenteric lymph node of swine infected with either *Salmonella enterica* serovar Typhimurium (ST) or *S.* Choleraesuis (SC) for 0, 8, 24, 48 or 504 hours post-inoculation (hpi). In total, 2,365 genes with statistical evidence for differential expression (DE; $p < 0.01$, $q < 0.26$, fold-change > 2) between at least two time-points were identified. Comparative Gene Ontology analyses revealed that a high proportion of annotated DE genes in both infections are involved in immune and defence responses. Hierarchical clustering of expression patterns and annotations showed that 22 of the 83 genes upregulated from 8-24 hpi in the SC infection are known NF-κB targets. The promoter sequences of human genes orthologous to the DE genes were collected and TFM-Explorer was used to identify a set of 72 gene promoters with significant over-representation of NF-κB DNA-binding motifs. All 22 known NF-κB target genes are in this list; we hypothesize that the remaining 51 genes are un-recognized NF-κB targets. Integration of these results and verification of putative target genes will increase our understanding of the porcine response pathways responding to bacterial infection.

INTRODUCTION

A major goal of genomics and bioinformatics research is the development of a "predictive" biology, or decreasing the uncertainty in describing individuals and populations and better prediction of the effects of genetic manipulations [1]. The data required for such depth of understanding is enormous and not yet available for any organism, and is especially complex for areas of biology that involve describing the output of multiple interacting molecules, cells and tissues. Immunology, which must account not only for proximate, stochastic and multiple intracellular responses to pathogens, but also understand long-range and short-range intercellular and intertissue communication, is a prime example. These interactions are interdependent and occur over multiple time scales, from minutes to years.

To begin the accumulation of such data in the porcine species, and to explore the important genes and pathways controlling resistance to intracellular pathogens, we performed transcriptional profiling of the mesenteric lymph node (LMN) in pigs inoculated with either *Salmonella enterica* serotype Typhimurium or *S.* Choleraesuis (SC). Transcriptional profiling can be used to identify the major regulatory pathways in this response [2], as well as to distinguish the responsible pathogen in patients presenting with similar etiology [3]. Large-scale transcript profiles can identify novel genes activating or suppressing the immune response to pathogens, especially for data sets with multiple time points throughout the infection response. Such data can also predict those genes whose genetic variability may control differential host response. This latter possibility is a major potential application of our data toward improving genetic resistance to food-borne pathogens and thus improve food safety. Here, our analyses of the porcine response to inoculation of either of two *Salmonella* bacterial strains are reported, emphasizing the differences in acute innate immune responses to each bacteria.

MATERIALS AND METHODS

Animals, tissue collection, RNA isolation, and Affymetrix data production

Piglets were weaned at 10 days of age and raised in isolation facilities to 7 weeks of age. Bacteriological culture of rectal swabs was performed twice to confirm that all pigs were free of *Salmonella* spp. Necropsy was performed on three control pigs per infection on day -3. On day 0, pigs in the infected group were intranasally challenged with 1 billion CFU of S. Choleraesuis χ3246 or S. typhimurium χ4232. At 8, 24, 48, and 504 hours post-infection (hpi), three randomly selected pigs were necropsied and mesenteric lymph nodes were collected. Total RNA was isolated (Qiagen RNeasy) and labeled (GeneChip® One-Cycle kit). Hybridization, washing and scanning was done using Affymetrix protocols and equipment.

Expression Data analysis

Data from Affymetrix cel files were converted to gene signal files by MAS5.0. Boxplot analysis using the natural log of the scaled median centered MAS5.0 data was used to check the uniformity of data (R2.0.1. software). To determine the expressed transcriptome of porcine MLN tissue, the number of genes showing a present call for all three replicates for at least one time point in each infection was calculated, removing duplicates across time points. Differentially expressed genes in each infection experiment were identified by the GLM procedure in SAS using $Y_i=\mu+T_i+\varepsilon_i$ where Yi is the natural log of the scaled median centered gene signal, T is the time-point fix effect, and ε is the random residual error. As the two infections were done separately, a combined statisical analysis was not performed. Thus comparisons across infections do not have direct statistical evidence.

qPCR

Real-time PCR was performed with 100 ηg cDNA (RNA equivalent) as previously described [2,4]. JMP 5.0 Software (Cary, NC) was used to perform the statistical analysis of the qPCR data. The effect of ST or SC infection on swine MLN mRNA expression was evaluated by one-way ANOVA that compared C_t values obtained from the negative control and post-infection samples. Fisher's LSD post-hoc test was applied to assess differences between groups of pigs at different time points post infection. $p \leq 0.05$ was considered statistically significant for all analyses.

Gene Ontology annotation of Affymetrix probe sets

A specific set of high level Gene Ontology (GO) terms which represent all host immune and regulatory response categories in biological process (GO slim) was created and GO analysis were performed for transcriptome and DE genes as described before using this GO-slim [2].

Clustering

After removing duplicates, genes which were found to be differentially expressed (DE, $p < 0.01$, fold change > 2 and $q < 0.26$) in at least one of the 10 possible time point pair-wise comparisons (8 h-C, 24 h-C, 48 h-C, 21 d-C, 24 h-8 h, 48h -8h, 21d-8h, 48 h-24 h, 21 d-24 h and 21 d-48 h) during ST infection (as well as these genes in the SC infection for comparison) were used to perform the hierarchical analysis and construct a heat map using the Gene Cluster 3.0 and tree view software (Stanford University, 2002).

Motif finding

The 1,500 bp 5′ and 500 3′ bp (relative to the annotated transcription start site) of the human orthologs for all annotated and upregulated DE porcine genes were obtained using in-house PERL scripts. Human orthologs were separated into different groups: an early set of 83 sequences (E83); one late set of 319 sequences (L319); five different randomly selected sets of 500 (A500, due to the 500-gene input limitation of TFM-Explorer) from the 560 upregulated and annotated genes across all time points. A list of genes known to be direct targets of NF-κB was obtained from two websites: http://www.nf-kb.org and http://bioinfo.lifl.fr/NF-KB/. The promoter sequences in these different groups were then searched for windows of sequence with over-representation of any of the available known NF-κB motifs defined by position-specific sequence matrices deposited in public TRANSFAC 7.0 (http://www.gene-regulation.com/pub/databases.html) or JASPAR (http://jaspar.genereg.net/) using the default parameters of TFM-Explorer [5].

Results

Summary of mesenteric lymph node (MLN) transcriptome and MLN genes differentially expressed due to Salmonella infection

When using a transcriptional profiling strategy, it is useful to determine the breadth of information obtained by the profiling tool used. Our results showed that 16,229 probe sets and 16,046 probe sets detected MLN expression, during ST infection and SC infection, respectively. Combining data shows that 16,599 probe sets detected expression in MLN during either *Salmonella* infection. For ease of discussion, the data in the following text will be described in terms of genes; in most cases one or more probe sets per gene actually provided the transcriptional data mentioned. Statistical analyses (see Methods) putatively identified many genes as differentially expressed (DE) across time points during infection or between infected and uninfected animals (data not shown). As the two infections were done at different times, it is not possible to provide statistical support for comparisons across infection, but an analysis across all time points within infection showed that 848 and 1,853 genes were identified as DE ($p < 0.01$, $q < 0.26$, fold-change > 2) between at least two time points during infection for the ST and SC experiments, respectively. There were 336 genes common to these two lists, indicating that 2,365 genes were shown to be DE in total.

Characterization of the transcriptional response to *Salmonella*

To identify the pathways affected by this differential response, a specific set of Gene Ontology terms associated with signal transduction and other processes relevant to immune response was enumerated (GO-slim; [2]). GO annotation of genes DE due to ST infection using this tool showed that DE genes annotated with "innate immune response," "immune response," "inflammatory response" and "defense response" were more highly represented as compared to those genes simply expressed in the MLN (transcriptome). To determine if this over-representation was primarily an induction or repression response, the DE genes were separately analyzed for GO terms. No category was more highly represented in the genes within the down-regulated list in comparison to the induced gene list, but both "immune response" and "cellular metabolism" terms were over-represented in the up-regulated genes. The SC infection showed very similar results to the ST experiment with regard to the induced gene list, although genes with the term "cell adhesion" were over-represented in the down-regulated group (Wang et al., manuscript submitted).

To further explore the type of response obtained, sets of genes with similar expression patterns across all time points were determined by using hierarchical clustering of genes DE in the ST infection. A heat map, constructed to more easily visualize the expression patterns, and the cluster hierarchy are shown in Figure 1. Many different co-expression patterns can be discerned, and a clear difference in the transcriptional response to each serovar is also evident. Sets of genes that are up- or downregulated only in the ST infection can be identified, and several clusters of genes with interesting patterns across infections are shown in Figure 1. Of especial interest are the 105 genes found in cluster 1 (bar graph 1), which did not show a large change at 8 hpi, but which had an increased expression level, from 8 hpi to 24 hpi, and then were down-regulated from 24 hpi to 48 hpi in the ST infection. In the SC infection, the expression of these genes gradually increased from 8 hpi to 24 hpi, reaching a peak response at 48 hpi. A majority of genes in this cluster are NF-κB target genes, cytokines and chemokines ([2] and Table 1). This sub-cluster indicates that the transcriptional response of these pathways in the host occurred earlier in the ST (at 24 hpi) than in the SC infection (48 hpi). Table 1 shows additional genes DE at specific time points relative to uninfected pigs that are connected to NF-κB in PubMed citations. Some of these genes are found in both ST and SC infections, especially between ST 24 hpi genes and SC 48 hpi genes, where 9 of the 13 genes up-regulated at 24 hpi in ST infected pigs were up-regulated at 48 hpi in SC infected pigs.

This profiling data was then validated by using quantitative real-time PCR (qPCR) for 21 genes. These genes included 18 known NF-κB target genes (*IL1A, IL15, CCL2, CCL3, CXCL5, PPBP, GBP1, GBP2, PTX3, IKBA, JUNB, NFKBIZ, CD14, ICAM1, TLR2, GZMB, TAP1* and *CCR5*), one T cell marker gene (*CD4*), and *CD163*. In addition, although we did not find any oligonucleotide set representing the *TGM3* gene found in the Affymetrix microarray, qPCR analysis of *TGM3* was conducted because it exhibited dramatic changes in expression (a fold change greater than 1,000) at 48 hpi in porcine lung during SC infection [6]. TREM1 was also selected

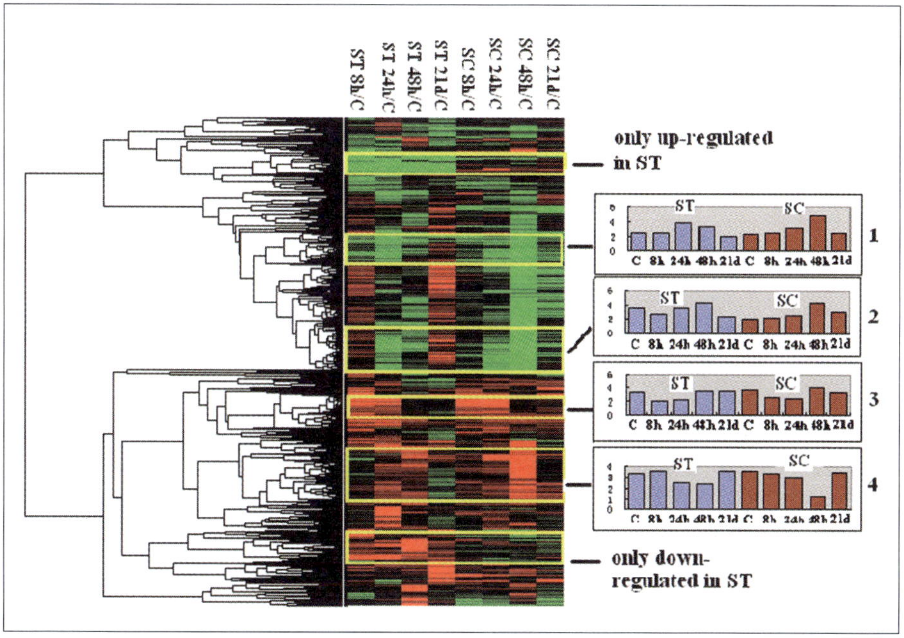

Fig. 1: Hierarchical clustering analysis of all genes showing differential expression among all possible comparisons in the ST infection ($p < 0.01$, $fc > 2$, $q < 0.24$).
The heat map was built by using Gene Cluster 3.0 software and the detailed four sub-clusters were constructed by using the centroid value from Gene Cluster 2.0. Red colour represents downregulation and green shows upregulation.

for qPCR validation due to its similar expression pattern to the known NF-κB target genes, although no reports have shown that *TREM1* is an NF-κB-regulated gene. The differential expression for all 20 genes with Affymetrix data, except for *CD4*, was confirmed by qPCR results, and the *TGM3* gene was shown to be strongly induced in both infections ([2] and data not shown).

Cell migrations cannot explain the differences in RNA levels in the MLN

During bacterial infection, T lymphocytes in the lymph nodes encounter antigen presenting cells, primarily macrophages and dendritic cells, carrying bacterial antigens from sites of infection. Cell migrations into or out of the lymph node could cause changes in the total RNA present in the MLN, and these RNA changes might be erroneously interpreted as a transcriptional response to infection within an immune cell type. To verify that the observed differences in gene expression during the infections are true transcriptional differences, the expression level of specific markers for T cells, macrophages, dendritic cells and granulocytes were evaluated (data not shown). No statistically significant changes in RNA levels for these marker genes, and thus no evidence of significant cell migration, was observed in either infection, indicating that the RNA differences observed are not primarily due to significant changes in the abundance of specific cell types in the MLN, but are instead likely to be representative of specific transcriptional or post-transcriptional responses within cells.

Table 1: Upregulated genes at each early time point connected to regulatory factor NF-κB in PubMed text-mining.

Time points	NF-κB neighbours	Total
ST 8 hpi	*DAF, TLR9*	2
ST 24 hpi	*BIRC3, DAF, EDN1, FGF2, GADD45B, IFNG, IL1B, IRF1, PPARD, SOD2, STAT1, TRAF3, VEGF*	13
ST 48 hpi	*CCR5, MAP3K, SAA1, STAT1*	4
SC 8 hpi	*EGR1, GRDD45B, JUN, KDR, MARP3K11*	5
SC 24 hpi	*BIRC3, CCL2, CEBPB, CEBPD, CXCL2, CYCS, EDN1, GADD45B, ICAM1, IER3, IL1A, IL1B, IRF1, JUN, JUNB, LIPG, MAP3K11, PTGS2, SELP, TGFB1, TNFRSF1A*	21
SC 48 hpi	*APP, BAX, BCL2, BIRC3, CASP3, CAST, CCL2, CCR5, CDKN1A, CEBPB, CEBPD, CEBPG, CFLAR, CSF3, CXCL1, CXCL2, CXCR4, CYCS, EDN1, ERBB3, FOS, GADD45B, GCLC, GRP58, HMOX1, ICAM1, IER3, IFNG, IGF1, IL1A, IL1B, IL1R1, IL6, IL8, IL10, IRF1, JAK2, JUN, JUNB, LIPG, MT2A, NFKBIA, OLR1, PDGFRA, PTGS1, PTGS2, RAC1, RXRA, SAA1, SELP, SOD2, STAT1, STAT3, STAT5A, TGFB1, TLR2, TLR4, TNF, TNFRSF5, TNFRSF1A, VEGF*	61

Pathway analysis and regulatory motif searches show known NF-κB target genes and putative novel NF-κB targets are abundant in the early response to SC infection

To better understand the large number of genes upregulated early in SC infection, two sets of these genes were investigated. We identified 83 annotated genes that were induced at either 8 or 24 hours post inoculation (called the Early set), 319 genes induced only at 48 hours post-inoculation (Late set), and 544 genes upregulated when all time points are compared (All set). Using available information (see Methods), we found that 22 Early genes (27%), 37 Late genes (12%) and 70 All genes (13%) are known NF-κB targets. As NF-κB is a well-studied inflammatory regulatory factor, these results indicated that co-regulation by NF-κB is a plausible mechanism for the co-induction of these genes, even though most are not recognized targets of NF-κB. To provide further evidence for this mechanism, and especially provide evidence for NF-κB regulation for the majority of these co-expressed Early and Late genes, the orthologous human promoter sequences were collected and TFM Explorer motif-finding software was used to identify those promoters with statistically significant over-representation of NF-κB recognition sequence motifs. As shown in Figure 2, TFM Explorer found most known NF-κB target genes for all groups (from 65 to 95%), indicating that the method is sufficiently sensitive. Interestingly, of the remaining genes with no published reports of direct NF-κB regulation, 51 of 61 genes in the Early, 182 of 283 in the Late, and 357 of 485 in the All group were found to contain NF-κB binding motifs.

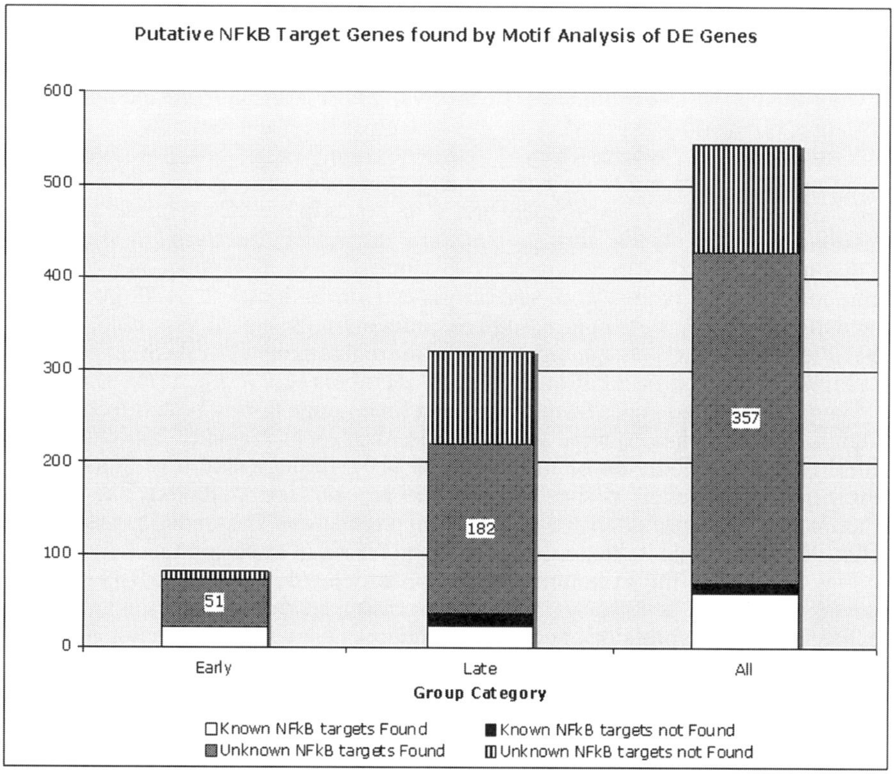

Fig. 2: Computational identification of NF-κB binding site motifs at human orthologs of up-regulated genes predicts putative NF-κB target genes responding to S. Choleraesuis.
Each bar graph shows for each gene group: a) the number of known genes with a promoter containing an NF-κB motif, b) the number of known target genes for which no motif was found in the promoter, c) the number of genes not known to be NF-κB targets for which an NF-κB motif was found, and d) promoters for which no motif was found. The number of putative novel NF-κB target genes identified by this approach is shown.

Discussion

Transcriptional profiling of the porcine response to *Salmonella* using the Affymetrix Genechip was very productive, as large numbers of genes responding in time specific and pathogen specific patterns were identified. A strong host transcriptional response at 48 hpi with *S.* Choleraesuis infection in particular was observed. In this experiment, many genes that showed large expression changes were found, such as cytokines, chemokines and heat shock proteins. These generally overlap with sets of genes that have been implicated in the host response to infection [7,8] indicating that this data can be integrated with similar results in other species. However, we also identified a number of genes that had not previously been shown to be involved in the response to bacterial infection, such as *STEAP4, PSTPIP2, LIPG, HK3, STXBP1, ITPR1, CA3, KRT17* and *PLN*. Even though the biological function of these genes in infection remains unclear, this data more fully describes the transcriptional response to SC in porcine MLN. Further, for both infections, DE genes with known immune function and those with unknown function add to the list of functional candidate genes for investigating the association between immune related traits and genetic variation. Polymorphisms of these candidates might be valuable markers for enhancing disease resistance, pig health, and food safety by molecular breeding methods.

A predominantly inflammatory, Th1-directed immune response was observed during infection with either serovar, although the peak gene expression induction was at 24 hours in the ST experiment and at 48 hours in the SC experiment. This is consistent with the results from porcine lung during SC infection [6] and qPCR results in porcine MLN in both ST and SC infection [2,9]. Apoptosis was also a commonly induced pathway, as several genes were induced at 24-48 hours post inoculation (*CASP1, TGM1*) in ST and SC infected pigs, but we observed, as with the Th1 response, that the magnitude of activation of this pathway in the SC experiment was higher and involved additional pathway members (*CASP3, CASP4, GZMB*).

Many upregulated genes from the earliest time points during both infections (8-48 hours post inoculation) were shown to be linked to, or regulated by, NF-κB, the central regulatory complex of inflammation [10], through Pathway Studio text-mining software as well as specific literature reports (see Methods). The weaker inflammatory response observed in the ST infected animals extended to the lack of feedback control gene induction; *NFKBIA* and *NFKBIZ* genes, induced by NF-κB binding to dampen the inflammatory response dependent on NF-κB [11], were induced only in the SC experiment animals. Further evidence that the molecular mechanism for co-expression of the large number of genes early in infection involved NF-κB was added by regulatory motif analysis of the orthologous human promoters of these genes. It was shown that a large majority of these co-expressed genes have NF-κB DNA binding motifs, indicating that they could be bound and directly regulated by NF-κB. Most of these co-expressed and NF-κB-motif containing genes have not been shown to be direct targets of NF-κB; these genes should be examined more closely to further explore the response to *Salmonella* in pigs and other mammals.

Conclusion

The confirmation of the microarray data by qPCR analysis of 21 genes shows that the Affymetrix platform efficiently and accurately measures transcriptional responses in gut associated lymph tissue, and identifies multiple pathways containing many known inflammaory genes, as well as many genes not previously implicated in innate resistance. Many genes were identified as responding to *Salmonella* infection within mesenteric lymph node, with clear differences between responses to SC versus ST infection. Significant numbers of genes are specific to the SC infection. Infection with ST resulted in upregulation of fewer genes for shorter times. qPCR analysis of DE genes was very successful in validating Affymetrix data, indicating that the profiling data was reliable. Cluster analysis identified an acute response cluster of co-expressed genes dominated by known NF-κB target genes, providing a mechanism for the observed co-regulation. Comparative bioinformatic analysis identified many putative novel NF-κB regulatory targets during infection that were co-expressed with known NF-κB targets and have NF-κB regulatory motifs in human flanking DNA sequences.

ACKNOWLEDGMENTS

We thank Tom Stabel for his collaboration in the production of these challenge populations. This project received support from the ISU Agriculture Experiment Station/Center for Integrated Animal Genomics, the USDA-ARS-NADC and USDA-ARS-BARC, and USDA-NRI 2004-35205-14202. OC is a recipient of a USDA-MGET Fellowship in Computational Biology USDA-CSREES-2001-52100-11506.

REFERENCES

1 Hood L, Heath JR, Phelps ME, Lin B: Systems biology and new technologies enable predictive and preventative medicine. Science 2004;306:640-643.

2 Wang Y, Qu L, Uthe JJ, Bearson SMD, Kuhar D, Lunney JK, et al: Global transcriptional response of porcine mesenteric lymph nodes to Salmonella enterica serovar Typhimurium. Genomics 2007;90:72-84.

3 Ramilo O, Allman W, Chung W, Mejias A, Ardura M, Glaser C, et al: Gene expression patterns in blood leukocytes discriminate patients with acute infections. Blood 2007; 109:2066-2077.

4 Dawson HD, Beshah E, Nishi S, Solano-Aguilar G, Morimoto M, Zhao A, et al: Localized multigene expression patterns support an evolving Th1/Th2-like paradigm in response to infections with *Toxoplasma gondii* and *Ascaris suum*. Infect Immun 2005;73:1116-1128.

5 Defrance M, Touzet H: Predicting transcription factor binding sites using local over-representation and comparative genomics. BMC Bioinformatics 2006;7:396.

6 Zhao SH, Kuhar D, Lunney JK, Dawson H, Guidry C, Uthe JJ, Bearson SM, Recknor J, Nettleton D, Tuggle CK: Gene expression profiling in Salmonella Choleraesuis-infected porcine lung using a long oligonucleotide microarray. Mammalian Genome 2006;17, 777-789.

7 Eckmann L, Kagnoff MF: Cytokines in host defense against Salmonella. Microbes Infect 2001;3:1191-1200.

8 Tukel C, Raffatellu M, Chessa D, Wilson RP, Akcelik M, Baumler AJ: Neutrophil influx during non-typhoidal salmonellosis: who is in the driver's seat? FEMS Immunol Med Microbiol 2006;46:320-329.

9 Uthe JJ, Royaee A, Lunney JK, Stabel TJ, Zhao SH: Porcine differential gene expression in response to *Salmonella enterica* serovars Choleraesuis and Typhimurium. Mol Immunol 2007;44:2900-2914.

10 Ghosh S, May MJ, Kopp EB: NFκB and Rel proteins: evolutionarily conserved mediators of immune responses. Annu Rev Immuno. 1998;16:225-260

11 Sun SC, Ganchi PA, Ballard DW, Greene WC: NF-kappa B controls expression of inhibitor I kappa B alpha: evidence for an inducible autoregulatory pathway. Science 1993;259:1912-1915.

Pr. Christopher K. Tuggle, 2255 Kildee Hall, Iowa State University, Ames, IA 50011 USA.
E-mail: cktuggle@iastate.edu

Pinard M-H, Gay C, Pastoret P-P, Dodet B (eds): Animal Genomics for Animal Health. Dev Biol (Basel). Basel, Karger, 2008, vol 132, pp 115-120.

EST-based Gene Indices for the Cattle Fever Tick, the Horn Fly, and the Screwworm Fly

F.D. Guerrero[1], M. Wang[2], V.M. Nene[3], S.E. Dowd[4], L. Saldivar[1]

[1] USDA-ARS Knipling-Bushland US Livestock Insects Research Lab, Kerrville, Texas, USA
[2] Lorus Therapeutics Inc, Toronto, Ontario, Canada
[3] University of Maryland School of Medicine, Baltimore, Maryland, USA
[4] USDA-ARS Livestock Issues Research Unit, Lubbock, Texas, USA

Key words: ESTs, gene index, G protein-coupled receptors, GO terms

Abstract: The southern cattle tick, *Rhipicephalus (Boophilus) microplus*, the horn fly, *Haematobia irritans*, and the New World screwworm, *Cochliomyia hominivorax*, are economically important parasites of cattle throughout the world. Understanding the biology and genomics of these pests is critical to developing novel control methods. We developed EST sequence databases for these parasites and used bioinformatic analysis to identify conceptual open reading frames and assigned Gene Ontology (GO) terms to those database transcripts that had informative BlastX hits. Bioinformatic analysis showed that a large fraction of the tick database entries have no clear matches in other sequenced genomes. Several putative tick G protein-coupled receptors (GPCRs) were identified that might serve as starting points to develop novel control technologies.

INTRODUCTION

Rhipicephalus *(Boophilus) microplus*, the tropical or southern cattle tick, is one of the most economically important tick vectors of pathogens that affect the global cattle population [1]. The tick transmits protozoan *(Babesia bovis* and *Babesia bigemina)* and bacterial *(Anaplasma marginale)* organisms that cause babesiosis and anaplasmosis, which can result in severe agricultural losses in milk and beef production and restriction in traffic of livestock. The United States (US) Department of Agriculture (USDA) led a campaign in the mid-20th century that eradicated the tick from the US [2]. The tick remains prevalent in Mexico and, since over a million cattle are imported annually into the US from Mexico, an extensive USDA quarantine program is in place to keep *Boophilus* ticks from re-establishing in the US.

The horn fly, *Haematobia irritans*, is considered to be the primary insect pest of cattle in the US. For pastured cattle production, annual economic losses due to horn fly damage have been estimated at $876 million [3]. Fly control is based upon the treatment of cattle with insecticides, and insecticide resistance has reduced the ability of cattle producers to control horn flies, thereby impacting the profitability of their operations.

The New World screwworm, *Cochliomyia hominivorax*, was a devastating pest of livestock and other animals, including humans, throughout the US, Mexico and Central America. The female fly deposits its eggs at a wound site and the larval stage feeds on living tissue. During periods of heavy fly infestation and oviposition, death of the affected animal will often ensue without human intervention. In the 1950s, annual losses to the US livestock industry topped $120 million [2]. Although eradicated from North America, the screwworm still is a pest in South America and the Caribbean. Reinfestation of North America is prevented by a sterile male release program conducted in Panama.

The development of expression sequence tag (EST) databases for these three cattle pests was undertaken to facilitate molecular studies to further the development of novel control methods to help maintain the profitability of livestock operations in the US.

Materials and Methods

The construction of an R. microplus normalized cDNA library, generation of ESTs, and assembly into the R. microplus gene index has been described [4]. Briefly, a single normalized cDNA library was synthesized from pooled RNA samples that had been purified from ticks subjected to various environmental exposures, including heat shock, cold shock, various acaricides, host odour, and infection with B. bovis. The acaricide exposure experiments utilized several strains of R. microplus that varied in their levels of susceptibility to pyrethroid, organophosphate and the formamidine amitraz. We also included RNA purified from eggs, nymphs, adults and dissected adult tick organs. Normalized cDNA libraries were also made from horn fly embryos and first instar larvae and screwworm embryos and second instar larvae. Approximately 8,000 EST sequences were obtained from each of these four libraries and the data combined into separate horn fly and screwworm EST databases, which were assembled into contigs for bioinformatic analysis. Protein domain analysis was performed by Reversed Position Specific (RPS)-Blast (ftp://ftp.ncbi.nih.gov/blast/documents/rpsblast.html) search against the Conserved Domain Database using a cut-off E-value < 1 × 10-8.

Results

Table 1 lists the number of ESTs and unique transcripts that are in our respective databases for the southern cattle tick, horn fly and screwworm. Acquisition and bioinformatic analysis of the 42,512 ESTs from the southern cattle tick has been previously reported [4,5]. These ESTs assembled into 9,403 assembled contigs with 4,240 ESTs remaining as unassembled sequences referred to as singletons. Thus, 13,643 unique tick transcripts are represented in this database. Approximately 8,400 and 8,300 ESTs were obtained from sequencing the horn fly embryo and 1st instar larval libraries, respectively. Each library consists of approximately 2,700 unique transcripts. The ESTs from these two libraries were combined and assembled into 2,850 assembled contigs and 1,052 singletons for a total of 3,902 unique transcripts. Analysis of the screwworm EST data is not yet complete, however, approximately 8,000 ESTs have been obtained from each library and these will be analyzed in similar fashion as the tick and horn fly libraries.

Since these cDNA libraries were normalized, strict quantitation of gene expression level is not possible with this data. However, Table 2 lists the most abundant protein domains found in the tick EST database. The four most common domains were the WD40, RNA recognition motif, P450, and trypsin-like serine protease.

During the bioinformatic analysis of the tick EST database, GO terms were assigned to the translated ESTs when possible [5; http://compbio.dfci.harvard.edu/tgi/cgi-bin/tgi/gimain.pl?gudb=b_microplus]. Table 3 lists those ESTs which were assigned the GO term 0004930 G-protein coupled receptor activity; assignments were based on Blast similarity to known GPCRs using an e-value cut-off of 1 X 10^{-25}.

The horn fly EST database has also been assembled and analyzed. Table 4 lists the ten most abundant ESTs in the database containing the combined embryo and 1st instar larval sequences. Again, these libraries were normalized and cannot be used to quantitatively analyze transcript abundance. However, the 16S mitochondrial ribosomal RNA sequence was the most abundant in this library, 5-fold more abundant than the second entry of Table 4. This RNA is involved in the translation complex of mitochondrial gene products and may have a role as an energy-independent protein chaparone involved in protein folding [6]. Following the 16S ribosomal RNA were transcripts that occurred between 31 and 45 times among the unassembled ESTs, including a fatty acid-binding protein, a vacuolar H+ ATPase G subunit, NADH dehydrogenase Fe-S protein 8, and H3 histone, among others.

Table 1: Gene Index statistics by species.

Arthropod	EST numbers	Unique transcripts
R. microplus*	42,512	9,403 assembled contigs
		4,240 singletons
H. irritans		
embryo	8,427	2,668
1st instar larvae	8,275	2,689
Combined	16,702	2,850 assembled contigs
		1,052 singletons
C. hominivorax		
embryo	8,000	
2nd instar larvae	8,000	

*Previously reported in [5].

Discussion

The 13,643 unique transcripts present in the tick EST database represent a significant fraction of the likely protein-coding transcripts of this organism. While the fraction of transcripts represented in the horn fly and screwworm EST databases is lower, these databases form a good foundation for molecular studies. The number of unique transcripts contained in each of these three databases will likely increase significantly, as more sequencing is taking place using 454 platform sequencing technology (data not shown).

Domain analysis showed that the two most commonly occurring domains, WD40 and RNA recognition motif, were involved in structural roles such as protein-RNA or protein-protein type interactions. WD40 domains are short ~40 amino acid motifs, often terminate in a Trp-Asp dipeptide, and are involved in a wide variety of functions such as G-protein coupling, RNA-processing, vesicular trafficking, and cell division

[7]. The RNA recognition motif domain provides an RNA-binding platform which can be used to regulate post-transcriptional gene expression [8]. The cytochrome P450 (P450) and serine protease domains are enzymatic in nature. P450 domains are present in mixed function oxidase enzymes, which are involved in the metabolism of xenobiotics, including pesticides. Their relative abundance in our tick database might be explained by our use of acaricide-treated tick larvae in several of the samples used to make up the pooled RNA, which served as the template for the normalized cDNA library synthesis. Finally, serine-type protease domains are critical to the tick with their role in digesting and processing the blood meal ingested during the larval, nymph, and adult stages of these organisms [9].

Up to half of all market drugs target G protein-coupled receptors (GPCRs) [10] and identification of tick- or arthropod-specific GPCRs would present novel targets for discovery or design of compounds with acaricidal activity. Thus the putative GPCRs (Table 3) identified by Blast similarity to known GPCRs from other organisms is a useful beginning for investigating the utility of tick coding regions as potential targets for designing novel control chemicals. Identification of a greater complement of tick GPCRs would likely benefit from bioinformatic approaches such as the Binary Topology Pattern method, which looks closely at transmembrane region topology patterns to classify GPCRs [11], because Blast analysis often returns few if any useful hits to tick coding regions. For example, Wang et al. [5] found that 66% of the unique transcripts from the tick EST database could not be assigned GO terms.

Table 2: Most abundant domains in tick EST database.

Domain	CDD accession #	# of hits	% of total hits
WD40	29257	58	1.78
RNA recognition motif	47687	44	1.35
P450	40168	39	1.20
Trypsin-like serine protease	29152	34	1.05
STKc kinase	29142	30	0.92
Peptidase M13	41481	29	0.89
Ubiquitin-conjugating enzyme E2, catalytic domain	29157	27	0.83
KU	29009	26	0.80
Serine proteinase inhibitor	29117	22	0.68
Carboxylesterase	40235	21	0.66

Previously reported in [5]

Table 3: Putative GPCRs in cattle tick EST database.

BmiGI ID	Top informative hit	E-value
TC9962	Frizzled 10	e-113
TC10217	Testis-specific lanthionine synthetase C-like protein	e-87
TC13135	Human putative GPCR	e-27
BEAA129TR	Metabotrophic glutamate receptor type 2	2 e-66
BEAD463TF	Calcitonin gene-related peptide receptor component	4 e-17

http://compbio.dfci.harvard.edu/tgi/cgi-bin/tgi/gimain.pl?gudb=b_microplu; based on assignment of GO term 0004930 G-protein coupled receptor activity)

Table 4: Most abundant ESTs in horn fly EST database.

Contig #	Top informative hit	# ESTs	E-value
165	16S ribosomal RNA, mitochondrion	229	0.0
36	Probable fatty acid-binding protein	45	2.6 e-29
1295	Vacuolar H+ ATPase G subunit	39	4.5 e-25
495	NADH dehydrogenase (ubiquinone) FE-S protein 8	39	9.0 e-88
70	H3 histone, family 3B	38	3.5 e-70
418	Peptidyl-prolyl cis-trans isomerase 5 precursor	35	3.4 e-69
64	*Hypothetical transcript from Drosophila melanogaster*	33	9 e-22
1039	Small ribonucleoprotein particle protein B	32	1.7 e-63
862	Oligomycin sensitivity-conferring protein	31	1.3 e-81
95	18S ribosomal RNA	31	0.0

Conclusion

EST databases for three major cattle pests have been synthesized and assembled as part of our research program aimed at developing novel control technologies for these pests. The cattle tick database likely contains a significant fraction of the gene coding regions of this organism, though deeper sequencing is expected to detect many low abundance transcripts. Although the horn fly and screwworm databases are smaller, further sequencing is ongoing, focusing on 454 platform methodology. Bioinformatic analysis identified a number of GPCRs as an initial phase of research aimed at finding putative targets for novel pesticide development.

References

1 de Castro JJ: Sustainable tick and tickborne disease control in livestock improvement in developing countries. Vet Parasitol 1997;71(2-3):77-97.

2 Graham OH, Hourrigan JL: Eradication programs for the arthropod parasites of livestock. J Med Entomol 1977;13(6):629-658.

3 Kunz SE, Murrell KD, Lambert G, James LF, Terrill CE: Estimated losses of livestock to pests, in Pimentel D (ed): CRC Handbook of Pest Management in Agriculture. CRC Press, Boca Raton, 1991, vol 1, 2nd ed. pp 69-98.

4 Guerrero FD, Miller RJ, Rousseau ME, Sunkara S, Quackenbush J, Lee Y, Nene V: BmiGI: a database of cDNAs expressed in *Boophilus microplus*, the tropical/southern cattle tick. Insect Biochem Mol Biol 2005;35(6):585-595.

5 Wang M, Guerrero FD, Pertea G, Nene VM: Global comparative analysis of ESTs from the southern cattle tick, *Rhipicephalus (Boophilus) microplus*. BMC Genomics 2007;8:368.

6 Sulijoadikusumo I, Horikoshi N, Usheva A: Another function for the mitochondrial ribosomal RNA: Protein folding. Biochemistry 2001;40(38):11559-11564.

7 Smith TF, Gaitatzes C, Saxena K, Neer EJ: The WD repeat: a common architecture for diverse functions. Trends Biochem Sci 1999;24(5):181-185.

8 Maris C, Dominguez C, Allain FH: The RNA recognition motif, a plastic RNA-binding platform to regulate post-transcriptional gene expression. Febs J 2005;272(9):2118-2131.

9 Mans BJ, Louw AI, Neitz AW: Evolution of hematophagy in ticks: common origins for blood coagulation and platelet aggregation inhibitors from soft ticks of the genus Ornithodoros. Mol Biol Evol 2002;19(10):1695-1705.

10 Flower DR: Modeling G-protein coupled receptors for drug design. Biochim Biophys Acta 1999;1422:207-234.

11 Inoue Y, Ikeda M, Shimizu T: Proteome-wide classification and identification of mammalian-type GPCRs by binary topology pattern. Comp Biol Chem 2004;28:39-49.

Dr. Felix D. Guerrero, USDA-ARS Knipling-Bushland US Livestock Insects Research Laboratory, 2700 Fredericksburg Road, Kerrville, TX 78028. USA.

E-mail: Felix.Guerrero@ars.usda.gov

Pinard M-H, Gay C, Pastoret P-P, Dodet B (eds): Animal Genomics for Animal Health. Dev Biol (Basel). Basel, Karger, 2008, vol 132, pp 121-130.

Comparison of Transcriptional Changes Associated with *E. acervulina* and *E. maxima* Infections using cDNA Microarray Technology

C.H. Kim[1], H.S. Lillehoj[1], Y.H. Hong, C.L. Keeler, Jr.[2]*

1 Animal Parasitic Diseases Laboratory, Animal and Natural Resources Institute, Agricultural Research Service, USDA, Beltsville, MD 20705, USA

2 Department of Animal and Food Sciences, College of Agriculture and Natural Resources, University of Delaware, Newark, DE 19716, USA

Key words: *Eimeria*; cDNA microarray; coccidiosis; mucosal immunity; chicken

Abstract: Enhanced understanding of host-pathogen interactions at local sites of infection will extend our knowledge of disease pathogenesis and will facilitate the development of novel preventive methodologies against many infectious diseases of economic importance. In the current study, a 9.6K avian intestinal intraepithelial lymphocyte cDNA microarray (AVIELA) was developed to compare the local transcriptional profiles following primary and secondary infections with two major *Eimeria* parasites, *E. acervulina* (EA) and *E. maxima* (EM), which infect the intestinal duodenum and jejunum, respectively. Gene Ontology analyses showed that EA infection primarily induced genes associated with lipid metabolism and intracellular trafficking whereas EM infection upregulated the genes involved in protein biosynthesis and metabolism, and downregulated apoptosis related genes. Following primary EA infection, there was a significantly enhanced expression of genes involved in the signal pathway of T cell activation and cytoskeletal regulation. Thus, the AVIELA array provides a valuable tool for investigating host-pathogen interactions in avian coccidiosis and allows for the comparison of the transcriptional regulations induced by species of *Eimeria* that infect different areas of the intestine.

INTRODUCTION

The apicomplexan protozoa *Eimeria* causes avian coccidiosis, the major parasitic disease of poultry. The growth and feed utilization of infected birds are seriously impaired by coccidiosis resulting in huge economical loss to the poultry industry worldwide [1]. Current disease control methods depend on the prophylactic medication and vaccination with live and attenuated parasite vaccines [2]. However, due to increasing regulations on the use of drugs in animal production, there is a timely need for the development of novel control strategies against this disease.

An understanding of the nature of protective immunity against coccidiosis is extremely important for the development of novel vaccines. *Eimeria* parasites undergo intracellular development in the epithelial cells in the gut, eliciting a wide spectrum of immune responses at the local level [3]. Intestinal mucosal surfaces are defended against enteric pathogens by the gut-associated lymphoid tissues (GALT), which mainly consist of intraepithelial lymphocytes (IELs) involved in defence against enteric pathogens [3]. Following infection with *Eimeria*, the number of gut IELs expressing the CD3, CD4, and CD8 T cell markers, as well as the αβ-T cell receptor (αβ-TCR) and γδ-TCR, were significantly increased [4]. However, only CD4+ cells remained elevated following secondary infection suggesting that this subpopulation of intestinal T lymphocytes is involved in local defence against coccidiosis.

Recent technological advances in functional genomics and systems biology are opening new doors for animal health and disease research. In particular, during the last five years various microarray platforms using oligonucleotides or cDNAs with high or low densities have become available for animal health research. Recently, our laboratory utilized several different microarray platforms to explore global transcriptional responses and protective immune mechanisms in avian coccidiosis [5,6]. Using a T lymphocyte cDNA microarray, the expression of several cytokine genes was examined following *E. acervulina* (EA) and *E. maxima* (EM) infections [5]. Furthermore, a 4.9K avian macrophage microarray (AMM) was used to profile transcriptional responses of macrophages infected with three major *Eimeria* species [6]. In an effort to investigate local gene regulation associated with coccidiosis in poultry, we developed an avian intestinal intraepithelial lymphocyte cDNA microarray (AVIELA) using 9,668 expression sequence tags (ESTs) that were selected from a previously cloned cDNA library developed from IEL of *Eimeria*-infected chickens [7]. In this report, it is demonstrated that this AVIELA array provides a valuable tool for examining local transcriptional responses following primary or secondary infections with *Eimeria* parasites.

Materials and Methods

Preparation of the AVIELA microarray

The avian intestinal intraepithelial lymphocyte (IEL) cDNA microarray (AVIELA) was constructed with 9,668 ESTs selected from the previously prepared 14,409 IEL clones [7]. The elements were organized into 24 duplicate subarrays, each as a 21 x 21 spot grid. Amplified and purified cDNA PCR products were spotted in duplicate after resuspension in 3X SSC with 0.01% SDS (1X = 150 mM NaCl, 15 mM sodium citrate, pH 7.0) on glass slides (Telechem International) using an OmniGrid Accent spotter (GeneMachines). The PCR products were fixed to the slides by UV cross-linking at 400 mJ and spotting quality was evaluated by SYBR-Green II staining (Invitrogen).

Animals, parasites, and experimental infection

Fertilized eggs of White Leghorn SPAFAS chickens (Charles River Laboratories, Wilmington, MA) were hatched at the Animal and Natural Resources Institute (Beltsville, MD). The chickens were divided into eight groups (12/group) for EA or EM infections. For primary infection, 3 groups of chickens at 3 weeks of age were orally inoculated with 1 x 104 sporulated oocysts/chicken. For secondary infection, a separate three groups received the same primary inoculation followed by oral infection at 6 weeks of age with 2 x 104 sporulated oocysts. The remaining two groups served as the day 0 uninfected controls.

RNA preparation and microarray hybridization

At 1, 2, and 3 days post-primary or post-secondary infection, and at day 0 for uninfected animals, IELs were isolated from the duodenum or jejunum from EA- or EM-infected chickens, respectively, as previously described [7]. IELs from each group were pooled, total RNA was prepared using TRIzol (Invitrogen) and the RNeasy Mini RNA Purification Kit (Qiagen), and aminoallyl cRNA was prepared using the Amino Allyl Message Amp II aRNA Amplification Kit (Ambion). Twenty μg of each aminoallyl cRNA was labeled with AlexaFluor 555 or AlexaFluor 647 dyes (Invitrogen). A simple loop design was employed [8] for hybridization, which was performed at 50°C overnight. The slides were scanned using a ScanArray Lite (PerkinElmer).

Data acquisition and analysis, and bioinformatics

Spot data were acquired using ScanArray Express software (PerkinElmer) and quality control (QC) was maintained by acquisition of data from spot intensities greater than the background plus 2 standard deviations. The data was normalized with block lowess using MIDAS 2.19 (http://www.tigr.org) and imported into GeneSpring GX 7.3 (Agilent) for expression and statistical analyses. Elements exhibiting greater than a 2.0-fold change during at least one of the three time points were analyzed by ANOVA ($P<0.05$ with Benjamini and Hochberg False Discovery Rate (FDR)) [9]. For identification of the AVIELA ESTs, a Blast search was applied and sequence files such as chicken genome and reference RNA were obtained from NCBI (http://www.ncbi.nlm.nih.gov). Chicken Entrez gene identities were mapped to Homologene ID for human, which were used for Gene Ontology (GO) analysis and pathway classification with the Panther database [10].

Results

To identify the transcriptional differences of genes modulated in primary or secondary infections with EA or EM, microarray data from 0 to 3 days post infection were extracted and normalized, followed by statistical analyses on elements with greater than 2-fold change. As shown in Figure 1, 87 and 26 of the AVIELA elements (i.e. genes and ESTs) were significantly upregulated in primary and secondary EA infections, respectively ($P<0.05$ with FDR). In EM infection, 50 and 41 of the array elements were elevated in primary and secondary infections, respectively. Significantly, there were 75 and 108 downregulated elements in primary and secondary EA infections, and 17 and 220 in primary and secondary EM infections, respectively. Fold change analysis indicated that upregulation of gene expression was dominant in primary EA infection, and downregulation was dominant in secondary EM infection.

Comparative functional classification of gene profiles provides the basic information on host-parasite interaction and pathogenic mechanisms during coccidiosis. The significantly changed host genes, excluding singleton ESTs from expression profiles in EA or EM infections, were mapped to biological process terms of GO definitions in the Panther database (http://www.pantherdb.org). As shown in Figures 2A and 2B, GO mapping results showed that the changes of biological processes were dependent on the total number of 2-fold changed genes in EA or EM infections. The genes involved in protein metabolism and development, differentiation, and proliferation were more induced in primary EM infection than other infections. Interestingly, only after primary and secondary EM infections, nucleotide metabolism related genes were upregulated whereas these genes were decreased following primary EA infection. The genes involved in immunity and defence, and apoptosis were not differentially induced between EA or EM infections, but they were more downregulated in EM infection.

To find statistically significant changes associated with the biological function of genes involved in EA and EM infections, GO and signal pathway information were analyzed using expression analysis tools from the Panther database. The Bonferroni correction ($P<0.1$) for multiple testing was applied to determine statistical significance of functional change. As shown in Table 1, four biological processes involving induced genes and one involving reduced genes were significantly changed during *Eimeria* infection. Primary EA infection was responsible for induction of the genes involved in lipid, fatty acid, and cholesterol metabolism. Secondary EA infection was associated with upregulation of intracellular trafficking genes. The genes involved in biosynthesis and metabolism of protein were significantly upregulated in primary EM infection. Furthermore, downregulated gene function was observed in the apoptosis process following primary EM infection. Significant differences in signalling pathways were observed with the genes involved in cytoskeleton regulation and T cell activation pathways in primary EA infection (Table 2).

Discussion

Using the AVIELA chip, we report that host transcriptional responses show significant changes in gene functions following primary and secondary EA and primary EM infections (Table 1). First, lipid metabolism related genes were elevated in primary EA infection. Lipids and fatty acids are important organic molecules that are needed for cell membrane synthesis and energy metabolism. Furthermore, these molecules are precursors of inflammatory mediators, suggesting an important role of lipid metabolism-related genes in *Eimeria* infection. *ALOX5AP* is required for production of leukotrienes, which are implicated in various types of inflammatory-related diseases [11]. Secondary EA infection was associated with enhanced expression of genes involved in intracellular trafficking. *GARNL1, CANX,* and *COPA* are involved in endocytosis/exocytosis and *MTMR1* in vesicle transport processes [10]. Generally, following secondary infection with *Eimeria*, nutrient uptake in the intestines is higher than with primary infection. Therefore, upregulation of these genes after secondary infection might be related to balancing gut functionality. Interestingly, 3 genes involved in apoptosis were downregulated in primary EM infection. *LGALS3* and *STAT3* have anti-apoptosis activity and suppression of these genes induced apoptosis [12,13]. Therefore, downregulation of *LGALS3* and *STAT3* might be related with acute phase pathogenesis by stimulation of host cell death during primary EM infection.

The significant expression changes in T cell activation pathway genes also were observed in primary EA infection (Table 2). During T cell development, the CD3D protein plays an important role in the transition of thymocytes from immature precursors to the mature $CD4^+$ or $CD8^+$ positive T cells [14]. Previously, we reported that the number of $CD8^+$ T cells was increased in the duodenum during primary EA infection [15]. Thus, elevated *CD3D* expression supports our previous findings. *RAC2* and *PRKCH* are involved in cell motility and transendothelial migration and their upregulation suggests enhanced activation of T cell migration toward the duodenum during primary EA infection.

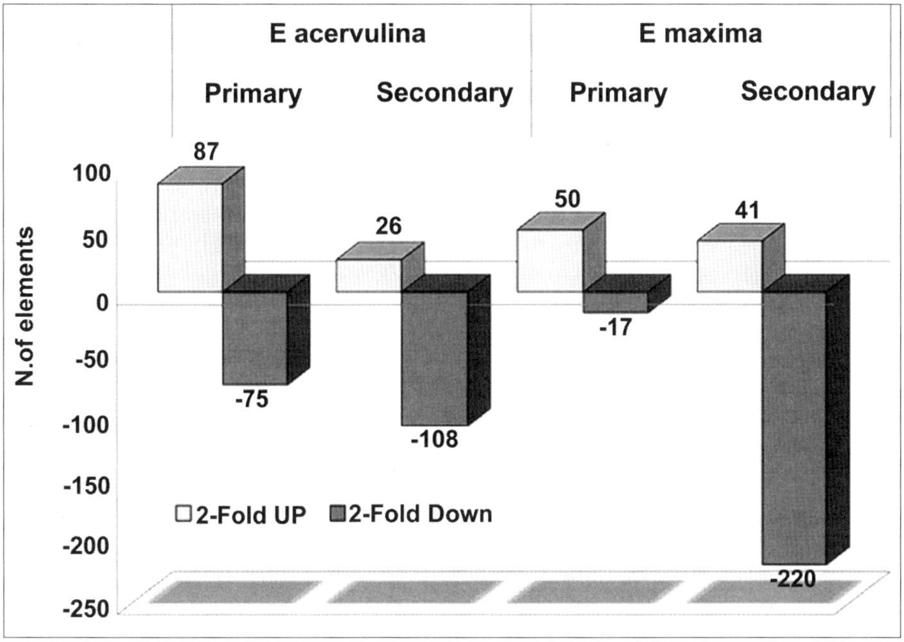

Fig. 1: Significantly up- or downregulated elements on the AVIELA array following primary or secondary infections with EA or EM parasites (P<0.05 with false discovery rate).

Table 2: Gene annotation and expression of significantly changed signaling pathways following EA or EM infections.

Significant condition	**Pathway**	**Regulation**	**Symbol**	**Description**
Primary EA	T cell activation	Up	*CD3D*	CD3d molecule, delta
		Up	*RAC2*	Ras-related C3 botulinum toxin substrate 2
		Up	*PRKCH*	protein kinase C, eta
		Down	*PIK3C2 B*	Phosphoinositide-3-kinase, class 2, beta polypeptide
	Cytoskeletal regulation	Up	*RAC2*	Ras-related C3 botulinum toxin substrate 2
		Up	*RICS*	Rho GTPase-activating protein
		Down	*RHOB*	ras homolog gene family, member B
		Down	*MYH9*	myosin, heavy chain 9, non-muscle

Fig. 2: Gene Ontology biological processes of upregulated (A) and downregulated (B) genes in primary and secondary infections with EA and EM. PEA; primary EA; SEA, secondary EA; PEM, primary EM; SEM, secondary EM infections.

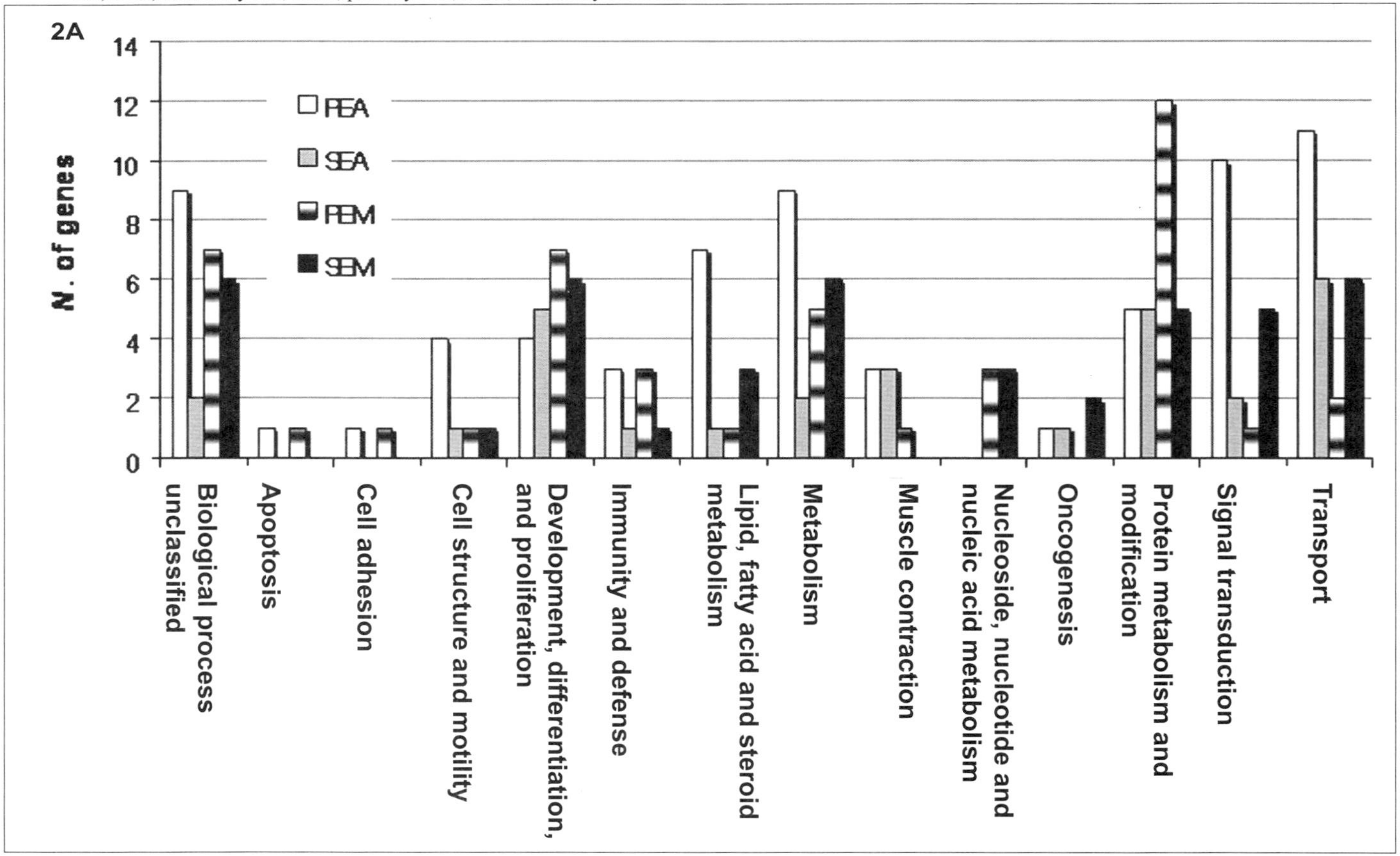

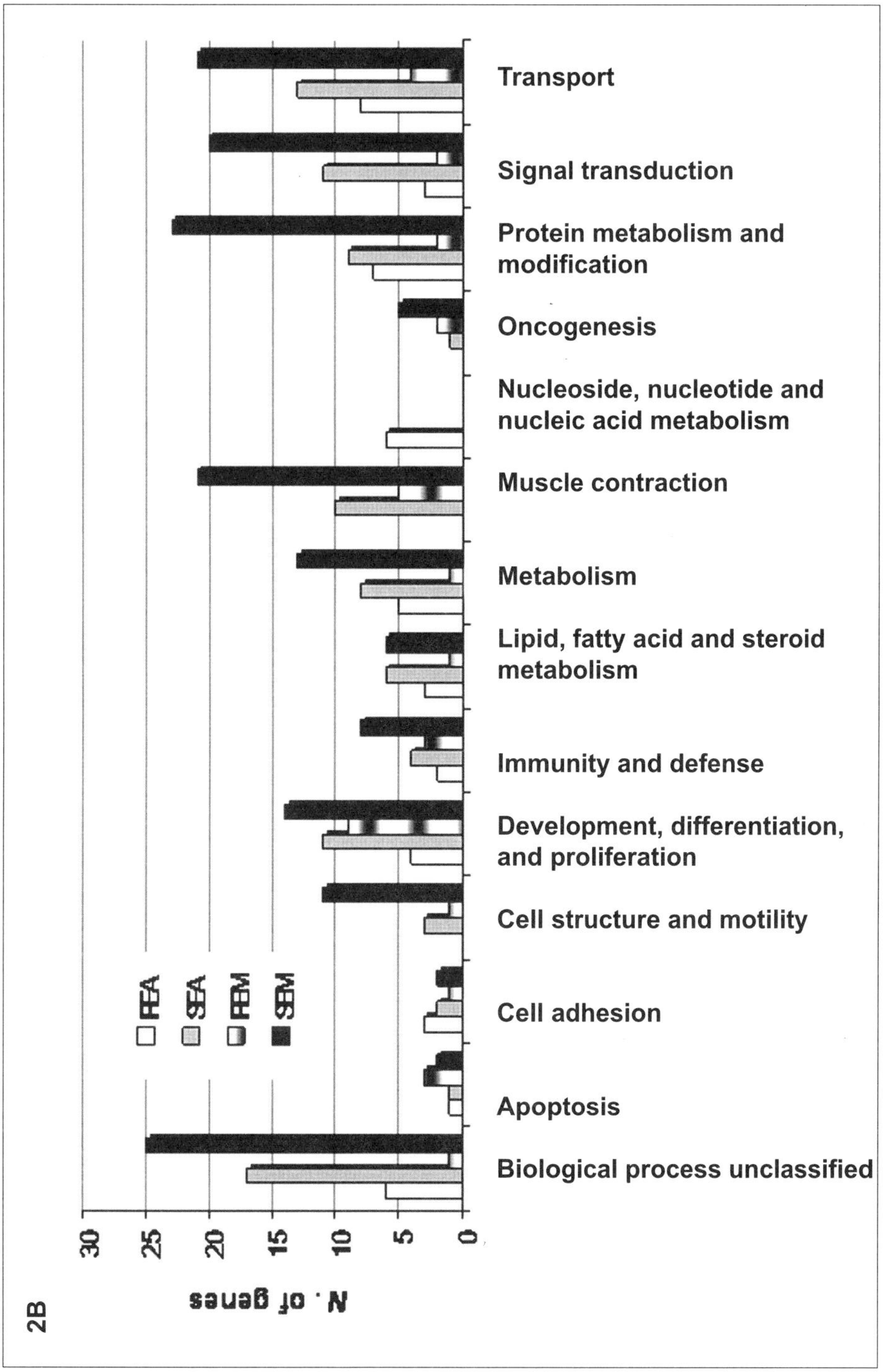
2B
N. of genes
0
5
10
15
20
25
30
FEA
SEA
FEM
SEM
Transport
Signal transduction
Protein metabolism and modification
Oncogenesis
Nucleoside, nucleotide and nucleic acid metabolism
Muscle contraction
Metabolism
Lipid, fatty acid and steroid metabolism
Immunity and defense
Development, differentiation, and proliferation
Cell structure and motility
Cell adhesion
Apoptosis
Biological process unclassified

Table 1: Gene annotation and expression of significantly changed Biological Processes of GO definitions following EA or EM infections.

Significant condition	Biological processes in GO terms	Regulation	Symbol	Description
Primary EA	Lipid, fatty acid and steroid metabolism	UP	*CYP2C9*	cytochrome P450, family 2, subfamily C, polypeptide 9
			ANXA5	annexin A5
			EPHX1	epoxide hydrolase 1, microsomal (xenobiotic)
			HMGCS1	3-hydroxy-3-methylglutaryl-Coenzyme A synthase 1
			ALOX5AP	arachidonate 5-lipoxygenase-activating protein
			HPGD	hydroxyprostaglandin dehydrogenase 15-(NAD)
			ELOVL1	elongation of very long chain fatty acids-like 1
Secondary EA	Intracellular trafficking	UP	*GARNL1*	GTPase activating Rap/RanGAP domain-like 1
			CANX	calnexin
			COPA	coatomer protein complex, subunit alpha
			MTMR1	myotubularin related protein 1
Primary EM	Protein biosynthesis and metabolism	UP	*TRIM2*	tripartite motif-containing 2
			RPS17	ribosomal protein S17
			RPL4	ribosomal protein L4
			RPL17	ribosomal protein L17
			CPM	carboxypeptidase M
			RPL3	ribosomal protein L3
			RPS3	ribosomal protein S3
			RPS15A	ribosomal protein S15a
			RPL9	ribosomal protein L9
			RPL18A	ribosomal protein L18a
			PPIB	peptidylprolyl isomerase B (cyclophilin B)
	Apoptosis	Down	*LGALS3*	lectin, galactoside-binding, soluble, 3 (galectin 3)
			STAT3	signal transducer and activator of transcription 3
			CYCS	cytochrome c, somatic

CONCLUSION

During primary or secondary infection with EA or EM, profound functional changes in host transcriptional responses occurred in the gut. Following primary EA infection, genes associated with lipid/fatty acid/steroid metabolism, T cell activation, and cytoskeletal regulation were modulated, implicating their involvement in the local host response to the protozoa. The 9.6K AVIELA array provides a valuable tool in the analysis of local host transcriptional regulation in avian coccidiosis and the application of this technology to the study of other enteric pathogens will expand our knowledge of the innate host defence system in poultry.

ACKNOWLEDGEMENTS

This work was supported by funds from the National Research Initiative of the USDA Cooperative State Research, Education and Extension Service, grants # 2002-35204-12368 and # 2004-35204-14798.

REFERENCES

1 Shirley MW, Ivens A, Gruber A, Madeira AMBN, Wan K, Dear PH, et al: The *Eimeria* genome projects: a sequence of events. Trends Parasitol 2004;20(5):199-201.

2 Dalloul RA, Lillehoj HS: Recent advances in immunomodulation and vaccination strategies against coccidiosis. Avian Dis 2005;49(1):1-8.

3 Lillehoj HS, Trout JM: Avian gut-associated lymphoid tissues and intestinal immune responses to *Eimeria* parasites. Clin Microbiol Rev 1996;9(3):349-360.

4 Hong YH, Lillehoj HS, Lillehoj EP, Lee SH: Changes in immune-related gene expression and intestinal lymphocyte subpopulations following *Eimeria maxima* infection of chickens. Vet Immunol Immunopathol 2006;114(3-4): 259-272.

5 Min W, Lillehoj HS, Kim S, Zhu JJ, Beard H, Alkharouf N, et al: Profiling local gene expression changes associated with *Eimeria maxima* and *Eimeria acervulina* using cDNA microarray. Appl Microbiol Biotechnol 2003;62(4):392-399.

6 Dalloul RA, Bliss TW, Hong YH, Ben-Chouikha I, Park DW, Keeler CL, et al: Unique responses of the avian macrophage to different species of *Eimeria*. Mol Immunol 2007;44(4):558-566.

7 Min W, Lillehoj HS, Ashwell CM, van Tassell CP, Dalloul RA, Matukumalli LK, et al: Expressed sequence tag analysis of *Eimeria*-stimulated intestinal intraepithelial lymphocytes in chickens. Mol Biotechnol 2005;30(2): 143-150.

8 Townsend JP: Multifactorial experimental design and the transitivity of ratios with spotted DNA microarrays. BMC Genomics 2003;4(1):41.

9 Benjamini Y, Hochberg Y: Controlling the False Discovery Rate: a Practical and Powerful Approach to Multiple Testing. J R Statis S B 1995; 57(1):289-300.

10 Mi H, Guo N, Kejariwal A, Thomas PD: PANTHER version 6: protein sequence and function evolution data with expanded representation of biological pathways. Nucleic Acids Res 2007;35:D247-52.

11 Helgadottir A, Manolescu A, Thorleifsson G, Gretarsdottir S, Jonsdottir H, Thorsteinsdottir U, et al: The gene encoding 5-lipoxygenase activating protein confers risk of myocardial infarction and stroke. Nat Genet 2004;36(3):233-239.

12 Fukumori T, Oka N, Takenaka Y, Nangia-Makker P, Elsamman E, Kasai T, et al: Galectin-3 regulates mitochondrial stability and antiapoptotic function in response to anticancer drug in prostate cancer. Cancer Res 2006;66(6):3114-3119.

13 Niu G, Shain KH, Huang M, Ravi R, Bedi A, Dalton WS, et al: Overexpression of a dominant-negative signal transducer and activator of transcription 3 variant in tumor cells leads to production of soluble factors that induce apoptosis and cell cycle arrest. Cancer Res 2001;61(8):3276-3280.

14 de Saint Basile G, Geissmann F, Flori E, Uring-Lambert B, Soudais C, Cavazzana-Calvo M, et al: Severe combined immunodeficiency caused by deficiency in either the delta or the epsilon subunit of CD3. J Clin Invest 2004;114(10):1512-1517.

15 Lillehoj HS: Analysis of Eimeria acervulina-induced changes in the intestinal T lymphocyte subpopulations in two chicken strains showing different levels of susceptibility to coccidiosis. Res Vet Sci 1994;56(1):1-7.

Dr. Hyun Soon Lillehoj, Animal Parasitic Diseases Laboratory; Animal and Natural Resources Institute, USDA-ARS, Bldg. 1040, BARC-East, Beltsville, MD 20705.

E-mail: hlilleho@anri.barc.usda.gov

Pinard M-H, Gay C, Pastoret P-P, Dodet B (eds): Animal Genomics for Animal Health. Dev Biol (Basel). Basel, Karger, 2008, vol 132, pp 131-140.

Differential Immune-related Gene Expression in Two Genetically Disparate Chicken Lines during Infection by *Eimeria maxima*

D.K. Kim[1], Y.H. Hong[1], D.W Park[1], S.J. Lamont[2], H.S. Lillehoj[1]

[1] Animal Parasitic Diseases Laboratory, Animal and Natural Resources Institute, United States Department of Agriculture, Beltsville, MD 20705

[2] Department of Animal Science, Iowa State University, Ames, IA 50011

Keywords: *Eimeria maxima*, chicken, infection, gene expression

Abstract: Two M5.1 and M15.2 *B* complex congenic lines of Fayoumi chickens were evaluated for body weight loss and faecal oocyst counts as parameters of avian coccidiosis. M5.1 chickens exhibited resistance to *E. maxima* compared with M15.2. To correlate the differential responses of the M5.1 and M15.2 lines to *E. maxima* infection with cellular immune responses, the expression levels of mRNAs encoding 14 immune-related molecules were measured by quantitative RT-PCR in intestinal intraepithelial lymphocytes (IELs) and splenocytes at 0, 3, 4, and 5 days following parasite infection. Intestinal IELs from M5.1 chickens expressed higher levels of transcripts encoding interferon gamma (*IFNG*), interleukin-1beta (*IL1B*), I*L6, IL8, IL12, IL15, IL17A*, inducible nitric oxide synthase (*iNOS*), and lipopolysaccharide-induced tumour necrosis α factor (*LITAF*), and lower levels of mRNAs for *IFNA, IL10, IL17D*, NK-lysin (*NKL*), and tumour necrosis factor superfamily 15 (*TNFSF15*) at 3 days post infection, compared with the M15.2 line. In the spleen, *E. maxima* infection was associated with higher expression levels of *IFNA*, and *IL15* and lower levels of *IL6, IL17D,* and *IL12* in M5.1 compared to M15.2 birds. Using an intestinal IEL cDNA microarray, the differential dynamics of gene expression in the gut of M5.1 and M15.2 chickens following experimental coccidiosis were evident. In particular, the genes encoding lymphotactin and parathymosin were expressed at significantly higher levels in M5.1 compared with M15.2 line chickens. In conclusion, genetic determinants within the chicken major histocompatibility complex (MHC) *B* complex influence resistance to *E. maxima* infection by controlling the local and systemic expression of immune-related cytokine and chemokine genes.

INTRODUCTION

Conventional disease control strategies against avian coccidiosis rely on prophylactic chemotherapy and vaccination, but both methods have serious drawbacks [1]. Genetic selection of disease resistant broiler lines offers an alternative to achieve this goal that may ultimately lead to the elimination of drugs in commercial poultry production.

The Fayoumi breed, which originated in Egypt, was imported to the United States in 1954 primarily because of its reported resistance to the avian leukosis. Subsequently, it was shown to develop a robust protective response against *E. tenella* [2]. Derived from the original Fayoumi breed, the M5.1 and M15.2 congenic pair of lines share an identical genetic background and differ only in the microchromosome bearing the major histocompatibility complex (MHC). Lines M5.1 and M15.2 are highly inbred, each with an inbreeding coefficient of 0.99, and are genetically distant from the broiler and Leghorn lines [3]. The present study was conducted to determine whether or not the major histocompatibility complex (MHC) *B* complex haplotypes expressed by M5.1 and M15.2 chickens influence protective immunity to avian coccidiosis, and to compare the expression of genes encoding immune-related molecules in the two lines. Additionally, comparative microarray analysis between two lines was performed to assess their systematic gene expression profile in the gut following *E. maxima* infection.

MATERIALS AND METHODS

Birds and experimental *Eimera* infection

Chicks were bred and maintained in the Poultry Genetics Program at Iowa State University (Ames, IA). *B* complex genotypes were confirmed by serological testing. Chickens (5/group) were orally inoculated at 3 weeks of age with 1.0 x 10^4 sporulated oocysts/chicken of *E. maxima*. Faecal oocyst numbers between days 6 to 9 post-infection and body weight gain between days 0 and 9 post-infection were determined as previously described [4,5]. All experiments were approved by the Animal and Natural Resources Institute IACUC.

RNA extraction, cDNA synthesis and quantitative RT-PCR

Following euthanization, intestines and spleens were taken at days 0, 3, 4, and 5 post-infection and jejunum IELs and splenocytes were prepared [4, 5]. Total RNA was isolated using Trizol (Invitrogen, Carlsbad, CA) and reverse-transcribed using the StrataScript first strand synthesis system (Stratagene, La Jolla, CA).

Amplification and detection of transcripts were carried out using equivalent amounts of total RNA with the Mx3000P system and Brilliant SYBR Green QPCR master mix (Stratagene). Standard curves were generated using log10 diluted standard RNA and levels of individual transcripts were normalized to those of GAPDH by the Q-gene program [6]. Each analysis was performed in triplicate.

Microarray analysis

The aminoallyl-labeled RNA from jejunum IELs from each group of 3 chickens was prepared using the Amino Allyl Message Amp II aRNA Amplification Kit (Ambion, 18 Austin, TX). For microarray hybridization, the reference design was employed. The samples were hybridized to the Avian Intestinal Intraepithelial Lymphocyte cDNA Microarray (AVIELA) constructed from 9,668 EST clones from the previously prepared 14,409 IEL clones [7]. After serial rinse, the microarray data acquisition, processing, and analysis were performed as previously described [8].

RESULTS

Assessment of disease parameters

To compare susceptibility to avian coccidiosis between Fayoumi M5.1 and M15.2 *B* haplotype congenic lines, body weight gain and faecal oocyst shedding were measured following oral infection with *E. maxima* oocysts. As shown in Figure 1, *Eimeria* infected M15.2 birds exhibited an approximately 20% decrease in body weight gain between 0 and 9 days post-infection compared with uninfected controls, and this difference was statistically significant ($P < 0.01$). In contrast, infected M5.1 birds displayed a tendency towards increased body weight gain compared with the uninfected controls, although this difference was not statistically significant ($P > 0.05$). When compared between infected chickens, M5.1 weight gains were significantly increased compared with the M15.2 line ($P < 0.05$). Additionally, M5.1 chickens had significantly reduced numbers of faecal *E. maxima* oocysts compared with M15 (Fig. 2).

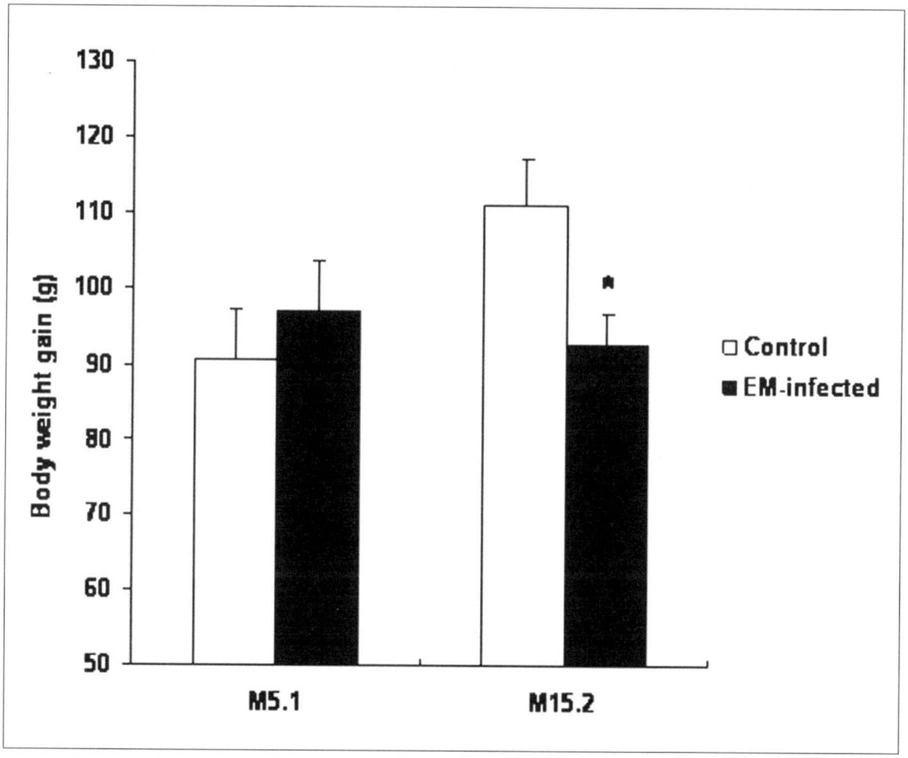

Fig. 1: Body weight gain of M5.1 and M15.2 chickens between 0 and 9 days post-infection with *E. maxima*. Each bar represents the mean ± S.D. values for each group (N = 5) (*: $P < 0.01$).

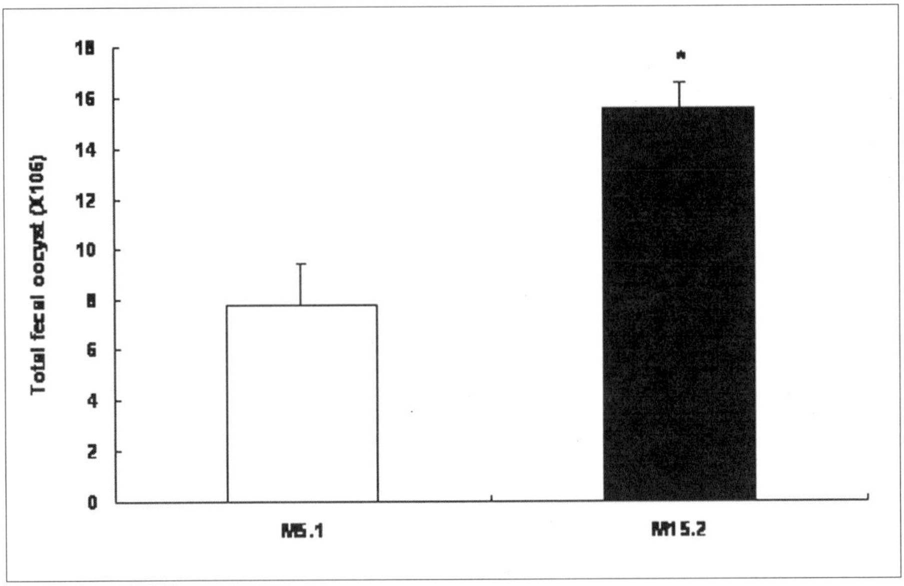

Fig. 2: Faecal oocyst numbers of M5.1 and M15.2 chickens between 6 and 9 days post-infection with *E. maxima*. Each bar represents the mean ± S.D. values for each group (N = 5) (*: $P < 0.01$).

Immune-related gene expression

In general, mRNAs encoding the proinflammatory cytokines *IFNA, IL1B, IL6, IL17A*, and *IL17D* were upregulated in intestinal IELs at 3, 4, and/or 5 days following *E. maxima* infection of M5.1 and/or M15.2 chickens (Table 1). Transcripts encoding *IL1B* and *IL6* were higher in M5.1 compared with M15.2 at day 3, whereas those for *IFNA* and *IL17D* were expressed at greater levels in M5.1 at day 4. Interestingly, while *IL17A* levels were undetectable at 0, 3 and 4 days post-infection in both chicken lines, its expression was markedly increased by 576-fold at day 5, but only in the M5.1 line. In the spleen, *IL6* and *IL17A* were downregulated following *E. maxima* infection and M15.2 had generally higher levels of *IL6, IL17A*, and *IL17D* mRNAs during the period examined. Finally, the proinflammatory CXC chemokine IL8 was downregulated following *E. maxima* infection in M5.1, but upregulated in M15.2, compared with uninfected controls (day 0).

Transcripts for the Th1/Th2-related molecules *IFNG, IL10, IL12,* and *iNOS* were generally upregulated in intestinal IELs of both M5.1 and M15.2 chickens at 3-5 days post-infection compared with uninfected controls (Tables 2 and 3). In the spleen, the levels of *IFNG, IL10, IL12*, and *IL15* mRNAs were decreased, particularly at day 3 in the M15.2 line, while the levels of these transcripts were relatively constant in M5.1.

The levels of *LITAF* transcripts were relatively unchanged in both M5.1 and M15.2 in IELs, but significantly decreased in splenocytes of M15.2 at days 3-5 compared with controls (Table 3). Similarly, *NKL* expression was downregulated in both chicken lines in spleen cells following *E. maxima* infection. By contrast, mRNAs encoding *TNFSF15* were higher in IELs and splenocytes from both lines.

Table 1: The quantitative analysis of mRNA levels of proinflammatory cytokines and chemokines in jejunum IELs and splenocytes of *E. maxima* infected M5.1 and M15.2 chickens (mean ± S.D, N = 3).

Tissue	Line		*IFNA*	*IL1B*	*IL6*	*IL8*	*IL17A*	*IL17D*
IEL	M5.1	Control	1.10E-3±1.15E-4	5.49E-8±2.06E-8	2.88E-7±2.20E-8	6.44E-3±2.03E-4	3.27E-10±6.90E-11	7.57E-6±5.10E-7
		3-dpi	1.41E-3±5.72E-5	3.81E-7±3.89E-8	1.58E-6±1.07E-7	2.85E-1±1.68E-2	1.32E-9±7.02E-11	2.89E-5±5.58E-6
		4-dpi	2.40E-3±2.04E-4	1.97E-7±1.86E-8	5.34E-7±1.98E-7	1.25E-1±9.26E-3	1.37E-10±1.23E-11	3.46E-5±2.51E-6
		5-dpi	2.28E-3±1.78E-4	3.33E-7±3.42E-8	8.57E-8±7.38E-9	8.57E-2±7.07E-3	1.85E-7±2.11E-8	4.91E-6±2.98E-7
	M15.2	Control	7.65E-4±1.42E-4	3.25E-7±3.46E-9	4.79E-7±7.15E-8	5.37E-3±1.16E-4	2.79E-10±3.00E-13	6.67E-6±7.55E-7
		3-dpi	1.95E-3±1.50E-4	2.51E-7±2.88E-8	6.67E-7±1.22E-7	4.22E-2±1.14E-3	2.28E-10±3.19E-11	3.33E-5±3.98E-6
		4-dpi	1.52E-3±2.08E-4	5.15E-7±1.70E-8	1.67E-7±1.18E-9	7.10E-2±1.04E-2	1.27E-10±1.54E-11	6.82E-6±7.75E-7
		5-dpi	9.99E-4±8.29E-5	3.26E-7±2.92E-8	2.04E-6±4.51E-7	1.65E-1±8.15E-3	9.45E-11±1.37E-11	2.64E-5±2.18E-6
Spleen	M5.1	Control	5.95E-2±6.85E-3	1.88E-6±1.00E-7	7.18E-5±1.32E-6	1.98E-1±1.85E-2	8.42E-11±1.15E-11	1.30E-5±4.02E-6
		3-dpi	3.52E-2±1.76E-3	1.71E-6±7.42E-8	1.87E-5±1.82E-6	3.17E-2±1.39E-3	2.35E-11±1.33E-12	5.51E-6±1.41E-6
		4-dpi	8.33E-2±8.54E-3	2.59E-6±2.78E-7	1.62E-5±2.88E-6	3.09E-2±1.15E-3	1.04E-11±6.79E-12	5.65E-6±3.47E-7
		5-dpi	1.78E-2±1.46E-3	3.98E-6±3.07E-8	2.23E-5±3.39E-6	7.58E-2±7.62E-3	8.64E-12±1.95E-12	1.83E-5±1.72E-6
	M15.2	Control	6.39E-2±6.07E-3	2.42E-6±3.79E-7	1.28E-4±1.03E-5	3.80E-2±1.18E-3	2.64E-9±2.55E-10	1.02E-4±1.84E-5
		3-dpi	3.81E-2±2.86E-3	3.19E-6±1.02E-7	1.08E-4±5.81E-6	4.90E-2±1.94E-3	3.57E-11±1.12E-11	9.25E-5±1.47E-5
		4-dpi	5.53E-2±4.90E-3	6.90E-6±1.03E-6	5.99E-5±9.18E-6	7.21E-2±4.25E-4	7.80E-11±2.41E-11	4.59E-4±2.87E-5
		5-dpi	4.40E-2±4.61E-3	2.47E-6±1.90E-7	4.08E-5±7.28E-6	6.16E-2±6.63E-3	9.59E-12±1.70E-12	4.30E-4±5.14E-5

Table 2: The quantitative analysis of mRNA levels of Th1/ Th2 cytokines in jejunum IELs and splenocytes of *E. maxima* infected M5.1 and M15.2 chickens (mean ± S.D, N = 3).

Tissue	Line		*IFNG*	*IL10*	*IL12*	*IL15*
IEL	M5.1	Control	6.72E-5±7.37E-6	1.50E-5±5.74E-7	7.53E-15±5.50E-15	8.00E-1±3.98E-2
		3-dpi	1.10E-3±6.35E-5	8.50E-6±3.55E-6	1.50E-14±1.29E-14	8.43E-1±3.11E-2
		4-dpi	1.70E-3±6.54E-5	2.79E-5±4.33E-6	7.88E-15±4.24E-15	6.81E-1±6.85E-2
		5-dpi	4.44E-4±2.93E-5	1.78E-5±2.95E-6	4.77E-15±4.39E-15	4.06E-1±2.97E-2
	M15.2	Control	3.97E-5±3.08E-6	5.89E-6±1.41E-6	5.95E-12±5.64E-13	4.30E-1±3.39E-2
		3-dpi	6.18E-4±1.80E-5	1.05E-5±1.34E-6	2.97E-12±1.82E-13	4.84E-1±4.33E-2
		4-dpi	2.01E-3±4.44E-5	4.28E-5±4.39E-6	3.80E-12±4.39E-13	4.53E-1±4.99E-2
		5-dpi	1.94E-4±2.76E-5	4.70E-6±5.67E-7	6.09E-12±8.05E-13	5.02E-1±8.34E-2
Spleen	M5.1	Control	1.24E-3±1.13E-4	3.38E-5±1.00E-5	3.93E-12±5.48E-13	4.10E-1±2.16E-2
		3-dpi	1.25E-3±2.13E-5	1.50E-5±1.36E-6	2.21E-12±2.43E-13	3.73E-1±2.90E-2
		4-dpi	1.36E-3±9.93E-5	2.45E-5±3.11E-6	2.66E-12±5.16E-13	5.99E-1±1.43E-2
		5-dpi	1.15E-3±7.63E-5	1.55E-5±2.36E-6	1.55E-12±2.70E-13	5.02E-1±1.96E-2
	M15.2	Control	2.01E-3±6.80E-5	1.02E-4±1.34E-5	5.95E-12±5.64E-13	8.83E-1±1.08E-1
		3-dpi	5.02E-4±2.28E-5	2.58E-5±3.34E-6	2.97E-12±1.82E-13	1.66E-1±1.97E-2
		4-dpi	4.36E-4±2.60E-5	2.00E-5±1.19E-6	3.80E-12±4.39E-13	1.86E-1±7.29E-3
		5-dpi	6.00E-4±4.31E-5	3.35E-5±2.92E-6	6.09E-12±8.05E-13	2.65E-1±2.34E-2

Table 3: The quantitative analysis of the gene expression of iNOS, LITAF, NK-lysin, and TNFSF15 in jejunum IELs and splenocytes of *E. maxima* infected M5.1 and M15.2 chickens (mean ± S.D, N = 3).

Tissue	Line		*iNOS*	*LITAF*	*NK-lysin*	*TNFSF15*
IEL	M5.1	Control	5.99E-5±9.99E-6	2.65E-1±2.54E-2	1.06E+0±1.43E-1	7.34E-4±9.45E-5
		3-dpi	1.89E-4±3.15E-5	3.46E-1±3.86E-2	1.95E+0±1.71E-1	1.15E-3±9.38E-5
		4-dpi	1.61E-4±2.96E-6	2.35E-1±2.65E-2	1.97E+0±1.81E-1	2.53E-3±2.20E-4
		5-dpi	4.91E-5±5.61E-6	2.78E-1±1.84E-2	4.49E-1±3.54E-2	1.31E-3±1.63E-4
	M15.2	Control	4.21E-5±3.22E-6	3.32E-1±3.14E-2	2.43E+0±2.14E-1	4.09E-4±2.38E-5
		3-dpi	7.45E-5±3.32E-6	3.36E-1±5.68E-2	4.28E+0±4.91E-1	1.73E-3±1.37E-4
		4-dpi	1.29E-4±5.26E-6	3.98E-1±6.43E-2	2.96E+0±1.90E-1	1.38E-3±1.47E-4
		5-dpi	2.68E-5±5.07E-6	2.60E-1±3.92E-2	3.00E+0±5.85E-1	2.58E-3±3.31E-4
Spleen	M5.1	Control	3.02E-4±1.68E-5	2.27E-1±2.94E-2	2.72E+0±2.80E-1	2.44E-4±1.06E-5
		3-dpi	3.03E-4±2.17E-5	1.81E-1±2.32E-2	9.93E-1±5.22E-2	2.55E-4±6.11E-6
		4-dpi	7.67E-4±5.65E-5	2.41E-1±1.33E-2	1.21E+0±8.18E-4	3.86E-4±2.31E-5
		5-dpi	8.27E-4±6.59E-5	2.51E-1±2.24E-2	6.92E-1±6.65E-2	5.30E-4±4.29E-5
	M15.2	Control	1.41E-4±8.87E-6	4.76E-1±8.44E-2	3.25E+0±3.74E-1	1.42E-4±1.24E-5
		3-dpi	1.88E-4±1.70E-5	2.37E-1±1.06E-3	1.31E+0±2.15E-1	1.75E-4±1.12E-5
		4-dpi	7.30E-4±5.32E-5	1.81E-1±1.92E-2	1.97E+0±2.15E-1	3.80E-4±3.07E-5
		5-dpi	2.56E-4±7.65E-6	2.19E-1±1.60E-2	1.78E+0±3.33E-1	1.75E-4±1.25E-5

Microarray analysis

The AVIELA microarray was used to compare the transcript levels in intestinal IELs of M5.1 and M15.2 line chickens from uninfected and *E. maxima* infected birds. When compared to uninfected controls and using a cut-off of > 2.0-fold alteration, M5.1 demonstrated altered expression of 1 (downregulated), 12 (6 upregulated, 6 downregulated), and 18 (5 upregulated, 13 downregulated) mRNAs at 3, 4, and 5 days post-infection respectively. In the case of the M15.2 line, altered expression was observed in 6 (3 upregulated, 3 downregulated), 29 (11 upregulated, 18 downregulated), and 32 (8 upregulated, 24 downregulated) transcripts at the three time points compared with uninfected controls. Comparison between infected M5.1 and M15.2 chickens revealed alterations in 32 (22 upregulated, 10 downregulated), 98 (55 upregulated, 43 downregulated), and 92 (59 upregulated, 33 downregulated) mRNAs at the 3 time points. Interestingly, of the genes with known chromosomal locations, the largest number of elements exhibiting > 2.0-fold altered expression levels between M5.1 and M15.2 were located on chromosome 1. Two genes were selected to illustrate the kinetics of their corresponding transcript levels following *E. maxima* infection. Lymphotactin, a chemokine produced by activated $CD8^+$ T cells, NK cells, and γδ T cells, was upregulated > 2.0-fold at days 4 and 5 in M5.1, but not in 15.2 chickens (Fig. 3). In contrast, parathymosin, a small nuclear acidic protein of unknown function, was downregulated at days 4 and 5 in M15.2, but not in 5.1 chickens (Fig. 4).

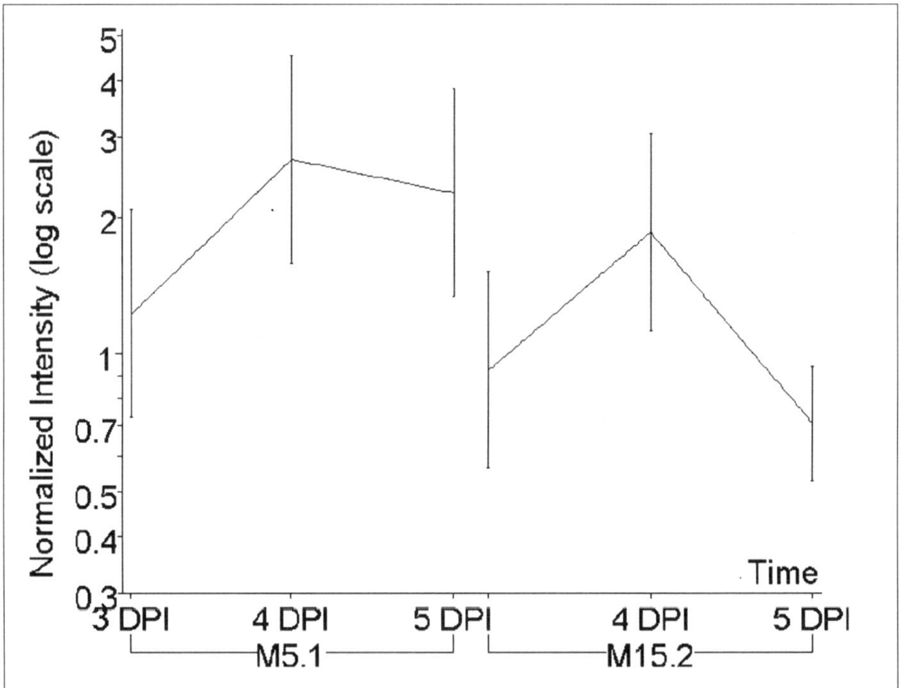

Fig. 3: Signal intensity of lymphotactin transcript levels in intestinal IELs of M5.1 and M15.2 chickens at 3, 4, and 5 days post-infection (dpi) with *E. maxima* measured using the AVIELA microarray and normalized with the levels in uninfected controls. Each data point represents the mean ± S.D. values for each group (N = 4).

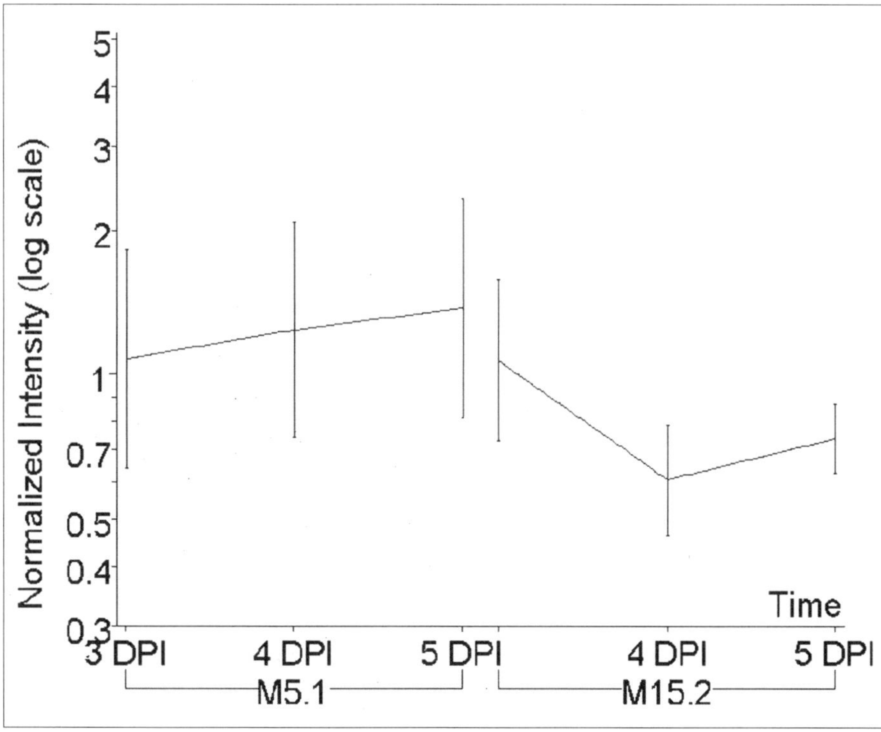

Fig. 4: Signal intensity of parathymosin transcript levels in intestinal IELs of M5.1 and M15.2 chickens at 3, 4, and 5 days post-infection (dpi) with *E. maxima* measured using the AVIELA microarray and normalized with the levels in uninfected controls. Each data point represents the mean ± S.D. values for each group (N = 4).

Discussion

The results of body weight gain and faecal parasite numbers during experimental *E. maxima* infection suggest that M5.1 chickens are more disease-resistant compared with the M15.2 line. Analysis of various immune-related molecules in the gut and spleen indicated that the enhanced resistance of M5.1 over M15.2 may be explained, in part, by their differential patterns of expression. Transcripts for the proinflammatory cytokines *IL1B* and *IL6* were significantly upregulated in intestinal IELs of M5.1 chickens at 3 days post-infection, compared with M15.2 chickens. It is tempting to speculate that differential expression of *IL1B* and *IL6* in M5.1 vs. M15.2 birds may underlie, to a large degree, their difference in susceptibility to *E. maxima* infection. The early robust *IL8* response in M5.1 may confer the ability of this strain to limit *Eimeria* infection in the gut, while the belated response in M15.2 may be too late to counteract parasite dissemination and intestinal destruction.

IL10 exerts inhibitory effects on proinflammatory cytokines as exemplified by the fact that chicken *IFNG* and IL12 were downregulated following expression of IL10 [9, 10]. The kinetics of *IFNG* and *IL10* expression were identical both in IELs and spleen cells of *E. maxima* infected chickens, suggesting that *IFNG* expression was not affected by *IL10* levels. *iNOS* expression was reported to be induced by *IFNG* in chicken macrophages and was highly upregulated during experimental *E. tenella* infection [11]. The temporal patterns of *iNOS* and *IFNG* expression were identical in M5.1 and M15.2 IELs, i.e. biphasic responses with maximum levels at 3 or 4 days following parasite infection.

In conclusion, early and high expression of pro-inflammatory cytokine and chemokine genes, including *IL1B, IL6,* and *IL8*, combined with lower levels of *IL10*, were associated with enhanced resistance to *E. maxima* infection in the Fayoumi M5.1 breeding line compared with the M15.2 strain. Moreover, Th1-related cytokines (*IFNG, IL12,* and *IL15*) as well as other mediators of innate immunity (*iNOS*) were also early markers associated with the resistant M5.1 line. Also noted were strain specific patterns of *IFNA* and *IL8* expression in comparison with previous studies, and most of the genes examined exhibited tissue specific expression between IELs and splenocytes. Transcriptional profiling data showed more gene expression changes in M15.2 than M5.1, however, most differences between the two chicken lines were exhibited at day 4 following *E. maxima* infection.

References

1 Dalloul RA, and Lillehoj HS: Recent advances in immunomodulation and vaccination strategies against coccidiosis. Avian Dis 2005;49:1-8.

2 Pinard-van der Laan MH, Monvoisin JL, Pery P, Hamet N, Thomas M: Comparison of outbred lines of chickens for resistance to experimental infection with coccidiosis (*Eimeria tenella*). Poult Sci 1998;77:185-191.

3 Zhou H, Lamont SJ: Genetic characterization of biodiversity in highly inbred chicken lines by microsatellite markers. Anim Genet 1999;30:256-264.

4 Lee S, Lillehoj HS, Park DW, Hong YH, Lin JJ: Effects of Pediococcus- and Saccharomyces-based probiotic (MitoMax) on coccidiosis in broiler chickens. Comp Immunol Microbiol Infect Dis 2007;30:261-268.

5 Dalloul RA, Lillehoj HS, Shellem TA, Doerr JA: Enhanced mucosal immunity against *Eimeria acervulina* in broilers fed a Lactobacillus-based probiotic. Poult Sci 2003;82:62-66.

6 Muller PY, Janovjak H, Miserez AR, Dobbie Z: Processing of gene expression data generated by quantitative real-time RT-PCR. Biotechniques 2002;32:1372-1374, 1376, 1378-1379.

7 Min W, Lillehoj HS, Ashwell CM, van Tassell CP, Dalloul RA, Matukumalli LK, Han JY, Lillehoj EP: Expressed sequence tag analysis of *Eimeria*-stimulated intestinal intraepithelial lymphocytes in chickens. Mol Biotechnol 2005;30:143-150.

8 Dalloul RA, Bliss TW, Hong YH, Ben-Chouikha I, Park DW, Keeler CL, Lillehoj HS: Unique responses of the avian macrophage to different species of *Eimeria*. Mol Immunol 2007;44:558-566.

9 Rothwell L, Young JR, Zoorob R, Whittaker CA, Hesketh P, Archer A, Smith AL, Kaiser P: Cloning and characterization of chicken IL-10 and its role in the immune response to *Eimeria maxima*. J Immunol 2004;173:2675-2682.

10 Groux H, Powrie F: Regulatory T cells and inflammatory bowel disease. Immunol Today 1999;20:442-445.

11 Laurent F, Mancassola R, Lacroix S, Menezes R, Naciri M: Analysis of chicken mucosal immune response to *Eimeria tenella* and *Eimeria maxima* infection by quantitative reverse transcription-PCR. Infect Immun 2001;69:2527-2534.

Dr. Hyun Soon Lillehoj, Animal Parasitic Diseases Laboratory, Animal and Natural Resources Institute, Building 1040, BARC-East, US Department of Agriculture, Beltsville, MD 20705, USA.

E-mail: Hyun.Lillehoj@ars.usda.gov

Pinard M-H, Gay C, Pastoret P-P, Dodet B (eds): Animal Genomics for Animal Health. Dev Biol (Basel). Basel, Karger, 2008, vol 132, pp 141-151.

Transcriptional Response of Chicken Macrophages to *Salmonella enterica* Serovar Enteritidis Infection

S. Zhang[1], H. S. Lillehoj[2], C.-H. Kim[2], C. L. Keeler[3], Jr., U. Babu[4], M. Z. Zhang[1]

1 Department of Pathobiology/Population Medicine, College of Veterinary Medicine, Mississippi State University, Pearl, MS, USA
2 Animal Parasitic Diseases, USDA-ARS, BARC E, Beltsville, MD, USA
3 Department of Animal and Food Sciences, University of Delaware, Newark, DE, USA
4 Center for Food Safety and Applied Nutrition, US Food and Drug Administration, Laurel, MD, USA

Keywords: ranscriptional response, chicken macrophage, *Salmonella enteritidis*

Abstract: The transcriptional profiles of chicken macrophages (HD11) infected with *Salmonella enterica* serovar Enteritidis (SE) were analyzed by using an avian macrophage microarray and real time RT-PCR. Out of 4,906 array elements interrogated, 269 genes exhibited a 2 fold change ($P < 0.001$) over a 24 h time-course. Genes coding for proinflammatory cytokines, CC and CXC chemokines, and chemokine ligand were upregulated; whereas genes associated with transcription, cell adhesion and proliferation were downregulated. Most transcriptional changes occurred at 5 hours post-inoculation (hpi), with more genes downregulated than upregulated. At 5 hpi, the levels of gallinacin 1, lymphotactin, *RhoA*, and *MHCIB2M* transcripts were significantly decreased. In contrast, the levels of *Cdc42* and *MHCIIBLB2* mRNA were elevated. Infection of HD11 cells with mutant SE strains carrying an inactivated type three secretion system (TTSS1 or TTSS2) induced significantly higher levels of *CCL4, K203*, lymphotactin, and *RhoA* than wild type SE. In conclusion, chicken macrophage genes belonging to diverse functional classes were transcriptionally modulated by SE and selective modulation of host innate responses involved the effectors of TTSS1/2.

INTRODUCTION

Salmonella enterica induced enterocolitis is a common cause of bacterial foodborne disease worldwide. Among more than 2,300 serovars, Enteritidis (SE) remains the second etiologic agent of nontyphoidal salmonellosis, accounting for 18.6% of all cases in the United States (US) [1]. In recent years, a definite upward trend in SE incidence has been observed in the US. The number of SE cases increased 25% in 2005 and 28% in 2006 over the average incidences from 1996 to 1998 [2]. Poultry meat and egg products have been identified as the major sources of human SE cases. The organism colonizes the intestine of adult chickens and the oviduct of laying hens without causing overt clinical signs while contributing to carcass and egg contamination [3]. To effectively control SE contamination of poultry products, research should aim at understanding the immunological and molecular mechanisms underlying SE infection and colonization of chickens.

Host protective immunity involves both innate and acquired responses. Being the first line of defense, macrophages serve as professional phagocytes and secrete various cytokines and immune mediators critical to the clearance of microbial infections [4]. However, *Salmonellae* have evolved the ability to combat macrophage-mediated killing by surviving and replicating within specialized vacuoles or the phagosomes of macrophages [5]. The organism possesses two virulence associated type three secretion systems (TTSS1 and TTSS2), which are responsible for the invasion of epithelial cells and survival in macrophages, respectively. At the present time, the complex interplay between SE and chicken macrophages remains to be understood. The newly developed avian macrophage-specific cDNA microarray (AMM) has made it possible to monitor the expression of thousands of gene transcripts in infected macrophages [6]. In this study, the transcriptional response of chicken macrophages to SE infection was investigated using the AMM system. In addition, the role of the TTSS1/2 in modulating immune gene expression during SE infection was also analyzed.

MATERIALS AND METHODS

Bacterial strains

SE100, a spontaneous nalidixic acid (Nal)-resistant mutant of SE 338, was selected by the serial passage method. TTSS1 and TTSS2 mutants, designated as SE103 and SE106, were constructed by introducing the suicide vector pEP185.2 into the open reading frames of *SipA* and *PipB*, respectively, as described [7]. The SE strains were cultured in LB broth containing 50 μg/ml of Nal for 18 h at 37°C with shaking at 200 rpm. Bacteria in the stationary phase of growth were resuspended in HBSS, and placed on ice prior to inoculation.

Macrophage culture and infection

HD11, a chicken macrophage cell line, was maintained as monolayers at 39°C in RPMI 1640 supplemented with 10% foetal bovine serum, 2% chicken serum, 2 mM glutamine, 50 U of penicillin G, and 50 μg/ml of streptomycin in a 5% CO2 humidified atmosphere. The HD 11 cells were inoculated with SE strains at an MOI of 10, and incubated for 1 h at 39°C. Following incubation, HD11 cells were washed three times with HBSS and treated with 100 μg/ml of gentamicin to remove and kill extracellular bacteria. The infected cells were maintained at 39°C in PRMI 1640 supplemented with 5% foetal bovine serum, 2% chicken serum, 2 mM glutamine, and 50 μg/ml of gentamicin.

RNA extraction

At 2, 5, and 24 hours post-inoculation (hpi), total RNA was prepared from infected and uninfected HD11 cells using TRIzol (Life Technologies). PolyA RNA was isolated using the Poly(A) Tract mRNA Isolation System IV (Promega).

cDNA microarray hybridization and data analysis

The AMM cDNA microarray was constructed as described previously [6]. Poly-A RNA from each sample (1.5 μg) was amplified into amino allyl modified RNA (aRNA) using the Ambion Amino Allyl messageAmp II aRNA Amplification kit (Ambion). aRNA (20 μg) was fluorescently labeled with AlexaFluor 555 (Cy3) or AlexaFluor 647 (Cy5). Hybridizations (60 μl) were performed in Mica hybridization chambers (GeneMachines) at 50°C overnight. After hybridization, slides were rinsed in 0.5 X SSC, 0.01% SDS at room temperature and then washed for 15 min in 0.2 X SSC, 0.2% SDS at 50°C, three times for 1 min in 0.2 X SSC at room temperature, and finally three times for 1 min in water at room temperature. Washed slides were scanned with an ArrayWoRx scanner (Applied Precision) using Cy3 and Cy5 filters. Spot and background intensities were acquired using SoftWoRx tracker (Applied Precision) and data analysis was performed using GeneSpring v7.0 (Silicon Genetics) as described previously [6, 8]. Background intensity was determined using the GeneSpring cell method. Abnormal spots (dust, bubbles) and spots with intensities not greater than background +2 standard deviations were flagged as abnormal. Elements that were not represented by at least two replicate spots on each slide were removed from further analysis. Elements exhibiting greater than 2 fold changes in signal intensity during at least one time point were analyzed by ANOVA using the Benjamini and Hochberg's false discovery rate (FDR) multiple testing correction with a P-value of < 0.001 to determine which biologically significant changes were also statistically significant. Statistical analysis was performed using the Significance Analysis of Microarray (SAM) software as described.

EST annotation and Gene Ontology analysis

All sequence data files were obtained from NCBI (http://www.ncbi.nlm.nih.gov). EST elements on the array were identified to reference RNA and protein sequences and UniGene sequence databases using NCBI Blast (version 2.2.13). Entrez gene and Homologene databases of chicken, human, and mouse from NCBI were used to gather gene information (gene identification number, symbol, and name) and Gene Ontology terms to biological processes. Python (version 2.2.4) and SQL scripts were used to process and manipulate Blast results and data processing, respectively.

Real time reverse transcription PCR (RRT-PCR)

RRT-PCR was conducted using MultiScribe reverse transcriptase (Invitrogen) and the DNA labeling dye SYBR Green (Applied Biosystems). The DNA sequences of chicken b-actin, and selected immune genes tested were obtained from NCBI (Table 1). Real-time PCR amplification of b-actin transcripts were used to normalize the cDNA concentrations of different samples. The changes in immune gene expression in infected HD11 cells were determined using the comparative C_T ($2^{-\Delta\Delta C_T \pm S}$) method as described previously [9].

Results

Gene expression in SE100-infected HD11 cells

Over a 24 h time-course, 269 out of 4,906 array elements (number of elements spotted on the chips) exhibited a 2 fold change with a statistical significance ($P < 0.001$). Of those 269 elements, 86 genes (32%) were upregulated and 183 (68%) were downregulated (Fig. 1). The kinetic analysis of gene expression also demonstrated that most transcriptional changes occurred at 5 hpi, with more gene downregulated than upregulated (Fig 1). Table 2 shows a partial list of the genes significantly modulated by SE100 at 5 hpi. Gene Ontology analysis of the transcripts revealed significant alterations in diverse functional processes, including apoptosis, cell adhesion, cytoskeleton, growth and differentiation, immune response, metabolism, transcription, translation, and transport (Fig. 2).

Table 1: Primers used for RT-PCR.

Name	Sequence	Amplicon size
BActin-F	TGCGTGACATCAAGGAGAAG	
BActin-R	GACCATCAGGGAGTTCATAGC	111 bp
iNOS-F	AGTTTGAAATCCAGTCGTGTTAC	
iNOS-R	AATATGTTCTCCAGGCAGGTAG	87 bp
MIP-1B-F	CAGACTACTACGAGACCAACAG	
MIP-1B-R	GCATCAGTTCAGTTCCATCTTG	233 bp
K203-F	TCTTCTCATCGCATCCTTCTG	
K203-R	ATCTTGTGTGTTATGTAAGTGGTG	93 bp
K60-F	GATGATGGGCAAGGCTGTAG	
K60-R	CTGGCATCGGAGTTCAATCG	107 bp
Lymphotactin-F	CCTGGTCATAGTCTGGCTTG	
Lymphotactin-R	AATACTTCTGATGTCCACTTTCTG	124 bp
MHC I beta2M-F	AACGTCCTCAACTGCTTCGT	
MHC I beta2M-R	GAAGGTCCAGTCGTCCTTGA	126 bp
RhoA-F	GCGCGCACTTTTTGTGTAAG	
RhoA-R	TCAGGGAGCTTAGAAAACTG	148 bp

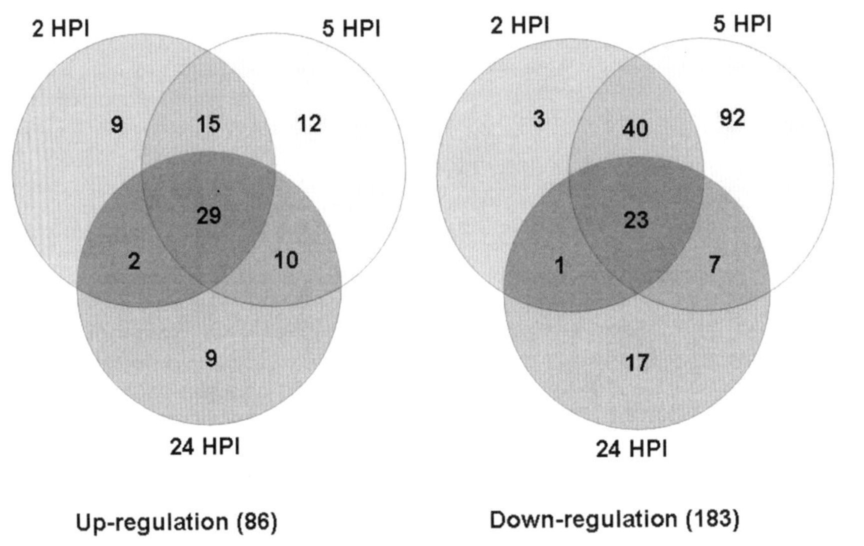

Fig. 1: Van Diagram of up- or downregulated elements (> 2 fold changed with P < 0.001) at 2, 5, and 24 hours post infection with SE100. Host cell adhesion and cytoskeleton

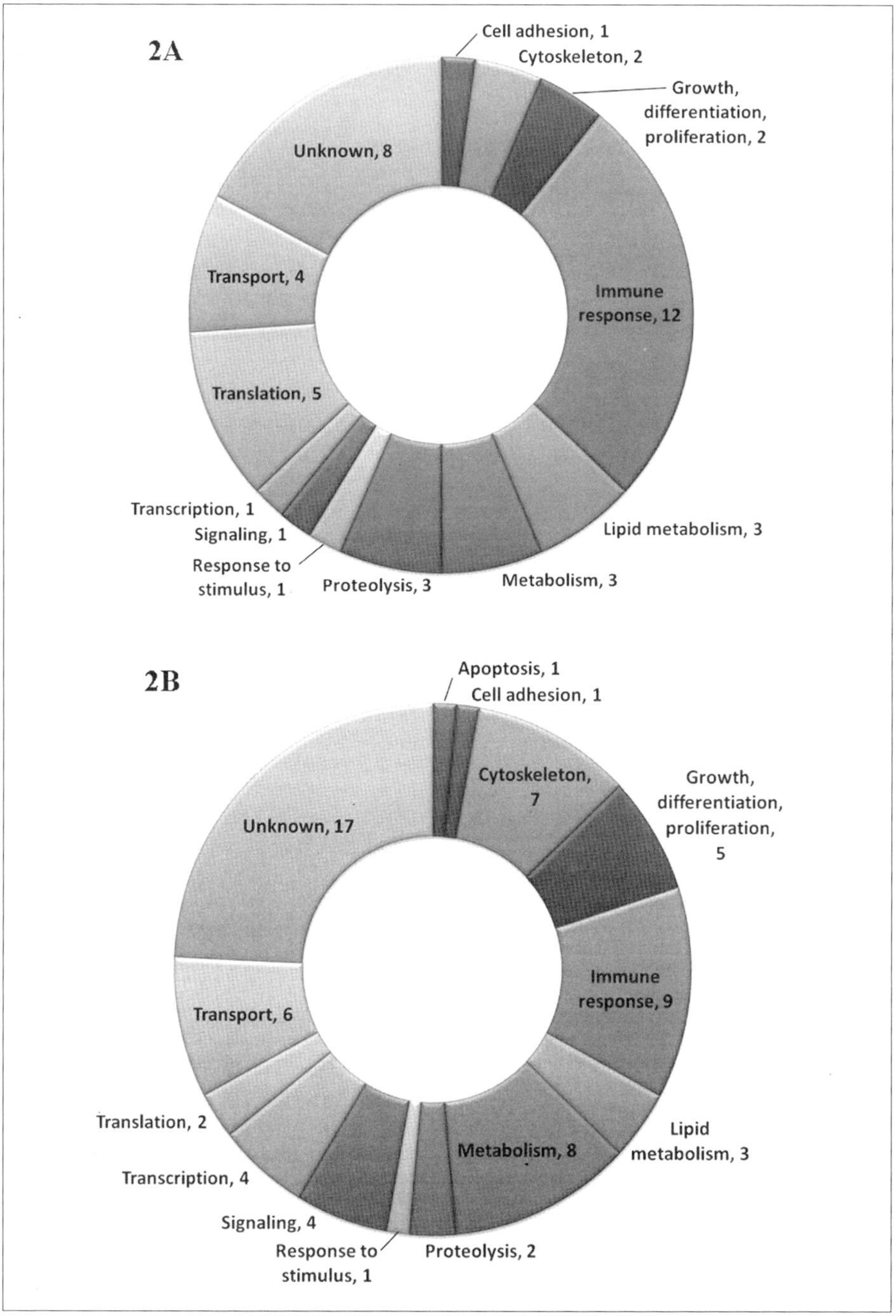

Fig. 2: Gene Ontology analysis of HD11 transcripts exhibiting greater than 2 fold changes ($P < 0.001$) at 5 hours post infection with SE100.

Table 2: A partial list of HD11 transcripts with known functions at 5 h post-inoculation with SE100.

Gene Name	Fold Change	Biological Processes and Detail Function	GenBank Acc ID	Gene Symbol
LPSj_P03	3.5	Cell adhesion	CK615099	PPFIBP1
LPSl_M18	2.9	Cytoskeleton, cell movement	CK613571	TUBB1
LPSh_G24	2.2	Cytoskeleton	CK614752	PLS3
IFNc_A13	5.5	Regulation of proliferation, positive	CK610285	IGF1R
LPSk_K11	3.6	Cell division, signal transduction	CK613423	CDC42
LPSh_I18	991.2	Immune response, inflammation	CK612928	CCL5
LPSm_K08	217.4	Inflammation	CK613680	CCL20
LPSm_I12	39.4	Immune response	CK615433	K60
LPSj_H24	26.9	Immune response	CK613244	ICOS
LPSm_L10	25.9	Humoral immune response	CK613692	IL6
IFNk_L14	16.4	Inflammation	CK611404	CCL4
LPSj_A20	15.4	Humoral immune response	CK613166	CD83
IFNi_J22	10.9	Inflammation	CK611118	IL1B
LPSm_M19	4.0	T cell activation	CK613706	CD82
LPSj_G10	3.9	Antigen processing	CK615011	MHCII BLB2
LPSm_I24	-13.7	Cell adhesion	CK615438	CTNND1
Controld_L19	-7.6	Cytoskeleton, actin binding	CK607331	TB4X
pmp1c.pk005.d07	-4.9	Cytoskeleton	BQ484493	COL1A1
pmp1c.pk007.n14	-4.7	Cytoskeleton, microtubule depolymerization	BQ485012	STMN1
IFNd_C13	-4.5	Cytoskeleton, actin binding	CK610406	PLS3
pmp1c.pk007.k07	-3.5	Cytoskeleton	BQ485000	PLEKHC1
LPSc_I09	-3.1	Cytoskeleton, actin binding	CK614064	PFN1
Control_H14	-2.5	Microtubule depolymerization, negative	CK606825	MID1IP1
pgm2n.pk004.a5	-11.2	Growth, differentiation, proliferation	BM487268	LOC776531
pgm2n.pk005.o23	-9.5	Growth, differentiation, proliferation	BM487890	FZD9
pmp1c.pk006.i03	-4.3	Growth, differentiation, proliferation	BQ484760	LLGL2
Control_C23	-4.1	Regulation of cell cycle	CK607529	CSN6
LPSf_I14	-3.8	Regulation of proliferation, negative	CK614488	PHB
LPSh_N04	-3.6	Cell division, signal transduction	CK612976	HRAS
IFNh_N10	-3.5	Regulation of growth, negative	CK609312	SERTAD2
LPSc_C10	-3.4	Proliferation	CK614012	EMP1
IFNm_D09	-2.7	Growth, differentiation, proliferation	CK611610	YWHAQ
Control_C14	-10.4	NF-κB activation	CK607525	RHOA
IFNg_B01	-4.1	Inflammatory response	CK609062	NFRKB
IFNc_K10	-3.3	Immune response	CK608618	TNFSF13B
Controlc_M15	-2.9	Antigen processing and presentation	CK607201	MHC I B2M
pat.pk0021.b7		Chemokine activity	AI980332	Lymphotactin
pmp1c.pk007.c06	-2.6	Antigen processing and presentation	BQ484906	PSME1
IFNl_D01	-2.6	T-cell signaling	CK611459	CMIP
pmp1c.pk002.a02	-2.4	Inflammatory response	BQ484156	CEBPB

Host cell adhesion and cytoskeleton

At 5hpi, SE100 had disparate effects on the genes related to cell adhesion and cell cytoskeleton. For example, the expression of PPFIBP1, a protein that interacts with protein-tyrosine phosphatase receptor type F (PTPRF) was upregulated. In contrast, catenin, a subunit of multiprotein cell-cell adhesion complex was downregulated. In addition, SE100 activated *Tubb1*, a gene coding for tubulin beta1, but repressed the expression of genes whose protein products are involved in actin binding (THYMOSINB4, PROFILIN1) and microtubule depolymerization (STATHMIN1, MID1 interacting protein 1).

Cell division, differentiation, and proliferation

Numerous genes involved in cell differentiation, growth, and proliferations were downregulated, except *Cdc42* and *Igf1r* that encode a small GTPase and an insulin-like growth factor 1 receptor, respectively. Among the downregulated genes, some (*Loc776531, Fzd9, Llgl2, Csn6, Hras,* and *Ywhaq*) promote cell differentiation and proliferation while others (*Phb* and *Sertad2*) function as negative regulators.

Immune response genes

SE100 significantly altered the expression of immune response genes, including those encoding proinflammatory cytokines, chemokines, MHC antigens, and cluster differentiation (CD) molecules. Among SE-induced chemokines, CCL4 (MIP1B), CCL5 (RANTES), and CCL20 (MIP3A) are responsible for recruiting leucocytes, especially eosinophils, basophils, and lymphocytes to the site of inflammation [11,12]. K60, a functional homolog of CXCL8 (IL8), is a potent chemotactic factor for heterophilic granulocyte [12]. Although SE100 elicited the expression of CC and CXC chemokines, it suppressed the transcription of lymphotactin (C chemokine) and gallinacin1. The expression of MHCIIBLB2 was upregulated, but HLADMB and MHCIB2 microglobulin, as well as PSME1 (proteasome activator 28 α protein) were downregulated. Additional genes downregulated by SE100 were *RhoA* and *NFRKB.* RHOA is a small GTPase with multiple biological functions, including activation of NFRKB that binds to the related kappa B site and promotes the transcription of IL2 receptor alpha.

SipA/PipB-associated transcriptional changes

To determine whether the effector proteins of TTSS1/2 are involved in modulating host immune responses, the expression levels of several chemokines, RhoA, and MHCIB2M in HD11 cells infected with SE100 (wild type), SE103 (*SipA*), or SE106 (*PipB*) were compared (Fig. 3). Data from RTPCR analyses indicated that SE103 and SE106 elicited higher levels of CCL4 (MIP1B), K60, and K203. SE106 showed a less inhibitory effect on the transcription of lymphotactin, compared to SE100. In sharp contrast to SE100, which suppressed the transcription of *RhoA*, SE103 and SE106 upregulated the expression of *RhoA*. Neither SE103 nor SE106 played a significant role in SE100-associated repression of *MHCIB2M.*

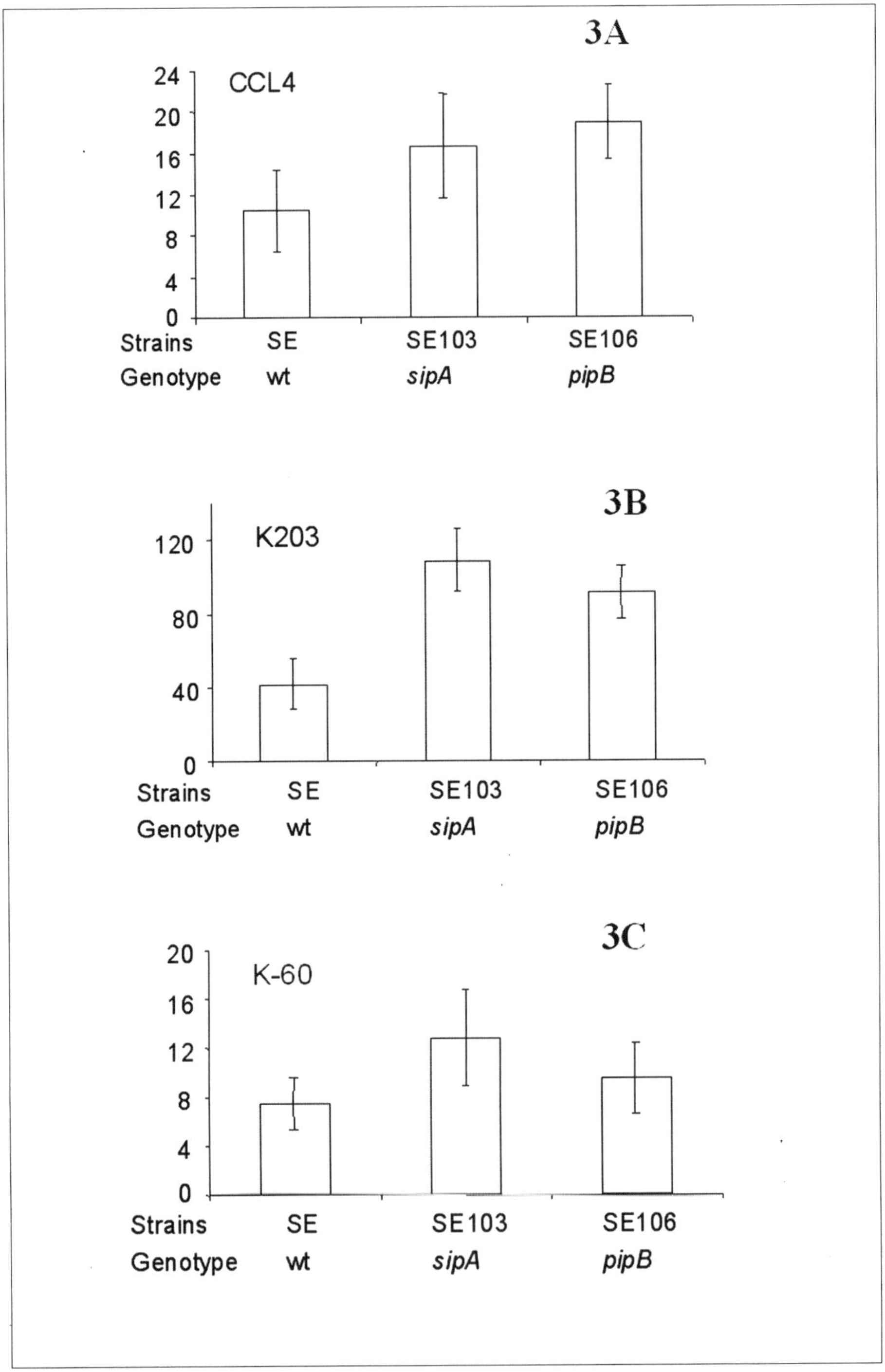
3A
CCL4
24
20
16
12
8
4
0
Strains
SE
SE103
SE106
Genotype
wt
sipA
pipB
3B
K203
120
80
40
0
Strains
SE
SE103
SE106
Genotype
wt
sipA
pipB
3C
K-60
20
16
12
8
4
0
Strains
SE
SE103
SE106
Genotype
wt
sipA
pipB

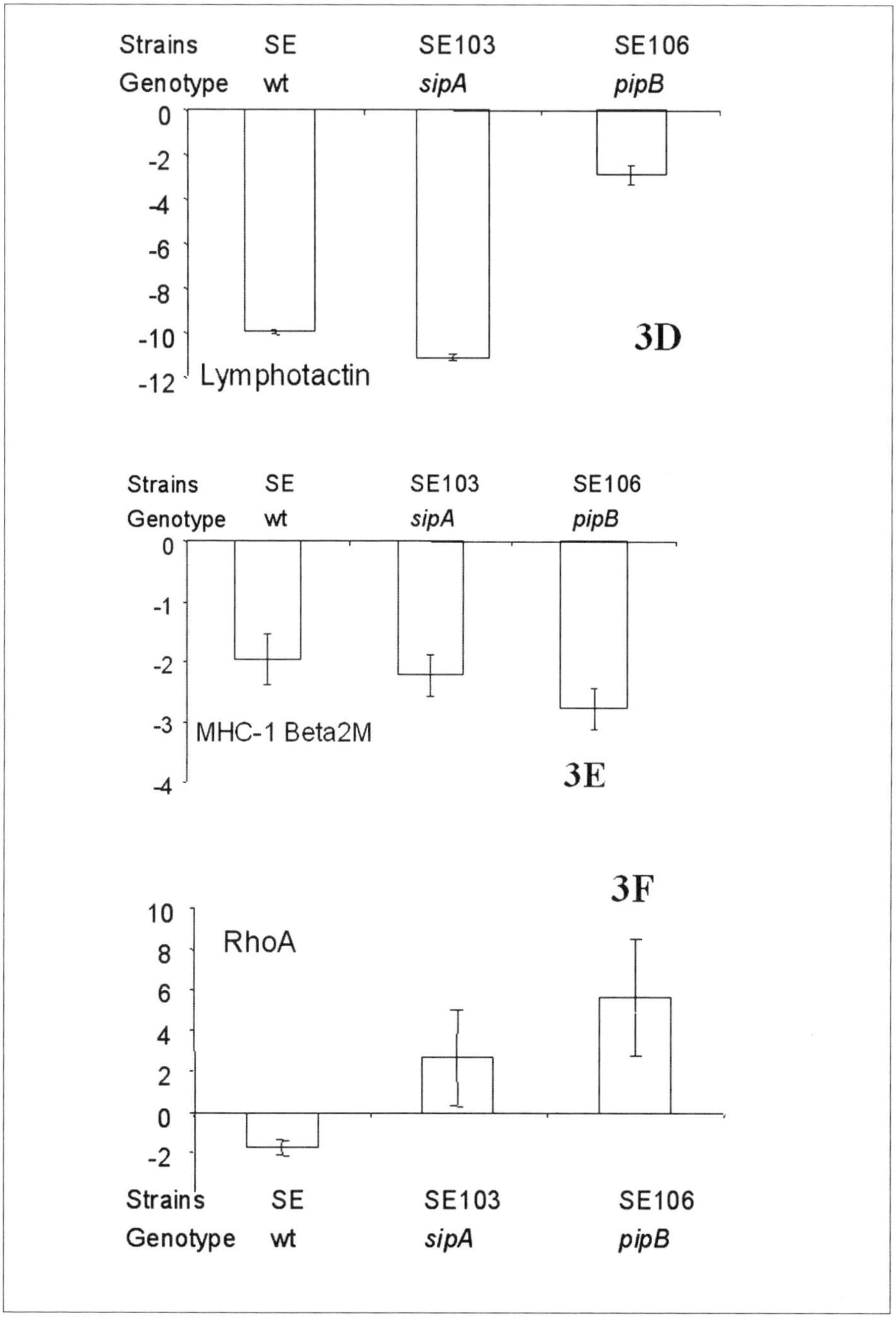

Fig. 3: RT-PCR analysis of immune gene transcription in HD11 following infection with SE100 (wt) and its isogenic TTSS1/2 mutant SE103 (SIPA) and SE106 (PIPB).

Discussion

The AMM was used to study the transcriptional response of chicken macrophage to SE infection. Although host response to SE was rapid and persisted throughout the course of the infection, most transcriptional changes occurred at 5 hpi, which is consistent with previous findings using mammalian cell models [9]. As expected, HD11 cells launched a proinflammatory response, as evidenced by the elevated expression of CC and CXC chemokines as well as cytokine IL1, which is critical to the clearance of infections. In this study, the genes involved in acquired immunity were transcriptionally modulated by the infection. Induction of IL6 and MHCIIBLB2 and repression of MHCIB2M and PSME1 (PA28A) suggest that SE promotes a humoral response over a cell-mediated response in chicken hosts.

It is noteworthy to point out that SE100 induced the transcription of *Cdc42*, but repressed *RhoA* and *Hras*. CDC42, RHOA, and HRAS are molecular switches in the regulation of diverse cellular functions, including signal transduction, cytoskeleton organization, and membrane trafficking. It is known that *Salmonella* TTSS1 effectors (SOPB and SOPE) activate CDC42, thereby facilitating actin rearrangement and bacterial internalization [13]. Data from this study provides first evidences that SE can also induce the transcription of CDC42. We also found that SE100 had an inhibitory effect on host cell proliferation, signal transduction, and transcription regulation. Although the exact biological consequences of such downregulation remain to be unravelled, it is likely that persistence of SE in the chicken host involves a mechanism of selective modulation of certain innate response genes by bacterial virulence factors.

In this study, SE103 (*SipA*) and SE106 (*PipB*) elicited higher transcription levels of certain chemokines and *RhoA*, suggesting a role for *SipA* and *PipB* in suppressing host innate responses. In conclusion, our data revealed that host genes belonging to diverse functional classes were transcriptionally modulated by SE. Further studies are necessary to gain insight into how TTSS1/2 transcriptionally modulates host cellular pathways.

References

1 CDC: Salmonella Surveillance: Annual Summary, 2005. Atlanta, Georgia: US Department of Health and Human Services, CDC, 2007.

2 FoodNet, CDC: *Salmonella Enteritidis* Update. Atlanta, Georgia: US Department of Health and Human Services, CDC, 2007.

3 Gast RK, Guraya R, Guard-Bouldin J, Holt PS, Moore RW: Colonization of specific regions of the reproductive tract and deposition at different locations inside eggs laid by hens infected with *Salmonella enteritidis or Salmonella heidelberg*. Avian Dis 2007;51:40-44.

4 Eckmann L, Kagnoff MF: Cytokines in host defense against *Salmonella*. Microbes Infect 2001;3:1191-200.

5 Brumell JH, Rosenberger CM, Gotto GT, Marcus SL, Finlay BB: SifA permits survival and replication of *Salmonella typhimurium* in murine macrophages. Cell Microbiol 2001;3:75-84.

6 Bliss TW, Dohms JE, Emara MG, Keeler CL, Jr: Gene expression profiling of avian macrophage activation. Vet Immunol Immunopathol 2005;105:289-299.

7 Zhang S, Santos RL, Tsolis RM, Stender S, Hardt WD, Baumler AJ, Adams LG: The *Salmonella enterica* serotype Typhimurium effector proteins SipA, SopA, SopB, SopD, and SopE2 act in concert to induce diarrhea in calves. Infect Immun 2002;70:3843-3855.

8 Dalloul RA, Bliss TW, Hong YH, Ben-Chouikha I, Park DW, Keeler CL, Lillehoj HS: Unique responses of the avian macrophage to different species of Eimeria. Mol Immunol 2007;44:558-566.

9 Zhang S, Adams LG, Nunes J, Khare S, Tsolis RM, Baumler AJ: Secreted effector proteins of *Salmonella enterica* serotype Typhimurium elicit host-specific chemokine profiles in animal models of typhoid fever and enterocolitis. Infect Immun 2003;71:4795-4803.

10 Broxmeyer HE: Regulation of hematopoiesis by chemokine family members. Int J Hematol 2001;74:9-17.

11 Sick C, Schneider K, Staeheli P, Weining KC: Novel chicken CXC and CC chemokines. Cytokine 2000;12:181-186.

12 Olson TS, Ley K: Chemokines and chemokine receptors in leukocyte trafficking. Am J Physiol Regul Integr Comp Physiol 2002:283: R7-R28.

13 Friebel A, Ilchmann H, Aepfelbacher M, Ehrbar K, Machleidt W, Hardt WD: SopE and SopE2 from *Salmonella typhimurium* activate different sets of RhoGTPases of the host cell. J Biol Chem 2001;276:34035-34040.

Dr. Shuping Zhang, DVM, PhD, Mississippi Veterinary Research and Diagnostic Laboratory, Department of Pathobiology and Population Medicine, College of Veterinary Medicine, Mississippi State University, PO Box 97813, Pearl, MS 39288, USA.

E-mail: zhang@cvm.msstate.edu

Pinard M-H, Gay C, Pastoret P-P, Dodet B (eds): Animal Genomics for Animal Health. Dev Biol (Basel). Basel, Karger, 2008, vol 132, pp 153-159.

Unique Co-expression of Immune Cell-related Genes in IBDV Resistant Chickens Indicates the Activation of Specific Cellular Host-response Mechanisms

E. Koren[1], H. Zhou[2], A. Cahaner[3], E.D. Heller[3], J. Pitcovski[1], S.J. Lamont[3]

1 MIGAL, Kiryat Shmona, Israel
2 Department of Animal Science, Iowa State University, Ames, IA, USA
3 Department of Animal Sciences, Faculty of Agricultural, Food and Environmental Quality Sciences, The Hebrew University of Jerusalem, Rehovot, Israel

Key words: infectious bursal disease virus, microarray, macrophage, gene, IL- 8, interferon-gamma

Abstract: Infectious bursal disease virus (IBDV) causes highly contagious, immunosuppressive disease that leads to high mortality in young chickens. The purpose of this study was to look for the genetic regulation of the immune acute immune response to IBDV in our selected lines. Chicks of a F2 generation of two lines divergently selected for early high (HH) or low (LL) antibody (Ab) response *to Escherichia coli* vaccination were challenged with virulent IBDV. Viral load in infected bursae was used to determine resistant (R) and susceptible (S) birds. By using a 13K chicken cDNA microarray, and pooled spleen mRNA of R, S and non-challenged, control (C) chicks, several genes were identified with differential expression associated with host resistance to IBDV. These genes were also subjected to RT-PCR on individual samples to verify the results obtained from microarrays. The major finding was the co-upregulation of seven genes – *ETS2, H963, RGS1, ABIN-2, CREM/ICER, DUSP1* and *CXCR4* – in several R, but not S or C individuals, and characterized by a high correlation of expression levels. Resistance also generally coincided with reduced transcript levels of acute-phase serum amyloid A (A-SAA) and increased levels of IL-8. Based on reported functions of these genes, these findings suggest that resistance was mediated by the activation of specific cellular mechanisms, indicated by increased activity of splenic macrophages and T-lymphocytes 3 days post-challenge.

Introduction

Infectious bursal disease virus (IBDV) causes highly contagious, immunosuppressive disease that leads to high mortality rates in chickens, mainly at the ages of 3 to 6 weeks. Mortality is high and the surviving chickens exhibit growth retardation and high susceptibility to other diseases because of the persistent immunosuppression.

The acute phase of the disease, in which bursal follicles are depleted of B cells and the bursa becomes atrophic, lasts for about 5-7 days. The peak of viral load in the bursa occurs at around 3-5 days post infection (PI).

The experimental population of this study is an F2 generation of two lines of chickens divergently selected for 16 generations for high (HH) or low (LL) antibody (Ab) response to *Escherichia coli* vaccination at 10 days of age. The Ab response in the HH line is also higher than in the LL line after vaccination with sheep red blood cells (SRBC) [1], Newcastle disease virus [2], and IBDV [3].

The goal of this research was to identify genes associated with resistance to IBDV. Gene-expression technologies were used to compare the two phenotypic groups: IBDV-resistant (R) and IBDV-susceptible (S) to each other and also to a non-challenged control group (C).

Materials and Methods

Experimental animals and IBDV challenge

The animal study was approved by the Ethics Committee of the Hebrew University of Jerusalem (Approval No. AG 20273-07). Male and female F2 chicks (N = 183) from the *E. coli*-selected lines were individually identified with wing-bands post-hatch, and grown under routine management conditions. One hundred and sixty chicks were divided into four groups with equal representation of genders and weights, and housed in four different isolators, for IBDV-challenge. The remaining 23 were grown under the same conditions in a separate isolator and served as non-challenged controls (C). At the age of 21 days, destined birds were challenged with KS strain of IBDV by 1 eye drop (approximately 10^5 EID_{50} - 50% egg infective dose per chick). At age 24 d (3 days post challenge), all chicks were weighed, and euthanized. Spleens and bursae were harvested.

Tissue processing

From spleens, a small slice (3-5 mm thick) was cut and immediately placed in a 1.5 ml-vial containing 1 ml of RNAlater buffer (Ambion). The tubes were kept 24 hours at 4°C and then stored at -20°C until RNA extraction. Whole bursae were taken from all birds, weighed and kept frozen (at -20°C) until processing.

ELISA for viral load

A capture-ELISA was performed, modified from the protocol described by Hudson et al., (2002) [4]. Briefly, ChαIBDV antiserum (described above) was diluted 500-fold in coating buffer (pH 9.6) and adsorbed to the plate overnight at room temperature – 100 µl per well. Then, 100 µl of 4% bovine haemoglobin (MP Biomedicals, Eschwege, Germany) solution in coating buffer was added to each well for 1 h of incubation at 37°C for blocking. Next, plates were washed with PBS containing 0.05% Tween 20, and 1:126 diluted bursae extractions were added. This dilution was obtained based on preliminary tests which yielded a standard curve with a linear range from 1:80 to 1:640 dilutions (8 µl of bursa lysate per 1 ml of dilution buffer – PBS containing 0.01% BSA). After 2 h of incubation at 37°C, plates were washed again. 100 µl/well of the primary antibody – goat anti VP2 (described above), was diluted 500-fold in 2% bovine haemoglobin solution in PBS, was added and incubated for 1 h. after incubation the plates were washed with PBS-Tween and secondary antibody – rabbit anti goat lgG- peroxidase - was added.. Plates were again incubated for 1 h, washed as previously and a substrate solution, o-phenylenediamine dihydrochloride (Sigma) was added. Colour was measured by ELISA READER (Lumitron) at OD_{450nm}.

Data analysis: All plates contained the same positive and negative controls. The positive control was the same bursa lysate used for challenge, diluted 100-fold. The negative control used for the correction was the bursa lysate from a SPF non-challenged bird. Substrate was added in intervals and read several times at different stages of color development. The data used for analysis was obtained from readings in which controls had similar values. The following formula was used for correction of the raw values:

$$\left(\frac{X_{ij} - N_j}{P_j - N_j}\right) \times 100$$

where X_{ij} is the raw O.D value for chicken i in plate j and N_j and P_j are the negative and positive controls (respectively) in that plate. This formula corrected for possible plate-bias by including the controls, and also corrected for differences in the substrate incubation times (between plates) and colour development. Next, the average score for the infected bursae in each plate was calculated (overall, 4 plates were used for infected samples) as well as the total average of all challenged birds. Finally, each score of challenged chicks was added or subtracted the difference between the average in its own plate and the overall average. The 20 chicks with the lowest scores were considered most resistant (R) and the 20 chicks with the highest scores were considered most susceptible (S). Basic statistical parameters were calculated for these two groups, as well as the control (C) group and the overall challenged group.

Microarray assay

Total RNA was extracted from preserved spleen tissues with TRIzol (Gibco BRL, Paisley, UK) according to manufacturer's instructions. Two µg of total RNA was amplified using MessageAmp II kit (Ambion, Inc., Austin, TX, U.S) to produce antisense-RNA (aRNA). Five µg of aRNA were then reverse-transcribed using SuperScript III (Invitrogen), incorporating aminoallyl-dUTP into the newly synthesized cDNA. After cleaning, each cDNA sample was coupled with either Cy3 or Cy5 dyes and then quenched.

Microarrays and hybridization

The arrays used in this study are "*Chicken 13K*" cDNA arrays acquired from the Fred Hutchinson Cancer Research Center (FHCRC) Seattle, WA, USA.

Scanning and quantification

Arrays were scanned using ScanArray 5000 and its corresponding software (PerkinElmer Life Sciences). Scanning resolution was 10 µm. Images were quantified using ImaGene 5 software (BioDiscoveries, Inc.).

Data analysis

Data were normalized using the locally weighted scatterplot smoothing (lowess) method, implemented in the statistical software package R [5].

Quantitative real-time PCR

mRNA levels were quantified by real-time PCR as described by Kaiser et al. [6] using QuantiTect™ SYBR® Green RT-PCR (Qiagen). The PCR mixture consisted of 50 ng/µl total RNA, 12.5 µl QuantiTect SYBR Green master mix, primers, 0.25 µl Quantitect RT mix and RNAse-free water to a final volume of 25 µl.

Results

ELISA for viral load

Adjusted scores for viral load varied from -10 to 133.5 (Fig. 1). The average score was 91.8 for challenged chicks and 1.0 for the non-challenged (p<0.01). Of the challenged birds, the 20 birds with the highest scores (116.5 – 133.5) were considered as IBDV-susceptible (S) and the 20 lowest scored birds (6.6 – 61.3) were considered as IBDV-resistant (R), and these birds were used for analysis of gene expression.

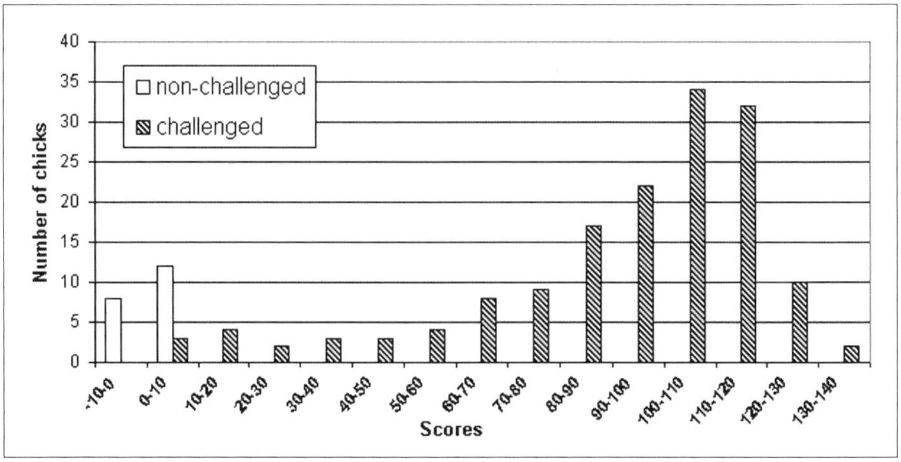

Fig. 1: Distribution of viral load scores among challenged and non-challenged chicks.

Microarray

A total of 219 genes were identified as differentially expressed. Eventually, 11 genes which had one specific hit within the full chicken genome sequence and a predicted protein, were selected to be analyzed by RT-PCR. Eight of these genes – coding for Ets2, H963, RGS1, ABIN-2, CREM/ICER, CACNG4, DUSP1 and CXCR4 – had similar microarray expression profiles: their transcription was higher in R chicks than either C or S chicks. The gene coding for SAA appeared to be down-regulated specifically in R birds and not in S birds. The gene coding for H5 was expressed at lower levels in R compared to both S and C, with lower levels in S compared to C. The gene coding for BCHE was expressed in significantly higher levels in R birds than S birds, with intermediate levels in C birds.

Microarray measurements were performed on RNA pools, and therefore RT-PCR measurements were conducted on individual samples from the same birds, to verify previous results and to provide additional insights on individual variation in gene expression.

Quantitative RT-PCR

Expression levels of 11 genes were measured using RT-PCR in 26 birds (8 from C group, 9 from R, and 9 from S). Levels of *IL*-8 mRNA were also quantified since it is known to be up-regulated in response to IBDV challenge in splenic macrophages as well as *IFN-gamma*, which is believed to play a role in that response.

Seven genes – coding for 2Ets, H963, RGS1, ABIN-2, CREM/ICER, DUSP1 and CXCR4 – were very similar in their mRNA expression variation profiles: variance of R-group chicks was higher than that of S chicks (p<0.01) for all genes, and higher than C group chicks for all but two, H963 (p= 0.074) and Ets2 (p=0.106) (Table 1).

SAA had significantly lower variance in expression levels in R chicks compared both to S and C groups (p< 0.01). H5 was significantly lower in expression in challenged birds (R and S) than non-challenged (C) (p<0.01).

IL-8 had high correlations with all seven co-expressed genes in the R-group chicks (average R^2 = 0.489). The mRNA expression levels of *IFN-G* were significantly higher in the S group in comparison to C group (p= 0.03) with R chicks exhibiting intermediate levels not different from S and C.

Table 1: Correlations between the seven upregulated genes among R group chicks (Table 1**a**) and among S and C groups chicks (Table 1**b**). Values above the diagonal line are R^2 and values below the diagonal line are the 2-tailed significance for the relevant R^2.

Table 1a

	H963	**ABIN-2**	**RGS1**	**ICER***	**DUSPI**	**ETS2**	**CXCR4**
H963		0.938	0.936	0.920	0.758	0.954	0.975
ABIN-2	<0.01		0.912	0.959	0.738	0.860	0.954
RGS1	<0.01	<0.01		0.967	0.914	0.921	0.922
ICER*	<0.01	<0.01	<0.01		0.855	0.895	0.933
DUSPI	0.002	0.003	<0.01	<0.01		0.737	0.709
ETS2	<0.01	<0.01	<0.01	<0.01	0.003		0.959
CXCR4	<0.01	<0.01	<0.01	<0.01	0.004	<0.01	

Table 1b

	H963	**ABIN-2**	**RGS1**	**ICER***	**DUSPI**	**ETS2**	**CXCR4**
H963		0.116	0.696	0. 139	0.115	0.378	0.381
ABIN-2	0.182		0.136	0.076	0.221	0.425	0.299
RGS1	<0.01	0.146		0.405	0.061	0.304	0.294
ICER*	0.141	0.284	0.006		0.153	0.015	0.002
DUSPI	0.183	0.057	0.339	0.120		0.215	0.121
ETS2	0.009	0.005	0.022	0.644	0.061		0.697
CXCR4	0.008	0.023	0.025	0.866	0.169	<0.01	

Discussion

By using a 13 K chicken cDNA microarray, and spleens from an IBDV challenge experiment, several genes with differential expression associated with host resistance to IBDV were identified. These genes were also subjected to RT-PCR on individual samples, which generally verified the results obtained from pooled samples on microarrays. A striking finding was the co-upregulation of seven genes - *2Est, H963, RGS1, ABIN-2, CREM/ICER, DUSP1* and *CXCR4* – in several resistant, but no susceptible or control individuals. Reported functions of these genes are described below. Most of the co-upregulated genes are members of an integrated network of cell-mediated immunity. Therefore, the identification of these genes gives insight into the genetic pathways and the gene interactions that contribute to host resistance to IBDV. Based upon the described functions of these genes, we propose the mechanistic gene network presented in Figure 2.

Measurements were obtained 3 days PC because this is a known peak of IBDV's acute phase [7-10]; therefore, resistance at that point of time was likely induced by innate responses rather than acquired ones. The reported functions of the co-expressed genes suggest that the resistance mechanisms may involve the intense formation of GCs in the spleen, specifically inducing macrophage survival and differentiation, as well as T cell activation, maturation and proliferation. The intense splenic activity of these cell types may occur partially on account of B lymphocyte activity, based on the microarray results of Ig light-chain probes. In addition, the acute phase chemo-attractant SAA was down-regulated, possibly through the inhibition of the pro-inflammatory cytokines IL-6, IL-1β and TNF-α and/or the inhibition of the transcription factor NF-κB. The downregulation of SAA may create a gradient of inflammatory signals that leads to the migration of activated lymphocytes, primarily T cells, and macrophages, from the GCs in the spleen towards the bursa [11-13]. It is likely that the main factor, which mediated resistance in the R chicks that exhibited the co-upregulation of these genes, was the rapid initiation of cellular processes. These may also involve the transition of the cellular response from primarily being macrophage-mediated to T-cell-mediated.

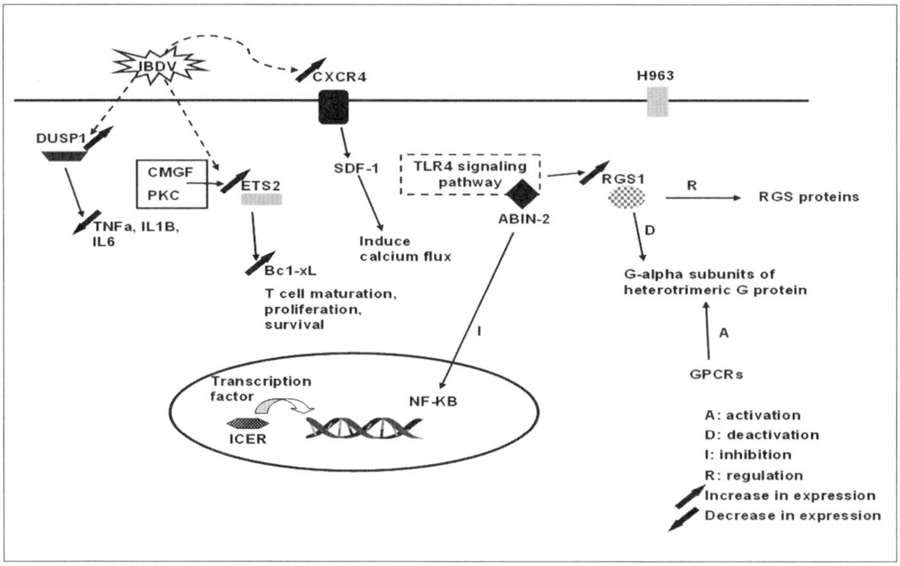

Fig. 2: Proposed gene network three days post challenge with IBDV in our chicken population.

References

1 Heller ED, Leitner G, Friedman A, Uni Z, Gutman M, Cahaner A: Immunological parameters in meat-type chicken lines divergently selected by antibody response to *Escherichia coli* vaccination. Vet Immunol Immunopathol 1992;34:159-172.

2 Leitner G, Uni Z, Cahaner A, Gutman M, Heller ED: Replicated divergent selection of broiler chickens for high or low early antibody response to *Escherichia coli* vaccination. Poultry Sci 1992;71:27-37.

3 Pitcovski J, Cahaner A, Heller ED, Zouri Z, Gutter B, Gotfrried Y, Leitner G: Immune response and resistance to infectious bursal disease virus of chicken lines selected for high or low antibody response to *Escherichia coli*. Poultry Sci 2001;80:879-884.

4 Hudson J, Hoerr CF J, Parker SH, Ewald SJ: Quantitative measures of disease in broiler breeder chicks of different major histocompatibiliy complex genotypes after challenge with infectious bursal disease virus. Avian Dis 2002;46:581-592.

5 Yang Y, Dudoit HS, Luu P, Lin DM, Peng V, Ngai J, Speed TP: Normalization for cDNA microarray data: a robust composite method addressing single and multiple slide systematic variation. Nucleic Acids Res 2002;30:e15-19.

6 Kaiser P, Hughes S, Bumstead N: The chicken 9E3/CEF4 CXC chemokine is the avian orthologue of IL8 and maps to chicken chromosome 4 syntenic with genes flanking the mammalian chemokine cluster. Immunogenetics 1999;49:673-684.

7 Tanimura N, Sharma JM: Significance of bursa of Fabricius as target organ in infectious bursal disease of chickens. Infect Immun 1998;27:364-367.

8 Sharma JM, Kim IJ, Rautenschlein S, Yeh HY: Infectious bursal disease virus of chickens: pathogenesis and immunosuppression. Dev Comp Immunol 2000;24:223-235.

9 Kim IJ, Karaca K, Pertile TL, Erickson SA, Sharma JM: Enhanced expression of cytokine genes in spleen macrophages during acute infection with infectious bursal disease virus in chickens. Vet Immunol Immunopathol 1998;61:331-341.

10 Kim IJ, You SK, Kim H, Yeh HY, Sharma JM: Characteristics of bursal T lymphocytes induced by infectious bursal disease virus. J Virol 2000;74:8884-8892.

11 Hirai K, Funakoshi T, Nakai T, Shimakura S: Sequential changes in the number of surface immunoglobulin-bearing B lymphocytes in infectious bursal disease virus-infected chickens. Avian Dis 1981;25:484-496.

12 Rautenschlein S, Yeh HY, Njenga MK, Sharma JM: Role of intrabursal T cells in infectious bursal disease virus (IBDV) infection: T cells promote viral clearance but delay follicular recovery. Arch Virol 2002;147:285-304.

13 Jungmann A, Nieper H, Muller H: Apoptosis is induced by infectious bursal disease virus replication in productively infected cells as well as in antigen-negative cells in their vicinity. J Gen Virol 2001;82:1107-1115.

Dr. E. Dan Heller, Department of Animal Sciences, Faculty of Agricultural, Food and Environmental Quality Sciences, The Hebrew University of Jerusalem, Rehovot 76100, Israel.

E-mail: dheller@agri.huji.ac.il

Pinard M-H, Gay C, Pastoret P-P, Dodet B (eds): Animal Genomics for Animal Health. Dev Biol (Basel). Basel, Karger, 2008, vol 132, pp 161-167.

Physiological Interactions between the Endocrine and Immune Systems Shown in Gene Analysis in Pigs

C. Phatsara[1], E. Jonas[1], S. Yammuen-art[1], H. Buschbell[1], D. Tesfaye[1], S. Ponsuksili[2], E. Tholen[1], H. Juengst[1], K. Schellander[1], K. Wimmers[2]

[1] Institute of Animal Science, University of Bonn, Germany
[2] Research Institute for the Biology of Farm Animals, Dummerstorf, Germany

Keywords: DUMI resource population, haemolytic complement activity, growth hormone axis

Abstract: A number of observations support the concept of important physiological interactions between the endocrine and immune systems. It could be confirmed that hormones secreted by the neuroendocrine system play an important role in communication and regulation of the cells of the immune system. Among protein hormones, this has been most clearly documented for prolactin (PRL), growth hormone (GH), and insulin-like growth factor 1 (IGF-1). A number of traits of the immune response in a Duroc x Berlin Miniature pig family (DUMI) were examined. The haemolytic complement activity in the classical complement pathway (CH50) and the alternative pathway (AH50) was examined at eight different time points in pigs that were vaccinated with different vaccines. Single nucleotide polymorphism (SNP) genotyping was employed to genotype the DUMI F2 animal for growth hormone (*Gh*), growth hormone releasing hormone (*Ghrh*), pituitary-specific transcription factor 1 (*Pit1*), and prolactin receptor (*Prlr*) loci, and also a microsatellite within insulin-like growth factor 1 (*Igf1*). Using a family-based association test (FBAT) program, a highly significant association of *Gh, Pit1*, and *Prlr* to AH50 ($p \leq 0.01$) and a significant association of *Pit1* to CH50 ($p \leq 0.05$) were found. Using the SAS system for mixed model, a highly significant association of Gh, and *Igf1* to AH50 and CH50 ($p \leq 0.01$) was detected, while *Prlr* and *Ghrh* had a highly significant association ($p \leq 0.01$) with CH50 only.

Introduction

The pituitary gland plays a key role in the regulation of growth, differentiation and function of all cells in the body, including the immunocytes. Immune reactions are generated through the proliferation of antigen-specific lymphocyte clones. Hormones like growth hormone (GH) enable lymphocytes to respond to antigen, which is delivered as an adherence signal in the context of the major histocompatibility surface molecules of antigen-presenting cells. Hormones secreted or regulated by the pituitary gland regulate every level of immune activity, including the competence of lymphocytes in responding to immune/inflammatory stimuli, signal transduction, gene activation, the production and activity of cytokines and other immune effector functions [1]. Insulin-like growth factor 1 (IGF1) is a member of the IGFs, a peptide family that plays an important role in mammalian growth and development. Insulin-like growth factor coupled proteins (IGFBP) bind with high affinity to both IGF1 and IGF2. These as well as the IGF receptor are necessary for the normal growth and normal development of the organism. Gene deletion of these elements shows how important IGF1 is for prenatal and postnatal development, whereas IGF2 is only used within the prenatal stage [2]. IGF1 is strongly connected to growth hormone and mediates many growth-promoting effects.

In recent years, many efforts have been undertaken to elucidate the complex interactions between mediators of the endocrine system and the immune system. The main effector of GH is IGF1, an endocrine mediator of growth and development under physiological conditions. Besides this important function, IGF1 also plays a prominent role in the regulation of immunity and inflammation. IGF1 is involved in innate as well as acquired immunity and host defence [3].

The binding of GH to the stoma cells induces the secretion of IGF1. In different studies, the influence of the GH on mannan-binding lectin (MBL) levels was summarized as well as the possible clinical consequences of this new link between the endocrine and the immune system [4]. The release of GH in the pituitary gland is stimulated by growth hormone releasing hormone (GHRH) from the hypothalamus through activation of a specific G-protein coupled receptor, the GHRH receptor (GHRHR). GHRHR expression in pig is regulated by a number of signals, which are different in a number of species [5].

The pituitary-specific transcription factor 1 (PIT1) regulates, together with the paired-like homeodomain transcription factor (PROP1), the differentiation of pituitary cells into somatotrophs, which synthesize and release GH. It is also known that growth hormone binds to the prolactin (PRLR), this being the basis of the induction of lactation by growth hormone. Also the PRLR is required for the development of mature lymphocytes and for the maintenance of immunocompetence [1]. GH and PRL influence granulopoiesis, erythropoiesis and immune function in vitro and in vivo in animal models, as mentioned in the review by Velkeniers et al. [6].

The results of the interaction between the endocrine and immune systems in different studies, together with our previous investigations of the *Mbl1* and *Mbl2* genes including additional quantitative trait loci (QTL) studies in two pig populations [7,8], led to the study of the association of polymorphisms within the *Igf1, Gh, Ghrh, Prlr* and *Pit1* genes to innate immunity traits in the pig.

Materials and Methods

This study included 115 F2 animals from a reciprocal cross between Duroc and Berlin Miniature pigs (DUMI population) and their parents comprising 9 half-sib families derived from 9 sows and one boar. Animals were reared at the Frankenforst Institute of Animal Science, University of Bonn research farm. Pigs were immunized with Mycoplasma, Aujeszky's and PRRS virus vaccines at different ages. Blood samples were taken pre- and post-immunization. In vivo haemolytic complement activity was measured in the classical and the alternative pathways. For classical complement activity (CH50), sensitized sheep red blood cells

(SRBC) were used to measure the haemolytic complement activity, and substituted by rabbit red blood cells for alternative complement activity (AH50). Haemolytic activity was expressed as the titre that lysed 50% of the erythrocytes (CH50 U/ml, AH50 U/ml) [7,8].

A short tandem repeat (STR) of *Igf1*, and SNPs of *Gh, Ghrh, Pit1* and *Prlr* reported previously [9-14], were used for genotyping and association study. The STR of the *Igf1* gene (GenBank Acc.No. X64400) [9] was amplified by PCR reaction. Electrophoresis was performed with the LiCor Sequencer System (LI-COR model 4200 automated DNA sequencer) using 12% SequaGel in 1×TBE buffer at 1500 V, 50 mA, 50 W and 50°C temperature. Sequence data was analyzed by OneDScan, version 4.10, software (Li-COR Biotechnology). A pair of primers designed by Nielsen and Larsen [10] was used to genotype the *Gh* gene (GenBank Acc.No. M17704). PCR was performed using 50 ng of genomic DNA, 50 μM of each dNTP, 2 ng of each primer, 0.5 U of Taq polymerase, and 1.5 mM of MgCl2 in a final volume of 25 μl (as a standard PCR protocol). After amplification of each reaction, 5 μl of the PCR product (605 bp) was digested with ApaI restriction enzyme. Samples were incubated overnight at 37°C. Fragments were visualized in 2.5% agarose gels, stained with ethidium bromide and photographed under UV light. The *Ghrh* genotype was investigated using the PCR-RFLP method. Primers designed by Moody et al. [11] in bovines and by Baskin and Pomp [12] in the pig were used (GenBank Acc.No. U90275). A 455 bp fragment spanning exon 3 of the *Ghrh* locus was amplified using standard PCR protocol as describe above. The PCR product was digested using AluI restriction enzyme. The *Pit1* gene was genotyped by the PCR-RFLP method. Primers used for *Pit1* amplification were as previously described by Yu et al. [13] (GenBank Acc.No. U00793). After PCR amplification, 5 μl of the PCR amplicon were digested using 2 U of RsaI restriction enzyme by incubation overnight at 37°C, and the bands were visualized on 2.5% agarose gels stained with ethidium bromide. Primers for genotyping *Prlr* were adapted from Vincent et al. [14] (GenBank Acc.No. U96306). PCR amplification was performed using the standard PCR protocol. PCR-RFLP was performed by digestion of the 457 bp fragment using AluI enzyme.

The family-based association test (FBAT, ver1.4) (http://www.biostat.harvard.edu/ ~fbat/fbat.htm) was used to perform the association analysis of the genotypes determining the values of CH50 and AH50 at different time points. Further analysis was also performed using a mixed model of the Statistical Analysis Software (SAS) program (version 9.1). The model used is described below:

$$Y_{ijkl} = \mu + sex_i + genotype_j + time_k + genotype_j{*}time_k + ANIMAL_l + e_{ijkl}$$

where:

Y_{ijkl} = phenotype (AH05 and CH50, respectively);

μ = population average;

sex_i = fixed effect of the sex (male or female);

$genotype_j$ = fixed effect of the genotype (allele1*allele2);

$time_k$ = effect of the repeated measurement of Y;

$ANIMAL_l$ = individual effect of the animal and

e_{ijkl} = residual error.

Results

Allele and genotype frequencies

Different alleles and genotypes of *GH, IGF1, GHRH*, and *PRLR* were observed that segregated into 9 half-sib families for a total of 115 animals. Three different genotypes (AA, AB and BB) were found segregating in the studied animals for *GH* and *Ghrh* polymorphisms, while only one homozygous and one heterozygous genotype (AA and BB) were found in these animals for *Prlr* and *Pit1* polymorphisms. For *Igf1* microsatellite genotypes, 5 different alleles (i.e. 222, 224, 231, 233 and 236) were detected, generating 12 genotype combinations and segregating in the studied animals. Almost all F1 animals were found to be heterozygous for *Gh, Igf1* and *Ghrh*, but not for *Prlr* and *Pit1*genes.

Association analysis

Using a family-based association test (FBAT), a highly significant linkage and association of AH50 with *Gh, Pit1, Prlr* ($p \leq 0.01$) were found, and a significant association of CH50 with *Pit1* ($p \leq 0.05$) was also found. Using SAS for mixed models, significant results were found for the association of *Igf1* and *Gh* with AH50 (Table 1). With the same model, significant results were observed for the association of *Igf1, Gh, Ghrh* and *Prlr* with CH50 (Table 1). There was no interaction between genotype and time-point observed in the analysis. Least square means and standard errors (LSM±SE) for the genes significantly associated with the complement activity traits (AH50 and CH50) are shown in Table 2. The allele combination 231/231 for the *Igf1* microsatellite had the lowest AH50 value, while the combination 224/231 had the highest. For *Gh* polymorphism, allele B had the highest AH50 value. In relation to CH50, it can be clearly seen that animals with genotype BB of the *Ghrh* polymorphism had a higher CH50 value than other genotypes. For the *Igf1* gene, animals with the 233/233 genotype had a higher CH50 value than other genotype combination, while the 224/231 combination had the lowest CH50 value. Among *Gh* polymorphisms, the heterozygous animals (AB) had a higher CH50 value than homozygous animals (AA and BB). It was also found that *Prlr* genotype BB led to a higher CH50 value when compared to another genotype found in the population.

Table 1: Analysis of variance of haemolytic complement activity (AH50, CH50) and gene polymorphisms.

Gene/ locus	Trait	Sex	Time	Genotype	Animal
***Igf1*/ STR**	AH50	n.s.	**	**	n.s.
	CH50	*	**	**	n.s.
***Gh*/ ApaI**	AH50	**	**	**	**
	CH50	**	**	**	**
***Ghr*h/ AluI**	AH50	n.s.	**	n.s.	**
	CH50	n.s.	**	**	**
***Pit1*/ RsaI**	AH50	n.s.	**	n.s.	n.s.
	CH50	n.s.	**	n.s.	**
***Prlr*/ AluI**	AH50	*	**	n.s.	**
	CH50	*	**	**	**

(n.s.) not significant
(*) significant with $p \leq 0.05$
(**) significant with $p \leq 0.01$

Table 2: Least square means (LSM) and standard errors (SE) of haemolytic complement activity (AH50 and CH50) associated with different gene polymorphisms.

Gene/ locus	Genotype	Trait	
		AH50	**CH50**
***Igf1*/ STR**	222/222	45.36 ± 4.59c,d,e	66.32 ± 6.71a,b
	222/224	66.07 ± 2.62a,b	45.86 ± 5.83d
	222/231	42.89 ± 3.99d,e	54.95 ± 6.50 b,c,d
	222/233	65.10 ± 2.77a,b	48.81 ± 6.35 c,d
	224/224	61.96 ± 1.64a,b,c	50.08 ± 5.80 b,c,d
	224/231	77.98 ± 4.38a	43.47 ± 7.55 d
	224/233	58.46 ± 1.26b,c,d	57.22 ± 5.67 b
	224/236	57.01 ± 2.34b,c,d,e	46.89 ± 6.06 d
	231/231	39.10 ± 4.42e	61.93 ± 6.66 a,b
	231/233	64.37 ± 4.21a,b,c	51.14 ± 7.43 b,c,d
	233/233	56.25 ± 1.43b,c,d,e	67.03 ± 5.74 a
	233/236	57.56 ± 2.42b,c,d,e	44.31 ± 6.09 d
***Gh*/ ApaI**	AA	40.14 ± 5.74b	20.71 ± 7.95c
	AB	61.72 ± 1.55a	68.46 ± 2.19a
	AB	66.15 ± 1.76a	58.85 ± 2.53b
***Ghrh*/ AluI**	AA	65.30 ± 2.06	55.33 ± 2.68b
	AB	61.83 ± 2.14	43.57 ± 2.09c
	BB	61.17 ± 1.67	76.15 ± 2.10a
***Pit1*/ RsaI**	AA	62.72 ± 0.81	62.84 ± 0.81
	AB	62.38 ± 2.73	60.85 ± 2.80
***Prlr*/ AluI**	AA	46.39 ± 12.45	28.55 ± 14.23b
	AB	61.33 ± 1.08	60.55 ± 1.36a

a,b,c,d,e indicate significant differences ($p \leq 0.01$)

Discussion

Linkage studies in the DUMI populations (F2 and backcross) revealed a number of QTL detected on different chromosomes [8,15]. On pig chromosome 5 (SSC5), QTL for different traits of the immune response were detected on the p-arm of the chromosome, but not in the position of the *Igf1* gene (the microsatellite within this gene was used in the linkage study). Different QTL were detected in the chromosomal region of *Gh* on SSC12 in the F2 population for AH50. This might be evidence that the *Gh* locus might be linked to an interesting gene, although it may not be the dominant gene affecting the traits. The most interesting results of the QTL studies were found on chromosome 17, where the *Ghr*h gene is located in pig.

These QTL were detected in similar regions as some of the gene locations, and most of the genes are described as playing an important role for the immune system. It was demonstrated that the treatment with GH, but not IGF1, significantly increases the MBL concentrations, which is known to play an important role in the immune response [4]. Hansen et al. [4] also suggest that the clinical consequences of this new link between the endocrine and the immune system remain to be elucidated. Long-standing hyper- and hyposecretion of PRL and GH does not induce any significant alteration of the immune system [6]; however, the results of the study from Velkeniers et al. [6] suggest that a greater understanding of the interactions between the endocrine and immune system may serve to benefit patients.

Conclusion

The results revealed association and/or linkage of several genes of the endocrine system with traits of the immune system. Further studies are needed to analyze possible linkage disequilibrium between relevant causative genes, pleiotropic effects, and the functional basis of links between the endocrine and the immune system. The results presented here promote studies on interactions between the endocrine and immune systems.

References

1 Berczi I: Pituitary hormones and immune function. Acta Paediatrica 1997:86:70-75.

2 Butler AA, Le Roith D: Minireview: Tissue-specific versus generalized gene targeting of the igf1 and igf1r genes and their roles in insulin-like growth factor physiology. Endocrinology 2001:142:1685-1688.

3 Heemskerk VH, Daemen M, Buurman WA. Insulin-like growth factor-1 (igf-1) and growth hormone (gh) in immunity and inflammation. Cytokine Growth Factor Rev 1999:10:5-14.

4 Hansen TK, Thiel S, Dall R, Rosenfalck AM, Trainer P, Flyvbjerg A, Jorgensen JOL, Christiansen JS: Gh strongly affects serum concentrations of mannan-binding lectin: Evidence for a new igf-i independent immunomodulatory effect of gh. J Clin Endocrinol Metab 2001:86:5383-5388.

5 Luque RM, Kineman RD, Park SJ, Peng XD, Gracia-Navarro G, Castano JP, Malagon MM: Homologous and heterologous regulation of pituitary receptors for ghrelin and growth hormone-releasing hormone. Endocrinology 2004:145:3182-3189.

6 Velkeniers B, Dogusan Z, Naessens F, Hooghe R, Hooghe-Peters EL. Prolactin, growth hormone and the immune system in humans. Cell Mol Life Sci 1998:54:1102-1108.

7 Phatsara C, Jennen DGJ, Ponsuksili S, Murani E, Tesfaye D, Schellander K, Wimmers K: Molecular genetic analysis of porcine mannose-binding lectin genes, mbl1 and mbl2, and their association with complement activity. Int J Immunogenet 2007:34:55-63.

8 Wimmers K: Kandidatengen- und QTL-Analyse für Merkmale der humoralen Immunabwehr beim Schwein. Habilitationschrift Rheinischen Friedrich-Wilhelms-Universität Bonn 2002.

9 Kirkpatrick BW: Identification of a conserved microsatellite site in the porcine and bovine insulin-like growth factor-i gene 5' flank. Anim Genet 1992:23:543-548.

10 Nielsen VH, Larsen NJ: Restriction-fragment-length-polymorphisms at the growth-hormone gene in pigs. Anim Genet 1991:22:291-294.

11 Moody D, Pomp D, Barendse W: Rapid communication: Restriction fragment length polymorphism in amplification products of the bovine growth hormone-releasing hormone gene. J Anim Sci 1995:73:3789.

12 Baskin LC, Pomp D: Rapid communication: Restriction fragment length polymorphism in amplification products of the porcine growth hormone-releasing hormone gene. J Anim Sci 1997:75:2285.

13 Yu TP, Schmitz CB, Rothschild MF, Tuggle CK: Expression pattern, genomic cloning and rflp analyses of the swine pit-1 gene. Anim Genet 1994:25:229-233.

14 Vincent AL, Wang L, Tuggle CK, Robic A, Rothschild MF: Prolactin receptor maps to pig chromosome 16. Mamm Genome 1997:8:793-794.

15 Phatsara C: Analyses of immune competence traits and their association with candidate genes in pigs. Dissertation Rheinische Friedrich-Wilhelms-Universität Bonn 2007.

Dr. Klaus Wimmers, Research Institute for the Biology of Farm Animals, Wilhelm-Stahl-Allee 2, 18196 Dummerstorf, Germany.

E-mail: wimmers@fbn-dummerstorf.de

Pinard M-H, Gay C, Pastoret P-P, Dodet B (eds): Animal Genomics for Animal Health. Dev Biol (Basel). Basel, Karger, 2008, vol 132, pp 169-174.

Effect of Porcine Reproductive and Respiratory Syndrome Virus on Porcine Alveolar Macrophage Function as Determined Using Serial Analysis of Gene Expression (SAGE)

L.C. Miller[1], G.P. Harhay[2], K.M. Lager[1], T.P.L. Smith[2], J.D. Neill[1]

1 National Animal Disease Center-USDA-ARS, Ames, Iowa, USA
2 United States Meat Animal Research Center-USDA-ARS, Clay Center, Nebraska, USA

Key words: SAGE, PRRSV, alveolar macrophages

Abstract: Porcine reproductive and respiratory syndrome virus (PRRSV) is a major pathogen of swine worldwide and causes considerable economic loss. The main target of infection is the porcine alveolar macrophage (PAM). Infection of PAMs by PRRSV causes significant changes in their function by mechanisms that are not understood. We have employed Serial Analysis of Gene Expression (SAGE) to examine the global expression of genes in PRRSV-infected PAMs. Total cellular RNA was prepared from in vitro mock-infected and PRRSV strain VR-2332-infected PAMs at 0, 6, 12, 16 and 24 hours after infection, and subjected to SAGE analysis to obtain >100,000 tags per time point. These sequences were processed to account for sequencing error before generating tag:count lists. These lists were deposited into a modified Identitag database for mapping to porcine and PRRSV genes. Identified unique mRNA tags were analyzed for their identity and relative abundance. Examination of the SAGE data indicated that there were changes in gene expression occurring in the PRRSV-infected PAMs over time post-infection. More than 400 unique tags with significantly altered expression levels were identified ($p< 0.01$ with Bonferroni correction). The validity and kinetics of expression of SAGE identified genes were evaluated using real-time RT-PCR.

INTRODUCTION

Porcine reproductive and respiratory syndrome virus (PRRSV), the causative agent of porcine reproductive and respiratory syndrome (PRRS) in swine, is a member of the *Arteriviridae* family in the order *Nidovirales*. PRRSV causes highly significant losses to the swine industry worldwide [1] as a result of both reproductive failure (late-term abortions and stillbirths) in pregnant sows and respiratory disease (pneumonia) in nursery and grower/finishing pigs [2]. The primary cellular target of PRRSV is the alveolar macrophage of the lung. Thus, porcine alveolar macrophages (PAMs) have a central role in PRRSV infections. Identifying macrophage function specific pathways affected by PRRSV replication can lead to novel gene targets for the control of PRRSV infection. Serial Analysis of Gene Expression (SAGE) is a powerful technique that allows for one to obtain detailed and profound quantitative and qualitative information regarding the gene expression profile, without previous knowledge of the sequence of genes analyzed [3]. SAGE was used to obtain a global picture of gene expression in normal and PRRSV infected PAMs over 24 hours post-infection.

Materials and Methods

Primary PAMs were isolated, cultured, and infected, as previously described [4]. Briefly, PAMs were harvested from clinically healthy, PRRS-negative gilts 6-8 weeks of age. Animals were humanely euthanized, following animal care and use protocols, and PAMs harvested under aseptic conditions. Viability of PAMs was determined by trypan blue dye exclusion and cells were tested by PCR for porcine circovirus and *Mycoplasma* spp. [5,6]. Aliquots of PAMs were frozen and stored in liquid nitrogen until used. Typical yields were 10^8-10^9 PAMs of >95% viability. Immediately prior to use, PAMs were thawed and cultured at 37°C, 5% CO2 in Dulbecco's Modified Eagles Media with 5% foetal bovine serum (FBS; Gibco-Invitrogen, Carlsbad, CA) and 1% antibiotic/antimyotic (Gibco-Invitrogen) for 2 hours. PRRSV strain VR-2332 [7] stock was propagated in MARC-145 cells [8]. To achieve near synchronization of cell infection, flasks containing adherent PAMs were infected at a multiplicity of infection (MOI) of 10 in chilled media and permitted to adhere at 4°C for 1 hour to allow for virus binding, but not entry into the cell. After 1 hour, pre-warmed media was added and the cells placed at 37°C, 5% CO_2 with humidification.

Total cellular RNA was prepared from *in vitro* mock-infected and PRRSV strain VR-2332-infected PAMs at 0, 6, 12, 16 and 24 hours after infection, and subjected to SAGE analysis [3]. The mRNA was converted to double stranded cDNA with the inclusion of the primer biotin-5'-T_{18}-3' using the Thermostable RT-PCR System (Invitrogen, Carlsbad, CA). The cDNA was then cleaved with *Nla*III and the 3'-terminal cDNA fragments were bound and purified using streptavidin-coated magnetic beads (Invitrogen, Carlsbad, CA). Oligonucleotides containing recognition sites for *Bsm*FI were ligated onto the cDNA fragments and the linked cDNA was released from the beads by digestion with *Bsm*FI (*Bsm*FI cuts DNA 14 bp 3' of its recognition sequence, thus generating unique sequence tags of 10 bp in length). The released tags were ligated to one another, concatomerized and cloned into the *Sph*I site of the vector pZero 1.0 (Invitrogen, Carlsbad, CA). Colonies were screened by PCR using the M13 forward and reverse primers to identify clones containing inserts greater than 300 bp (>25 tags) in length. Clones were amplified and sequenced using our high-throughput sequencing pipeline with an ABI 3730 automated sequencer and ABI chemistry (Applied Biosystems Inc., Foster City, CA) at USMARC.

Validation of the results and corroboration of the altered gene expression levels were analyzed by real-time reverse transcription-PCR (RT-PCR). Real-time RT-PCR was done in 25 µl reaction volumes using the Invitrogen SYBR green qRT-PCR reaction mix (Invitrogen, Inc., Carlsbad, CA) per supplier's specifications. Primer sequences are available upon request. All primers were used at 100 nM final concentration. PCR cycling conditions were 95°C for 15 minutes, and 40 cycles of 94°C for 10 seconds, 60°C for 30 seconds and 72°C for 30 seconds using an Opticon 2 fluorescent thermocycler (MJ Research, Waltham, MA). Analysis of amplification products was done by melt curve where the PCR reactions were heated from 50 to 94 C at a rate of 0.5°C/second. For real-time PCR and SAGE tag count comparisons, beta-2-microglobulin (*B2M*) served as the internal control where the number of SAGE tags for *B2M* derived from this library and the amplification curve from real-time PCR were considered equal and were the basis for comparison of all genes analyzed. To quantify RNA abundance, the comparative threshold cycle (C_T) method was used, which compares the relative amount of target sequence with the reference value chosen and presents the result as relative to the reference value (expression level of mock-infected PAMs). The comparative C_T method ($\Delta\Delta C_T$) is the most practical method if the target and reference have similar dynamic ranges [9]. Calculations for the target transcript quantitation begins with the difference (ΔC_T) between the C_T values of the target and reference. This is followed by transforming these values to absolute values, using the formula: Comparative expression level = $2^{-\Delta\Delta C_T}$.

Results

Total cellular RNA was prepared from in vitro mock-infected and PRRSV strain VR-2332-infected PAMs at 0, 6, 12, 16 and 24 hours post-infection, and subjected to SAGE analysis to obtain >100,000 tags per time point. Identified unique mRNA tags were analyzed for their identity and relative abundance. A tag has a quality score (derived from PHRED, a base calling program for DNA sequence traces [10]) and a sequence, ten base pairs in length, that can be used to identify the corresponding transcript. Tags were mapped to transcripts and genes by exact regular expression matching to sequences in GenBank, Harvard Gene Index, Pig Expression Database (Japan), and the USMARC EST databases. Tag abundance was corrected for sequencing error using R and sagenhaft [11]. Relative abundance was calculated based upon the number of times a tag was represented in a given SAGE library [12]. Multidimensional statistical tests [13] were applied to determine which changes in tag abundance were significant. Examination of the preliminary SAGE data indicated that there were changes in gene expression occurring in the PRRSV-infected PAMs based on more than 400 unique tags with significantly altered expression levels identified ($p<0.01$ with Bonferroni correction). SAGE libraries were compared to identify common or differential patterns of expression. Our interest was in expression changes that affect PAMs function; particularly in regard to innate immunity, antigen presentation, and intra- and extracellular signaling.

The SAGE libraries yielded, thus far, 280,903 sequenced tags, which were used to generate a table, representing genes expressed in mock and PRRSV-infected PAMs. The derived catalogue of expressed genes represents a first attempt to generate comprehensive analysis of the PRRSV-infected PAM expression profile (available in Genbank). The wealth of information obtained allowed for the detection of genes involved in normal porcine alveolar macrophage physiology, as well as possible target genes of PRRSV infection. In general, tag frequency in a typical SAGE experiment follows a Gaussian distribution [14,15]. Table 1 summarizes the general statistics of these libraries. Tags with a frequency of 1 were not considered for quantitative purposes, because these were likely to represent artefacts of sequencing or of the SAGE procedure [16].

The validity and kinetics of expression of SAGE identified genes were evaluated using real-time RT-PCR (Fig. 1). Control genes that did not respond to PRRSV infection were also included in the analysis as internal controls. Differential expression of the most abundantly expressed tag, corresponding to the ferritin light polypeptide gene (*FTL*), was confirmed by real-time RT-PCR. The increase in the *FTL* transcript abundance, as well as a cytochrome P450 (*CYP3A*) and thioredoxin (*TXN*), was found to be a result of cell culture. The pro-inflammatory cytokines *IL1A* and *CCL4* declined in transcript abundance in accordance with the findings of Lopez-Fuertes et al. (2000) [17]. Real-time PCR amplification of transcripts known to be up-regulated following PRRSV infections (*MX1, RHIV1*, and *USP*) showed 40-, 7.5- and 14.5-fold increases at 24 hours PI, respectively. The transcripts encoding RNA helicase *RHIV1* and *MX1* were induced between 0 and 12 h post-PRRSV-infection in accordance with Zhang et al. (1999) [18].

Table 1: Summary of SAGE analysis of transcripts dysregulated by PRRSV.

Library[a]	SAGE Tags Total	Unique	Known Transcripts	Unknown Transcripts	No Match
T0 Mock	46,421	13,583	11832 (25%)	5031 (11%)	29557 (64%)
T6	66,486	23,547	17110 (26%)	5277 (8%)	44094 (66%)
T12	66,194	26,187	15458 (23%)	4977 (8%)	45759 (69%)
T16	36,446	13,650	9244 (25%)	2254 (6%)	24949 (68%)
T24	65,362	18,668	16687 (26%)	5812 (9%)	42863 (66%)
Total	280,903	95,604	70331 (25%)	23351 (8%)	187222 (67%)

[a]The libraries are T0 Mock, a total SAGE tag count of porcine alveolar macrophages mock-infected; T6, a total SAGE count of porcine alveolar macrophages infected with PRRSV 6 hours post-infection; T12, a total SAGE count of porcine alveolar macrophages infected with PRRSV 12 hours post-infection; T16, a total SAGE count of porcine alveolar macrophages infected with PRRSV 16 hours post-infection; and T24, a total SAGE count of porcine alveolar macrophages infected with PRRSV 24 hours post-infection.

Discussion

SAGE libraries were generated defining the transcription of non-infected PAMs and PAMs-infected with PRRSV that will be a publicly available resource. Macrophage transcript abundance that varied in correlation with PRRSV replication was identified. This research may identify genes that respond to specific stages of the virus life cycle such as virus binding and entry versus replication. Responding genes were identified that have not yet been represented in sequence databases. We periodically BLAST these sequences against sequence databases until matches are identified. Further study of the genes, their transcript abundance, protein level, and protein function will enhance our understanding of the interaction of PRRSV with the porcine macrophage. Possible outcomes may include the identification of virulence mechanisms, the development of next generation diagnostic assays and more rational vaccine design to more effectively limit viral replication and spread.

Acknowledgements

The authors would like to acknowledge Drs. M. E. Kehrli, Jr. and T. L. Nicholson for helpful discussions and reviewing of the manuscript. We would like to thank W. Boatwright, S. Pohl and R. Steeves for technical assistance, and S. Ohlendorf for secretarial assistance in preparation of the manuscript. This work was supported in part by a grant from the National Pork Board to J.N. and L.M. (06-122).

Mention of trade names or commercial products in this article is solely for the purpose of providing specific information and does not imply recommendation or endorsement by the U.S. Department of Agriculture.

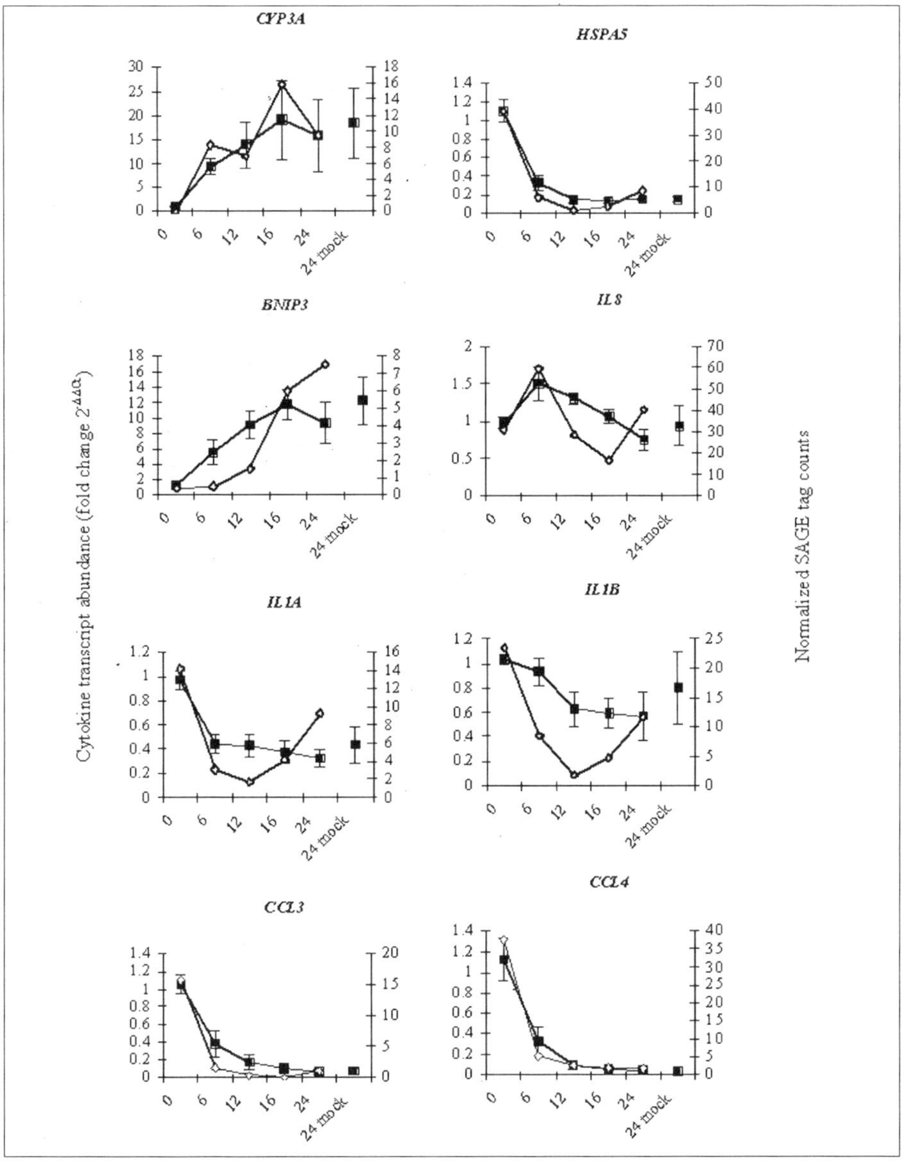

Fig. 1: Real-time RT-PCR validation of SAGE results.

Validation by real-time RT-PCR of RNA from the in vitro PRRSV-infected PAM samples used in the SAGE analysis of selected transcripts showing differential expression. Real-time RT-PCR transcript abundance results (filled squares; left *y* axis) and SAGE tag counts normalized to total tags per library (open diamonds; right *y* axis) are shown for transcripts: Cytochrome P450 (*CYP3A*; TC281236); heat shock 70kDa protein 5 (*HSPA5*; BW956325); BCL2/adenovirus E1B 19kDa interacting protein 3 (*BNIP3*; AK240169); interleukin-8 (*IL8*; NM_213867); interleukin 1, alpha (*IL1A*; NM_214029); interleukin 1, beta (*IL1B*; NM_214055); chemokine (C-C motif) ligand 3 (*CCL3*; NM_001009579); chemokine (C-C motif) ligand 4 (*CCL4*; TC288791). The '24 mock' result shows the real-time RT-PCR fold change in transcript abundance as a result of being in cell culture.

References

1 Rossow KD: Porcine reproductive and respiratory syndrome. Vet Pathol 1998;35(1):1-20.

2 Zimmerman JJ, Yoon KJ, Wills RW, Swenson SL: General overview of PRRSV: a perspective from the United States. Vet Microbiol 1997;55(1-4):187-196.

3 Velculescu VE, Zhang L, Vogelstein B, Kinzler KW: Serial analysis of gene expression. Science 1995;270:484-487.

4 Chiou MT, Jeng CR, Chueh LL, Cheng CH, Pang VF: Effects of porcine reproductive and respiratory syndrome virus (isolate tw91) on porcine alveolar macrophages in vitro [In Process Citation]. Vet Microbiol 2000;71(1-2):9-25.

5 Opriessnig T, Yu S, Gallup JM, Evans RB, Fenaux M, Pallares F, et al: Effect of vaccination with selective bacterins on conventional pigs infected with type 2 porcine circovirus. Vet Pathol 2003;40(5):521-529.

6 Stakenborg T, Vicca J, Butaye P, Imberechts H, Peeters J, De Kruif A, et al: A multiplex PCR to identify porcine mycoplasmas present in broth cultures. Vet Res Commun 2006;30(3):239-247.

7 Allende R, Kutish GF, Laegreid W, Lu Z, Lewis TL, Rock DL, et al: Mutations in the genome of porcine reproductive and respiratory syndrome virus responsible for the attenuation phenotype. Arch Virol 2000;145(6):1149-1161.

8 Kim HS, Kwang J, Yoon IJ, Joo HS, Frey ML: Enhanced replication of porcine reproductive and respiratory syndrome (PRRS) virus in a homogeneous subpopulation of MA-104 cell line. Arch Virol 1993;133(3-4):477-483.

9 Pfaffl MW: A new mathematical model for relative quantification in real-time RT-PCR. Nucleic Acids Res 2001;29(9):e45.

10 Ewing B, Green P: Base-calling of automated sequencer traces using phred. II. Error probabilities. Genome Res. 1998;8(3):186-194.

11 Beissbarth T, Hyde L, Smyth GK, Job C, Boon WM, Tan SS, et al: Statistical modeling of sequencing errors in SAGE libraries. Bioinformatics 2004;20 Suppl 1:i31-39.

12 Madden SL, Wang CJ, Landes G: Serial analysis of gene expression: from gene discovery to target identification. Drug Discov Today 2000;5(9):415-425.

13 Romualdi C, Bortoluzzi S, D'Alessi F, Danieli GA: IDEG6: a web tool for detection of differentially expressed genes in multiple tag sampling experiments. Physiol Genomics 2003;12(2):159-162.

14 Audic S, Claverie JM: The significance of digital gene expression profiles. Genome Res 1997;7(10):986-995.

15 Ruijter JM, Van Kampen AH, Baas F: Statistical evaluation of SAGE libraries: consequences for experimental design. Physiol Genomics 2002;11(2):37-44.

16 Yamamoto M, Wakatsuki T, Hada A, Ryo A: Use of serial analysis of gene expression (SAGE) technology. J Immunol Methods 2001;250(1-2):45-66.

17 Lopez-Fuertes L, Campos E, Domenech N, Ezquerra A, Castro JM, Dominguez J, et al: Porcine reproductive and respiratory syndrome (PRRS) virus down- modulates TNF-alpha production in infected macrophages. Virus Res. 2000;69(1):41-46.

18 Zhang X, Shin J, Molitor TW, Schook LB, Rutherford MS: Molecular responses of macrophages to porcine reproductive and respiratory syndrome virus infection. Virology. 1999;262(1):152-162.

Dr. Laura C. Miller, Virus and Prion Diseases of Livestock Research Unit, National Animal Disease Center, Agricultural Research Service, USDA, P.O. Box 70, 2300 Dayton Avenue, Ames, IA 50010, USA.

E-mail: laura.miller@ars.usda.gov

Pinard M-H, Gay C, Pastoret P-P, Dodet B (eds): Animal Genomics for Animal Health. Dev Biol (Basel). Basel, Karger, 2008, vol 132, pp 175-183.

Immediate Early Responses of Avian Tracheal Epithelial Cells to Infection with Highly Pathogenic Avian Influenza Virus

L. Sarmento[1], M. Pantin-Jackwood[1], D.R. Kapczynski[1], D.E. Swayne[1], C.L. Afonso[1]

[1] Southeast Poultry Research Laboratory, ARS-USDA, Athens, GA, USA

Key words: avian influenza, microarrays, epithelial cell

Abstract: Highly pathogenic (HP) avian influenza viruses (AIV) present an ongoing threat to the world poultry industry. In order to develop new AIV control strategies it is necessary to understand the underlying mechanism of viral infection at mucosal respiratory sites. Chicken and duck tracheal epithelial cells systems (TEC) were developed to study early host responses to AIV infection on TEC. Infection of 2 week-old chickens and ducks with the highly pathogenic AIV H5N1 Ck/Hong Kong/220/97 and Egret/Hong Kong/757.2/02 viruses together with TEC early responses to infection suggest the induction of differential innate immune responses. Growth curves indicated that although chicken and ducks TEC supported viral replication and re-infection, the capacity of the two viruses to replicate was not equal. A 42K probes chicken microarray system used to characterize differences in gene expression between chicken tracheal epithelial cells infected with these two highly pathogenic AIV identified expression of virus-specific molecular markers. The existence of dissimilar patterns of host gene expression as early as six hours post infection suggests that the differential growth characteristics of the two highly pathogenic AIV in tracheal epithelial cells is preceded by distinct host responses.

INTRODUCTION

Outbreaks of highly pathogenic avian influenza (HPAI) of serotype H5N1 have originated in Asia and spread through several Middle Eastern, African, and European countries in birds. Since 2002, these outbreaks have been characterized by unusually high pathogenicity, with some isolates capable of causing significant morbidity and mortality in wild birds and in mammals. Infection of chickens with highly pathogenic avian influenza virus (HPAIV) CK/HK/220/97 and Egret/HK/757.2/02 causes 100% mortality [1]. Although differences in pathogenicity are normally minor in chickens, in ducks these isolates cause a range of effects from minor respiratory disease to 100% mortality [1]. Avian influenza virus (AIV) targets the epithelial cells of the upper respiratory tract [2,1]. While *in vivo* systems are best for studying disease response, primary cell culture systems provide a valuable intermediate between animal studies and the use of cell lines to study under highly controlled conditions the mechanisms of infection and the immediate host responses. Recent progress in tissue culture methods has facilitated the isolation of epithelial cells from other species including mouse [3], and human [4]. Isolation conditions and specialized media that support mammalian epithelial cell culture have been previously described [5,6]. In order to identify early differences in host responses to infection with AIV of different pathogenicity, and to compare viral growth characteristics, we developed avian tracheal epithelial cells and microarray systems.

Materials and Methods

Viruses and reagents

The highly pathogenic H5N1 AIV A/CK/HK/220/97 and A/Egret/HK/757.2/02 were used in this study. Virus stocks were propagated in SPF embryonating chicken eggs. All experiments including work with animals, were performed in biosecurity level-3 Ag (BSL-3-Ag) facilities at the Southeast Poultry Research Laboratory (SEPRL), Agricultural Research Service, United States Department of Agriculture (USDA), and all personnel were required to wear a powered air protection respirator with high efficiency particulate air (HEPA)-filtered air supply (3M™, St. Paul, MN). All tissue culture and chemical reagents were obtained from Sigma Chemical Company (St. Louis, MO) unless otherwise noted.

Pathogenicity studies in chickens and ducks

Two week-old SPF White Leghorn chickens (*G. gallus domesticus*) (from SEPRL flocks) and two week-old White Pekin ducks (from Metzer Farms, CA) were inoculated intranasally (IN) to determine the pathogenicity of the two H5N1 viruses, as previously described [7]. The birds were housed in self-contained isolation cabinets that were ventilated under negative pressure with HEPA-filtered air and maintained under continuous lighting. Feed and water were provided *ad libitum*. General care was provided as required by the Institutional Animal Care and Use Committee, as outlined in the *Guide for the Care and Use of Agricultural Animals in Agricultural Research and Teaching* [8]. Groups of 10 birds, were inoculated with 0.1 ml of an inoculum containing 10^6 EID_{50}/ml of either virus and clinical signs of disease evaluated for up to 10 days.

Tissue dissociation and epithelial cell culture

Seventeen day-old embryonating SPF Leghorn chicken eggs or 24 day-old embryonating duck eggs were used to harvest embryos and the intact tracheas were collected in phosphate buffered saline (PBS) using sterile technique. Tracheas were cut into 1 cm long sections and rinsed twice in PBS and 0.1% DL-dithiothreitol (DTT), and once each in PBS and Dulbecco's Modified Eagle's Medium (DMEM). The sections from 20 tracheas were incubated for 2 hours at 37°C in 20 ml dissociation solution (Minimum Essential Medium (MEM) containing 1% penicillin/streptomycin antibiotic, 0.2% type XIV protease from *Streptomyces griseus*, and 0.02% DNase). The reaction was stopped by addition of 5 ml foetal bovine serum (FBS) and cells were dispersed by repeated pipetting. The cell mixture was filtered on a 40 nm nylon cell strainer (BD Biosciences) and cells harvested by centrifugation at 200 x *g*. The cell pellet was resuspended in DMEM and applied to a 60/40/20% discontinuous Percoll gradient. The cell band was collected and pelleted by centrifugation at 290 x *g*, washed twice with PBS, and resuspended in a 1:1 solution of DMEM and bronchial epithelium growth media (BEGM, Cambrex, East Rutherford, NJ), supplemented with 1% each of L-glutamine and antibiotics (as previously described), with 5% BIT-9500 serum substitute (Stem Cell Technologies, Vancouver, BC). Twenty-four hours prior to tracheal tissue dissociation, 24 well tissue culture plates (BD, Franklin Lakes, NJ) were coated with 170µl per well of 0.01% Type I collagen as per product specifications. Following tissue dissociation, cells were stained with trypan blue and viable cell counted using a hemacytometer and adjusted to a concentration of 1 x 10^6 cells/ml. Two hundred μl of cell suspension were plated into each well of a collagen-coated 24-well plate and allowed to attach for at least 24 hours in the presence of BEGM.

Growth curves

At 24 hours post-plating, 1 x 10^5 TECs were inoculated with CK/HK/220/97 and Egret/HK/757.2/02 AIV at an MOI of 0.005. Following adsorption for 1 hour at 37ºC, non-adsorbed viruses were removed and a solution of DMEM and bronchial epithelium growth media (BEGM, Cambrex, East Rutherford, NJ), supplemented with 1% each of L-glutamine and antibiotics (as previously described), and 5% BIT-9500 serum substitute (Stem Cell Technologies, Vancouver, BC) were added. Cell supernatants were collected at 0, 12, 24, 36, 48 and 60 hours post-infection (HPI). Serial dilutions (1:10) of each supernatant were used to infect chicken embryo fibroblasts. To determine the titres of virus in cell culture supernatants, three independent experiments were performed. Virus titres were determined by cytopathic effect (CPE) and reported as the 50% tissue culture infectious dose (TCID50). The TCID50 assay end point was calculated by the method of Reed and Muench [9]. Titres obtained from the three independent experiments were analyzed statistically by one-way ANOVA.

RNA isolation, reverse transcription, labeling, and microarray hybridization

Six hours post infection, RNA was isolated from TECs infected with AIV CK/HK/220/97 and Egret/HK/757.2/02 at an MOI of 10 with TRIzol Reagent (Invitrogen, Carlsbad, CA) followed by LiCl precipitation. Ten μg of total RNA was reverse transcribed to cDNA and labeled with Cy3 or Cy5 using the Amino Allyl cDNA Labeling Kit (Ambion, Austin, TX). An array composed of 42,034 probes (65 nucleotides) and 6 additional probes containing AIV genes was designed in collaboration with Agilent Technologies (Santa Clara, CA) from coding sequences of the entire chicken genome. The genome was annotated by BLASTN analysis of the DNA sequences against TIGR and GenBank non-redundant databases followed by BLASTX analysis of these sequences against translated protein of the RefSeq Release 20, Human and Mouse databases. A total of 22,364 probes were annotated with a probability E value of < 1.0e-4. Labeled cDNAs were hybridized for 17 hours at 65°C with gene expression hybridization solution (Agilent, Santa Clara, CA). Slides were disassembled from their hybridization chambers in Gene Expression Wash Buffer 1 (Agilent) then washed for 1 minute in Gene Expression Wash Buffer 1, and for 1 minute in Gene Expression Wash Buffer 2 (Agilent) as per product specifications.

Microarray data collection and analysis

Differential expression measurements of infected TEC were based on simultaneous two-colour hybridizations and scanning with a GenePix 4200A scanner and the GenePix Pro 7.0 data acquisition and analysis software (Axon Instruments, Molecular Devices, Sunnyvale, CA) was used for all normalization and statistical analysis of the output. Four slides from four independent experiments were analyzed. The intensity ratio of expression for each gene was calculated by dividing the measured virus Egret/HK/757.2/02 values (test channel) by the intensity of the virus CK/HK/220/97 (control channel). All output with control channel values of less than 300.0 were not considered. Intensity-dependent normalization was used to reduce the ratios to the residual of the lowess fit of the intensity-versus ratio curve.

Results

Intranasal (IN) administration of CK/HK/220/97 and A/Egret/HK/757.2/02 AIV resulted in 100% mortality of the 2 week-old chickens, whereas no mortality was observed in ducks inoculated with CK/HK/220/97. In contrast, A/Egret/HK/757.2/02 induced 100% mortality in ducks (Table 1). Because of the observed differences in virulence in ducks between these two AIV isolates, they were compared further.

Previously, tracheal epithelial cells (TECs) were successfully isolated from embryonic chicken separated from other cell types by centrifugation in a Percoll gradient and characterized using microarray comparison of gene expression [10]. Here the use of this procedure has been extended to duck tracheal cells (Fig. 1). To determine if chicken and duck TECs were capable of supporting viral growth with productive infection and virus dissemination, cells were infected 24 hours post-plating with Egret/HK/757.2/02 and CK/HK/220/97 AIV at an MOI of 0.005. Cell supernatants were collected at specified time points and titrated in cells by CPE as previously described. Results show that TECs from both avian species support the growth of both viruses *in vitro* with titres rapidly increasing and reaching up to 5 log differences in EID_{50}/ml, in comparison to time zero (Fig. 2). Notably, the two viruses grow at different rates, with the less pathogenic virus (CK/HK/220/97) displaying an enhanced replication efficiency (2-3 log increase) in both duck and chicken cells. Despite intra-viral differential growth capacity, there were no significant differences in the titres of individual viruses in chicken vs duck TECs comparisons, suggesting that TEC from both avian species posses a comparable capacity to support viral replication (Fig. 2).

Table 1: Mortality and mean death time (MDT) of Ck/Hong Kong/220/97 and Egret/Hong Kong/757.2/02 in 2 week-old chickens and ducks.

	Chickens[a]	**Ducks**[a, b]
Ck/Hong Kong/220/97	10/10 (1.5) [c]	0/10 (0)
Egret/Hong Kong/757.2/02	10/10 (2.0)	10/10 (5.5)

[a]Birds were inoculated intranasally with 105 EID50 of the viruses.
[b]Number of dead ducks/number of inoculated or exposed ducks.
[c]Mean death time in days.

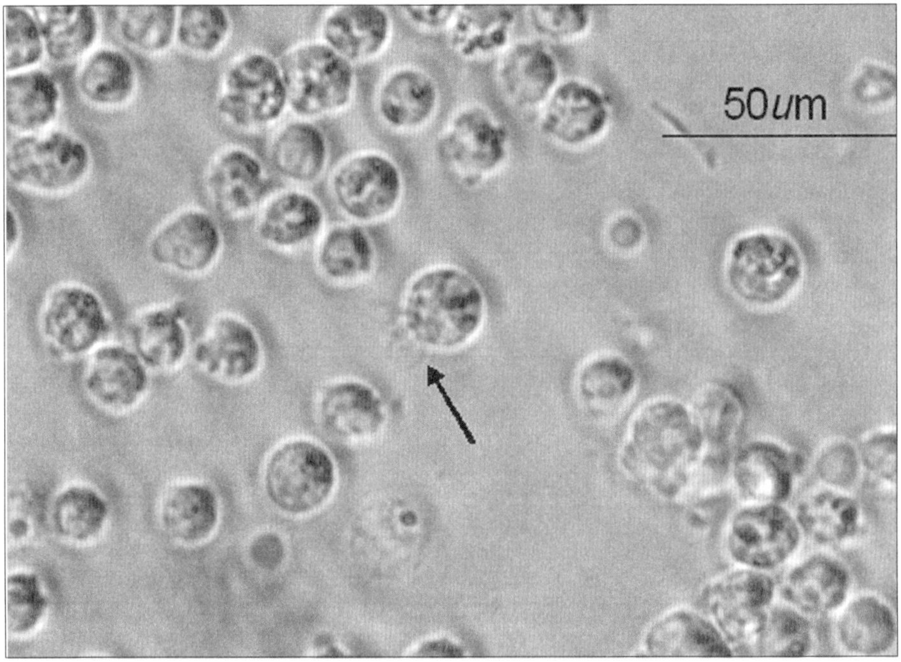

Fig. 1: Morphological characteristics of freshly isolated duck tracheal epithelial cells. Visible cilia on a freshly isolated tracheal cell is indicated by the arrow.

A

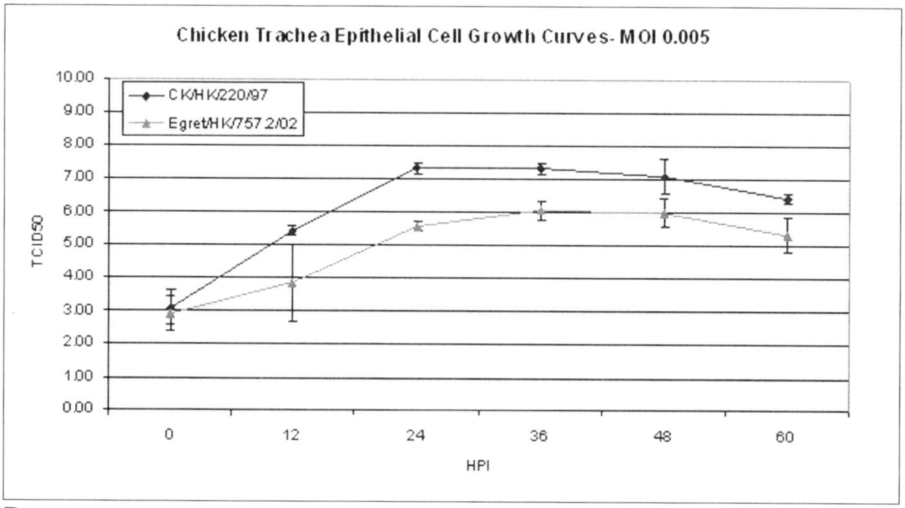

B

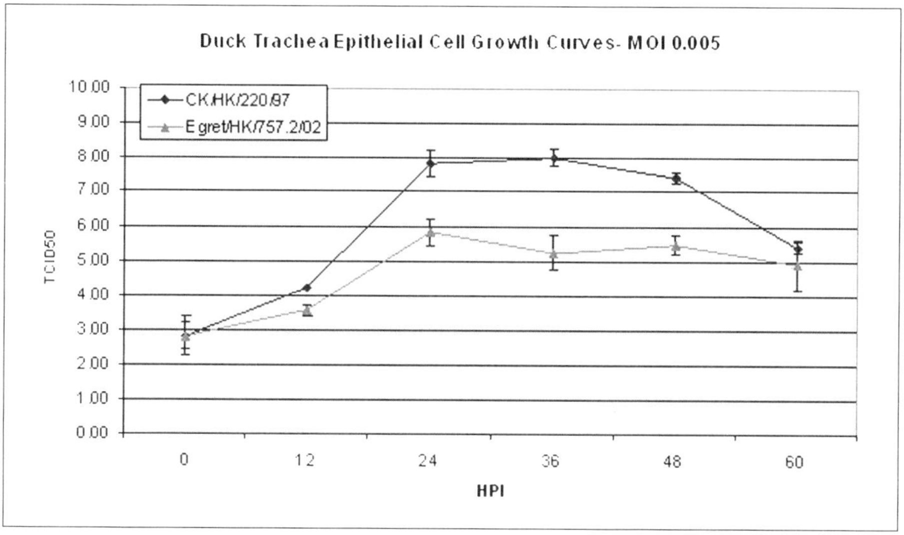

Fig. 2: AIV growth curves in chicken and duck TECs.
Twenty-four hours post-plating, chicken (A) or duck (B) TECs were infected with an MOI of 0.005 with AIV H5N1 strains Ck/Hong Kong/220/97 and Egret/Hong Kong/757.2/02. Cell supernatants were collected at 0, 12, 24, 36, 48, and 60 HPI. Serial dilutions (1:10) of each supernatant were titrated on chicken or duck embryonic fibroblast cells. Infection was assessed by CPE. Virus titres (TCID50/ml) were determined by the Reed and Munch method and log values represented in the vertical axis.
* = $p < 0.05$

Early gene expression was characterized by microarray analysis following inoculation of chicken TECs with either Ck/Hong Kong/220/97 or Egret/Hong Kong/757.2/02. While 2,018 genes had statistically reproducible values, the majority of the genes (97%) did not display significant changes in gene expression. Using a probability of < 0.05 and fold increases of >2, 21 genes were identified that were differentially up-regulated by the Egret/HK/757.2/02 virus while 32 genes displayed higher levels in the Ck/Hong Kong/220/97 infection (Table 2). The Egret/HK/757.2/02 upregulated genes were mainly novel genes of unknown function. Egret/HK/757.2/02 upregulated genes of known function included the neurotrypsin precursor motopsin; the solute carrier family 13; member 4 (Na(+)/sulfate cotransporter SUT-1), solute carrier family 6, the arachidonate 5-lipoxygenase (5-lipoxygenase) (5-LO) and the CD27 binding Siva protein that inhibits activation of NF-κB and promotes T cell receptor mediated apoptosis [11-14]. Genes preferentially expressed by the Ck/Hong Kong/220/97 virus (negative ratios) also included several unknown genes, as well as multiple ribosomal proteins including L6, L7a, L21, L24, L27a, L32, L38, LP1, S2, S3a, S16, S26, SA associated protein, and the protein kinase C inhibitor Wpkci, among others.

Table 2 a: Microarray analysis of gene expression in trachea epithelial cell cultures. Chicken TECs were infected at an MOI of 10 with Egret/Hong Kong/757.2/02 (Cy5) or Ck/Hong Kong/220/97 (Cy3). RNA was isolated at 6 HPI, reverse transcribed to cDNA, amplified, labeled, and analyzed by microarray. Four slides from four independent experiments were analyzed as previously described. The numbers of differentially expressed genes (>2.0 fold up or down) between Egret/Hong Kong/757.2/02 (Cy5) or Ck/Hong Kong/220/97 (Cy3) were compared.

Ratio	p-value	GenBank	RefSeq	RefSeq Annotation
4.368	0.00789	BU422905	XP_698127.1	unknown
3.617	0.00971	TC202470	ZP_01294054.1	unknown
3.108	0.0241	TC206120	XP_859097.1	unknown
2.503	0.0325	CR390650	ZP_01373703.1	unknown
2.495	0.05	TC219612	XP_416432.1	unknown
2.471	0.0366	BU285118	XP_001220935.1	unknown
2.453	0.0165	TC226829	NP_660403.1	unknown
2.446	0.037	TC206061	NP_989963.1	unknown
2.371	0.00178	BU220555	XP_001141101.1	unknown
2.299	0.0201	CR387588	YP_292828.1	unknown
2.286	0.0088	BU335121	XP_426502.1	unknown
2.159	0.0218	CR389003	XP_420637.1	neurotrypsin precursor
2.136	0.0105	CR353818	ZP_00657814.1	unknown
2.128	0.0178	CR338786	XP_416360.1	Solute carrier family 13
2.115	0.0198	BU406049	XP_367292.1	unknown
2.071	0.0439	CR390015		unknown
2.06	0.0213	BX933269	XP_421397.1	CD27 binding Siva protein
2.055	0.0141	TC205915	XP_307890.2	unknown
2.018	0.029	TC225467	XP_421641.1	arachidonate 5-lipoxygenase (5-lipoxygenase)
2.017	0.0281	TC223985	XP_419789.1	unknown
2.009	0.0142	AJ851469	NP_001003311.1	solute carrier family 6
0.497	0.0276	AF148455	NP_990019.1	protein kinase c inhibitor Wpkci
0.496	0.0332	L48915	ZP_01001036.1	unknown
0.491	0.0304	BX932372	XP_417127.1	ribosomal protein L21

Table 2 b

0.491	0.0304	AF334677	NP_989533.1	claudin 3 tight junction protein
0.483	0.0208	TC207241	YP_133820.1	unknown
0.479	0.0237	NM_001006685	NP_001006686.1	heat shock protein 70
0.478	0.0251	BX929997	XP_872162.1	60S ribosomal protein L38
0.478	0.0249	AF421549	NP_001012296.1	fizzy/cell division cycle 20 related
0.477	0.0245	CO758431	XP_668857.1	unknown
0.476	0.0243	TC207164	XP_001099771.1	ribosomal protein S3a
0.476	0.0242	X68200	XP_724982.1	unknown
0.476	0.0243	X13876	NP_990653.1	ribosomal protein large P1
0.476	0.0243	X06580	XP_415105.1	polyubiquitin with 3 Ub domains
0.476	0.0242	U25954	NP_990729.1	tumor protein translationally-controlled
0.476	0.0243	M33331	NP_001009.1	ribosomal protein S15
0.476	0.0244	CR524410	XP_422734.1	unknown
0.476	0.0244	CR390230	XP_997995.1	unknown
0.476	0.0243	CR386179	XP_414845.1	40S ribosomal protein S2
0.476	0.0242	CR353354	XP_429924.1	similar to thymosin-like 8.
0.476	0.0244	CR352709		ribosomal protein L24
0.476	0.0243	BX931089	NP_989830.1	ribosomal protein L27a and transcription factor Crx
0.476	0.0243	BX929563	XP_414453.1	60S ribosomal protein L32
0.476	0.0243	BX264617	XP_418414.1	prostate stem cell antigen
0.476	0.0242	AY448022	XP_418817.1	p40 ribosomal SA associated protein
0.476	0.0244	AY032864	NP_989483.1	ribosomal protein L6
0.476	0.0243	AJ851769	NP_001012582.1	tumor-associated calcium signal transducer 1
0.476	0.0241	AJ719518	NP_956319.1	ribosomal protein S26 [Danio rerio]
0.475	0.0246	CR524103	XP_415762.1	uroplakin 3B isoform b uroplakin IIIb
0.475	0.0241	BX930915	XP_517208.2	cell division control protein septin 11
0.475	0.0241	AB221488	NP_001004379.1	ribosomal protein L7a
0.447	0.00919	CN232184	XP_001213247.1	unknown
0.419	0.00639	BU474360	XP_730698.1	unknown
0.365	0.00079	TC213118	XP_001114281.1	UDP-GlcNAc

Discussion

The establishment of a chicken TEC *in vitro* system provides an invaluable new tool in the field of avian diseases. The ability to identify molecular changes using an *in vitro* model that represents the mucosal sites of early AIV infection *in vivo* will facilitate future studies of the mechanism of early host gene responses at the initial site of infection. This system will allow studies to be performed under highly controlled conditions, using large number of viruses and minimal resources. This advantage is even more significant in the case of evaluating infection in ducks because experimental animals are not as genetically uniform and readily available as the chicken.

Wild ducks are natural reservoirs of AIV and epidemiologic evidence and experimental infections show that domestic ducks are also susceptible to infection [15-17]. However until recently, AI viruses did not cause severe disease or death in ducks [15-16]. In recent years, a number of Asian HPAI H5N1 viruses have been found to be pathogenic in ducks, however this pathogenicity appears to be dependent on the viral strain and the species of ducks tested. Here the use of microarrays to examine host responses under tightly controlled conditions (e.g. TEC) suggests that virus specific responses are induced very early during infection. Future comparison of chicken vs duck responses will likely facilitate the identification of duck specific responses and may lead to the recognition of disease resistance genes.

The induction of a large number of ribosomal genes by Ck/Hong Kong/220/97 is notable as it suggests that an increased translational activity may occur in cells infected with this isolate and may explain the increased capacity of the virus to replicate. In contrast, the induction of other genes that may have systemic function, such as arachidonate 5-lipoxygenase, was observed with the Egret/HK/757.2/02. Five-lipoxygenase (also known as 5-LO or Alox5), is responsible for the production of leukotrienes. These are naturally produced eicosanoid lipid mediators, which affect the inflammatory response with different roles such as chemotactic effects on migrating neutrophils (LTB4) and vascular permeability. Egret/HK/757.2/02 also induced the Siva protein, which inhibits activation of NF-κB and promotes T cell receptor-mediated apoptosis. Although it is premature to speculate how the expression of these genes will affect the outcome of infection, overall these studies have shown the suitability of the TEC model for replicating HPAI viruses and the usefulness of the microarray analysis for measuring early host responses following infection with avian influenza virus.

Acknowledgements

The authors would like to thank Kristin Kaffuto, Jeremy Volkering, and Denise Brinson for their excellent technical assistance. This work was funded by USDA CRIS project numbers 6612-32000-049 and 6612-32000-048.

References

1 Swayne DE, Pantin-Jackwood M: Pathogenicity of avian influenza viruses in poultry. Dev Biol (Basel) 2006; 124:61-67.

2 Swayne DE: Understanding the complex pathobiology of high pathogenicity avian influenza viruses in birds. Avian Dis 2007; 51:242-249.

3 Ibricevic A, Pekosz A, Walter MJ, Newby C, Battaile JT, Brown EG, Holtzman MJ, Brody SL: Influenza virus receptor specificity and cell tropism in mouse and human airway epithelial cells. J Virol 2006; 80:7469-7480.

4 Matrosovich MN, Matrosovich TY, Gray T, Roberts NA, Klenk HD: Neuraminidase is important for the initiation of influenza virus infection in human airway epithelium. J Virol 2004; 78:12665-12667.

5 Yamaya M, Finkbeiner WE, Chun SY, Widdicombe JH: Differentiated structure and function of cultures from human tracheal epithelium. American Journal of Physiology 1992; 262:L713-L724.

6 Davidson DJ, Kilanowski FM, Randell SH, Sheppard DN, Dorin JR: A primary culture model of differentiated murine tracheal epithelium. Am J Physiol Lung Cell Mol Physiol 2000; 279:L766-L778.

7 Lee CW, Senne DA, Linares JA, Woolcock PR, Stallknecht DE, Spackman E, Swayne DE, Suarez DL: Characterization of recent H5 subtype avian influenza viruses from US poultry. Avian Pathol 2004; 33:288-297.

8 Craig J, Dean VWF, Havenstein GB, Kruger KK, Nestor KE, Purchase GH, Siegel PB, van Wicklen GL: Guidelines for poultry husbandry. 1999.

9 Reed LJ, Muench H: A simple method of estimating fifty percent endpoints. Am J Trop Med Hyg 1938; 27:493-497.

10 Zaffuto KM, Estevez CN, Afonso CL: Primary chicken tracheal cell culture system for the study of infection with avian respiratory viruses. Avian Pathol 2008;37(1):25-31.

11 Chu F, Borthakur A, Sun X, Barkinge J, Gudi R, Hawkins S, Prasad KV: The Siva-1 putative amphipathic helical region (SAH) is sufficient to bind to BCL-XL and sensitize cells to UV radiation induced apoptosis. Apoptosis 2004; 9:83-95.

12 Gudi R, Barkinge J, Hawkins S, Chu F, Manicassamy S, Sun Z, Duke-Cohan JS, Prasad KV: Siva-1 negatively regulates NF-kappaB activity: effect on T-cell receptor-mediated activation-induced cell death (AICD). Oncogene 2006; 25:3458-3462.

13 Prasad KV, Ao Z, Yoon Y, Wu MX, Rizk M, Jacquot S, Schlossman SF: CD27, a member of the tumor necrosis factor receptor family, induces apoptosis and binds to Siva, a proapoptotic protein. Proc Natl Acad Sci USA 1997; 94:6346-6351.

14 Yoon Y, Ao Z, Cheng Y, Schlossman SF, Prasad KV: Murine Siva-1 and Siva-2, alternate splice forms of the mouse Siva gene, both bind to CD27 but differentially transduce apoptosis. Oncogene 1999; 18:7174-7179.

15 Li KS, Guan Y, Wang J, Smith GJ, Xu KM, Duan L, Rahardjo AP, Puthavathana P, Buranathai C, Nguyen TD, Estoepangestie AT, Chaisingh A, Auewarakul P, Long HT, Hanh NT, Webby RJ, Poon LL, Chen H, Shortridge KF, Yuen KY, Webster RG, Peiris JS: Genesis of a highly pathogenic and potentially pandemic H5N1 influenza virus in eastern Asia. Nature 2004; 430:209-213.

16 Pantin-Jackwood MJ, Swayne DE: Pathobiology of Asian highly pathogenic avian influenza H5N1 virus infections in ducks. Avian Dis 2007; 51:250-259.

17 Senne DA, Suarez DL, Stallnecht DE, Pedersen JC, Panigrahy B: Ecology and epidemiology of avian influenza in North and South America. Dev Biol (Basel) 2006; 124:37-44.

Dr. Claudio L. Afonso, SEPRL, ARS-USDA, SAA, 934 College Station Road, Athens, GA 30605.
E-mail:claudio.afonso@ars.usda.gov

Pinard M-H, Gay C, Pastoret P-P, Dodet B (eds): Animal Genomics for Animal Health. Dev Biol (Basel). Basel, Karger, 2008, vol 132, pp 185-195.

Porcine Genome-wide Gene Expression in Response to Tetanus Toxoid Vaccine

S. Ponsuksili[1], E. Murani[2], K. Wimmers[2]

Research Institute for the Biology of Farm Animals (FBN Dummerstorf),
[1] Research Group `Functional Genomics
[2] Research Unit `Molecular Biology´, Dummerstorf, Germany

Key words: early immune response, swine, microarray, functional annotation, real-time PCR

Abstract: The very early in vivo response to immune stimuli was studied using tetanus toxoid as a model antigen known to induce Th1 and Th2 responses. Eighteen weaning piglets were vaccinated subcutaneously with tetanus toxoid. Leukocyte RNAs were isolated from samplings before and 2, 4, 8, and 24 hours after vaccination. After competitive hybridization of a 13297 porcine 70-mer oligo microarray (Qiagen-Operon NRSP8), subsequent image analysis and normalization, the data was analysed using analyses of variance (ANOVA) to identify genes regulated due to vaccination (ANOVA, $p \leq 0.05$; corresponding to false discovery $q \leq 0.12$). Of 8,240 probes representing genes expressed in leukocytes, 1,289 genes showed differential expression. Results were exemplarily confirmed by real-time PCR. Holistic expression data was clustered to six prominent groups of genes with similar changes in expression patterns using a k-means algorithm. To get more insight into functional and structural components and impact of the genes represented in each cluster, the EASE score was used to identify Gene Ontology functional categories. The results showed that in vivo significant changes of expression profiles of peripheral blood mononuclear cells (PBMCs) occurred very early after immune stimuli. These alterations concerned genes of pathways related to immune response as well as other metabolic and regulatory pathways including ATP/energy metabolism, transcription, structural molecule activity, biosynthesis, and metabolism. The analysis reveals new functional candidate genes for traits related to immune responsiveness and also provides new insight into the interaction of immune response, and metabolic and endocrine status. This will facilitate a better understanding of the relationship between immune and performance traits.

Introduction

Genetic improvement of the health and resistance status of farm animals contributes to higher product safety and quality and has a positive impact on production costs and animal welfare. Because major health problems in intensive as well as animal welfare friendly production systems arise from multifactorial diseases, there is a need for animals that exhibit optimal endogenous protection against many pathogens including intra- or extra-cellular microorganisms. Selection for high and low immune responsiveness has been successfully demonstrated to improve protection and responsiveness to numerous pathogens and growth performance [1]. Identification of genes controlling disease resistance and immune competence will facilitate the genetic improvement of animal health.

Here we report on a microarray study of the early leukocyte transcriptional response to immunisation in the pig. Expression profiling reveals insight into global in vivo gene regulation and facilitates creating lists of functional candidate genes for immune competency.

Material and Methods

Animals and vaccination

Three male and three female weaning piglets derived from each of three litters (n=18) of a German Landrace herd were vaccinated subcutaneously with 1 ml (30 IU) tetanus toxoid (Equilis Tetanus-Vaccine, Intervet). Blood samples were obtained immediately prior to and 2, 4, 8, and 24 hours after vaccination. Peripheral blood mononuclear cells (PBMCs) were isolated by density gradient centrifugation on Histopaque (Sigma) from 4 ml of blood.

RNA and target preparation and microarray hybridization

For each sampling time point, three RNA pools of six individuals, one male and one female per litter, were used to reduce biological variation. Therefore, total RNA was isolated using TRI Reagent (Sigma) first and subsequently mRNA was isolated from total RNA using magnetic oligo-(dT)-beads (Dynal Dynabeads). Pools of 1 µg of mRNA were labelled with Cy3 or Cy5 using the CyScribe Post-Labelling Kit (GE Healthcare). Labelled cDNAs were hybridised to the set of 13,297 pig-specific 70-mer oligonucleotides (Qiagen-Operon NRSP8) at 42°C for 14–18 hours using the Pronto Universal Hybridization kit (Corning B.V. Life Science). Hybridized slides were washed once with 1 × SSC, 0.2% (w/v) SDS for 10 minutes at 42°C and twice with 0.1 × SSC, 0.2% (w/v) SDS. Finally, slides were dipped into distilled water and air-dried.

Data analysis

Microarrays were scanned using the 428 Array Scanner (Affymetrix). Spots were quantified and quality filtering was applied using ImaGene™5 software version 5.6 (BioDiscovery. Inc.). The subtracted signal medians of genes that passed filtering were analysed further using the TIGR Microarray Data Analysis System (MIDAS) and the TIGR Multiexperiment Viewer (MEV) [2]. T-tests were conducted for each comparison and p-values were converted to q-values (QVALUE) [3]. To compare all stages of transcriptional response, ANOVA ($p \leq 0.05$) and hierarchical clustering were performed. The Functional Annotation Chart and Functional Annotation Clustering tools from the Database for Annotation, Visualization, and Integrated Discovery (DAVID) bioinformatics resources [4] were used to identify functional gene categories that were significantly overrepresented ($p \leq 0.05$) in particular Gene Ontology terms, KEGG, BIOCARTA, and BBID pathways.

Quantitative real-time RT-PCR (qRT-PCR)

Differential expression data of selected genes obtained from porcine oligonucleotide chips were validated by real-time PCR carried out in a LightCycler instrument (Roche Diagnostics). First, all individual RNA samples were reverse transcribed to cDNA and then gene specific PCR was performed using the LightCycler DNA Master SYBR Green I (Roche Diagnostics). For all assays standard curves were generated by amplifying serial dilutions of specific PCR products. Normalisation of variation in RT-PCR efficiency and initial RNA input was performed by using the *RPL32* gene as an internal standard. All assays were done in duplicate.

Results

Blood from eighteen weaning piglets was taken just before vaccination (0 hour) and 2, 4, 8, and 24 hours after vaccination. mRNA isolated from PBMC collected at 0 hour was used to set up three RNA pools of six individuals each that were reverse transcribed and labelled with Cy3. These pools were used as a reference for comparisons with the corresponding three pools for each of the time points, 2, 4, 8, and 24 hours after vaccination, derived from mRNA from the same individuals used for the reference pools. These RNA pools were reverse transcribed and labelled with Cy5 and used for hybridisation. Filtering of the spots based on the criteria of spot consistency, minimum intensity levels, and minimum signal-to-background ratio successfully discharged Arabidopsis gene, which were on the array as negative controls, from analysis. The background corrected values of median intensity of each spot were used for further analysis. In total, 11 array hybridizations were analysed. For the comparison of all time points for transcriptional response to vaccination, the gene expression patterns of porcine PBMC before and after immunisation were analysed using ANOVA. Of the 13,297 probes on the chip and 8,240 probes representing genes expressed in PBMC, 1,289 genes showed differential expression between any of the five time points of analysis prior to and after immunisation ($p \leq 0.05$; $q \leq 0.12$). Functional categories of genes that were overrepresented cover those related to cytoskeleton, cell structure, ~organization and ~motility, programmed cell death and apoptosis, and nucleotide/RNA metabolism and binding, as well as categories related to immune response like lymphocyte activation, positive regulation of T cell activation, and immunoglobulin binding (Table 1). A number of genes of the IL2 signalling pathway and IL12/STAT4 dependent signalling pathway of Th1 development (BIOCARTA) were significantly regulated.

Pair wise comparisons of expression profiles of PBMCs before and after vaccination

Pair wise comparisons of expression patterns of the reference time point before vaccination and each of the subsequent time points revealed 241 genes differently expressed before *vs* 2 hours post-vaccination (p.v.) with 170 down- and 71 upregulated genes. At 4 hours p.v., 72 and 22 genes were down- and upregulated. The number of regulated genes thereafter increased to 165, including 97 down- and 68 upregulated genes, at 8 hours p.v., and to 170 genes, including 89 up- and 81 downregulated genes at 24 hours p.v.. Assignment of the differentially regulated genes to GO functional categories revealed that these changes in expression profiles cover genes assorted to pathways related to response to stimuli/immune response as well as other metabolic and regulatory pathways related to cell growth and maintenance, cell communication, morphogenesis and other physiological and cellular processes (Table 2).

Table 1: Functional annotation of genes regulated in PBMCs during the early immune response to immunisation.

	Category	Term	Count	P-Value
cytoskeleton, cell structure and ~motility	GOTERM_BP_ALL	regulation of actin polymerization and/or depolymerization	9	0.04
	GOTERM_BP_ALL	regulation of actin filament length	9	0.04
	BIOCARTA	mCalpain and friends in cell motility	6	0.05
	BIOCARTA	agrin in postsynaptic differentiation	8	0.05
	GOTERM_BP_ALL	negative regulation of actin filament depolymerization	6	0.05
	GOTERM_BP_ALL	actin filament depolymerization	6	0.05
	GOTERM_BP_ALL	actin filament capping	6	0.05
	GOTERM_BP_ALL	barbed-end actin filament capping	6	0.05
	GOTERM_BP_ALL	regulation of actin filament depolymerization	6	0.05
immune response	GOTERM_MF_ALL	immunoglobulin binding	6	0.001
	GOTERM_MF_ALL	IgG binding	4	0.03
	GOTERM_BP_ALL	positive regulation of T cell activation	5	0.04
	GOTERM_BP_ALL	lymphocyte activation	11	0.05
	BIOCARTA	IL12 and STAT4 dependent signalling pathway in Th1 development	6	0.07
	BIOCARTA	IL2 signalling pathway	7	0.07
programmed cell death, apoptosis	GOTERM_BP_ALL	positive regulation of programmed cell death	20	0.03
	GOTERM_BP_ALL	positive regulation of apoptosis	20	0.03
	GOTERM_BP_ALL	induction of programmed cell death	18	0.04
	GOTERM_BP_ALL	induction of apoptosis	18	0.04
nucleotide metabolism	GOTERM_MF_ALL	RNA binding	81	0.002
	GOTERM_BP_ALL	RNA metabolism	62	0.02
	GOTERM_MF_ALL	nucleotide binding	167	0.03
	GOTERM_MF_ALL	helicase activity	20	0.04
	GOTERM_MF_ALL	ATP-dependent helicase actvity	15	0.04
	GOTERM_MF_ALL	RNA helicase activity	9	0.05
physiological processes	GOTERM_MF_ALL	manganese ion binding	17	0.02
	GOTERM_BP_ALL	small GTPase mediated signal transduction	30	0.02
	GOTERM_MF_ALL	protein binding	366	0.03
	GOTERM_BP_ALL	negative regulation of protein metabolism	12	0.04
	GOTERM_BP_ALL	negative regulation of cell organization and biogenesis	8	0.04

Table 2: Functional annotation of differentially expressed genes obtained by pair-wise comparisons of PBMC expression profiles at 2, 4, 8, and 24 hours after *vs* 0 hours before immunisation.

Category	**Term**	**Count**	**P-Value**
2 vs 0 hours			
GOTERM_BP_ALL	response to external stimulus	10	0.02
GOTERM_BP_ALL	response to wounding	8	0.04
GOTERM_MF_ALL	ATP binding	21	0.04
GOTERM_MF_ALL	adenyl nucleotide binding	21	0.05
GOTERM_MF_ALL	G-protein coupled receptor activity	6	0.02
GOTERM_BP_ALL	G-protein coupled receptor protein signalling pathway	8	0.02
4 vs 0 hours			
GOTERM_MF_ALL	calcium ion binding	9	0.04
GOTERM_BP_ALL	transport	22	0.04
GOTERM_BP_ALL	vesicle-mediated transport	7	0.05
8 vs 0 hours			
GOTERM_BP_ALL	regulation of lymphocyte activation	3	0.04
GOTERM_BP_ALL	interferon-gamma biosynthesis	2	0.05
GOTERM_BP_ALL	regulation of cell activation	3	0.05
GOTERM_BP_ALL	intracellular signalling	9	0.04
GOTERM_BP_ALL	biosynthesis	11	0.04
GOTERM_BP_ALL	deoxyribonucleotide biosynthesis	2	0.05
GOTERM_BP_ALL	DNA metabolism	11	0.05
GOTERM_MF_ALL	L-ascorbic acid binding	3	0.02
24 vs 0 hours			
GOTERM_BP_ALL	immune response	17	0.002
GOTERM_BP_ALL	defence response	17	0.004
GOTERM_BP_ALL	humoral defence mechanism (sensu Vertebrata)	5	0.02
GOTERM_BP_ALL	response to biotic stimulus	17	0.008
GOTERM_BP_ALL	response to external stimulus	10	0.03
GOTERM_BP_ALL	humoral immune response	5	0.05
BIOCARTA	BCR signalling pathway	4	0.02
GOTERM_BP_ALL	regulation of physiological process	43	0.03
GOTERM_BP_ALL	regulation of cellular physiological process	41	0.04
GOTERM_BP_ALL	morphogenesis	14	0.005
GOTERM_BP_ALL	cellular morphogenesis	9	0.02

Time course analysis of expression profiles of PBMC before and after vaccine

Data of expression levels at different time points was clustered using a k-means algorithm to reveal prominent groups of genes with similar changes in expression patterns (Fig. 1). From this analysis, six clusters could be distinguished. Cluster 1 covered 369 genes that were consistently slightly downregulated at 2, 4, and 8 hours and showed consistently slight up-regulation at 24 hours compared to 0 hour in each of the three microarrays analysed 2, 4, 8 hours *vs* 0 hour post-vaccination and both microarrays analysed 24 hours *vs* 0 hours post-vaccination. Cluster 2 represented 172 genes with a similar expression pattern but more pronounced differential regulation. Cluster 3 consisted of 151 genes that were not regulated at 2 and 4 hours but were downregulated at 8 hours and even more so at 24 hours. Cluster 4 represented 108 genes that were upregulated at 2 and 4 hours and showed slight or non-regulation at 8 and 24 hours. Genes upregulated at 2, 4 and particularly at 8 hours but downregulated at 24 hours were assigned to cluster 5 (179 genes). Cluster 6 covered 310 genes with an expression pattern opposite to those merged in cluster 1, i.e., that were slightly upregulated at 2, 4, and 8 hours but downregulated at 24 hours.

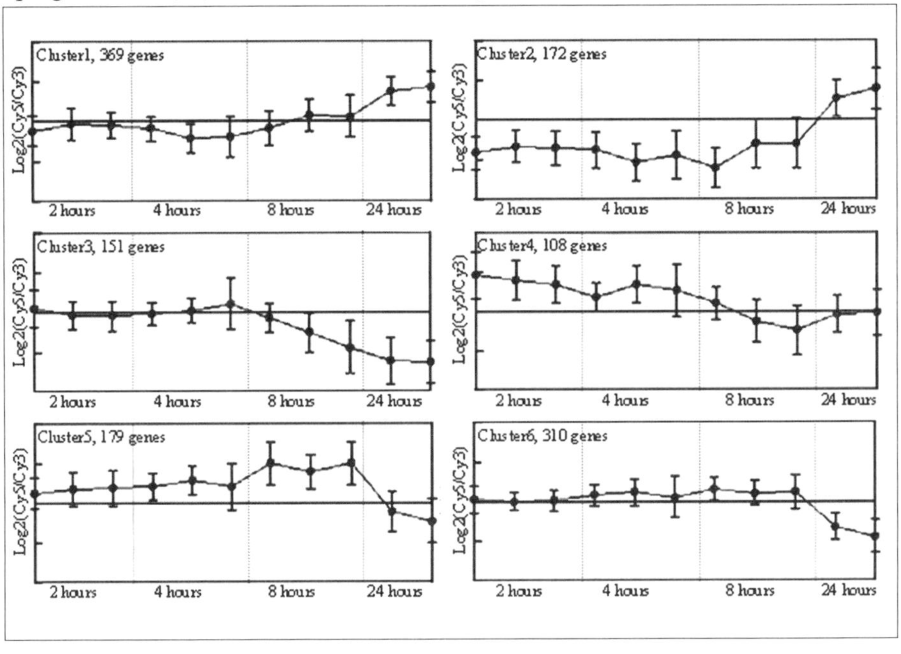

Fig. 1: Clusters derived from time course analysis of gene expression profiles of PBMC at five sampling times before and after immunisation obtained by analysis of variance. Data of comparisons between post-vaccination expression profiles at four sampling times compared to pre-vaccination expression profiles of PBMC were clustered to reveal prominent groups of genes with similar changes in expression patterns using a k-means algorithm. The cluster number and the number of genes is denoted at the top left of each cluster. For each cluster, the log2(post-/pre-vaccination) ratio is on the y-axis. The 11 microarrays analysed (three for each of the comparisons 2, 4, and 8 *vs* 0 hours; two at 24 *vs* 0 hours) are represented along the x-axis. Data points represent means of the log2(post-/pre-vaccination) ratio of the number of genes as denoted for each cluster; error indicators show standard deviations.

Table 3: Functional annotation of differentially expressed genes obtained by clustering of genes showing similar PBMC expression profiles during the early response to immunisation by k-means algorithm.

Category	**Term**	**Count**	**P-Value**
Cluster 1			
GOTERM_BP_ALL	amino acid metabolism	12	0.02
GOTERM_BP_ALL	biosynthesis	43	0.045
GOTERM_BP_AL	cellular biosynthesis	39	0.049
GOTERM_MF_ALL	immunoglobulin binding	4	0.003
GOTERM_MF_ALL	RNA binding	27	0.014
Cluster 2			
GOTERM_BP_ALL	physiological process	112	0.006
GOTERM_BP_ALL	cellular physiological process	106	0.017
Cluster 3			
GOTERM_BP_ALL	induction of apoptosis	5	0.044
GOTERM_BP_ALL	induction of programmed cell death	5	0.044
GOTERM_BP_ALL	nucleic acid metabolism	35	0.05
Cluster 4			
GOTERM_MF_ALL	transferase activity, transferring glycosyl groups	5	0.022
GOTERM_MF_ALL	manganese ion binding	4	0.029
GOTERM_MF_ALL	transferase activity, transferring hexosyl groups	4	0.029
Cluster 5			
GOTERM_BP_ALL	negative regulation of cellular biosynthesis	3	0.043
GOTERM_BP_ALL	negative regulation of biosynthesis	3	0.05
GOTERM_BP_ALL	macromolecule metabolism	63	0.05
Cluster 6			
GOTERM_BP_ALL	actin cytoskeleton organization and biogenesis	12	0.001
GOTERM_BP_ALL	regulation of protein metabolism	13	0.012
GOTERM_BP_ALL	intracellular signalling cascade	32	0.017
GOTERM_BP_ALL	regulation of biological process	77	0.022
GOTERM_BP_ALL	regulation of physiological process	69	0.033
GOTERM_BP_ALL	regulation of cellular process	71	0.033
GOTERM_BP_ALL	regulation of cellular physiological process	67	0.037
GOTERM_BP_ALL	protein biosynthesis	25	0.038
GOTERM_BP_ALL	regulation of progression through cell cycle	18	0.039
GOTERM_BP_ALL	regulation of cell cycle	18	0.039
BIOCARTA	EGF signalling pathway	5	0.027

To get more insight into the functional and structural components and the impact of the genes represented in each cluster, DAVID tools were used to identify functional categories that were significantly overrepresented in each of the defined k-means clusters. Functional annotation charts displaying Gene Ontology terms of enriched genes ($p \leq 0.05$), shown in Table 3, cover a broad spectrum of biological functions and structures. This corresponded to results of pair-wise analysis between expression patterns obtained before and after immunisation that showed transcriptional response and general functional categories.

Quantitative real-time PCR

Transcript abundance of three genes was determined by real-time PCR using individual samples from 18 animals obtained at the five time points prior to and after vaccination. Genes were selected from the groups of genes that were markedly up- or downregulated early after immunisation (clusters 2 and 5) and the group of genes that showed later regulation according to the microarray analyses (cluster 6). *CSPG6* (chondroitin sulphate proteoglycan 6), *XRCC1* (x-ray repair cross complementing protein 1), and *MUTYH* (muty homolog) showed strong regulation in the microarray experiments that was confirmed by real-time PCR. These are genes acting on nucleic acids that might be involved in DNA repair and mitosis. For five other genes, differential regulation at particular time points after vaccination was confirmed by real-time PCR (Fig. 2). In general, a 1.5 fold change in transcript abundance shown by microarray analysis was confirmed by real-time PCR.

Discussion

Specific disease resistance and global defence capability are heritable but difficult to measure and to address by conventional breeding. The measurement of parameters of immune functions was successfully applied in experimental populations towards improved global defence capability. Map-based and function-driven approaches were applied towards the identification of loci affecting resistance to infectious diseases. Genome scans revealed QTL for cellular und humoral immune traits including antibody titres segregating on chromosomes 1, 4, 5, and 6 in an experimental cross of Wild Boar and Yorkshire [5-7]. In an experimental population based on Duroc and Berlin Miniature pigs, QTL for antibody titres were detected on SSC 1, 2, 6, and 7. QTL for humoral innate immune traits (haemolytic complement activity and acute phase response) were detected on all autosomes except SSC 13 [8]. Function-driven attempts cover analyses focused on direct candidate genes identified based on the knowledge of physiology, biochemistry or pathology. Candidate genes that have been analysed include genes of the MHC, T cell receptors, complement factors, acute phase proteins, immune globulins and cytokines.

Approaches analyzing expression profiles of particular subsets of cells and/or individuals with certain phenotypes by microarray analyses detect functional networks controlling the trait of interest and functional candidate genes. In order to follow-up the transcriptional response of PBMC to immunisation, pigs were vaccinated with tetanus toxiod and blood was repeatedly sampled before and after vaccination. Tetanus toxoid is a non-ubiquitous antigen known to promote humoral and cellular, TH1 and TH2 immune responses and thus a good model antigen [9,10]. The progress of an infectious disease largely depends on the very early interaction between host

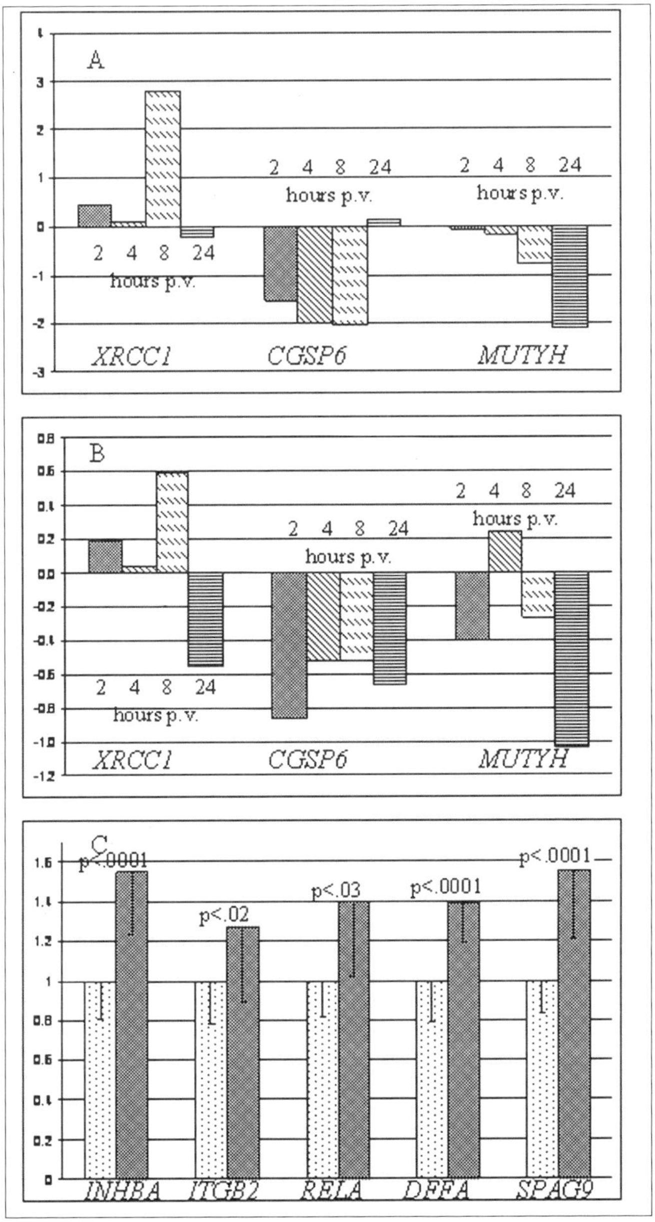

Fig 2: Levels of expression of genes in PBMCs regulated after vaccination according to microarray analysis and confirmed by real-time RT-PCR. A and B: Expression pattern of three genes during the early immune response as determined by real-time RT-PCR (A) and microarray analysis (B). Y-axes show log2 (post-/pre-vaccination) ratio. C: Significant regulation of genes at 8 hours after compared to 0 hours before vaccination. Y-axis shows level of expression relative to 0 hours (=1).

and pathogen. Therefore the early in vivo response was addressed here. In vitro experiments have shown a number of genes that are regulated in cultured PBMC after immune or mitogenic stimulation [11-14]. However, an intense cross-talk of cells and tissues in vivo affects the PBMC transcriptome during the establishment of an immune response. In order to target genes that are candidate genes for immune competency, these endogenous effects need to be taken into account. Genes found to be differentially expressed among the five time points prior to and after immunisation covered general Gene Ontology functional categories like cell structure and motility, nucleotide metabolism, physiological processes, programmed cell death and apoptosis, as well as categories related to the immune response. Using the DAVID Functional Annotation Clustering tool, the first and second clusters showing most significant overrepresentation covered genes related to actin filament polymerization/depolymerization. Clusters 3 and 4 included genes relevant to immune cell activation. Analysis of temporal regulation of gene expression by pair-wise comparison of measurements before and after immunisation revealed that after 2 hours genes were regulated that are relevant to the modulation of responses to external stimulus. At 4 hours, the functional categories of transport and cation binding were enriched. At 8 and 24 hours, genes related to the immune response made up the clusters with the highest enrichment scores. When clustering genes according to a similar course of expression along the five time points, it became obvious that genes showing marked downregulation after 8 hours were related to apoptosis and programmed cell death; genes showing downregulation only after 24 hours control cell structure. Genes upregulated after 24 hours were involved in immunoglobulin binding. Gene showing up- and downregulation soon after immunisation belong to functional categories of physiological process and regulation of biosynthesis. Genes showing co-regulation with immune genes, that were expected to show up in this analysis, but belonging to other functional categories need to be addressed further in order to clarify to what extent their differential expression is caused by the establishment of an immune response or whether their expression might have impact on the immune response.

Conclusion

The results showed that significant in vivo changes of expression profiles of PBMC occurred very early after immune stimuli. These alterations concerned genes of pathways related to the immune response but also other metabolic and regulatory pathways including ATP/energy metabolism, transcription, structural molecule activity, biosynthesis, and metabolism. Analysis revealed new functional candidate genes for traits related to immune responsiveness and also provided new insight into the interaction of the immune response and metabolic and endocrine status. This will facilitate partly understanding the relationship between immune response and performance traits.

References

1 Mallard BA, Wilkie BN, Kennedy BW, Quinton M: Use of estimated breeding values in a selection index to breed Yorkshire pigs for high and low immune and innate resistance factors. Anim Biotech 1992;3:257-280.

2 Saeed AI, Sharov V, White J, Li J, Liang W, Bhagabati N et al: TM4: a free, open-source system for microarray data management and analysis. Biotechniques 2003;34:374-378.

3 Storey JD, Tibshirani R: Statistical significance for genomewide studies. Proc Natl Acad Sci USA 2003;100:9440-9445.

4 Dennis G, Sherman BT, Hosack DA, Yang J, Gao W, Lane HC, Lempicki RA: DAVID: Database for Annotation, Visualization and integrated discovery. Genome Biol 2003;4:R60.

5 Edfors-Lilja I, Wattrang E, Marklund L, Moller M, Andersson-Eklund L, Andersson L et al: Mapping quantitative trait loci for immune capacity in the pig. J Immunology 1998;161:829-835.

6 Edfors-Lilja I: Genetic analysis of immune capacity in the pig. Proceedings of the International Symposium on Candidate Genes for Animal Health, Rostock-Warnemünde, Germany. Arch Tierz Special Issue 1999;42:4-10.

7 Edfors-Lilja I, Wattrang E, Andersson L, Fossum C: Mapping quantitative trait loci for stress induced alterations in porcine leukocyte numbers and functions. Anim Genet 2000; 31:186-193

8 Wimmers K: QTL and gene discovery for immune response and disease resistance traits in the pig. Proceedings of the 3rd Symposium on Genetics of Animal Health, Ames, USA, AgBiotechNet 2005:58-71.

9 ElGhazali GE, Paulie S, Andersson G, Hansson Y, Holmquist G, Sun JB et al: Number of interleukin-4- and interferon-gamma-secreting human T cells reactive with tetanus toxoid and the mycobacterial antigen PPD or phytohemagglutinin: distinct response profiles depending on the type of antigen used for activation. Eur J Immunol 1993;23:2740-2745.

10 Tierney TJ, Simpson-Morgan MW: The immune response of foetal calves. Vet Immunol Immunopathol 1997;57:229-238.

11 Regnström K, Ragnarsson E, Artursson P: Gene expression after vaccination of mice with formulations of diphtheria toxoid or tetanus toxoid and different adjuvants: identification of shared and vaccine-specific genes in spleen lymphocytes. Vaccine 2003;21:2307-2317.

12 Regnström K, Ragnarsson E, Rydell N, Sjöholm I, Artursson P: Tetanus antigen modulates the genes expression profile of aluminium phosphate adjuvant in spleen lymphocytes in vivo. Pharmocogenomics J 2002;2:57-64.

13 Diaz-Mitoma F, Alvarez-Maya I, Dabrowski A, Jaffey J, Frost R, Aucoin S et al: Transcriptional analysis of human periphereal blood mononuclear cells after influenza immunization. J Clin Virol 2004;31:100-112.

14 Cortez KJ, Lyman CA, Kottilil S, Sup Kim H, Roilides E, Yang J et al: Functional genomics of innate host defence molecules in normal human monocytes in response to Aspergillus fumigatus. Infect Imm 2006;74:2353-2365.

PD. Dr. Klaus Wimmers, Research Unit `Molecular Biology´ FBN-Dummerstorf, Wilhelm-Stahl-Allee 2, 18196 Dummerstorf. Germany.

E-mail: wimmers@fbn-dummerstorf.de

Pinard M-H, Gay C, Pastoret P-P, Dodet B (eds): Animal Genomics for Animal Health. Dev Biol (Basel). Basel, Karger, 2008, vol 132, p 197.

SESSION IV

The Search for Markers to Select for Desirable Animal Traits

Pinard M-H, Gay C, Pastoret P-P, Dodet B (eds): Animal Genomics for Animal Health. Dev Biol (Basel). Basel, Karger, 2008, vol 132, pp 199-204.

Detection and Confirmation of a Major QTL Affecting Resistance to Infectious Pancreatic Necrosis (IPN) in Atlantic Salmon (*Salmo Salar*)

R.D. Houston[1], A. Gheyas[1,2], A. Hamilton[2], D.R. Guy[2], A.E. Tinch[2], J.B. Taggart[3], B.J. McAndrew[3], C.S. Haley[1], S.C. Bishop[1]

[1] Division of Genetics and Genomics, The Roslin Institute and Royal (Dick) School of Veterinary Studies, Roslin BioCentre, Midlothian EH25 9PS, UK
[2] Landcatch Natural Selection Ltd., Alloa, Clackmannanshire FK10 3LP, UK
[3] Institute of Aquaculture, University of Stirling, Stirling FK9 4LA, UK

Key Words: Atlantic salmon, disease resistance, recombination, QTL, marker-assisted selection

Abstract: Infectious pancreatic necrosis (IPN) is a viral disease currently presenting a major problem to the aquaculture of Atlantic salmon (Salmon salar), during both the freshwater and seawater stages of production. Genetic variation in resistance to IPN has previously been demonstrated and the purpose of this study was to determine whether this variation includes loci of major effect. The initial QTL detection methodology utilized the limited recombination seen in male salmon to detect QTL in ten large full-sib families, using a genome-wide scan of two to three markers per linkage group. QTL were then positioned by adding additional markers to the significant linkage groups in a female-based analysis. The most significant QTL was mapped to LG 21, and further confirmation of the LG 21 QTL is provided in an analysis of the QTL flanking markers in an additional nine full-sib families from the same population. The size of QTL effect is such that the QTL flanking markers can be immediately applied in marker-assisted selection programmes to improve the resistance of salmon populations to IPN, thus reducing mortality due to the disease.

Introduction

Infectious pancreatic necrosis (IPN) is a highly contagious viral disease that impacts on farmed salmon during two specific windows of the life cycle [1]. The first window of susceptibility to IPN is when salmon are newly hatched as fry in freshwater, and the second window is during a short period following the transfer of fish from freshwater to seawater [1]. The causative agent for IPN is a double-stranded, non-membraned RNA birnavirus (IPNv), of which several subtypes have been characterised [1]. While IPN outbreaks can potentially be prevented in the freshwater environment through stringent biosecurity, the emergence and persistence of virulent IPNv strains in the marine environment currently presents a significant threat to the salmon farming industry worldwide.

Both experimental tank IPN challenges and data collected from field outbreaks have suggested that there is substantial host genetic variation in resistance to IPN [2,3], with most families showing consistent levels of resistance across different challenge sites. Therefore, selective breeding for IPN resistance is potentially a key strategy in the effective control of IPN. Selective breeding programmes may be considerably enhanced by the identification and utilisation of suitable molecular genetic markers for quantitative trait loci (QTL) affecting IPN resistance in salmon.

Mapping QTL in salmon is complicated by the unique structure and properties of their genome. The male salmon has severely limited recombination during meiosis in comparison to the female. As a result, the power to detect QTL in the male is greatly increased, but the trade-off is a reduced ability to position the QTL within a linkage group [4]. This phenomenon can be exploited in a QTL mapping strategy that effectively utilises the low male recombination to detect QTL using relatively few markers dispersed throughout the genome. The significant QTL detected can then be fine mapped on the linkage group by genotyping further markers and tracking segregation through the female parent, thus using the additional recombination to position the QTL.

The purpose of this study was to utilise the aforementioned QTL mapping strategy to detect and position QTL affecting resistance to IPN in several commercial Atlantic salmon families, as previously described by our group [5], and to confirm the population-wide effect of the most significant QTL by analysing a second sample of families from the same population.

Materials and Methods

Detailed materials and methods for the genome scan have been described previously [5]. Briefly, a selection of ten full sib families (total n = 604), with intermediate levels of mortality from a field post-seawater transfer IPN epidemic, were genotyped for 2-3 microsatellite markers on each of the 29 Atlantic salmon linkage groups. The linkage arrangements between the markers were checked for consistency with existing linkage maps, and a linkage map built for this population. A linear-regression QTL analysis was applied using a sire-based analysis for the QTL detection, and subsequently a dam based analysis with additional markers for positioning the QTL on the significant linkage group. The relevant significance thresholds were determined by permutation tests, while the confidence intervals for the QTL were determined by a bootstrap analysis using QTL express.

To confirm the most significant QTL, which was located on LG 21, and to obtain improved estimates of QTL effect and frequency, a further nine full sib families (n = 386), also of intermediate mortality from the same cohort were genotyped for two polymorphic flanking markers for the QTL (BHMS217 and Alu333). These data were analysed as above, both in a sire and dam-based analysis with updated significance thresholds determined by permutation. Also, a combined analysis of all 19 families was performed for the two markers.

Results

As previously reported [5], in the first stage sire-based genome scan, two genome-wide significant QTL were detected on LG 21 and LG 26, with a suggestive QTL on LG 19 (Table 1). For the most significant QTL on LG 21, four sires showed evidence of segregation, and the mean additive effect on mortality for those sires was 0.41 (SE 0.13). This QTL also reached genome-wide significance in the dam-based analysis, whereas the QTL on LG 26 and LG 19 showed suggestive significance (Table 1). Five additional markers were genotyped on LG 21 to position the QTL in the significant families, and the estimated QTL position was narrowed to a confidence interval of approximately 10 cM in these ten families (Fig. 1).

Table 1: Details of the significant and suggestive QTL from the genome scan of the initial ten families (sample 1) (results adapted from [5]), and the analysis of LG 21 in the second sample of nine families (sample 2) showing the significance levels and estimated percentage of within-family variation explained (PVE).

Genome scan sample 1 (10 families n=604)				
Linkage Group	Sire F ratio	Dam F ratio	Significance level	Estimated PVE* (%)
21	4.77	3.57	Genome-wide	20.9
26	3.74	2.54	Genome-wide	14.3
19	2.32	1.45	Chromosome-wide	6.0
LG 21 Sample 2 (9 families n=386)				
21	2.67	7.00	Genome-wide	34.4

*calculated based on both the sire and dam analysis as described in [5].

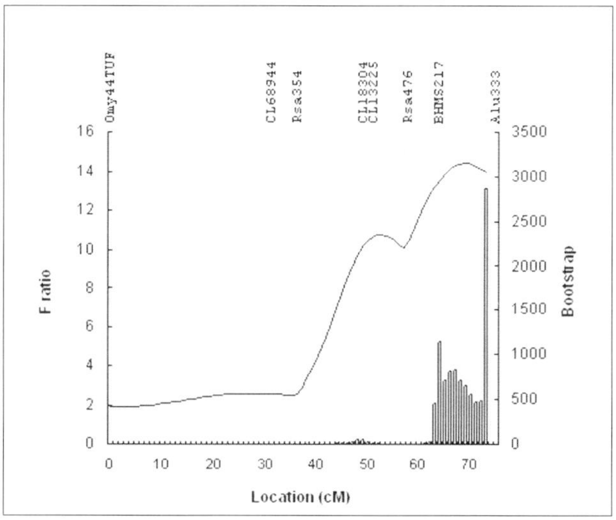

Fig. 1: The likelihood profile for linkage group 21 from the dam-based QTL positioning based on the original 10 families, adapted from [5].

In the nine additional families that were selected from the same year group, the BHMS217 and Alu333 markers were sufficient to detect a QTL significant at the genome-wide level in the dam-based analysis, and at the suggestive level in the sire-based analysis (Table 1). When all 19 families were analysed together for these two markers, the evidence for the QTL was even stronger, with F = 3.35 in the sire-based analysis and F = 7.10 in the dam-based analysis, compared to a genome-wide significance level of F = 2.50. In total, 6 sires and 6 dams were shown to be segregating for the QTL (absolute T value>2) (Table 2). Therefore, approximately 30% of the chosen parental fish were heterozygous for the QTL, implying that the QTL is segregating at moderate frequencies in this commercial population. The mean effect (standard error) on IPN mortality in the segregating parents was 0.49 (0.14) (Table 2), and the overall estimated percentage of within-family variation explained by the LG 21 QTL across the whole dataset was 31.5%.

Table 2: The LG 21 QTL effect on mortality and associated absolute T values in segregating individual parents across both family samples.

Parent (L*M = sire, L*F = dam)	QTL effect estimate (standard error) *	Absolute T value
L3M3076	0.57 (0.13)	4.5
L2M0167	0.50 (0.12)	4.1
L2M0243	0.26 (0.12)	2.1
L3M3080	0.26 (0.13)	2.0
L3M0068	0.53 (0.16)	3.7
L2M0144	0.35 (0.16)	2.1
L2F0056	0.73 (0.14)	5.3
L2F0067	0.35 (0.12)	2.8
L2F0133	0.32 (0.13)	2.6
L2F0695	0.43 (0.13)	3.4
L2F0850	0.65 (0.19)	3.4
L2F0041	0.89 (0.16)	5.7
Mean effect (sire)	0.41	
Mean effect (dam)	0.56	
Mean effect (overall)	0.49	

* The estimate of effect is expressed in terms of mortality proportion.

Two full-sib families were detected where both sire and dam were heterozygous for the QTL. Therefore, to gain an improved indication of the size of the QTL effect within full-sib families, the predicted QTL genotypes of the offspring were calculated based on segregation of the microsatellite marker alleles (Table 3). This showed that offspring inheriting resistance alleles from both parents had a mean mortality of just 4%, compared to offspring inheriting susceptibility alleles from both parents with a mean mortality of 79 %. Heterozygous offspring had mortality levels between these extremes, although there were fewer fish classified as heterozygous due to both parents of one family carrying the same microsatellite alleles.

Table 3: The differences in IPN mortality levels between alternative LG 21 QTL genotypes within two full-sib families where both parents were segregating for the QTL*.

		Sire QTL haplotype	
		Resistant	Susceptible
Dam QTL Haplotype	Resistant	**4% mortality** (n = 23)	**62% mortality** (n = 13)
	Susceptible	**22% mortality** (n = 9)	**79% mortality** (n = 28)

* The number of heterozygotes in this table is lower than expected since one family had parental marker genotypes such that it was not possible to determine which allele originated from which parent for heterozygotes. Overall, the mortality rate for heterozygotes in that family was 16% (N = 19), although there were substantially more survivor than mortality samples in this particular family (31 survivors, 9 mortalities).

Discussion

This study has demonstrated that major QTL affecting resistance to IPN are segregating in commercial Atlantic salmon populations. The largest and most significant QTL was located on LG 21, and this QTL has now been confirmed in a second sample of full sib families from the population. The mean additive effect on mortality for all segregating parents is 0.49, highlighting the large effect of this QTL, although such estimates of additive effects tend to be inflated in QTL studies. Further evidence for the within-family effect of the QTL comes from families where both parents are segregating for the QTL. In two such families, the contrast in mortality between homozygous resistant and homozygous susceptible offspring was 66%.

The detection of this major QTL was based on a novel two-stage genotyping and analysis strategy, where the low recombination rate was utilized in males to give a higher power of QTL detection, with the increased recombination in females used to position the QTL [4,5]. By taking a second sample of families from the same cohort, we have now confirmed this QTL, and have a clearer estimate of the frequency of segregation and the within-family QTL effects. Since 12 out of the 38 parents are segregating for the QTL, this indicates a heterozygous frequency close to 30%, which demonstrates the large scope for improvement in population-wide IPN resistance by selecting for the favourable QTL allele. It should be noted, however, that the families chosen for analysis are not fully representative of the overall population, since they were deliberately chosen for intermediate mortality to maximize the chances of QTL segregation.

To utilize this QTL immediately in marker-assisted selection (MAS), it is necessary to know the phase relationship between the marker allele and the QTL allele. One method of establishing this phase relationship is to genotype siblings of the selection candidates where the siblings have undergone an IPN challenge (as currently occurs in salmon selective breeding programmes). It is then possible to analyse the segregation of the QTL flanking marker alleles from parents to dead and surviving siblings of the selection candidates to identify putative "resistant" and "susceptible" parental haplotypes. The selection candidates themselves can then be genotyped to determine whether they have inherited the "resistant" or susceptible" haplotype and this information can then be used in conjunction with traditional phenotype-based selection to make within-family selection decisions, thereby improving the genetic resistance to IPN in the population. A pilot study of this method is underway in the LNS breeding program. In parallel, the effect of the QTL on other economically important traits is currently being assessed.

While the QTL can be utilized immediately with the flanking markers, it is desirable and interesting to move towards the identification of the causal gene(s). Genomics resources in Atlantic salmon lag behind those in some terrestrial species, although molecular marker maps are improving in density and an extensive physical map is available [6]. Investigation of the gene-expression response to experimental IPN infection of fish carrying alternative QTL alleles may be a fruitful route towards identifying which genes and pathways are important in the underlying physiological effect of the QTL. Integration of such data with improved structural genomics resources will lead towards the specific genes, and polymorphisms, underlying the QTL.

Acknowledgements

The authors gratefully acknowledge Professor William Davidson (Simon Fraser University, Canada), Dr. Bjorn Høyheim (Norwegian College of Veterinary Medicine, Norway) and Dr. Karim Gharbi (University of Glasgow, UK) for assisting with the provision of microsatellite markers for the project. We acknowledge funding from the British Biotechnology and Biological Sciences Research Council (BBSRC), the European Animal Disease Network of Excellence for Animal Health and Food Safety (EADGENE) and a Knowledge Transfer Partnership project (KTP001367).

References

1. Roberts RJ, Pearson MD: Infectious pancreatic necrosis in Atlantic salmon, *Salmo salar* L. J Fish Dis 2005;28:383-390.
2. Guy DR, Bishop SC, Brotherstone S, Hamilton A, Roberts RJ, McAndrew BJ, et al: Analysis of the incidence of infectious pancreatic necrosis mortality in pedigreed Atlantic salmon, *Salmo salar* L., populations. J Fish Dis 2006;29:637-647.
3. Wetten M, Aasmundstad T, Kjøglum S, Storset A: Genetic analysis of resistance to infectious pancreatic necrosis in Atlantic salmon (*Salmo salar* L.). Aquaculture 2007;272:111-117.
4. Hayes BJ, Gjuvsland A, Omholt A: Power of QTL mapping experiments in commercial Atlantic salmon populations, exploiting linkage and linkage disequilibrium and effect of limited recombination in males. Heredity 2006;97:19-26.
5. Houston RD, Haley CS, Hamilton A, Guy DR, Tinch AE, Taggart JB, et al: Major QTL Affect Resistance to Infectious Pancreatic Necrosis in Atlantic Salmon (Salmo salar). Genetics 2008; in press.
6. Ng SH, Artieri CG, Bosdet IE, Chiu R, Danzmann RG, Davidson WS, et al: A physical map of the genome of Atlantic salmon, *Salmo salar*. Genomics 2005;86:396-404.

Dr. Ross Houston, Genetics and Genomics, Roslin Institute (Edinburgh), Midlothian EH25 9PS; UK.
E-mail: ross.houston@bbsrc.ac.uk

Pinard M-H, Gay C, Pastoret P-P, Dodet B (eds): Animal Genomics for Animal Health. Dev Biol (Basel). Basel, Karger, 2008, vol 132, pp 205-212.

Identifying Genes for Intestinal Nematode Resistance using Transcriptional Profiling

O.M. Keane[1,2], *K.G. Dodds*[3], *A.M. Crawford*[3], *J.C. McEwan*[3]

1 AgResearch Molecular Biology Unit, Department of Biochemistry, University of Otago, Dunedin, New Zealand
2 Presently: Smurfit Institute of Genetics, Trinity College, Dublin, Ireland
3 AgResearch Invermay Agricultural Centre, Mosgiel, New Zealand

Key words: sheep, microarray, gastrointestinal nematode, MHC II, DQA1

Abstract: Gene expression was compared between resistant and susceptible Perendale lambs that had either never been exposed to gastrointestinal nematode challenge (had a naïve immune system with respect to parasites) or had been naturally challenged on pasture with nematodes. Only a small number of genes were differentially expressed between the naïve resistant and susceptible animals, but many genes were differentially expressed between the resistant and susceptible challenged animals. The differentially expressed genes were involved in a variety of biological processes, most notably the immune response, the stress response and gene regulation via chromatin remodelling. The transcriptional profiling experiments also detected gene expression differences in the *Ovar-DQA1* gene between resistant and susceptible challenged animals. A null allele of this gene was demonstrated to be associated with susceptibility to gastrointestinal parasites in some, but not all populations. This allele is not thought to be causal for susceptibility.

INTRODUCTION

Gastrointestinal nematodes are a serious cause of morbidity and mortality in grazing ruminants. Infection with these parasites leads to clinical disease and reduced productivity. This results in significant losses for farmers, particularly those in developing countries where personal wealth is closely tied with the number of livestock owned and livestock productivity. Parasites of economic importance include *Trichostrongylus*, *Teladorsagia*, *Cooperia* and *Haemonchus* species. In the developed world, livestock infected with gastrointestinal nematodes have been successfully treated for the past 40 years with anthelmintic drenches. This method of treatment is becoming less useful, however, as anthelmintic resistant nematodes evolve. Alternative strategies of nematode control are urgently required.

Resistance to gastrointestinal nematodes is thought to be primarily acquired via a Th2 mediated immune response. Resistance to gastrointestinal nematodes has been shown to be a heritable trait [1]. Importantly, resistance to one species of nematode also provides protection against other species. Therefore, a potential method of parasite control is to select for resistant animals. This strategy is currently in use in both Australia and New Zealand where the Nemesis and WormFEC programmes selectively breed for nematode resistant animals. This selection is based on the phenotype of the faecal egg count (FEC). This phenotype has low to moderate heritability ($h^2 = 0.2 - 0.3$) and a preferable method of selection would be to select animals based on genotype. Many studies have been carried out to map genes linked or associated with gastrointestinal nematode resistance, but few candidate genes have emerged from these studies, as resistance to nematodes is likely to be extremely complex and polygenic [2,3]. We decided to take a complementary approach and see if genes associated with nematode resistance could be identified using transcriptional profiling. This approach has been previously demonstrated to identify genes responsible for phenotypic variation [4].

Materials and Methods

Duodenum tissue was collected from resistant and susceptible Perendale selection line lambs that were either exposed to natural gastrointestinal nematode challenge or had never been exposed to gastrointestinal nematode challenge. Phenotypic faecal egg count data was also collected. The resistant and susceptible line animals differ in their faecal egg counts by approximately 5-fold [1]. RNA was extracted from the tissue, quantified and hybridized to ovine cDNA microarray slides as described previously [5]. Details of the animals used are summarized in Table 1. At each time point (age), gene expression in four randomly selected, unrelated resistant and four randomly selected, unrelated susceptible animals was compared in a 4 x 4 factorial experimental design as described previously [6]. Gene expression results for the naïve animals were analysed independently, but the results of the three experiments using challenged animals were combined into a single meta-analysis. The data for each slide was normalized according to the procedure of Baird et al [7]. For each expression sequence tag (EST), the normalized data for all slides was combined and ESTs were counted as differentially expressed if they lay outside the 95% confidence limit for the log ratio of the T value. Differentially expressed ESTs were confirmed by resequencing or reamplification from the original clone. All public ovine ESTs were assembled into contigs using CAP3 after an initial clustering step using BLAST. All contigs were annotated using BLASTN against the human RefSeq database and an expectation value cut-off of 1×10^{-10}. Results from the individual microarray experiments are shown in Table 2. All microarray information has been submitted to NCBI Gene Expression Omnibus and the experiment series are GSE3738 and GSE5492. Differential gene expression was confirmed for a subset of the genes using quantitative RT-PCR or Northern blotting as described previously [5,8]. Biological themes associated with the differentially expressed genes were based on the annotated human RefSeqs and were found using EASEonline, Protein Annotators' Assistant and Biorag.

Table 1: Animals used in the microarray experiments.

Expt No	Year born	Average Age / days	Sex	No of R† animals	No of S‡ animals	Ovine array used	Immune status
1	2002	84	M	4	4	11k	Challenged
2	2001	175	M	4	4	11k	Challenged
3	2000	276	M	4	4	20k	Challenged
4	2003	84	F	4	4	20k	Naïve

† R = resistant
‡ S = susceptible

Table 2: Summary of results from four individual microarray experiments.

Expt No	Year born	No. of ESTs in R animals	No. annotated	No. of ESTs in S animals	No. annotated
1	2002	0	0	0	0
2	2001	8	7	89	78
3	2000	204	177	557	500
4	2003 (naïve)	40	16	60	30
1-3	2000-2002 (challenged)	81	51	408	272

Animals were genotyped for the presence or absence of *Ovar-DQA1* using PCR with the primers DQA1exon2F (5′ AATGGAGCCCACAATGTTTG 3′) and DQA1exon2R (5′ TGTCCTTGGAAGGAGA-GAGTTT 3′). Animals were assigned a *DQA1*$^{-/-}$ (null/null) genotype if no PCR product could be detected. Specific *Ovar-DQA1* alleles were determined by sequencing the PCR products and comparing the sequence of bases 1-154 of exon 2 of *DQA1* with that from known *DQA1* genotypes in the published literature. Animals for which only a single DQA1 sequence was obtained were randomly assigned *DQA1*$^{-/+}$ or *DQA1*$^{+/+}$ genotypes in accordance with Hardy-Weinberg equilibrium. Peddrift, a programme that calculates exact probabilities of allele frequency difference in selection lines taking founder effect and genetic drift into account, was used to test for allele frequency differences between the selection lines [9].

Results

Gene expression was compared in four resistant and four susceptible lambs that had never been exposed to gastrointestinal nematodes and had a naïve immune system with respect to intestinal parasites. Only a small number of genes were differentially expressed between resistant and susceptible animals [5]. This indicates that many of the gene products required to resist infection are most likely induced in response to infection and are not constitutively expressed. Genes differentially expressed in naïve animals had a variety of biological functions and the most significant (P <0.01) Biological Process and Molecular Function Gene Ontology (GO) terms associated with them are listed in Table 3. Genes more highly expressed in resistant animals tended to have immune specific functions while genes more highly expressed in susceptible animals tended to be involved in the host response to stress and pathogens and response to stimulus.

Gene expression in resistant and susceptible lambs that had been exposed to nematodes was also compared [8]. Only two genes that were more highly expressed in naïve susceptible animals were also more highly expressed in challenged susceptible animals and these genes were *EVE1* and *TFF3*. No genes were more highly expressed in both naïve and challenged resistant animals. This indicates that infection with gastrointestinal nematodes alters gene expression in both resistant and susceptible animals.

Many genes were differentially expressed between resistant and susceptible animals that had been exposed to nematodes and these genes had a variety of functions [8]. The most significant Gene Ontology terms associated with genes more highly expressed in challenged resistant and susceptible animals are listed in Table 3. Many of the genes more highly expressed in susceptible animals have heat shock response, gene regulation or chromatin remodelling functions. This indicates that gene expression is being altered in these animals in response to infection. Many of the genes more highly expressed in resistant animals had immune, fatty acid metabolism or free radical scavenging functions. The most significant Gene Ontology terms associated with the list of genes more highly expressed in resistant animals related to Major Histocompatibility Complex Class II (MHC II) functions. MHC II encodes cell surface expressed glycoproteins that present exogenously derived peptides to T helper cells. These genes are therefore good candidate genes to be involved in nematode resistance. There have also been many reports in the literature of association between parasite resistance and markers in the MHC region [10-13]. MHC is encoded on sheep chromosome 20 and this region has been shown to be highly polymorphic in many species, including sheep [14]. Three MHC II genes (*Ovar-DQA1*, *Ovar-DQB1* and *Ovar-DRA*) were confirmed by quantitative RT-PCR to be more highly expressed in resistant than susceptible animals and this is illustrated in Figure 1. Interestingly, *Ovar-DRB1* was not found to be differentially expressed, although it is a marker in this gene that has been shown to be associated with resistance to internal parasites in many breeds and populations [12,13]. However, more animals would need to be tested to confirm that *Ovar-DRB1* expression is not associated with nematode resistance.

Table 3: Gene Ontology terms significantly associated with genes differentially expressed between resistant and susceptible Perendale selection line lambs.

Immune Status	Line	GO term	P value
Naive	Susceptible	Response to biotic stimulus	0.00012
		Response to stimulus	0.00028
		Response to external stimulus	0.00067
		Defence response	0.00072
		Response to stress	0.00124
		Organismal physiological process	0.00425
		Immune response	0.00486
		Response to pest/pathogen/parasite	0.00956
Challenged	Susceptible	heat shock protein activity	0.000614
		chromatin assembly/disassembly	0.00149
		peptidyl-prolyl cis-trans isomerase activity	0.00173
		cis-trans isomerase activity	0.00214
		chromosome organization and biogenesis	0.00231
		nuclear organization and biogenesis	0.00266
		protein folding	0.00269
		establishment and/or maintenance of chromatin architecture	0.00314
		DNA packaging	0.00454
		chaperone activity	0.00584
		chloride channel activity	0.0061
Challenged	Resistant	antigen presentation, exogenous antigen	1.6×10^{-5}
		antigen processing, exogenous antigen via MHC class II	1.6×10^{-5}
		MHC class II receptor activity	4.2×10^{-5}
		antigen processing	1.1×10^{-4}
		antigen presentation	1.3×10^{-4}
		fatty acid metabolism	0.00524
		oxidoreductase activity, acting on CH-OH group of donors	0.00239

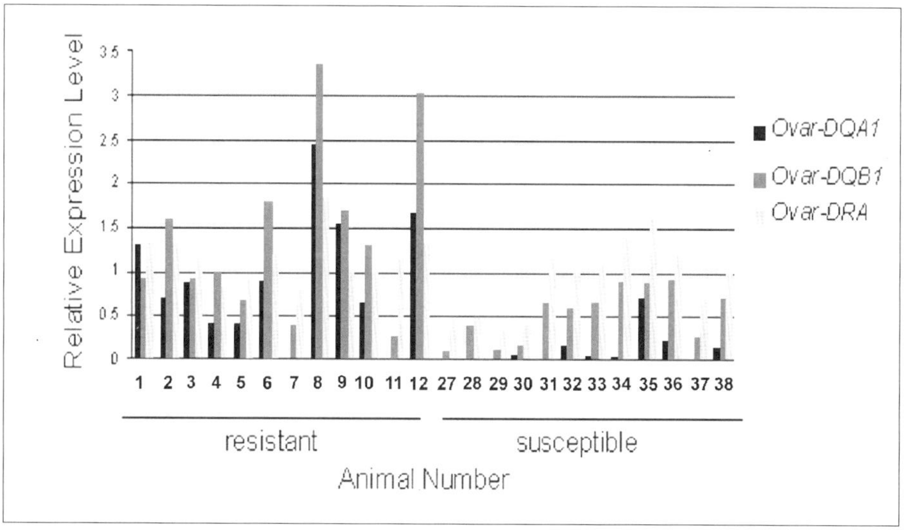

Fig. 1: Relative expression of three MHC Class II genes in the challenged resistant and susceptible Perendale selection line lambs.

As can be seen in Figure 1, some of the animals had very low or non-detectable *DQA1* expression. Null alleles of *DQA1* have been reported to be common in sheep populations [14]. To determine if the animals not expressing *DQA1* did not have *DQA1* in their genome *Ovar-DQA1* was amplified from the genome of 52 animals from our Perendale selection lines. These animals were comprised of 26 resistant animals and 26 susceptible animals and included all animals used in the expression array studies. Fourteen of the animals lacked *DQA1*, but *DQA1* null alleles were much more common in the susceptible selection line than the resistant line (4/26 resistant animals were *DQA1*$^{-/-}$; 10/26 susceptible animals were *DQA1*$^{-/-}$; Table 4). While this allele frequency difference between resistant and susceptible selection line animals was nominally significant, it was not significant when calculated using Peddrift which adjusts the error term for founder effects and genetic drift (P = 0.345). Another selection line, consisting of resistant and susceptible Romney animals that have been divergently selected since 1979 and differ in FEC by 40-fold was also genotyped for the null allele of *DQA1*. In this case there was a significant *DQA1* null allele frequency difference between the resistant and susceptible selection line animals (P = 0.003; Table 4) using the same test.

As the *DQA1* null allele appeared to be associated with parasite susceptibility in at least some populations, we decided to screen commercial populations in order to see if *DQA1* genotype was associated with nematode susceptibility in these populations. In total, 432 commercial progeny tested sires that had breeding values for FEC were genotyped for the presence or absence of *Ovar-DQA1*. These animals consisted of Perendales, Romneys and Coopworths. The *DQA1* null genotype was only associated with susceptibility to internal nematodes in one of the populations (Perendales; Table 4) indicating that the null allele is unlikely to be responsible for parasite susceptibility directly, but may be linked to another allele that contributes to resistance status.

Table 4: Association of Ovar-*DQA1* null allele (*DQA1*$^{-}$) with parasite susceptibility in different sheep populations.

Breed/Line	No of animals genotyped	No of *DQA1*$^{-/-}$ animals	*DQA1*$^{-}$ frequency	P value
Perendale R selection line†	26	4	0.39	0.345
Perendale S selection line†	26	10	0.62	
Perendale commercial‡	113	9	0.28	0.008
Romney R selection line†	50	3	0.24	0.003
Romney S selection line†	50	33	0.81	
Romney Commercial‡	203	15	0.27	0.93
Coopworth Commercial‡	116	1	0.09	0.24

† P value represents association between selection line and *DQA1* null genotype.
‡ P value represents association between estimated breeding value for FEC2 (log of the faecal egg count at 9 months of age) and *DQA1* null genotype.

Discussion

Transcriptional profiling detected many genes that were more highly expressed in resistant or susceptible animals when they were compared to each other. Gene expression was compared in two groups of immunologically distinct populations. One group had never been exposed to gastrointestinal nematodes and had a naïve immune system with respect to nematodes. Only a small group of genes were differentially expressed between resistant and susceptible animals of this group. These genes had a variety of biological functions although genes more highly expressed in resistant animals tended to have immune related functions while genes more highly expressed in susceptible animals tended to have stress response functions. The second group consisted of animals that had been exposed to natural gastrointestinal nematode challenge and many genes were differentially expressed between resistant and susceptible animals of this group. This indicates that many of the gene products required to respond to and control infection are differentially induced or suppressed upon infection. Genes more highly expressed in challenged resistant animals had immune-specific, fatty acid metabolism and free radical scavenging functions while genes more highly expressed in challenged susceptible animals had chromatin remodelling and heat shock response functions. Many more genes were more highly expressed in the duodenum of susceptible challenged animals than resistant challenged animals. The increased expression of these genes in susceptible animals may be the result of infection rather than the cause of susceptibility. Therefore much work remains to be done to distinguish cause from effect in order to develop a useful genetic marker of resistance.

Of the three MHC class II genes more highly expressed in challenged resistant animals, it was noted that one of these genes, *Ovar-DQA1* had non-detectable expression in a number of animals, particularly susceptible animals. This was due to a null allele of *Ovar-DQA1* in these animals. The null allele had a much higher frequency in the susceptible line from two independent selection experiments, however, the allele frequency difference was only significant in the Romney selection line when calculated using Peddrift. The null allele was also found to be significantly associated with a high faecal egg count in commercial Perendale sires, however, there was no association between the *DQA1* null allele and faecal egg count in commercial Romney or Coopworth sires. Although the *DQA1* null allele is a good candidate for nematode susceptibility, the finding that it is associated with nematode susceptibility in some but not all sheep populations indicates that it is not causal for *susceptibility. The* DQA1 gene is located immediately upstream of the *Ovar-DRB1* gene on sheep chromosome 20. Many groups have demonstrated an association between *Ovar-DRB1* alleles and nematode resistance [12,13]. Therefore, the association between *DQA1* and susceptibility could be due to an alternative allele of the resistance enhancing allele detected in other studies. This allele may be in linkage disequilibrium with the null allele of *Ovar-DQA1* in only some populations leading to the observation that *Ovar-DQA1* null is associated with susceptibility to gastrointestinal nematodes in some populations only. Animals lacking *Ovar-DQA1* have been shown to carry a *DQA2*-like gene instead [14]. Alleles of this gene may, therefore, play a role in determining resistance status. Alternatively the null allele of *DQA1* could be in linkage disequilibrium with another gene, either within the MHC region or outside it, that confers resistance. The level of DQA1 expression on the surface of antigen presenting cells may also contribute to resistance status as *DQA1* was still more highly expressed in resistant than susceptible animals even after *DQA1* null animals were excluded from the analysis.

Conclusion

These were the first transcriptional profiling experiments looking at gene expression in the duodenum of sheep resistant and susceptible to gastrointestinal nematodes. These experiments have detected many genes differentially expressed between the animals and these genes are candidate genes for resistance to internal parasites. Much work remains to be done, however, to distinguish genes that contribute to resistance status from those that are merely differentially expressed due to the difference in parasite load between the resistant and susceptible animals. The experiments also confirmed a candidate gene for resistance, namely *Ovar-DQA1*. A null allele of this gene was shown to be associated with susceptibility to gastrointestinal nematodes in some populations. This demonstrated the usefulness of examining gene expression in the search for the underlying genotype responsible for phenotype.

Acknowledgements

We would like to thank Gordon Greer, Nadia McLean, Roger Wheeler, Chris Morris, Wendy Bain and Mary Wheeler for carrying out all animal husbandry, Theresa Wilson and Dianne Hyndman for developing and preparing the microarray slides used, David Baird for normalizing the microarray data, and Meredith Roberts-Thomson for genotyping the Romney selection line for *DQA1*.

References

1 Morris CA, Wheeler M, Watson TG, Hosking BC, Leathwick DM: Direct and correlated responses to selection for high or low faecal nematode egg count in Perendale sheep. NZ J Agric Res 2005;48:1-10.

2 Crawford AM, Paterson KA, Dodds KG, Diez-Tascon C, Williamson PA, Roberts Thomson M, et al: Discovery of quantitative trait loci for resistance to parasitic nematode infection in sheep: I. Analysis of outcross pedigrees. BMC Genomics 2006;7(1):178.

3 Davies G, Stear MJ, Benothman M, Abuagob O, Kerr A, Mitchell S, et al: Quantitative trait loci associated with parasitic infection in Scottish blackface sheep. Heredity 2006;96(3):252-258.

4 Abzhanov A, Kuo WP, Hartmann C, Grant BR, Grant PR, Tabin CJ: The calmodulin pathway and evolution of elongated beak morphology in Darwin's finches. Nature 2006;422:563-567.

5 Keane OM, Zadissa A, Wilson T, Hyndman DL, Greer GJ, Baird DB, et al: Gene expression profiling of Naïve sheep genetically resistant and susceptible to gastrointestinal nematodes. BMC Genomics 2006;7:42.

6 Diez-Tascon C, Keane OM, Wilson T, Zadissa A, Hyndman DL, Baird DB, et al: Microarray analysis of selection lines from outbred populations to identify genes involved with nematode parasite resistance in sheep. Physiol Genomics 2007;21:59-69.

7 Baird D, Johnstone P, Wilson T: Normalisation of microarray data using a spatial mixed model analysis which includes splines. Bioinformatics 2004;20:3196-3205.

8 Keane OM, Dodds KG, Crawford AM, McEwan JC: Transcriptional profiling of *Ovis aries* identifies *Ovar-DQA1* allele frequency differences between nematode resistant and susceptible selection lines. Physiol Genomics 2007; 30:253-261

9 Dodds KG, McEwan JC: Calculating exact probabilities of allele frequency differences in divergent selection lines. Proc Assoc Advmt Anim Breed Genet 1997;12:556-560.

10 Buitkamp J, Filmether P, Stear MJ, Epplen JT: Class I and class II major histocompatibility complex alleles are associated with faecal egg counts following natural, predominantly Ostertagia circumcincta infection. Parasitol Res 1996;82(8):693-696.

11 Paterson S, Wilson K, Pemberton JM: Major histocompatibility complex variation associated with juvenile survival and parasite resistance in a large unmanaged ungulate population. Proc Natl Acad Sci USA 1998;95(7):3714-3719.

12 Sayers G, Good B, Hanrahan JP, Ryan M, Angles JM, Sweeney T: Major histocompatibility complex DRB1 gene: its role in nematode resistance in Suffolk and Texel sheep breeds. Parasitology 2005;131(Pt 3):403-409.

13 Schwaiger FW, Gostomski D, Stear MJ, Duncan JL, McKellar QA, Epplen JT, et al: An ovine major histocompatibility complex DRB1 allele is associated with low faecal egg counts following natural, predominantly *Ostertagia circumcincta infection.* Int J Parasitol 1995;25(7):815-822.

14 Snibson KJ, Maddox JF, Fabb SA, Brandon MR: Allelic variation of ovine MHC class II DQA1 and DQA2 genes. Anim Genet 1998;29(5):356-362.

Dr. Orla Keane, 228Ryevale Lawns, Leixlip, Co. Kildare, Ireland.

E-mail: orlakeane@yahoo.com.au

Pinard M-H, Gay C, Pastoret P-P, Dodet B (eds): Animal Genomics for Animal Health. Dev Biol (Basel). Basel, Karger, 2008, vol 132, pp 213-218.

High Resolution, Advanced Intercross Mapping of Host Resistance to *Salmonella* Colonization

J. R. Hasenstein, A.T. Hassen, J.C.M. Dekkers, S.J. Lamont

Department of Animal Science and Center for Integrated Animal Genomics, Iowa State University, Ames, Iowa, USA

Key words: Salmonella, chicken, QTL, SNP, advanced intercross line

Abstract: Availability of a dense single-nucleotide polymorphism (SNP) map in chickens has allowed for whole-genome QTL mapping for disease resistance. In this study, two F8 advanced intercross lines of broiler by Fayoumi and broiler by Leghorn chickens, and a dense, genome-wide SNP panel were used to map genomic regions associated with host resistance to bacterial colonisation. One week after inoculation of day-old chicks with *Salmonella enteritidis* (SE), caecum and spleen tissues were collected to quantitate the bacterial load. Of 2,733 genotyped SNPs, 875 were homozygous and the remaining SNPs with a minor allele frequency of > 0.2 were individually tested for association with SE burden by utilising a Chi-square log-likelihood test between models with and without SNP genotype. Using a Q-value of 25%, calculated utilising 1,207 SNPs with Chi-square P-Value < 1.0 to control for false discovery, 21 SNPs identifying 19 genes were significantly associated with SE bacterial levels. Ten genes were in pathways associated with immune response to Salmonella (toll-like receptor signaling, apoptosis, and MAPK signaling), further supporting their involvement in host resistance pathways. In addition to identifying new candidate genes for bacterial resistance, the trait-associated SNPs may be useful in marker assisted selection programmes for disease resistance.

INTRODUCTION

A major safety concern in animal production and the processing of food products is the avoidance of pre-harvest and post-harvest pathogenic contamination. *Salmonella* is a major food poisoning agent in the United States, the European Union, and many developing nations. Consumers are also concerned about possible emergence of antibiotic-resistant bacteria due to subtherapeutic use of antibiotics in livestock. Genetic selection programmes may enhance vaccine efficacy and host resistance to disease, while lowering dependence on antibiotics.

Sequencing of the chicken genome [1] and the simultaneous identification of millions of single-nucleotide polymorphisms (SNPs) [2] has allowed for the development of large SNP panels which can be used to identify genomic regions containing the quantitative trait loci (QTL) controlling complex biological traits. Typically, marker-trait association studies for livestock have been performed on either F2 crosses or backcrosses which possess a large amount of linkage disequilibrium (LD). To more accurately position the QTL location, reductions in LD may be obtained through advanced intercross lines (AIL) or recombinant inbred lines [3].

The specific objective of this study was to analyse SNPs for associations with Salmonella bacterial burden in poultry in two F8 advanced intercross lines utilising an SNP panel of over 3,000 SNPs. This works contributes to the overall goal of identification of DNA markers for phenotypic variation in host response to bacterial pathogens.

MATERIALS AND METHODS

Two advanced intercross lines were established from crosses between highly inbred (>99%) Leghorn (Leg) and Fayoumi (Fay) dams and a single commercial outbred Broiler (Br) sire [4]. Two separate AIL were maintained with random mating and avoidance of sib mating. Sixty-seven Br x Fay and 66 Br x Leg F8 birds from two hatches were used. Numbers of birds per hatch and sex were similar. Chicks were inoculated intraoesophagally with pathogenic SE phage type 13a at 1 day of age and euthanized at one week. The spleen and one caecum were removed aseptically and the bacterial counts in these tissues were quantified as representatives of enteric and visceral organs [5]. Bacterial counts were transformed to natural logarithms to achieve normal distributions.

Genomic DNA was typed by Illumina (San Diego, CA) on a chicken Genotyping BeadChip, which generated 2,733 successful SNP genotypes. The SNPs with minor allele frequency (MAF) <0.2, within line, were removed from the study. Associations between individual SNP genotypes and the SE bacterial count of each tissue were determined through comparisons of two (full and reduced) linear mixed models using the SAS programme [6].

Full Model: $Y_{ijklmnpqrs} = \mu + hatch_i + sex_j + room_k + necropsy\ day_l + line_m + SNP\ genotype_n + SNP\ genotype*line_p + sire(line)_q + dam(sire, line)_r + e_{ijklmnpqrs}$

where $Y_{ijklmnpqrs}$ is the response variable from each individual F8 bird (natural logarithms of spleen or caecal bacterial count). The reduced model was the same as the full model, with the removal of all terms containing SNP genotype. Hatch, sex, room, necropsy day, line, SNP genotype, and SNP genotype*line were set as fixed effects, while sire(line) and dam(sire, line) were fitted as independent random effects. A Chi-square analysis was conducted comparing the full and reduced model to test for significance:

$$X^2 = -2[\log L(\text{Reduced Model}) - \log L(\text{Full Model})]$$
$$P = [1\text{-probCHI}(X^2, df)],\ df = s\text{-}n$$

where s equals number of line*genotype classes and n equals number of lines. Sex, hatch, room, and necropsy day were not significant and were excluded from the final analysis.

With multiple tests being conducted, q-values were calculated on the Chi-square results to estimate the minimum false discovery rate (FDR), represented as the expected proportion of false positives among the tests found to be significant [7]. A threshold level of 25% FDR was set and q-values were calculated utilising the R statistical programme Q-value. Each significant SNP was examined for an interaction of genotype with line.

RESULTS

Of the 3,072 SNPs on the panel, 2,733 were successfully genotyped. Initially, SNPs were removed from analysis if MAF < 0.2 and, subsequently, if the difference between the full and reduced models resulted in a Chi-squared P-value of 1.0 (no difference between models indicating no genotype effect). A total of 1,207 (44.1%) SNPs passed these two steps. After correcting for multiple comparisons by utilising q-values, and selecting an FDR of 25%, a total of 21 SNPs were significant for SE bacterial levels in the birds (Table 1), and none had significant interaction with line.

Table 1: SNPs and genes associated with SE bacterial burden in AIL chicken caecum and spleen.

SNP Name	Chr	SNP Position (Mb)	Organ	X^2 P-Value	Gene Name	Description
snp-7-85-3009-S-1	1	53.5	Caecum	0.009	*TST*	Mitochondrial matrix enzyme encoded by nucleus
snp-5-453-22528-S-2	2	45.5	Caecum	0.012	near *ARPP-21*	Protein phosphatase type 1 regulator activity
snp-5-477-15705-S-2	2	46.1	Caecum	0.012	*MLH1*	Cancer gene, involved in DNA mismatch repair
snp-0-517-22931-S-1	3	78.4	Caecum	0.002	*UBE2J1*	Recognizes degradation signals characteristic of the target protein
snp-1-1110-22847-S-1	5	34.3	Caecum	0.005	near *MEIS2*	Role in cell cycle control, proliferation, and apoptosis
snp-1-1005-6873-S-1	5	36.9	Caecum	0.006	*NUPBL*	Involved in regulation of centrosome duplication
snp-54-156-6817-S-2	7	20.6	Caecum	0.009	*STK39*	Stimulates apoptotic signaling induced by *TRAIL*
snp-13-440-8821-S-2	7	24.2	Caecum	0.004	*TNS1*	Negatively regulates activation of *NFκ-B*
snp-109-86-3227-S-3	10	17.9	Caecum	0.015	near *IGF1R*	Transmembrane tyrosine kinase, mediates apoptosis
snp-572-1-2764-S-2	13	14.1	Caecum	0.016	near *MAPK9*	Influences both antigen-induced activation and apoptosis
snp-24-25-5353-S-2	18	0.4	Caecum	0.015	*MYH13*	Involved in myosin's catalytic activity and fast speed contractions
snp-50-62-8294-S-2	24	0.8	Caecum	0.017	near *TIRAP*	Extracellular matrix protein involved in signal transduction and apoptosis of dendritic cells
snp-50-198-20693-S-1	24	2.8	Caecum	0.009	near *ATP12A*	Phosphorylation protein activated in response to hypotonic stress
snp-5-388-49809-S-2	2	43.2	Spleen	0.003	*KIF15*	Component of intracellular signal transduction
snp-5-453-22528-S-2	2	45.5	Spleen	0.008	near *ARPP-21*	Protein phosphatase type 1 regulator activity
snp-5-477-15705-S-2	2	46.1	Spleen	0.007	*MLH1*	Cancer gene, involved in DNA mismatch repair
snp-0-583-46062-S-1	3	81.0	Spleen	0.010	*IBTK*	Activated by LPS and is required for LPS-induced TNF production
snp-3-177-39024-S-2	4	36.3	Spleen	0.020	*FAM13A1*	Involved in myosin's catalytic activity and slow speed contractions
snp-10-586-6490-S-2	8	14.6	Spleen	0.015	*BCAR3*	Involved cell adhesion/migration and antiestrogen resistance
snp-25-120-19345-S-1	9	7.6	Spleen	0.020	near *SLC25A36*	Important in metal transport, especially iron
snp-114-104-1895-S-2	28	2.7	Spleen	0.007	*SF4*	Functions in pre-mRNA splicing mechanisms

Thirteen SNPs were associated with bacterial burden in the caecum and eight in the spleen. Each SNP accounted for approximately 3-4% of the phenotypic variation in bacterial burden. Nine SNPs were in exons, four in introns, and eight were located within 200 kb of the 3' or 5' region of the genes.

The 21 significant SNPs are in or near genes with a wide variation of molecular activity (Table 1). Six genes (*IBTK, STK39, TNS1, TIRAP, IGF1R,* and *MAPK9*) are involved in the immune response. Four genes (*KIF15, ARPP21, MEIS2,* and *UBE2J1*) function in intracellular signal messaging. The remaining nine genes are involved in cellular maintenance, DNA synthesis and processing, and membrane structure. *ARPP21* and *MLH1* were associated with Salmonella burden in both organs.

Discussion

Host response to SE in poultry is a polygenic trait for which several individual candidate genes have been identified [4,8-13]. These genes have generally been selected for study based upon prior knowledge of their function in similar systems. Thus, the confirmation of their role in response to Salmonella in poultry is valuable, but not surprising, because their study was biased in a positive fashion to previously identified genes. The current availability of complete genome sequences for livestock opens the opportunity for true gene-discovery experiments in which the identification of QTL for disease resistance is not confined to the investigation of genes previously associated with similar functions. This timely opportunity was the impetus for the current study. The major finding of this study was the association of 21 SNPs in 19 different genes with bacterial burden in the caecal content and spleen of two F8 AILs of chickens.

Several of the SNPs identified in the current study as having significant associations with bacterial burden levels were located in or near genes that regulate eukaryotic immune responses (Fig. 1). Tensin 1 (*TNS1*) negatively regulates activation of nuclear factor kappa-B driven (*NF-κB*) transcription, an important cellular transcription factor located downstream from other candidate genes for response to SE infection. Found upstream of *NF-κB* in the Toll-like receptor pathway, inhibitor of Bruton's tyrosine kinase (*IBTK*) has been linked to the immune response to lipopolysaccharide (LPS), a surface component of SE.

Apoptosis, which occurs in many cells infected with SE, is a morphological process of cellular self-destruction controlled through complex pathways of intracellular signals. The current study identified associations between bacterial burden and SNPs for serine threonine kinase 39 (*STK39*), toll-interleukin 1 receptor protein (*TIRAP*), insulin-like growth factor 1 receptor (*IGF1R*), and mitogen-activated protein kinase 9 (*MAPK9*). These genes are in the same pathways as genes previously identified (inhibitor of apoptosis protein-1 (*IAP1*), toll-like receptor 4 (*TLR4*), MD2 protein (*MD2*), and tumour necrosis factor related apoptosis inducing ligand (*TRAIL*) as having SNPs associated with SE bacterial burden levels in chickens (Fig. 1) [11,12,14,15]. The agreement between the findings of the current study and previous studies suggest that apoptosis signaling genes are critical in the innate immune response to SE in chickens.

Four genes identified in the current study are involved in signal transduction or cell signaling. Kinesin family member 15 (*KIF15*), cyclic AMP-regulated phosphoprotein 21 kD (*ARPP-21*), Meis homeobox 2 (*MEIS2*), and ubiquitin-conjugating enzyme E2 J1 (*UBE2J1*) are in pathways involving intracellular signaling for either cellular functions or protein degradation. These genes may start or participate in a signal cascade to induce the immune response to bacterial infections.

The current results demonstrate the usefulness of a genome-wide SNP study to locate regions of the genome associated with response to a pathogen challenge in livestock. Three unique lines (Fayoumi, Leghorn and broiler) were used in the development of the resource population, and their distinct genetics generated a high percentage of polymorphic loci in the genotyped SNPs. Additionally, the use of F_8 AIL reduced the linkage disequilibrium around the significant SNPs, thus providing more accurate fine mapping of the true location of the QTL. The fact that the high-density SNP array was able to detect novel candidate genes located within the same pathways as previously identified biological candidates demonstrates the value of the approach to discover new genes that are associated with disease related traits.

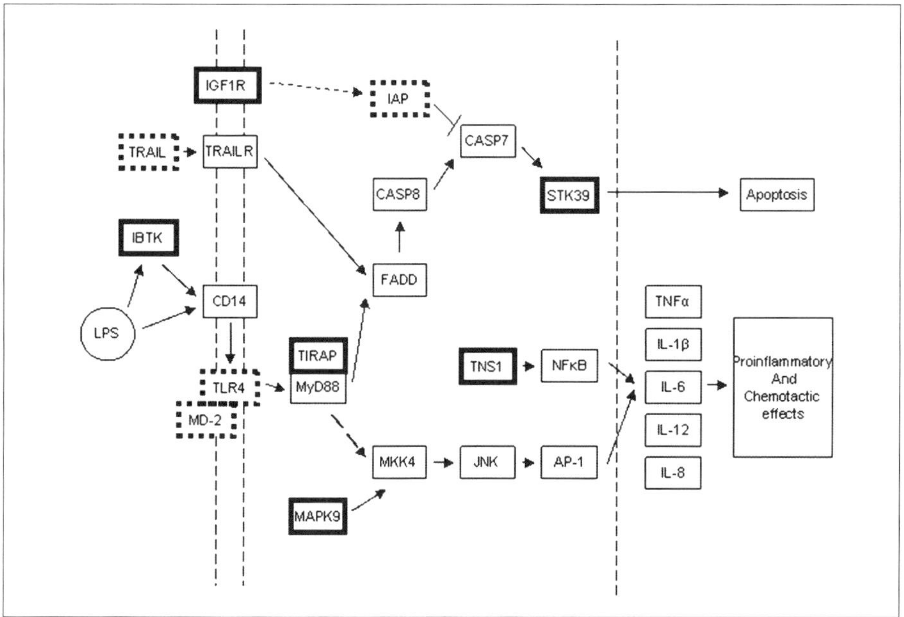

Fig. 1: Immune function genes and signaling pathways. Bold outlines = genes identified in the current study. Dashed outlines = genes previously identified with SNPs with significant associations in host response after SE challenge of related resource populations.

Conclusion

This study demonstrates the association of 21 SNPs with SE colonisation in the caecum and spleen of two F_8 advanced intercross lines of chickens. These SNPs provide markers that may be valuable used in marker-assisted selection programmes to enhance the immune response of young birds to SE. The results identified many new candidate genes for SE bacterial burden, and also validated previous studies that identified associations with genes involved in the toll-like receptor, apoptosis, and kinase pathways.

ACKNOWLEDGEMENTS

This paper of the Iowa Agriculture and Home Economics Experimental Station, Ames, Iowa, was supported by Animal Health and Iowa State University Center for Integrated Animal Genomics Funds, National Research Initiative Grant no. 2004-35205-14234 from the USDA Cooperative State Research, Education, and Extension Service, and Research Grant US-3408-03 from BARD, the Binational Agriculture Research and Development fund. The SNP assays were developed with the support of the USDA Agricultural Research Service and through the efforts of Drs. Cheng, Muir, Wong, Groenen, and Zhang funded by USDA-CSREES-NRICGP proposal no. 2004-05434. The authors gratefully acknowledge the Poultry Research Center crew at Iowa State University for managing the birds, and Michael Kaiser, Jennifer Cheeseman, Xianghai Ye, and Bill Larson for technical support.

REFERENCES

1. Hillier LW, Miller W, Birney E, Warren W, Hardison RC, Ponting CP, et al: Sequence and comparative analysis of the chicken genome provide unique perspectives on vertebrate evolution. Nature 2004;432:695-716.
2. Wong G, Liu B, Wang J, Zhang Y, Yang X, Zhang Z, et al: A genetic variation map for chicken with 2.8 million single-nucleotide polymorphisms. Nature 2004;432:717-722.
3. Darvasi A, Soller M: Advanced intercross lines, an experimental population for fine genetic mapping. Genetics 1995;141:1199-1207.
4. Liu W, Kaiser MG, Lamont SJ: Natural resistance-associated macrophage protein 1 gene polymorphisms and response to vaccine against or challenge with *Salmonella* enteritidis in young chicks. Poult Sci 2003;82:259-266.
5. Kaiser MG; Lamont SJ: Genetic line differences in survival and pathogen load in young layer chicks after *Salmonella enterica* serovar enteritidis exposure. Poultry Sci 2001;80:1105-1108.
6. SAS: SAS/STAT User's Guide. SAS Institute Inc., Cary, N.C. 2004.
7. Storey, JD, Tibshirani R: Statistical significance for genomewide studies. Proc Natl Acad Sci USA 2003;100:9440-9445.
8. Beaumont C, Protais J, Pitel F, Leveque G, Malo D, Lantier F, et al : Effect of two candidate genes on the *Salmonella* carrier state in fowl. Poult Sci 2003;82:721-726.
9. Hasenstein JR, Lamont SJ: Chicken gallinacin gene cluster associated with *Salmonella* response in advanced intercross line. Avian Dis 2007;51:561-567.
10. Girard-Santosuosso O, Menanteau P, Duchet-Suchaux M, Berthelot F, Mompart F, Protais J, et al: Variability in the resistance of four chicken lines to experimental intravenous infection with *Salmonella enteritidis* phage type 4. Avian Dis 1999;42:462-469.
11. Kramer J, Malek M, Lamont SJ: Association of twelve candidate gene polymorphisms and response to challenge with *Salmonella enteritidis* in poultry. Anim Genet 2003;34:339-348.
12. Malek M, Hasenstein JR, Lamont SJ: Analysis of chicken TLR4, CD28, MIF, MD-2, and LITAF genes in a *Salmonella enteritidis* resource population. Poult Sci 2004;83:544-549.
13. Tilquin P, Barrow PA, J. Marly J, Pitel F, Plisson-Petit F, Velge P, et al: A genome scan for quantitative trait loci affecting the *Salmonella* carrier-state in the chicken. Genet Sel Evol 2005;37:539-561.
14. Liu W, Lamont SJ: Candidate gene approach: potentional association of caspase-1, inhibitor of apoptosis protein-1, and prosaposin gene polymorphisms with response to *Salmonella* enteritidis challenge or vaccination in young chicks. Anim Biotechnol 2003;14:61-76.
15. Malek M, Lamont SJ: Association of INOS, TRAIL, TGF-beta2, TGF-beta3, and IgL genes with response to *Salmonella* enteritidis in poultry. Genet Sel Evol 2003;35 Suppl 1:S99-111.

Dr. Susan J. Lamont, Department of Animal Science, 2255 Kildee Hall, Iowa State University, Ames, IA 50011 USA.

E-mail: sjlamont@iastate.edu

Pinard M-H, Gay C, Pastoret P-P, Dodet B (eds): Animal Genomics for Animal Health. Dev Biol (Basel). Basel, Karger, 2008, vol 132, pp 219-223.

Predicting Genetic Merit for Mastitis and Fertility in Dairy Cattle using Genome Wide Selection and High Density SNP Screens

H.W. Raadsma[1,2], G. Moser[1], R.E. Crump[1,4], M.S. Khatkar[1,2], K.R. Zenger[1,2], J.A.L. Cavanagh[1,2], R.J. Hawken[1,3], M. Hobbs[1,2], W. Barris[1,3], J. Solkner [1,2], F.W.Nicholas[1,2], B.Tier[1,4]

1 Co-operative Research Centre for Innovative Dairy Products-CRC IDP
2 ReproGen – Centre for Advanced Technologies in Animal Genetics and Reproduction, Faculty of Veterinary Science, The University of Sydney, Camden, Australia
3 CSIRO Livestock Industries, Brisbane, Australia
4 AGBU, University of New England, Armidale, Australia

Key words: genome wide selection, SNP, dairy cattle, somatic cell count, fertility

Abstract: Two novel methods for genome wide selection (GWS) were examined for predicting the genetic merit of animals using SNP information alone. A panel of 1,546 dairy bulls with reliable EBVs was genotyped for 15,380 SNPs that spanned the whole bovine genome. Two complexity reduction methods were used, partial least squares (PLS) and regression using a genetic algorithm (GAR), to find optimal solutions of EBVs against SNP information. Extensive internal cross-validation was used to find the best predictive models followed by external validation (without direct use of the pedigree or SNP location). Both PLS and GAR provided both accurate fit to the training data set for somatic cell count (SCC) (max r = 0.83) and fertility (max r = 0.88) and showed an accuracy of prediction of r = 0.47 for SCC, and r = 0.72 for fertility. This is the first empirical demonstration that genome wide selection can account for a very high proportion of additive genetic variation in fitness traits whilst exploiting only a small percentage of available SNP information, without use of pedigree or QTL mapping. PLS was computationally more efficient than GAR.

INTRODUCTION

Consequently to the development of high-density large-scale single nucleotide polymorphism (SNP) genotyping platforms [1], we recently reported on the development of genome wide selection (GWS) in cattle [2-4]. This paper discusses the development of a GWS platform for dairy cattle with a specific focus on fertility and somatic cell count (SCC).

GENOME WIDE ANALYSES TO PREDICT GENETIC MERIT

Whole-genome analyses require methods that are capable of handling cases where the number of marker variables largely exceeds the number of individuals. A common approach to address this problem is to select a 'few' important markers to construct statistical models by testing multiple hypotheses about their gene action. It is well known that estimates of a subset of single-nucleotide polymorphism (SNP) effects or quantitative trait loci (QTL) are upwards biased [5-8]. Furthermore, SNP selection methods perform poorly on collinear markers, due to the sheer amount of SNPs and the co-segregation of alleles due to linkage disequilibrium as described by Khatkar et al [9].

Alternative statistical methods that treat SNP effects as random variables [10-12] avoid or reduce the impact of overestimation and collinearity but, as pointed out by Gianola et al [13], these methods require some strong assumptions, e.g. known distribution of SNP effects. The robustness of these methods is unknown and has only been shown to work in simulation.

GWS is performed to maximize the relationship between genetic merit derived from molecular data and true breeding value (TBV) or a proxy based on estimated breeding value (EBV). Other than the requirements that markers are located across the genome, no additional information, such as marker location or pedigree, is required to assemble a molecular breeding value (MBV). GWS thus differs substantially from QTL, candidate gene and high density marker association approaches, which all focus on estimation of single locus effects and assembling these in a multi-locus or multi-marker model to predict genetic merit.

DEVELOPMENT OF A GWS PLATFORM FOR DAIRY CATTLE

Selection of animals, SNP selection, SNP quality scoring and genome scanning and quality control (QC) of SNP genotype data were described by Raadsma et al [4], Khatkar et al [14] and Zenger et al [15]. In brief, 1,546 Australian progeny-tested Holstein Friesian dairy bulls were screened for 15,036 SNP markers. The sires were born between 1955 and 2001, with more than 96% of sires born after 1980. All sires had EBV data calculated from the industry genetic evaluation service by ADHIS (Australian Dairy Herd Improvement Scheme) for 42 traits associated with lactation performance, conformation, reproductive fitness and disease resistance. Over 85% of sires had a reliability greater than 0.70 for their EBV.

SNP complexity reduction

The challenge of dealing with over parameterized data sets where the number of SNP variables greatly exceeds the number of observations was dealt with by using elements of feature reduction or feature selection. Several approaches for analyzing high-dimensional whole-genome SNP data were used, including the method of supervised dimension reduction through partial least squares (PLS) as described by Moser et al [3]. Optimal search algorithms for exploring the parameter space using genetic algorithm regression (GAR) as described by Crump et al [2] were used for prediction of genetic merit based on MBV. Other methods such as principal component analyses or least angle regression were not used here, neither were non-statistical SNP reduction methods such as tag SNPs in defined haplotypes.

Prediction and validation of MBV

Model building was used in a subset of the 1,546 genotyped sires (the 'training' set), which was further divided into a number of other subsets for internal cross-validation to find the best predictive models. External validation was used to predict the genetic merit of a complementary group of sires (the 'test' set), without direct use of the pedigree or SNP location. Both PLS and GAR provided an accurate fit to the training data set for SCC (max $r = 0.83$) and fertility (max $r = 0.88$). Of greater interest is the accuracy of prediction in the test data set where EBV for SCC could be predicted with an accuracy of $r = 0.47$, and $r = 0.72$ for fertility. The relationship between MBV and EBV in both training and test data is shown in Figure 1a for fertility, and Figure 1b for SCC.

This is the first empirical demonstration that GWS can account for a very high proportion of additive genetic variation in fitness traits whilst exploiting only a small percentage of available SNP information, without use of pedigree or QTL mapping. However, for predictions regarding young cohorts of bulls based on derived predictions from the older population (forward prediction), a reduction in accuracy of prediction was apparent. It is unknown if pre-selection of the young bulls based on pedigree information, poor accuracy of early proofs, insufficient SNP information, or population structure affect the utility of this approach. However no loss or gain of efficiency was observed when 6,000 of the available SNPs were used in GWS development or panels were extended to 25k SNPs, respectively.

Deriving MBV from a population in which future predictions have to be made is immediately applicable to young sire and elite dam selection (as described by Tier et al [16]). Features of GWS can readily be incorporated with advanced reproductive technologies, leading to greatly increased rates of genetic gain and potential significant cost reduction, as breeding programmes move from progeny testing in sire selection to progeny validation. Use of MBV allows for screening of suitable germplasm from global sources, and may possibly extend to incorporate G x E and G x G and a numerator relationship matrix based on shared genome content in genetic evaluation. The critical subset of SNPs to give best prediction, termed here "molecular keys" for GWS, can readily be updated as new sires enter the industry. Alternatively, as genotyping costs continue to decrease and additional use of SNP marker information is envisaged, such as in quality assurance and as biomarker prospects for phenotype management, whole genome scans may be conducted on all animals which contribute to information management. It is unknown to what extent the same molecular key can predict MBV and therefore TBV in other breeds, but it warrants exploration. Similarly, the frequency by which the molecular relationships with MBV need updating is unknown; but it would be reasonable, in dairy cattle at least, for all new sires with significant progeny records to have a high density genome scan performed. Application of GWS may make QTL and gene discovery programmes for marker assisted selection (MAS) redundant and offer significant new promise.

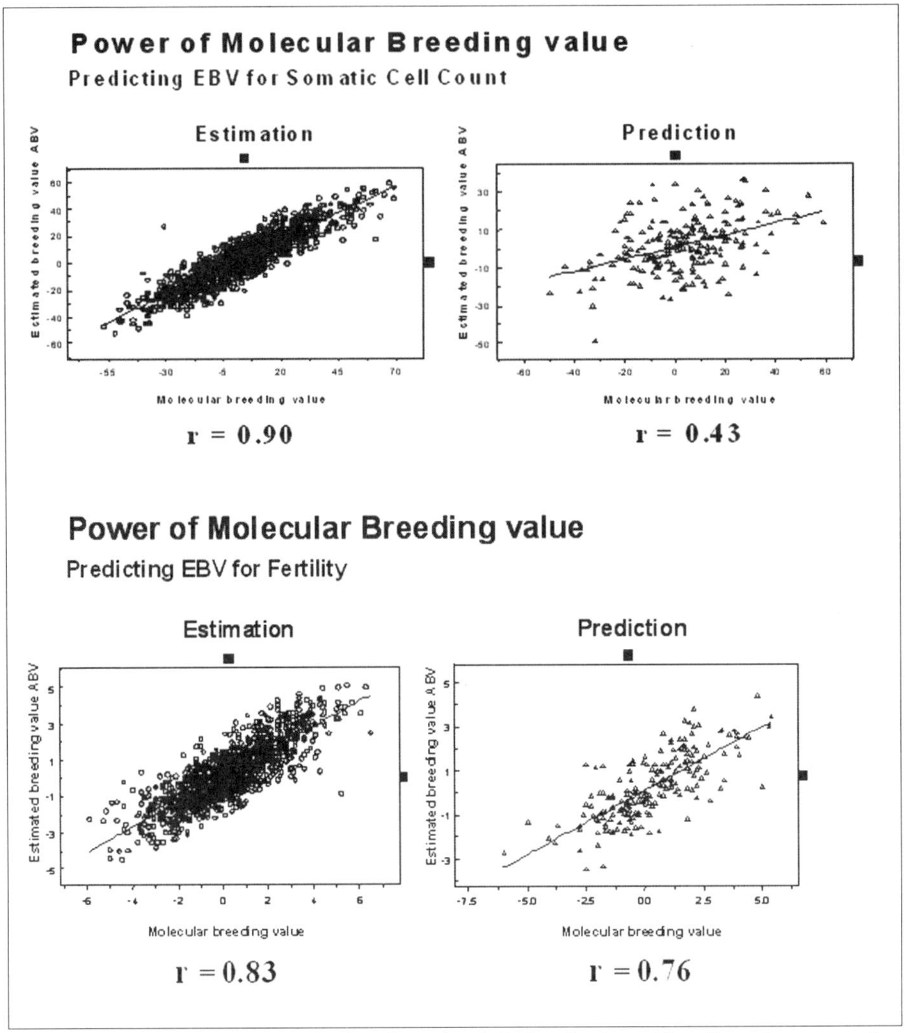

Fig. 1: Relationship between molecular breeding value (MBV) and estimated breeding value expressed as Australian Breeding Value (ABV) for estimation and prediction of (a) somatic cell score and (b) fertility.

ACKNOWLEDGMENTS

The programme was supported by the CRC for Innovative Dairy Products, and platform establishment grant from Dairy Australia.

REFERENCES

1 Hardenbol P, Yu F, Belmont J, Mackenzie J, Bruckner C, Brundage T, et al: Highly multiplexed molecular inversion probe genotyping: over 10,000 targeted SNPs genotyped in a single tube assay. Genome Res 2005,15(2):269-275.

2 Crump RE, Tier B, Moser G, Solkner J, Kerr RJ, Woolaston AF, et al: Genome wide selection in dairy cattle: use of genetic algorithms in the estimation of molecular breeding values. Proc Assoc Adv Anim Breed Genet 2007;304-307.

3 Moser G, Crump RE, Tier B, Solkner J, Zenger KR, Khatkar MS, et al: Genome based genetic evaluation and genome wide selection using supervised dimension reduction based on partial least squares. Proc Assoc Adv Anim Breed Genet 2007;227-230.

4 Raadsma HW, Zenger KR, Khatkar MS, Crump RE, Moser G, Solkner J, et al: Genome wide seleciton in dairy cattle based on high-density genome-wide SNP analysis: from discovery to application. Proc. Assoc Adv Anim Breed Genet 2007;231-234.

5 Melchinger AE, Utz HF, Schön CC: Quantitative trait locus (QTL) mapping using different testers and independent population samples in maize reveals low power of QTL detection and large bias in estimates of QTL effects. Genetics 1998;149:383-403.

6 Bost B, De Vienne D, Hospital F, Moreau L, Dillman C: Genetic and nongenetic bases for the L-shaped distribution of quantitative trait loci effects. Genetics 2001;157:1773-1787.

7 Xu S: Theoretical basis of the Beavis effect. Genetics 2003;165:2259–2268.

8 Weller JL, Shlezinger M, Ron M: Correcting for bias in estimation of quantitative trait loci effects. Genet. Sel Evol 2005;37:501-522.

9 Khatkar MS, Collins A, Cavanagh JAL, Hawken RJ, Hobbs M, Zenger KR, et al W: A first generation metric linkage disequilibrium map of bovine chromosome 6. Genetics 2006;174(1):79-85.

10 Fernando RL, Grossman M: Marker-assisted selection using best linear unbiased prediction. Genet Sel Evol 1989;21:467-477.

11 Meuwissen THE, Hayes BJ, Goddard ME: Prediction of total genetic value using genome-wide dense marker maps. Genetics 2001;15:1819-1829.

12 Gianola D, Perez-Enciso M, Toro MA: On marker-assisted prediction of genetic value: beyond the ridge. Genetics 2003;163:347-365.

13 Gianola D, Fermando R, Stella A: Genome assisted prediction of genetic value with semi-parametric procedures. Genetics 2006;173:1761-1776.

14 Khatkar MS, Zenger KR, Cavanagh JAL, Hawken RJ, Hobbs M, Barris W, et al: Extent of linkage disquilibrium and haplotyle locks in bocine genome using high density SNP markers. Proc Assoc Adv Anim Breed Genet 2007;215-218.

15 Zenger KR, Khatkar, MS, Tier B, Hobbs M, Cavanagh JAL, Solkner J, et al: QC analyses of SNP array data: experiences from a large population of dairy sires with 23.8 million data points. Proc Assoc Adv Anim Breed Genet 2007;123-126.

16 Tier B, Crump R, Moser G, Solkner J, Thomson PC, Woolaston A, et al:. Genome wide selection: issues and implication. Proc Assoc Adv Anim Breed Genet 2007;308-311.

Prof. Herman W. RAADSMA, Director, ReproGen- Centre for Advanced Technologies in Animal Genetics and Reproduction, Faculty of Veterinary Science, University of Sydney, PMB 3, Camden NSW 2570, Australia.

E-mail: raadsma@camden.usyd.edu.au

Pinard M-H, Gay C, Pastoret P-P, Dodet B (eds): Animal Genomics for Animal Health. Dev Biol (Basel). Basel, Karger, 2008, vol 132, pp 225-230.

On the Search for Markers of Tick Resistance in Bovines

L.C.A. Regitano[1], A.M.G. Ibelli[2], G. Gasparin[2], M. Miyata[2], A.L.S. Azevedo[4], L.L. Coutinho[3], R.L. Teodoro[4], M.A. Machado[4], M.V.G.B. Silva[4], L.C. Nakata[2], L.G. Zaros[3], T.S. Sonstegard[5], A.M. Silva[2], M.M. Alencar[1], M.C.S. Oliveira[1]

1 Embrapa Pecuária Sudeste, São Carlos, São Paulo, Brazil
2 PPGGev, Universidade Federal de São Carlos, São Carlos, São Paulo, Brazil
3 Universidade de São Paulo-ESALQ, Piracicaba, São Paulo, Brazil
4 Embrapa Gado de Leite, Juiz de Fora, Minas Gerais, Brazil
5 USDA, ARS, Bovine Functional Genomics Laboratory, Beltsville, Maryland, USA

Key words: *Rhipicephalus (Boophilus) microplus*, QTL, candidate gene, gene expression, cattle

Abstract: Genetic differences in susceptibility to ticks (*Rhipicephalus (Boophilus) microplus*) are considerable in bovines. Here, mapping, association and gene expression approaches were employed to further advance our understanding of the molecular basis of tick resistance. A *B. taurus* x *B. indicus* F_2 population was developed by Embrapa and 382 individuals were measured for parasitic load. Scanning of all chromosomes is in progress. Quantitative trait loci (QTL) for tick load were mapped to chromosomes 4, 5, 7, 10, 14, 18 and 23 out of the 20 chromosomes scanned and were dependent on the season in which the phenotype was scored. In the candidate gene approach, females from the genetic groups Nelore (NE - 184), Canchim x Nelore (CN - 153), Aberdeen Angus x Nelore (AN - 123) and Simmental x Nelore (SN - 120) were evaluated under natural infestation. Microsatellite markers close to the genes for interleukin 2 (*IL2*), interleukin 4 (*IL4*) and interferon gamma (*IFNG*) were analysed. Tick counts were associated with the marker for interleukin 4 ($P<0.05$) in three genetic groups. Differences in cytokine mRNA levels of naïve versus infested Nelore calves as well as between resistant versus susceptible cows from NE, CN and AN genetic groups were also investigated. Comparison of cytokines from infested and naïve animals showed downregulation of *IL2*. When resistant cows were compared to susceptible animals, *IL8* was downregulated. These results reinforce the multiloci nature of tick resistance and the need to consider QTL and environment interactions.

Introduction

The bovine tick *Rhipicephalus* (*Boophilus*) *microplus* and tick-borne diseases result in significant economic losses in cattle production in tropical and subtropical areas. The scenario becomes even worse with the selection of chemical resistant parasites and the increased public demand for residue free animal products. Host resistance to ticks is a complex trait influenced by a great number of environmental and physiological factors, such as temperature, humidity, gender and age [1]. Regardless, there is a genetic component of tick resistance with estimates of heritability varying from very low to high, according to evaluation method (artificial or natural challenge), study population and statistical method. Several studies have reported an association of genetic markers with tick count in different bovine breeds [1-3], suggesting that a portion of genetic control of tick resistance could be determined and used in the selection of resistant cattle.

Materials and Methods

For the interval analysis in the F_2 generation, a *Bos taurus* (Holstein) x *Bos indicus* (Gyr) population of 382 F_2 animals was developed from 1999 to 2005. The F_2 generation was evaluated for tick resistance by experimental challenge in two seasons, as previously described [4]. Microsatellite markers were selected to cover 20 chromosomes at 20 cM intervals and linkage maps were assembled. Significance of the fixed effects and covariables for each trait was evaluated by least square analysis of tick count transformed to $\log_{10}(n + 1)$, where n is the number of ticks. Quantitative trait loci (QTL) analysis was performed using regression interval mapping for F_2 families [5]. The model included the additive and dominance effects of the QTL and all fixed effects (sex, year in which the animal was evaluated, coat color and hair type) that were significant in the analysis of variance (ANOVA). Age of the animal (days) at evaluation was included as a covariable.

For the candidate gene experiment, females with different physiological status (calves, pregnant and open heifers, primiparous and pluriparous cows with and without a calf) were selected from the following genetic groups (GG): Nelore (NE – n=184), Canchim x Nelore (CN – n=153), Aberdeen Angus x Nelore (AN – n=123) and Simmental x Nelore (SN – n=120). They were evaluated under natural infestation conditions in seven to ten tick counts, from July 2003 to December 2004. Animals were raised and evaluated at the Southeastern Embrapa Cattle Station, located in São Carlos, São Paulo, Brazil. During the entire evaluation period, animals were maintained on pastures, with no parasite control. Ticks 4.5 mm or larger were counted on one side of the animals, and the data transformed to $\log_{10}(n + 1)$. The transformed tick count, adjusted for genetic group (GG), animal within GG, year-season (YS) and physiological status, was obtained for each animal that had at least seven observations. DNA samples from the evaluated animals were part of a DNA repository held at the Southeastern Embrapa Cattle Station. Microsatellite markers were chosen based on their proximity to candidate genes from the bovine immune system, *BL4* (interferon gamma), *IL4* (interleukin 4) and *BMS941* (interleukin 2) and genotyped using an ABI3100. Adjusted tick counts estimated for each animal were used in least square analysis considering the fixed effects of sire, number of observations (7 to 10) and marker genotype. Low frequency genotypes (<4 animals) were excluded from the analysis. Since GG had different allelic frequencies, association analysis was done within GGs.

The experiment for measuring cytokine mRNA levels of naïve versus infested Nelore was held at the Southeastern Embrapa Cattle Station in Brazil (22°01'S and 47°53'W). Ten calves from ten different mothers were born in isolated pens, and kept free of ticks along with their mothers under intensive chemical control. When calves were four months old, they were assigned to two groups, reference and challenge. Paternal half-sibs were assigned each to a different group. Animals from the challenge group (five calves) were left without chemical tick control for 30 days before they were artificially infested with aliquots of *R. microplus* larvae, spread on the back of each animal. On the ninth day after tick infestation, samples from skin and inguinal lymph node were surgically collected using local anaesthesia. Total RNA was isolated using TRIzol® reagent (Invitrogen) and synthesis of complementary DNA (cDNA) was performed from 5.0 μg total RNA using oligo(dT) primer and Superscript II (Invitrogen). RT-PCR primers were designed [6] and real-time PCR was performed using SYBR Green I in a LightCycler™ (Roche Diagnostics, Mannheim, Germany), with *RPL19* (ribosomal protein L-19) as housekeeping gene. To calculate the relative expression ratio between reference and challenge groups, the Relative Expression Software Tool (REST)© [7] was used.

To determine cytokine mRNA levels of resistant versus susceptible bovines, ten cows from each genetic group (5 resistant and 5 suceptible) were selected from the candidate gene experiment to represent the extreme phenotypes of each genetic group. Animals were kept under chemical tick control for three months and left without chemical tick control for 30 days before they were submitted to artificial infestation. Tick challenge, tissue collection, RNA extraction and RT-PCR were as described for the naïve versus infested experiment.

RESULTS

In the F_2 experiment, the average tick count value was 40 ± 72.4 for the rainy season, with eight animals presenting more than 200 ticks, out of 302. The maximum observed value was 792 ticks in one animal, and 25 animals were free of ticks during this season. In the dry season, an average of 33 ± 43.3 ticks per animal was observed. The maximum observed value in one animal was 412, four animals had more than 200 ticks, and ten were completely free of ticks, out of 338 evaluated. Year of evaluation and hair type were significant for tick counts obtained during the rainy season (P<0.01). For the dry season, tick count was affected by coat colour and hair type (P<0.01). Interval analysis of F_2 has covered 20 chromosomes so far. QTL have been detected on chromosomes 4, 5, 7, 10, 11, 14, 18 and 23 (Table 1). Dominance deviations were significant for QTL on chromosomes 4, 5, 7, 14 and 18. Different QTL were found for phenotypes scored during the rainy or dry season. Altogether, QTL explained 13.07% of the total phenotypic variation for tick count during the rainy season and 11.28% of the total phenotypic variation for tick count during the dry season.

Table 1: Summary of QTL for tick resistance identified in the Holstein x Gyr F_2 population.

Chromosome	Season[1]	Significance	Position (cM)	d[2]	$\%\delta^2P$[3]
4	Rainy	5%	98	yes	2.40
5	Rainy	5%	132	yes	1.70
7	Dry	10%	73	yes	1.90
10	Dry	1%	18	no	6.20
11	Rainy	5%	32	no	3.40
14	Dry	5%	25	yes	3.20
18	Rainy	1%	60	yes	1.97
23	Rainy*	5%	50	no	3.60

[1] Season in which ticks were counted.
[2] Dominance deviation.
3 Proportion of the phenotypic variance explained by the QTL.
* This QTL was not tested for dry season data.

In the candidate gene experiment, the difference in resistance among genetic groups was dependent on the year and month of evaluation, but in all cases NE cows were less infested than all the other genetic groups. Of the molecular markers investigated, only the microsatellite linked to the IL4 microsatellite marker was associated (P<0.05) with adjusted tick count in three of the genetic groups in the study (NE, CN and AN). Genotype means are presented in Table 2. Since most of the genotypic combinations were not represented, it was not possible to assess allele substitution effects. There was no clear pattern of genotypes associated with higher or lower tick loads within or among genetic groups.

Table 2: Means and standard deviations for microsatellite IL-4 genotypes within genetic groups.

	Nelore		½ Canchim ½ Nelore		½ Angus ½ Nelore	
Genotype	Mean[1]	± SE	Mean[1]	± SE	Mean[1]	± SE
81089	-	-	0.527[b]	± 0.266	-	-
85093	-	-	1.219[a]	± 0.142	-	-
89089	0.699[a]	± 0.104	1.131[a,c]	± 0.209	-	-
89091	0.266[a,b]	± 0.160	1.236[a]	± 0.184	-	-
89093	0.478[a,b]	± 0.112	1.201[c]	± 0.119	1.343[b]	± 0.096
89105	0.412[b]	± 0.090	0.586[a]	± 0.186	-	-
91091	0.472[a,b]	± 0.155	-	-	-	-
91093	0.383[a,b]	± 0.172	0.689[a,c]	± 0.241	1.772[a]	± 0.150
91097	-	-	1.033[a,c]	± 0.413	-	-
91101	-	-	1537[a]	± 0.296	-	-
91105	0.685[a]	± 0.095	-	-	-	-
93093	0.414[a,b]	± 0.116	-	-	1.621[a,b]	± 0.112
93097	-	-	0.977[a]	± 0.423	-	-
93105	0.513[a,b]	± 0.101	1.286[a,c]	± 0.160	1.500[a,b]	± 0.102
105105	0.422[a,b]	± 0.100	1.898[a,c]	± 0.373	-	-

[a;b;c] Values with different letters within columns are significantly different ($P<0.01$).

Analysis of cytokine mRNA abundance in challenged compared to naïve Nelore calves (Fig.1) showed downregulation of *IL2* (P<0.05) mRNA in challenged animals, while no effect of tick challenge was observed for *IL4*, *IL8*, *IL12p35*, *TNF-α* and *MCP-1*. When extremes of tick resistance were compared across all genetic groups (Fig.1), downregulation (P<0.05) of *IL8* was observed in resistant cows. When the same comparison was performed within the different genetic groups, downregulation was observed only in the AN group (P<0.01) (data not shown).

Discussion

The number of QTL found in the F_2 resource population suggests a multiloci nature for host resistance. Although tick resistance was evaluated by artificial challenge in order to minimize environmental effects, all the QTLs in question were dependent on the season in which the phenotype was scored. If tick counts were only performed during the rainy season, two QTL would not have been detected. This dependency could be related to seasonal variations in coat length and thickness, nutritional status, circadian rhythm or heat stress. The genotype-environment interaction could imply that animals selected from one environment would perform poorly in another. Moreover, the sampling problem associated with seasonal variation suggests the necessity for testing across many years.

Results for IL4 microsatellite association to ticks did not allow for the identification of a specific allele for resistance. This gene is located in the same region on chromosome 7 as a suggestive dominant QTL that was detected, between microsatellite markers IL4 and BM6117, in the F_2 experiment. Investigation of SNPs in the *IL4* gene could be helpful for confirming and understanding this association.

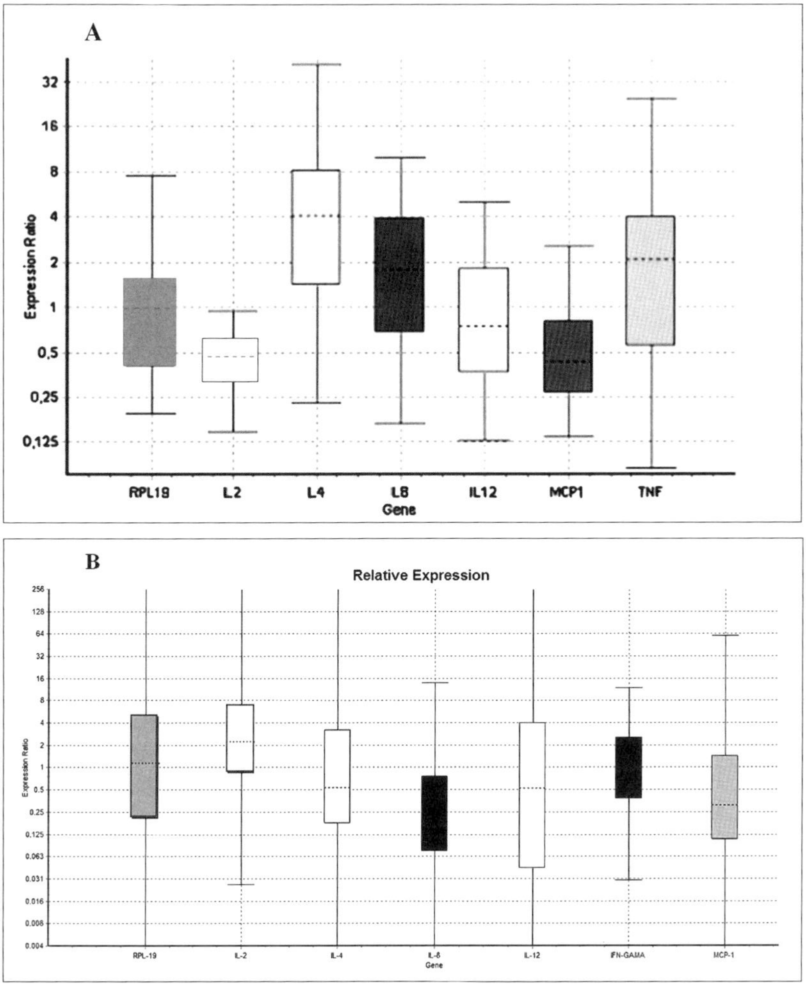

Fig. 1: Cytokine mRNA profile in lymph nodes. **A.** Relative expression of infested against naïve Nelore Calves. **B.** Relative expression of resistant compared to susceptible cows. Boxes represent the interquartile range, or the middle 50% of observations expressed as the n-fold differences between groups (A=Infested X Naïve, B= Resistant X Susceptible) in relation to the housekeeping gene. The dotted line represents the median gene expression. Whiskers represent the minimum and maximum observations (some are out of range not displayed).

The cytokine IL2 is a growth factor that guides the proliferation and differentiation of T cells, and is self-produced by activated T cells. In mice infested with *Rhipicephalus sanguineus* ticks, *IL2* and *IFNG* were found to be largely downregulated, whereas *IL4* and *IL10* were upregulated [8]. We have shown that *IL2* is downregulated after a first challenge in Nelore calves, a resistant breed, with no difference between resistant and susceptible cows, which suggests that this cytokine may not play a main role in host resistance. IL8 is a chemokine that act mainly as a chemo-attractant for leukocytes, recruiting monocytes, neutrophils and other effector blood cells to infection sites. It is produced by monocytes, macrophages, fibroblasts and keratinocytes. Salivary gland extracts (SGE) from several ixodid tick species have been shown to reduce IL8 levels through the presence of IL-8 binding proteins in SGE of some species [9]. According to our results, in *R. microplus* infested cows, the reduction would occur at the mRNA level and in a breed specific pattern. Unfortunately, no relationship between expression patterns and marker data could be drawn. Both *IL2* and *IL8* are on chromosome 6, which was not scanned in the F_2 experiment, and the microsatellite selected to mark *IL2* segregation in the candidate gene experiment turned out to be invalid, since it was selected based on the first build of the *Bos taurus* genome view, in which *IL2* was mapped to chromosome 17.

Conclusion

The marker data presented here suggests multiloci inheritance for tick resistance in bovines, with the predominance of non-additive effects and a strong genotype-environment interaction. Differences in gene expression of resistant cows compared to susceptible cows were breed-specific.

Acknowledgements

This work was financially supported by Prodetab and CNPq. We thank CNPq for providing scholarships to Gasparin and Miyata and fellowships to Coutinho, Alencar and Regitano.

References

1 Ashton GC, Seifert GW, Francis J: An association between serum amylase phenotype and tick infestation in cattle. Aust J Biol Sci 1968;21:303-308.

2 Stear MJ, Hetzel DJ, Brown SC, Gershwin LJ, Mackinnon MJ, Nicholas FW: The relationships among ecto- and endoparasite levels, class I antigens of the bovine major histocompatibility system, immunoglobulin E levels and weight gain. Vet Parasitol 1990;34:303-321.

3 Acosta-Rodriguez R, Alonso-Morales R, Balladares S, Flores-Aguilar H, Garcia-Vazquez Z, Gorodezky C: Analysis of BoLA class II microsatellites in cattle infested with *Boophilus microplus* ticks: class II is probably associated with susceptibility. Vet Parasitol 2005;127:313-321.

4 Gasparin G, Miyata M, Coutinho LL, Martinez ML, Teodoro RL, Furlong J, et al: Mapping of quantitative trait loci controlling tick [Rhipicephalus (Boophilus) microplus] resistance on bovine chromosomes 5, 7 and 14. Anim Genet 2007;38:453–459.

5 Haley CS, Knott SA, Elsen JM: Mapping quantitative trait loci in crosses between outbred lines using least squares. Genetics 1994;136:1195-1207.

6 Zaros LG, Bricarello PA, Amarante AFT, Coutinho LL: Quantification of bovine cytokine gene expresion using real-time RT-PCR methodology. Genet Mol Biol 2007;30:575-579.

7 Pfaffl MW, Horgan GW, Dempfle L: Relative expression software tool (REST©) for group-wise comparison and statistical analysis of relative expression results in real-time PCR. Nucleic Acids Res 2002;30:1-10.

8 Ferreira BR, Silva JS: Successive tick infestations selectively promote a T-helper 2 cytokine profile in mice. Immunology 1999;96:434-439.

9 Hajnicka V, Kocakova P, Slavikova M, Slovak M, Gasperik J, Fuchsberger N, et al: Anti-interleukin-8 activity of tick salivary gland extracts. Parasite Immunol 2001;23:483-489.

Dr. Luciana Correia de Almeida Regitano, Embrapa Pecuária Sudeste, São Carlos, São Paulo 13560970, Brazil.
E-mail: luciana@cppse.embrapa.br

Pinard M-H, Gay C, Pastoret P-P, Dodet B (eds): Animal Genomics for Animal Health. Dev Biol (Basel). Basel, Karger, 2008, vol 132, pp 231-237.

Detection of Germline and Somatic Copy Number Variations in Cattle

G.E. Liu[1], C.P. Van Tassell[1], T.S. Sonstegard[1], R.W. Li[1], L.J. Alexander[2], J.W. Keele[3], L.K. Matukumalli[1,4], T.P. Smith[3], L.C. Gasbarre[1]

1 USDA-ARS, ANRI, Bovine Functional Genomics Laboratory, BARC-East, Beltsville, MD, USA
2 USDA-ARS, LARRL, Ft Keogh Miles City, MT, USA
3 USDA-ARS, US Meat Animal Research Center, Clay Center, NE, USA
4 Bioinformatics and Computational Biology, George Mason University, Manassas, VA, USA

Key words: copy number variation (CNV), array comparative genomic hybridization (array CGH), cattle genome, structural variants

Abstract: As a complement to the Bovine HapMap Consortium project, we initiated a systematic study of the copy number variation (CNV) within the same cattle population using array comparative genomic hybridization (array CGH). Oligonucleotide CGH arrays were designed and fabricated to cover all chromosomes with an average interval of 6 kb using the latest bovine genome assembly. In the initial screening, three Holstein bulls were selected to represent major paternal lineages of the Holstein breed with some maternal linkages between these lines. Dual-label hybridizations were performed using either Hereford L1 Dominette 01449 or L1 Domino 99375 as reference. The CNVs were represented by gains and losses of normalized fluorescence intensities relative to the reference. The data presented here, for the first time, demonstrated that significant amounts of germline and fewer somatic CNVs exist in cattle, that many CNVs are common both across diverse cattle breeds and among individuals within a breed, and that array CGH is an effective tool to systematically detect bovine CNV. Selected CNVs have been confirmed by independent methods using real-time (RT) PCR. The strategy used in this study, based on genome higher-order architecture variation, is a powerful approach to generating resources for the identification of novel genomic variation and candidate genes for economically important traits.

INTRODUCTION

Recently it has become apparent that previously unappreciated genomic structural variation contributes significantly to individual health and disease in humans and rodents [1]. Genomic structural variation ranging from 1 kb to 5 Mb comprises copy number variations (CNVs, i.e., deletions and insertions ≥1 kb), as well as inversions and translocations spanning millions of nucleotides of heterogeneity within a genome. More than 1,500 CNVs have been detected in humans by experimental and in silico methods with dozens of these variations existing between two normal individuals [2]. These large-scale deletions and insertions are polymorphic (7-20% allele frequency) within the human population. Analyses of these variations have provided compelling evidence of linkage disequilibrium with flanking single nucleotide polymorphisms (SNPs) in humans. Segmental duplications have been suggested as a major catalyst of these genomic imbalances in the human genome [3]. However, the existence, frequency and impact of such CNVs in cattle are still completely unknown as no systematic study has yet been conducted.

Cytogenetic analysis of metaphase chromosomes using fluorescence *in situ* hybridization (FISH) allows for the screening of genomic structural variations including CNVs. However, these methods have significant limitations both in their resolution (detection of structural variations ≥3-4 Mb) and in their scalability. More recently, array comparative genomic hybridization (array CGH) has emerged as a technology that promises to circumvent both of these limitations [4]. These whole-genome arrays are based on long oligonucleotides representing a tiling path along each chromosome. The method has proven successful in detecting hundreds of CNVs [5]. These technologies, along with the existence of bovine genome reference sequence, make it possible for the first time to assess CNVs and their impact on bovine health in a systematic, cost-effective, and high-throughput fashion.

Materials and Methods

For sequence alignment and analyses, orthologous sequence identification, genomics sequence alignment and insertion deletion analyses were performed as previously described [6,7].

For oligonucleotide array construction, genomic DNA was purified from sperm, whole blood, and ear notch as previously described [8]. Three Holstein bulls (Hs1, Hs2 and Hs3) were selected to represent major branches of the Holstein breed with some maternal linkages between branches. DNA samples from the sequenced cow, Hereford L1 Dominette 01449 (whole blood: Dt), or its sire L1 Domino 99375 (semen: Do) were used as the reference in hybridization experiments. High-density oligonucleotide CGH arrays were designed and fabricated as previously described [5]. Unamplified genomic DNA (2.5 μg) was labelled with Cy3 (samples) or Cy5 (references), and hybridizations, data normalization, and segmentation were performed [5]. A set of conservative criteria (≥5 probes in a segment and a mean log2 ratio shift across segment ≥0.5) were used to define the final set of high confidence CNV calls [9]. Human orthologous gene annotation and overlap were determined based on net human-cow alignment using the UCSC Genome Browser.

For PCR validation, the relative copy numbers were determined by quantitative PCR (qPCR) using TaqMan and/or SYBR Green detection chemistries. Probe and primer sequences were as follows: control probe, 5' 6-FAM- CCTCCCTGTCCAGCACCAACTCCT-BHQ-1; forward primer, CTTTGAGACCC-CACGGACTG; reverse primer, GATGCGAAGAACTGACTCATTTGA; test probe, 5' 6-FAM-AAGTGTGAAGCCTTTTCCCTTGAC-BHQ-1; forward primer, AGACCAAAACAAACAAACAAAA; and reverse primer, TGGAGTTAATGAATGTAGTTAGTT. The averaged relative comparative threshold cycle (ddCT) values of triplicates were normalized against the control primers. Assuming that there were two copies of DNA in the control regions, the relative copy number for each test region was calculated as $2^{(1+ddCT)}$. A linear regression model was used to compare the copy number estimates from array CGH and qPCR.

Results and Discussion

Before array CGH experiments were initiated, two independent pieces of evidence were collected indicating the existence of CNVs in the cattle genome.

Direct high-quality sequence comparison between Holstein and Hereford

For direct high-quality sequence comparison between Holstein and Hereford breeds, a large-scale comparative genome analysis was previously performed on 11 Mb of high-quality finished genomic sequence from cattle, dog, and humans to study evolutionary changes that have shaped mammalian genomes [7]. These same 29 bacterial artificial chromosome (BAC) clones from the RP42 library (Holstein) in the above study were *in silico* mapped on version 3 of the bovine genome assembly. In order to provide high-quality large-scale genomic alignments, the analysis was limited to the completely finished BAC sequences. All regions were excluded if ambiguous bases were present in the bovine genome assembly. Most of these BACs are from the greater CFTR (cystic fibrosis transmembrane conductance regulator) region on bovine chromosome 4, comprising 1.8 Mb of genomic sequence. Insertion/deletion (indel) events were examined in Holstein and Hereford breeds as shown in Figure 1. SNPs and small indels (<1,000 bp) were more common (1 event per 700 bp) by count. However, large-scale events (≥1 kb) were predominant by length. In this case, the length ratio was about 2:1. The 1.1-kb sequence in RP42 was also mapped to bovine chromosome 29 in the bovine genome assembly corresponding to human clone 23560 mRNA, which is expressed in infant brain with unknown function. The nearest segmental duplication sequence was 1 Mb away in the human orthologous genomic region. Assuming our sample was representative, it was extrapolated that there will be ~1,500 – 2,000 such large-scale indels in the whole bovine genome, which correlates closely with human data [6]. However, unless the costs of genome sequencing drop dramatically, the scarcity of high-quality genomic sequence in cattle except for Dominette and Domino will prevent the application of this direct sequence comparison strategy.

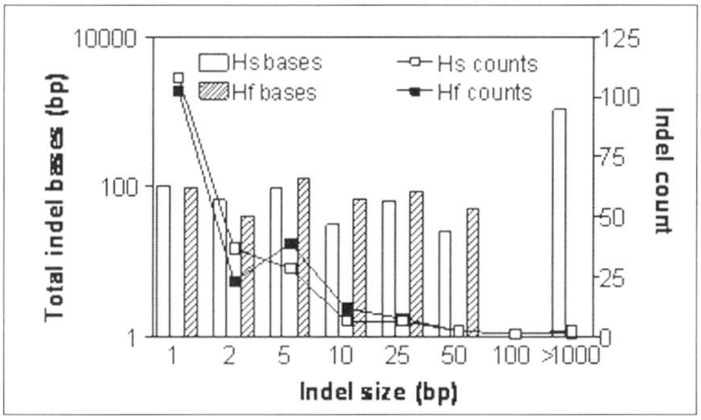

Fig. 1: Distribution of insertion/deletion (indel) events in Holstein (Hs) and Hereford (Hf) breeds. Indels were binned according to the indel size, and total indel basepairs (left Y axis in log scale) and indel counts (right Y axis in linear scale) were determined.

Clustered SNP genotyping errors

Another piece of evidence for the existence of CNVs comes from the bovine SNP data. From the human HapMap data, ~1,000 deletion structural variations (1-745 kb) were discovered by mapping clusters of SNP genotype errors, Mendelian transmission errors and/or sites that do not pass Hardy-Weinberg equilibrium [10,11]. An initial screen of the first 10,000 SNP data points from cattle has found clustered SNP genotyping errors as candidate deletion variations (unpublished results). However, the density of SNP markers is too low to achieve a high-resolution map of CNV across the entire genome. These limitations are the critical factors that motivated us to study CNV using array CGH.

Detection and validation of CNV in the cattle genome

The microarray for CGH contains 385,048 probes spanning the cattle genome with a median spacing of 5,779 bp. Semen DNA from three Holstein bulls (Hs1, Hs2, and Hs3) was tested against Dominette blood DNA or Domino semen DNA. DNA copy number gains and losses were evident in all Holsteins. By using a set of stringent criteria (see Materials and Methods), 25 high confidence CNVs from Holstein vs. Hereford comparisons were identified on 16 cattle autosomes (Table 1).

Table 1: CNVs from Holstein vs. Hereford comparisons.

Index	Chr	Start	End	Length	Calls	Genes
1	chr3	9,765,000	9,819,000	54,001	loss	CD1D, related to MHC
2	chr3	52,335,000	52,485,000	150,001	loss	Guanylate binding protein family
3	chr4	102,603,035	102,735,412	132,378	gain	T cell receptor family
4	chr5	44,292,274	44,321,089	28,816	loss	Lysozyme
5	chr5	57,103,490	57,454,046	350,557	loss	Olfactory receptor family
6	chr7	9,108,229	9,240,437	132,209	gain	Olfactory receptor family
7	chr9	79,373,243	79,465,247	92,005	gain	*RAET1* (related to MHC class I)
8	chr10	59,715,000	59,895,000	180,001	loss	T cell receptor family
9	chr10	72,789,344	72,818,105	28,762	gain	LOC440184 (unknown function)
10	chr12	66,033,000	66,393,000	360,001	loss	ATP-binding cassette sub-family
11	chr12	67,545,311	67,942,187	396,877	loss	ATP-binding cassette sub-family
12	chr13	61,381,259	61,410,203	28,945	gain	C20orf70 (unknown function)
13	chr14	12,828,345	13,131,000	302,656	loss	
14	chr14	13,605,000	13,635,000	30,001	loss	
15	chr14	16,910,763	17,008,636	97,874	loss	
16	chr15	5,823,000	5,913,000	90,001	loss	Progesterone receptor
17	chr15	50,310,000	50,370,000	60,001	gain	
18	chr17	67,499,437	67,574,186	74,750	gain	Zinc finger proteins, *PRAME* (preferentially expressed antigen in melanoma)
19	chr18	54,705,000	54,825,000	120,001	gain	CD33 antigen
20	chr18	58,275,000	58,311,000	36,001	loss	SLC7A3 (solute carrier family 7)
21	chr20	44,229,233	44,269,270	40,038	loss	
22	chr23	21,223,487	21,286,588	63,102	loss	MHC class II
23	chr25	1,395,000	1,545,000	150,001	gain	LAR protein-tyrosine phosphatase-interacting protein (liprin) family
24	chr25	28,215,000	28,251,000	36,001	gain	LOC283932 (unkonwn function)
25	chr29	9,000	171,000	162,001	gain	

Due to the lack of sequence and/or mapping uncertainty, chromosomes X, Y and Un chromosomes were not considered. The variable regions vary in size (ranging from 28,762 to 396,877 bp; with a mean of 127,879 bp and standard deviation of 110,989 bp) and number per animal (Hs1: 14; Hs2: 8; Hs3: 10). As expected, no segments were identified in the reference DNA self–self hybridizations (unpublished data). To validate CNVs detected by array CGH, quantitative PCR (qPCR) assays were used to measure relative copy number in test regions (Test) compared to control regions (Control) (Fig. 2). Copy number assessed by qPCR was concordant with array CGH in unaffected animals: Domino and Hs3. DNA from Hs1 with copy number loss yielded only background level amplification by PCR, implying that the target sequences were not present (null for both alleles). These results were confirmed using a second, independent qPCR assay within the same CNV. Copy number changes in Hs2 detected by array CGH were confirmed by qPCR using the CT method. Regression analysis demonstrated a linear relationship between copy numbers determined by both platforms (Fig. 2C).

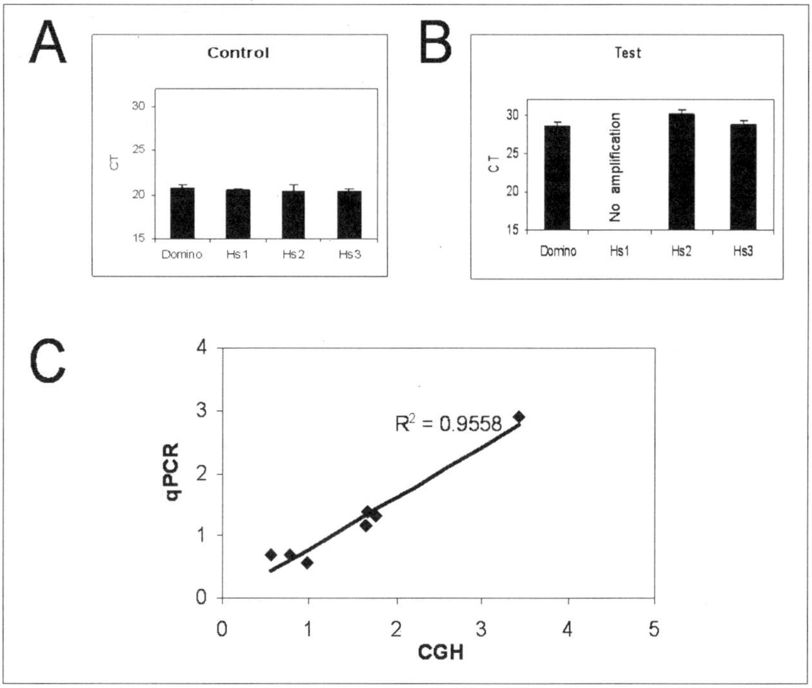

Fig. 2: Validation of CNV.
A region of copy number variation (loss) in chr5:57,103,490-57,454,046 (size ~360 kb) identified by whole-genome array CGH is confirmed by qPCR (A, B). Copy number estimates from array CGH and qPCR are highly concordant (C).

Observed differences in the copy number of genome segments between samples from two individuals could reflect germline differences or somatic variations. Therefore, control hybridizations were conducted between different tissues from each donor (sperm, skin and whole blood for Domino and skin and whole blood for Dominette), i.e. DNA samples from different tissues of the same donor were compared to each other. All comparisons of Domino DNA samples (blood:semen, skin:semen and blood:skin) showed no detectable somatic variation under the calling criteria used. In the comparison of Dt blood: Dt skin, only one somatic gain of 60 kb was observed in Dominette blood at chromosome 28:7,650,000-7,710,000, which contains the *HEATR1* (HEAT repeat containing 1) gene.

Since the bovine gene annotation is still in its early phase, a primary survey was performed on human orthologous genes derived from human syntenic regions. For the 25 high confidence CNVs reported here, 19 (76%) human syntenic regions contain or overlap at least one gene (Table 1). Immune related genes are overrepresented: CNVs 3 and 8 overlap with the T cell receptor alpha locus and CNVs 1, 7, 19 and 22 are related to the major histocompatibility complex (MHC). Another enriched gene function is nucleotide binding (CNVs 2, 10, and 11), which are commonly involved in biosynthesis and metabolism. Several genes in our analysis of bovine CNVs were also found to be overrepresented in an analysis of human CNV data (e.g., olfactory receptors, and zinc finger proteins).

Conclusion

The data presented here demonstrates, for the first time, that significant amounts of germline and fewer somatic CNVs exist in cattle, that many CNVs are common both across cattle breeds and among individuals within a breed, and that array CGH is an effective way to detect these bovine CNVs. Due to the limited sampling of this study, the CNV list reported here is probably not complete. It is possible that the current CNV calling criteria are too stringent, and that CNV events are underestimated. In addition, because the array design is based on the Dominette reference genome (Bta 3.1), sequences absent in Dominette and present in other animals will not be detected in this analysis. We are performing array CGH with more samples to investigate the frequency, pattern and extent of such CNVs and their possible impact on the improvement of health, well-being and productive efficiency in cattle.

References

1 Sebat J, Lakshmi B, Troge J, Alexander J, Young J, Lundin P, et al.: Large-scale copy number polymorphism in the human genome. Science 2004;305:525-528.

2 Redon R, Ishikawa S, Fitch KR, Feuk L, Perry GH, Andrews TD, et al: Global variation in copy number in the human genome. Nature 2006;444:444-454.

3 Eichler EE: Widening the spectrum of human genetic variation. Nat Genet 2006;38:9-11.

4 Pinkel D, Segraves R, Sudar D, Clark S, Poole I, Kowbel D, et al: High resolution analysis of DNA copy number variation using comparative genomic hybridization to microarrays. Nat Genet 1998;20:207-211.

5 Selzer RR, Richmond TA, Pofahl NJ, Green RD, Eis PS, Nair P, et al: Analysis of chromosome breakpoints in neuroblastoma at sub-kilobase resolution using fine-tiling oligonucleotide array CGH. Genes Chromosomes Cancer 2005;44:305-319.

6 Liu G, Zhao S, Bailey JA, Sahinalp SC, Alkan C, Tuzun E, et al: Analysis of primate genomic variation reveals a repeat-driven expansion of the human genome. Genome Res 2003;13:358-368.

7 Liu G, Matukumalli LK, Sonstegard TS, Shade LL, Van Tassell CP: Genomic divergences among cattle, dog and human estimated from large-scale alignments of genomic sequences. BMC Genomics 2006;7:140.

8 Sonstegard TS, Garrett WM, Ashwell MS, Bennett GL, Kappes SM, Van Tassell CP: Comparative map alignment of BTA27 and HSA4 and 8 to identify conserved segments of genome containing fat deposition QTL. Mamm Genome 2000;11:682-688.

9 Graubert TA, Cahan P, Edwin D, Selzer RR, Richmond TA, Eis PS, et al: A high-resolution map of segmental DNA copy number variation in the mouse genome. PLoS Genet 2007;3:e3.

10 Conrad DF, Andrews TD, Carter NP, Hurles ME, Pritchard JK: A high-resolution survey of deletion polymorphism in the human genome. Nat Genet 2006;38:75-81.

11 McCarroll SA, Hadnott TN, Perry GH, Sabeti PC, Zody MC, Barrett JC, et al: Common deletion polymorphisms in the human genome. Nat Genet 2006;38:86-92.

Dr. George E. Liu, Bovine Functional Genomics Laboratory, USDA-ARS, Building 200, Room 124B, BARC-East, Beltsville, MD 20705, USA.

E-mail: George.Liu@ars.usda.gov

Pinard M-H, Gay C, Pastoret P-P, Dodet B (eds): Animal Genomics for Animal Health. Dev Biol (Basel). Basel, Karger, 2008, vol 132, pp 239-245.

Dynamic Differential Regulation of Innate Immune Transcripts during the Infection of Alveolar Macrophages by the Porcine Reproductive and Respiratory Syndrome Virus

T. Ait-Ali[1], A.D. Wilson[1], D.G. Westcott[2], J.P. Frossard[2], M.A. Mellencamp[3], T.W. Drew[2], S.C. Bishop[1], A.L. Archibald[1]*

[1] Roslin Institute, Roslin, UK
[2] Department of Virology, Veterinary Laboratories Agency, Weybridge, Surrey UK
[3] Genus / PIC, Hendersonville, USA

Key words: pig, macrophage, innate, immunity, transcripts, PRRSV

Abstract: Porcine reproductive and respiratory syndrome virus (PRRSV), an arterivirus, is the etiologic agent of an infectious disease of that name, characterized by respiratory disorders, abortion in pregnant sows and high mortality in piglets, resulting in significant economic losses in the pig industry worldwide. In order to identify whether genetic differences in PRRSV response may exist in pigs, alveolar macrophages were used to assess the progression of the type-I interferon (IFN) transcript response in porcine alveolar macrophages infected by PRRSV. Our results suggest that a dynamic differential regulation of the type-I IFN and chemokine transcripts may operate during the first hours of infection with and entry of the virus in alveolar macrophages, and provide a compelling mechanism for the establishment of PRRSV replication in susceptible cells.

INTRODUCTION

More than 20 years after its emergence, the porcine reproductive and respiratory syndrome (PRRS) is still having major impacts on pig health and welfare. PRRS is characterized by respiratory disorders, abortion in pregnant sows and high mortality in piglets. In the United States alone, PRRS is costing the industry at least $600 million annually and probably accounts for a third of the annual loses from infectious diseases in pigs. The etiologic agent of this disease is the PRRS virus (PRRSV) an enveloped, single stranded, 15 kb positive RNA virus which is a member of the *Arteriviridae* family.

Since 1987 extensive information has been gained on the biology of PRRSV and the structure of its RNA genome. The virus has a specific tropism for bronchio-alveolar macrophages in which it replicates actively before spreading to other organs. Little is known about the host response during the early PRRSV interaction and replication. This report provides a glimpse of the host transcriptional response to PRRSV during the early hours of infection and identifies a possible mechanism involving a type-I interferon response for the initiation of PRRSV replication in alveolar macrophages.

Materials and Methods

Four 3-4 weeks old Pietrain piglets were studied. Bronchio-alveolar lavage fluid (BALF) cell isolation, preparation, characterization and *in-vitro* viral infection were described previously [1]. The H2 isolate of PRRSV was used for the infection studies [2]. Cells were harvested at 0, 2, 4, 8, 12, 16, 24 and 30 h post infection (PI) and RNA isolated as described earlier [1]. Samples were either assayed individually, or equal amounts of cDNA from each time point were pooled and then analysed. Real-time qPCR analyses were performed to assess the production of the PRRSV *ORF7* RNA as an indicator of virus replication in the cells. Levels of specific RNAs were assayed by real-time qPCR as described earlier with some modifications [1]. The cDNA of the four individual pigs was pooled and the corresponding cycle threshold (C_T) measured as previously. Table 1 shows the pairs of primers used in this study.

Results

Viral replication in alveolar macrophages

Fig. 1 shows that the accumulation of the PRRSV *ORF7* transcripts apparently occurs in three phases: an early phase (0-8 h) with a low accumulation of the *ORF7* mRNA up to about 1 Log, an intermediate phase (8-12 h) with a dramatic accumulation of the *ORF7* transcript up to about 3.5 Logs and a terminal phase (12-30 h) showing a further increase of 1 Log. Figure 1 also demonstrates that for each time point tested, pooled samples produced similar results to the mean of the four separate samples tested individually.

Innate immune transcript response

Figure 2A shows the changes in transcript levels for sixteen genes encoding important innate immune response proteins (Table 1) during PRRSV infection of alveolar macrophages. The same host response was measured following treatment with polyI/C, a well known inducer of a type-I interferon response (Figure 2B). Over the first 8-12 hours post infection, the levels of the majority of innate immune transcripts within PRRSV-infected macrophages clearly showed a transient reduction. Of those tested, the only exceptions were *TNFA* mRNA and some chemokine-encoding transcripts such as *IL8* and *IL1A*, which show a mild multiphasic pattern of accumulation during the course of the infection. For example, for the first 12 hours, the *IFNA* and *IFNB* transcript levels were concomitantly maintained at a low level in a comparable way as type-I interferon response transcripts such as *MX1*, *IFT1*, *UBL* or *UBP*. Interestingly, chemokine-encoding transcripts such as *CCL2* and the anti-inflammatory *IL10* were substantially repressed too, compared to the controls.

By 12 h post infection the levels of most of the transcripts were steadily increasing with the exception of the *CCL2* transcripts which were still kept below the level of the mock infected-T0 control.

In polyI/C-treated alveolar macrophages (Figure2B), 8 of the 16 transcripts levels tested showed a significant up-regulation from about 1 to 3 logs above the untreated control within 4 to 6 hours. Transcripts encoded *IFNA*, *IL1A*, *IL10*, *Casp2*, *Anx2*, *TRAF2* and *PACT*.

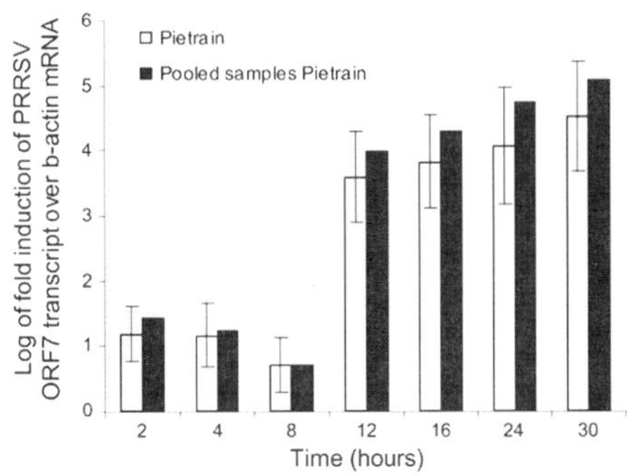

Fig. 1: In vitro H2 PRRSV infection of alveolar macrophages isolated from the 4 Pietrain pigs. Expression of PRRSV ORF7 transcript measured by qPCR over 30 hours. Bronchio-alveolar lung lavage (BALF) cells from four individuals were infected at a multiplicity of infection (MOI) of 1. Data were expressed as Log10 of the fold induction of the ORF7 transcript normalised to the t0 and the actin mRNA level during infection as described previously [1].

Table 1: List of primers used to detect the transcripts analysed in this study.

Gene product name and/or function / GenBank number	Primer sequences (5'-3') and (3'-5')
IP-10 (CXCL10 homolog) / CB469489	AAGAGAATATGGGCCCCTTG
	TTCCACTTTGGGACTTAATCG
MX1 (homology to GTP binding protein) / M65087	TCTCTCCATCCCTCTCTCCA
	CCACACAGGTGGCTGTAATG
IFIT1(interferon induced tetratricopeptide)/ CB469191	CTTTCGAATGATGCCGTTTC
	TTTCCACTATGGCCGATTTC
UBL(Ubiquitin-like protein protein degradation)/ CB471836	TAGGCTTGAGGGCATACTCC
	AAACAGCAGGTGTGTCACCA
UBP (deubiquitination enzyme)/ AF134195	CTGTGAGAACTGTGGCAGGA
	CAAGCTCTGGGGGAAATACA
PACT (P53-associated protein)/ CB476189	CCATCATTTTTCCGTTTGGT
	GATGCCGAGTCCATCACTTT
TNFalpha (tumour necrosis factor-alpha)/ X57321	CCACCAACGTTTTCCTCACT
	CCCAGGTAGATGGGTTCGTA
Actin / AY550069	GAGAAGCTGTGCTACGTCGC
	CCAGACAGCACTGTGTTGGC

IFNA (Interferon-a1)/ NM214393	ATGGCCCCAACCTCAGCCTTC
	TCACTCCTTCTTCCTGAGTCT
IFNB(Interferon-b)/) AY687281	ATGGCTAACAAGTGCATCCTCCAA
	TCAGTTCCGGAGGTAATCTGTAAG
IL1A (Interleukin(-1a)/X52731/)	ACTTCCTGGGGACGGCATGGATAAA
	GCATCATTCAGGATGCACTGGTGGT
IL10 (interleukin-10)/ NM214041	GCTGAAGACCCTCAGGCTGA
	TTGCTCTTGTTTTCACAGGGC
IL 8 (interleukin-8)/ X61151	TGGACCAGAGCCAGAGAGACTAG
	CCAGAACTGCAGCCTGCACAG
TRAF2 (TNFreceptor-associated factor)/ BG384589	ACCACAGAAGCCCATACCTG
	CCTTGGGTGGAACAGGATAA
Casp2 (apoptosis)/ AW347363	AATGAGGGAGCACATTCAGG
	CAGCAGGTTCAGAAGTTCC
Anx2 (annexin2)/ AY706383	CCAACCGCAGCAATGAACAG
	GGCCCTTCAATGGAGGCTTTC
CCL2 (monocyte chemoattractant protein-1)/ X79416	CACCAGCAGCAAGTGTCCTA
	TCCAGGTGGCTTATGGAGTC

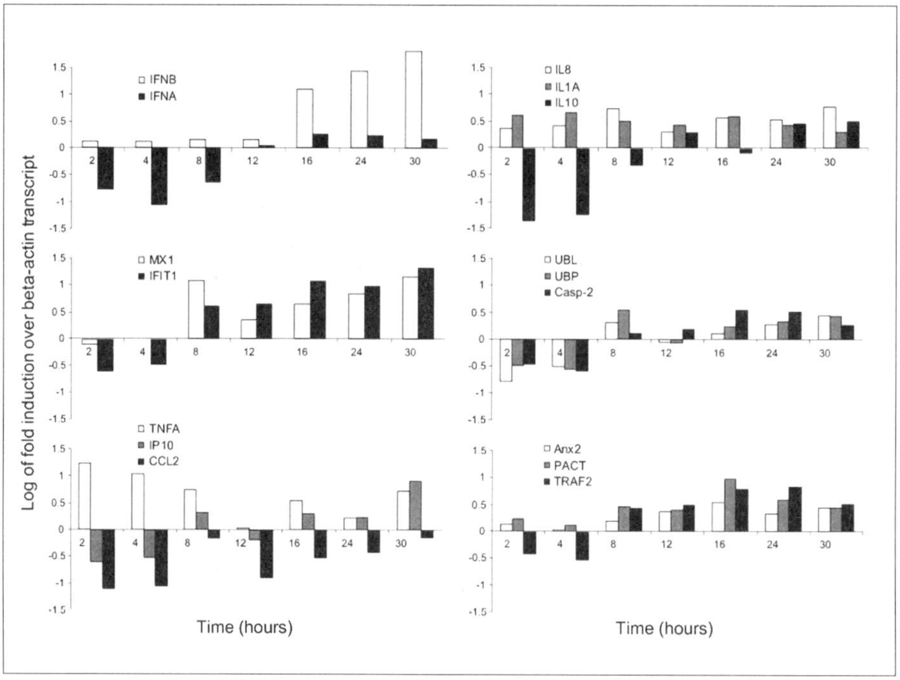

Fig. 2 A: qPCR analyses of host gene expression and cytokine levels during the infection of BALF cells by H2 PRRSV shown in Fig1.
A: The data for each transcript level were expressed as a Log_{10} of the fold induction normalised over t0 and the endogenous actin transcript level.

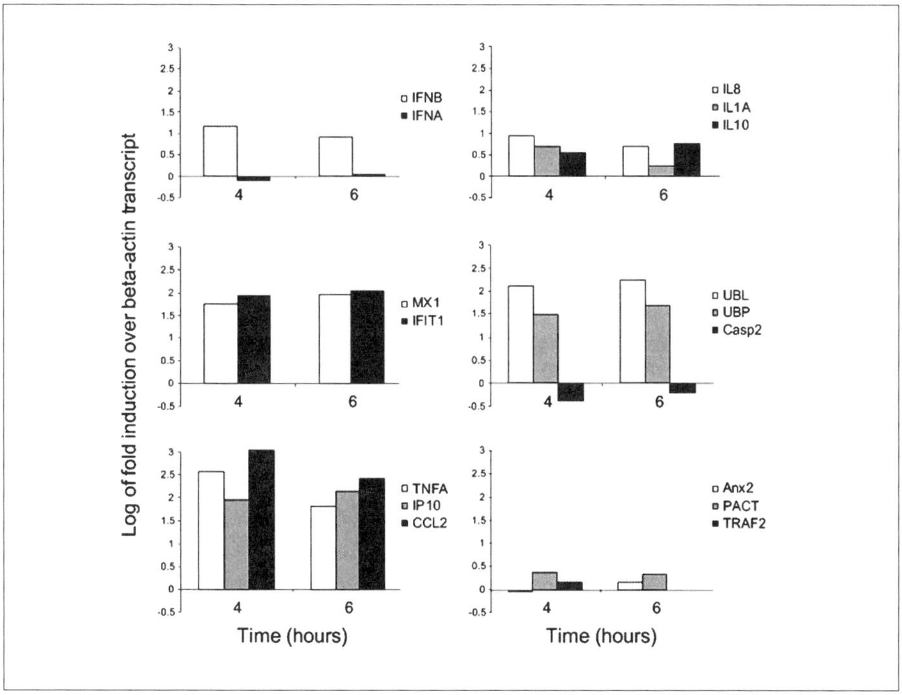

Fig. 2 B: qPCR analyses of host gene expression and cytokine levels during the infection of BALF cells by H2 PRRSV shown in Fig1.
B: Alveolar macrophages were treated with polyI/C (10 μg/ml) for 4 and 6 h. The data for each transcript level was expressed as a Log_{10} of the fold induction normalised over t0 and the endogenous β-actin transcript level.

Discussion

Innate immune responses provide the host with early, non-specific protection against infectious agents, including viruses, and help shape the subsequent adaptive immune responses [3]. Alveolar macrophages play a central role in the development of the primary defence mechanisms against pathogens, including viruses. Many viruses have evolved mechanisms to evade detection by the immune system and even to permit their replication in IFN-α/β-competent cells. Type-I IFNs, including IFN-α and IFN-β cytokines, have emerged as critical mediators of the host responses which are transcriptionally induced during viral infection. They control a large set of genes and exert potent and essential antiviral activities. The secretion of IFNs is not only essential to mount an innate antiviral response, but is equally crucial to trigger an adaptive immune response [4]. Little is known about the regulation of signalling pathways in PRRSV infected alveolar macrophages, prompting us to look further at the transcriptional regulation of the type-I IFN response during PRRSV infection in vitro. Three phases were identified during the productive replication of the H2 PRRSV. Due to the relatively low levels of viral RNA present immediately after infection, it is self-evident that the early phase (0-8 h) will play a critical role in determining the extent of the replication of the RNA genome 8 h post infection. Except for the limited accumulation of *TNFA*, *IL8* and *IL1A* transcripts during the early phase, most of the host mRNA levels tested were maintained at a significantly lower level than in the mock control, which may indicate a general inhibition of host innate immune transcription. The H2 PRRSV isolate used for the infection studies had only been passaged three times and thus was expected to be fully pathogenic, and indeed cytopathic effects were routinely detectable at 30 h post infection. Thus, the muted host response is unlikely to be attributable to a lack of pathogenicity of the virus [1]. Alternatively, the virus could specifically and temporarily reduce the accumulation of innate immune transcripts as a way of subverting the innate immune response and establishing a productive replication of its genome. The fact that substantial type-I interferon responses could be triggered in polyI/C-treated alveolar macrophages, 4 h post-treatment, indicates that interferon signalling components are present for a rapid response in these cells, and would support the possibility that PRRSV may have the ability to suppress the innate immune response very rapidly, and without the need of extensive replication of its genome. Interestingly, transcripts which are known to be regulated, at least in part, by type-I IFN signalling such as *MX1*, *IFIT1*, *UBL* and *UBP* were mostly downregulated concomitantly with the IFNB transcript accumulation during the early infection time (2-4 h post-infection), which favours the latter hypothesis [5]. It is still unclear which mechanism triggers this early innate immune suppression. IL-10, a potent suppressant of macrophage functions, may not play an immediate role during the early PRRSV response, since we have observed that the accumulation of *IL10* transcripts is downregulated. Furthermore, the level of transcripts encoding chemo-attractants such as *CCL2* remains very low and would predict that in a context of an in-vivo infection the attraction of immune cells to the site of infection may be temporarily inhibited.

Later during the viral infection, *IFNB* transcript accumulated substantially, 12 h post infection while *IFNA* mRNA was still kept at a very low level. This differential regulation between the two cytokines transcripts is intriguing. It is well recognized that the replication of the PRRSV genome requires the production of a dsRNA intermediate, also known to stimulate TLR3-dependent and -independent mechanisms, leading to the activation, at least in part, of NFκB, AP1 and IRF3 transcriptional factors and, finally triggers the production of type-I interferon [4]. Consistent with this view, *IFNB* transcript induction follows the concomitant up-regulation of the *PACT* and *TRAF2* transcript levels, protein products of which were shown to regulate NFκB via activation of the double-stranded RNA dependent protein kinase and a pathway involving other TRAF regulators, respectively [6,7]. Furthermore the induction of *MX1*, *IFIT1*, *UBL* and *UBP* transcript accumulations may also reflect the apparent release from restraint effected through IFNB cytokine accumulation. These results are in agreement with the demonstration that PRRSV activates the NFκB pathway through IκB degradation [8]. In contrast, the poor upregulation of the *IFNA* transcript induction is unclear, but it agrees with the observation that PRRSV is a poor inducer of IFN-α in vitro and in vivo [9]. Taken together, these observations suggest the possibility that PRRSV replication exercises differential control of the expression of *IFNA* and *IFNB* transcripts in alveolar macrophages which could be explained, at least in part, by the difference between the structure of *IFNB* and *IFNA* promoters [10]. Alternatively, a previous study has suggested that *IFN-α* suppression by PRRSV could take place at the post-transcriptional level [11]. Additional investigations are required to clarify this point.

This report provides the first indication that PRRSV infection delays the interferon response in alveolar macrophages, and contributes to the dissection of the specific PRRSV-induced signalling during innate immune response in infected alveolar macrophages.

Acknowledgements

We would like to thank Mr Laurence Hall and Dr Elisabeth Glass for fruitful discussions and comments on the manuscript. We are also grateful to Mr Norrie Russell for the graphic designs. This work was supported by the BBSRC grant EGA16307.

References

1 Ait-Ali T, Wilson AD, Westcott DG, Clapperton M, Waterfall M, Mellencamp MA et al: Innate immune responses to replication of porcine reproductive and respiratory syndrome virus in isolated Swine alveolar macrophages. Viral Immunol 2007;20(1):105-118.

2 Drew TW: A review of evidence for immunosuppression due to porcine reproductive and respiratory syndrome virus. Vet Res 2000;31(1):27-39.

3 Takaoka A, Yanai H: Interferon signalling network in innate defence. Cell Microbiol 2006;8(6):907-922.

4 Garcia-Sastre A, Biron CA: Type 1 Interferons and the Virus-Host Relationship: A Lesson in Detente. Science 2006;312(5775):879-882.

5 Zhang X, Shin J, Molitor TW, Schook LB, Rutherford MS: Molecular responses of macrophages to porcine reproductive and respiratory syndrome virus infection. Virology 1999;262(1):152-162.

6 Darnay BG, Haridas V, Ni J, Moore PA, Aggarwal BB: Characterization of the Intracellular Domain of Receptor Activator of NF-kappa B (RANK). Interaction with tumor necrosis factor receptor-associated factors and activation of NF-kappa B and c-jun N-terminal kinase. J Biol Chem 1998;273(32):20551-20555.

7 Huang X, Hutchins B, Patel RS: The C-terminal, third conserved motif of the protein activator PACT plays an essential role in the activation of double-stranded-RNA-dependent protein kinase (PKR). Biochem J 2005;366:175-186.

8 Lee SM, Kleiboeker SB: Porcine arterivirus activates the NF-kappaB pathway through I-kappaB degradation. Virology 2005;342(1):47-59.

9 Albina E, Carat C, Charley B: Interferon-alpha response to swine arterivirus (PoAV), the porcine reproductive and respiratory syndrome virus. J Interferon Cytokine Res 1998;18(7):485-490.

10 Levy DE, Marie I, Prakash A: Ringing the interferon alarm: differential regulation of gene expression at the interface between innate and adaptive immunity. Curr Opin Immunol 2003;15(1):52-58.

11 Lee SM, Schommer SK, Kleiboeker SB: Porcine reproductive and respiratory syndrome virus field isolates differ in in vitro interferon phenotypes. Vet Immunol Immunopathol 2004;102(3):217-231.

Dr. Tahar Ait-Ali, Roslin Institute, Roslin, EH25 9PS, UK.
E-mail: tahar.aitali@bbsrc.ac.uk

Pinard M-H, Gay C, Pastoret P-P, Dodet B (eds): Animal Genomics for Animal Health. Dev Biol (Basel). Basel, Karger, 2008, vol 132, pp 247-253.

Identification of Polymorphisms in Bovine *TLR2* and *CARD15*, Associations between *CARD15* Polymorphisms and Milk Somatic Cell Score in Canadian Holsteins, and Functional Relevance of SNP c.3020A>T

S.D. Pant, F.S. Schenkel, I. Leyva-Baca, B.S. Sharma, N.A. Karrow

Centre for Genetic Improvement of Livestock, Department of Animal and Poultry Science, University of Guelph, Guelph, Ontario, Canada N1G 2W1.

Keywords: somatic cell score, SCS, mastitis, CARD15/NOD2, SNP

Abstract: Toll-like receptor-2 (TLR2) and caspase recruitment domain 15 (CARD15) are important pattern recognition receptors that play a role in the initiation of the inflammatory and subsequent immune response. They have been previously identified as susceptibility loci for inflammatory bowel diseases in humans and are, therefore, suitable candidate genes for inflammatory disease resistance in cattle. The objective of this study was to identify single nucleotide polymorphisms (SNPs) in the bovine *TLR2* and *CARD15* and evaluate the association of these SNPs with health and production traits in a population of Canadian Holstein bulls. A selective DNA pool was constructed based on the estimated breeding values (EBVs) for somatic cell score (SCS). Gene segments were amplified from this pool in PCR reactions and the amplicons sequenced to reveal polymorphisms. A total of four SNPs, including one in intron 10 (c.2886-14A>G) and three in exon 12 (c.3020A>T, c.4500A>C and c.4950C>T) were identified in *CARD15*; none were identified in *TLR2*. Canadian Holstein bulls (n=338) were genotyped and haplotypes were reconstructed. Two SNPs, c.3020A>T and c.4500A>C, were associated with EBVs for health and production traits. The SNP, c.3020A>T for example, was associated with SCS EBVs ($p=0.0097$) with an allele substitution effect of 0.07 score. When compared to the most frequent haplotype Hap12(AC), Hap22(TC) was associated with increased milk ($p < 0.0001$) and protein ($p = 0.0007$) yield EBVs, and hap21(TA) was significantly associated with increased SCS EBV ($p = 0.0120$). All significant comparison-wise associations retained significance at 8% experimental-wise level by permutation test. The role of SNP c.3020A>T in MDP induced IL-1β expression was investigated by real-time quantitative reverse transcription-PCR (real-time quantitative RT-PCR). The induction of IL-1β by MDP was highly variable between individuals, and no association was observed between IL-1β expression and SNP c.3020A>T genotypes. In summary, the association study indicates that SNP c.3020A>T might play a role in the host response against mastitis; however, it is not the sole determinant of MDP induced IL-1β expression in blood leukocytes. Further detailed studies are needed to understand the functional implications of SNP c.3020A>T.

Introduction

Inflammatory diseases such as mastitis, inflammatory bowel disease, metritis and laminitis are economically important diseases for the dairy industry. Mastitis is the most common amongst them, alone accounting for losses in excess of $2 billion to the US dairy industry [1]. A number of therapeutic, prophylactic and management strategies have been proposed to minimize this complex disease. However, a widely accepted complimentary strategy is based on improving the host genetics through selective breeding. Since it is difficult to use a direct index to measure the mastitis phenotype, milk somatic cell score (SCS) is often used as an indirect index to select animals for breeding [2]. Milk SCS is a log2 score of the milk somatic cell count and has been genetically correlated with clinical mastitis ($r = 0.7$) [3]. It has been demonstrated in sheep that intramammary infections occur less frequently after one generation of breeding for low SCS [4]. Breeding strategies in dairy cattle are similarly based on selection for low estimated breeding values (EBVs) for SCS [5].

The etiology of mastitis is diverse and may include infections caused by bacteria, parasites, viruses and fungi. In order for an effective host immune response to occur against these pathogens, the host must possess receptors that recognize conserved molecular patterns associated with different classes of pathogens. These are referred to as pathogen associated molecular patterns (PAMPs). Receptors that recognize PAMPs are collectively referred to as pattern recognition molecules (PRMs) [6].

Caspase recruitment domain 15 (CARD15), also known as nucleotide oligomerization domain 2 (NOD2), is a cytosolic PRM capable of initiating inflammation following PAMP recognition. CARD15 is involved in intracellular recognition of muramyl dipeptide (MDP), the minimal bioactive structure of peptidoglycan, which is common to the cell wall of both Gram-positive and Gram-negative bacteria [7].

Phagosomal toll-like receptor-2 (TLR2) is involved in recognizing PAMPs associated with Gram-positive bacteria. Murine studies for example, have demonstrated that TLR2 is in involved in the early recruitment of neutrophils in response to intraperitoneal challenge with either *Staphylococcus aureus*, or peptidoglycan (PGN) from *S. aureus* [8]. Ruminant studies have demonstrated that TLR2 gene (designated as '*TLR2*') expression occurs in dermal and gut-associated tissues, and is highly induced during mastitis caused by *S. aureus* [9].

It has been shown that polymorphisms, particularly single nucleotide polymorphisms (SNPs), within the genes coding for different PRMs impair the ability of certain individuals to respond properly to infections [10]. It is important therefore, to understand the biology of such genetic variants, in order to identify and select animals with enhanced resistance to mastitis for breeding programmes.

The purpose of this study was to identify SNPs in the bovine *TLR2* and *CARD15*, evaluate the association of these SNPs with SCS and other production traits in a population of Canadian Holstein bulls, and to assess the functional relevance of SNP c.3020A>T in *CARD15*.

Material and Methods

The resource population consisted of 2,166 Holstein bulls selected on the basis of extreme EBV for either protein yield or SCS. A total of 338 semen samples of these bulls were selected within families on the basis of extreme EBVs for either protein yield or SCS for this study. The EBV's for SCS, udder depth, milk yield, protein yield and fat yield of these bulls were obtained from a national genetic evaluation database generated in August 2006 by the Canadian Dairy Network, Guelph, Ontario, Canada, and used in the association studies. Samples were kindly provided by Semex Alliance (Guelph, Ontario, Canada).

DNA was extracted from the semen samples using a slightly modified phenol chloroform procedure [11]. Twenty bulls with high and twenty bulls with low EBVs for SCS were selected from the previously selected 338 animals, to create a DNA pool containing equal amounts of DNA from each bull. This pool was used in PCR reactions to amplify the *CARD15* and *TLR2* exons, flanking introns and 2 kb promoter regions. Each PCR amplicon was sequenced in forward and reverse direction for SNP discovery. Polymorphisms were detected by scrutinizing the forward and reverse electropherograms generated from the sequences. A total of 338 animals were genotyped for the identified *CARD15* SNPs using the tetra-primer Amplification Refractory Mutation System (ARMS)-PCR procedure, as described by Ye et al. [12].

Statistical Analyses

Preliminary analyses revealed only two CARD15 SNPs (c.3020A>T and c.4500A>C) to be associated with any of the different traits analyzed; only these two SNPs were used for the detailed analysis. Data was analyzed using PROC REG (SAS Institute, Inc., Cary, NC), by the following model:

$$y_j = \mu + \sum_{i=1}^{2} \beta_i G_i + e_j$$

where: y_j = trait EBV for the jth animal; μ = overall mean, β_i = regression coefficient (allele substitution effect) for the ith SNP, e_j = random error and G_i = the genotype of the ith SNP. Mallow's criterion was used to select the final regression model. Type I experimental errors were controlled by implementing permutation test. The traits of interest were milk yield, fat yield, protein yield, udder depth, and SCS.

The haplotype probabilities were reconstructed using the program 'HAPROB' developed by Boettcher et al. [13]. Haplotype effects were estimated by regressing EBV's on haplotype probabilities, which are expressed as the expected number of copies of each haplotype. The following model was used for statistical analyses using SAS software (SAS Institute, 1999):

$$EBV_j = \mu + \sum_{i=1}^{4} \beta_i \, Hap_{ij} + e_j$$

where EBVj is trait EBV of the jth animal, βi is the linear regression coefficient for the ith haplotype, Hapij is the probability of the ith haplotype for the jth bull, and ej are random residual effects. A permutation test was implemented to control type I experimental-wise errors.

Real-time quantitative reverse transcription polymerase chain reaction (RT-PCR)

Fifteen Holstein cows, genotyped for SNP c.3020A>T (AA, AT and TT), were used for the functional study. Whole blood, 30 ml from each cow, was challenged with 10 mg/ml of MDP (Invivogen, San Diego, CA) at 37°C in 5% CO_2, and total RNA was extracted from the buffy coat leukocytes at 0 hours; and 1 and 3 hours post-challenge using Trizol (Invitrogen, Carlsbad, CA). RNA quality was checked by methylene blue staining and cDNA was prepared using stratascript reverse transcriptase (Stratagene, CA) Real-time RT-PCR was used to quantify *IL-1β* and the housekeeping gene *RPLP0* (ribosomal protein, large, P0). Real-time RT-PCR was performed using SYBR green (Invitrogen, Carlsbad, CA) and an ABI Prism 7000 sequence detection system (Applied biosystems, Foster City, CA). Standard curves were generated using serial dilutions of *IL-1β* and *RPLP0*, and the relative expression of *IL-1β* and *RPLP0* was determined.

Results

Investigation of the coding exon, flanking introns and promoter sequences of *TLR2* and *CARD15* revealed a total of four SNPs in *CARD15*, including one in intron 10 (c.2886-14A>G) and three in the exon 12 (c.3020A>T, c.4500A>C and c.4950C>T); and no SNP in *TLR2*

Statistical analyses revealed associations between CARD15 SNP c.3020A>T and EBVs for milk yield, protein yield, udder depth and SCS; and between SNP c.4500A>C and milk yield, fat yield and protein yield. Amongst all the possible regression models for SCS, a single SNP model including SNP c.3020A>T was found to be the best model, showing a significant association between this SNP and SCS (p=0.0097). The average allele substitution effect of this SNP for SCS was 0.068, with the allele 'T' increasing SCS over allele 'A' (21% of the SD for SCS EBV). All significant associations were retained at an 8% experimental-wise significance level by permutation test. A complete description of the average allele substitution effects is presented in Table 1.

Table 1: Average allele substitution effect of SNPs c.3020A>T and c.4500A>C on EBVs for milk yield, fat yield, protein yield, udder depth (UD) and somatic cell score (SCS).

	Trait				
	Milk (Kg)	**Fat (Kg)**	**Prot (Kg)**	**UD**	**SCS**
SNP	**Allele substitution effects ± SE (comparison-wise P level)**				
c.3020A>T	359.77± 107.70	-	9.90±3.20	-1.10±0.41	0.07±0.03
	(<0.0001++)		(0.0003++)	(0.0087+)	(0.0097+)
c.4500A>C	348.44±117.80	8.29±2.40	11.31±3.51	-	-
	(<0.0001++)	(0.0010++)	(<0.0001++)		

++Significant effect by permutation test at an experimental-wise significance level of 5%.
+Significant effect by permutation test at an experimental-wise significance level of 8%

Two *CARD15* SNPs (c.3020A>T and c.4500A>C) were used for haplotype reconstruction. The estimated haplotype frequencies were 42.6%, 26.1%, 19.8% and 11.4% for Hap12(AC), Hap21(TA), Hap22(TC) and Hap11(AA), respectively. The linear effects of each of the haplotypes were estimated by treating the effect of the most frequent haplotype (Hap12) as a control and contrasting the effects of the other haplotypes (Hap21, Hap22 and Hap11) against it (Table 2). Analysis revealed statistically significant differences between Hap21 and Hap12 for SCS and fat yield, and between Hap22 and Hap12 for milk yield and protein yield at different levels of significance. All significant associations were retained at 8% experimental-wise level by permutation test.

Table 2: Linear effects of CARD15 on EBVs for milk yield (milk), fat yield (fat), protein yield (prot), udder depth (UD) and somatic cell score (SCS), when comparing haplotypes Hap11, Hap22, and Hap21, against Hap12.

	Trait				
	Milk (Kg)	**Fat (Kg)**	**Prot (Kg)**	**UD**	**SCS**
Haplotypes	**Haplotype effects ± SE (P level)**				
Hap11(AA)	-325.2 ±175.30	-8.23±5.60	-12.68±5.20	-0.91±1.00	0.02±0.06
	(0.06)	(0.14)	(0.01)	(0.36)	(0.74)
Hap22(TC)	523.5 ±129.70	6.89±4.15	13.15±3.85	-1.71±0.74	0.05±0.05
	(<0.0001++)	(0.07)	(0.0007++)	(0.02)	(0.26)
Hap21(TA)	-57.5 ±81.62	-7.27±2.61	-3.10±2.42	-0.93±0.46	0.07±0.03
	(0.48)	(0.0057++)	(0.20)	(0.04)	(0.012+)

++Significant effect by permutation test at an experimental-wise significance level of 5%.
+Significant effect by permutation test at an experimental-wise significance level of 8%.

Real-time quantitative RT PCR did not reveal any significant differences between MDP induced IL-1B expression and SNP c.3020A>T genotypes (Table 3). However, IL-1B induction varied between individuals within and across each genotype.

Table 3: Mean and standard deviation (SD) of cycle threshold (ct) values of blood leukocyte IL-1β expression measured 0, 1 and 3 hours post-MDP challenge in vitro, by real-time RT-PCR relative to the RPLP0 housekeeping gene.

Genotype	Mean ct values ± SD		
	0 hour	1 hour	3 hours
AA	0.026±0.033	1.943±1.868	0.116±0.087
AT	0.073±0.099	1.025±0.658	0.081±0.076
TT	0.119±0.126	1.923±2.943	0.087±0.071

Discussion

TLR2 and CARD15 play a role in the initiation of inflammatory and immune responses to bacterial infections. A number of studies have reported that SNPs in PRMs of different species play an important role in contributing towards disease susceptibility. Therefore, the main objective of this study was to identify SNPs in the *TLR2* and *CARD15* and to explore associations between these SNPs and SCS, a trait directly related to udder inflammation. Out of a total of four SNPs identified in *CARD15*, only one SNP was found in the coding region (c.3020A>T). This SNP, located in exon 12, is a non-synonymous SNP coding for the terminal LRR domain of CARD15 receptor. While allele 'A' at this position codes for glutamine, allele 'T' codes for leucine. The association of the T allele with increased SCS and decreased udder depth, which predisposes animals to mammary infections, suggests that changes in the composition of the LRR domain of CARD15 may contribute to disease susceptibility. The terminal LRR domains, as in CARD15, are common to different PRMs and are responsible for PAMP recognition. There are several reports of polymorphisms in the human and mouse LRR coding gene segments that contribute to differential binding to several bacterial components [14]. In humans, CARD15 variants have been hypothesized to alter bacterial component recognition by altering the structure of the LRR domain of the adjacent region. Thus, we wanted to explore the potential functional relevance of SNP c.3020A>T by activating genotype specific blood leukocytes with MDP. MDP acts as a NOD2 ligand, which upon stimulation induces signalling pathways culminating in the production of pro-inflammatory cytokines including IL-1β. Therefore, real-time quantitative RT-PCR was carried out to investigate the effect of this SNP on MDP induced IL-1β expression in blood leukocytes. All individuals were induced by MDP challenge. IL-1β induction peaked at one hour post-challenge and was found to be highly variable amongst individuals. However, no association was found between MDP induced release of IL-1β and SNP c.3020>T genotypes. This suggests that NOD2 may not contribute significantly to MDP induced expression of IL-1β in blood leukocytes. A recent study, for example, has shown that MDP-induced release of IL-1β from bone marrow derived macrophages requires a number of proteins in addition to NOD2 [15]. Thus, there is a need to identify these proteins and characterize their function.

Conclusion

In conclusion, four SNPs in *CARD15* in a sample of Canadian Holstein bulls were discovered. Statistical analyses revealed that SNP c.3020A>T was associated with EBVs for SCS and udder depth and milk and protein yields, while SNP c.4500A>C was only associated with milk, fat and protein yields. The most common haplotypes for these two SNPs in the population differed significantly for their effect on SCS. Moreover, the most common haplotype carried alleles at both loci that are favourable for reducing SCS and increasing production EBVs. This implies that these two SNPs, together with other polymorphisms found to be associated with SCS in other studies, may potentially be used for genetic selection for mastitis resistance and production. A functional investigation of the SNP c.3020A>T did not reveal any direct effect on MDP induced expression of IL-1β. However, CARD15 is not the sole determinant accounting for the variability in MDP induced expression of IL-1β and different factors involved in this process need to be identified.

Acknowledgements

We would like thank Natural Science and Engineering Research Council, Dairygen, Canadian Bovine Mastitis Research Network and Ontario Ministry of Agriculture, Food and Rural Affairs for financial support. We would also like to thank Semex alliance for providing DNA samples for this study.

References

1 National Mastitis Council: Current concepts of bovine mastitis, 4th edition edn. Madison, WI; 1996.

2 Description of National Genetic Evaluation Systems for dairy cattle traits as applied in different Interbull member countries [http://www-interbull.slu.se/national_ges_info2/framesida-ges.htm]

3 Rupp R, Boichard D: Genetic parameters for clinical mastitis, somatic cell score, production, udder type traits, and milking ease in first lactation Holsteins. J Dairy Sci 1999;82(10):2198-2204.

4 Rupp R, Bergonier D, Dion S, Hygonenq MC, Aurel MR, Foulon E et al: Effects of SCC-Based selection for mastitis resistance: First results from a divergent selectioin experiment in sheep. . In: Eighth world congress on genetics applied to livestock production. Belo horizonte, MG, Brazil; 2006.

5 Mark T, Fikse WF, Emanuelson U, Philipsson J: International genetic evaluations of Holstein sires for milk somatic cell and clinical mastitis. J Dairy Sci 2002,85(9):2384-2392.

6 Schumann RR: Host cell-pathogen interface: molecular mechanisms and genetics. Vaccine 2004;22 Suppl 1:S21-24.

7 Girardin SE, Boneca IG, Viala J, Chamaillard M, Labigne A, Thomas G et al: Nod2 is a general sensor of peptidoglycan through muramyl dipeptide (MDP) detection. J Biol Chem 2003;278(11):8869-8872.

8 Mullaly SC, Kubes P: The role of TLR2 in vivo following challenge with Staphylococcus aureus and prototypic ligands. J Immunol 2006;177(11):8154-8163.

9 Goldammer T, Zerbe H, Molenaar A, Schuberth HJ, Brunner RM, Kata SR et al: Mastitis increases mammary mRNA abundance of beta-defensin 5, toll-like-receptor 2 (TLR2), and TLR4 but not TLR9 in cattle. Clin Diagn Lab Immunol 2004;11(1):174-185.

10 Schroder NW, Schumann RR: Single nucleotide polymorphisms of Toll-like receptors and susceptibility to infectious disease. Lancet Infect Dis 2005;5(3):156-164.

11 Winfrey MR, Rott MA, Wortman AT: Unraveling DNA: Molecular Biology for the Laboratory: Prentice Hall, New Jersey, USA; 1997.

12 Ye S, Dhillon S, Ke X, Collins AR, Day IN: An efficient procedure for genotyping single nucleotide polymorphisms. Nucleic Acids Res 2001;29(17):E88-88.

13 Boettcher PJ, Pagnacco G, Stella A: A Monte Carlo approach for estimation of haplotype probabilities in half-sib families. J Dairy Sci 2004;87(12):4303-4310.

14 Matsushima N, Tachi N, Kuroki Y, Enkhbayar P, Osaki M, Kamiya M et al: Structural analysis of leucine-rich-repeat variants in proteins associated with human diseases. Cell Mol Life Sci 2005;62(23):2771-2791.

15 Pan Q, Mathison J, Fearns C, Kravchenko VV, Da Silva Correia J, Hoffman HM et al: MDP-induced interleukin-1beta processing requires Nod2 and CIAS1/NALP3. J Leukoc Biol 2007;82(1):177-183.

Dr. Sameer Pant, Centre for Genetic Improvement of Livestock, Department of Animal and Poultry Science, University of Guelph, Guelph, Ontario, Canada N1G 2W1.

E-mail: sameerpant@gmail.com

Pinard M-H, Gay C, Pastoret P-P, Dodet B (eds): Animal Genomics for Animal Health. Dev Biol (Basel). Basel, Karger, 2008, vol 132, pp 255-259.

Associations between Polymorphisms in the Porcine Haptoglobin Gene and Baseline Levels of Serum Haptoglobin

A. B. Diack[1], C. D. Gladney[2], M. A. Mellencamp[2], M. J. Stear[1], P. D. Eckersall[1]

[1] Faculty of Veterinary Medicine, University of Glasgow, Glasgow, UK
[2] Genus Plc/PIC, De Forest, Wisconsin, USA

Key words: haptoglobin, polymorphisms, porcine SNPs

Abstract: Haptoglobin (Hp) is an acute phase protein that is a marker in blood for clinical and subclinical disease in the pig. The aim of this study was to identify single nucleotide polymorphisms (SNPs) in the Hp gene and analyse their influence on baseline serum levels. DNA samples and serum were collected from 345 boars. Of 13 SNPs identified, 5 were genotyped using PCR-RFLP and Pyrosequencing®. Serum Hp levels were measured using a biochemical assay. A general linear model was fitted with line and genotype as fixed effects. In addition, linkage disequilibrium (LD) was estimated between the 5 SNPs using r-square and D prime. Serum Hp concentrations in the population showed a skewed distribution with a mean of 0.34 g/L (range 0 – 2.65 g/L). Three SNPs were found to be associated with baseline Hp levels (p-value = 0.0093, 0.0051 and 0.0094). These 3 SNPs were also found to be in high LD with each other. This is the first study to find associations between polymorphisms in the porcine Hp gene and baseline Hp serum levels. The results have implications for breeding for resistance to infection.

INTRODUCTION

Biomarkers such as acute phase proteins (APPs) have been proposed for monitoring pig health and welfare. APPs are a group of proteins produced by the liver, following cytokine production resulting from inflammation, injury or infection. The function of APPs is to return the body to homeostasis [1]. Haptoglobin (Hp), a major porcine APP, binds haemoglobin released during erythrocyte destruction; this prevents iron loss and has a bacteriostatic effect [2]. Serum Hp concentrations are an effective indicator of respiratory disease, infection by pathogens such as *Streptococcus suis* and *Actinibacillus pleuropneumoniae* and can also be used to differentiate between animals with a high and low weight gain at slaughter [3-5]. Little is known about the genetic control of porcine APPs. The aim of this study was to identify single nucleotide polymorphisms (SNPs) in Hp and analyse their association with baseline serum Hp concentrations.

Materials and Methods

A population of 345 pure line high health boars from seven genetic lines was made available by Genus Plc (PIC). Each boar had an archived tissue sample taken at birth and one (n= 115) or two (n= 230) serum samples available for analysis. DNA was isolated from porcine tail tissue using DNeasy® 96 Tissue kits (Qiagen Inc., CA, USA). Hp concentrations were measured in duplicate using a biochemical assay [6]. To identify SNPs, PCR followed by a Big Dye V3.1 (Applied Biosystems, CA, USA) sequencing reaction with analysis on an ABI 3100 automated sequencer (Applied Biosystems) was carried out.

SNP genotyping was done by either PCR-RFLP or Pyrosequencing®. For PCR-RFLP, restriction enzymes were selected using NEBcutter V2.0 software. PCR products were digested using the appropriate restriction enzyme (NEB, MA, USA) and using the recommended buffer systems and incubation times. Following incubation, the products were placed on a 3% SFR agarose gel stained with ethidium bromide for 30 min at 150V. Genotypes were scored manually.

Pyrosequencing® primers were designed using PSQ Assay Design Software V1.0 and supplied by Integrated DNA Technologies Inc (Coralville, IA, USA). PCR was carried out and the resulting product immobilised onto Streptavidin Sepharose™ beads, washed (70% ethanol), denatured (0.2 M NaOH) and washed (10 mM tris-acetate, pH 7.6) before the beads were released into a sequencing primer/annealing buffer solution and heated to 80°C for 2 min. Once cooled to room temperature (RT), the Pyrosequencing® reaction was performed at RT using an automated Pyrosequencing® system PSQ 96 MA and the associated software (Biotage Inc. Uppsala, Sweden). The genotype scoring was done using PSQ 96 SNP software.

Statistical analysis using SAS 9.1 (The SAS Institute, NC, USA) was carried out on log-transformed data. A general linear model (GLM) with line and genotype as fixed effects was fitted.

Results

The Hp serum concentrations of the population (n=345) are shown in Fig.1. The population showed a skewed distribution with a mean of 0.34±0.34 g/L (range: 0.00 – 2.65 g/L) and the data log-transformed before further analysis.

Thirteen polymorphisms in the porcine Hp gene were identified by sequencing a pool of genomic DNA. Five non-coding SNPs were genotyped across a high health pig population (n=345). Analysis of the 5 SNPs showed that they were not in Hardy-Weinberg equilibrium and additionally were in high LD with one another (Table 1).

Table 1: Linkage disequilibrium measurements for five single nucleotide polymorphisms (SNPs) within the porcine haptaglobin (Hp) gene.

Locus 1	Locus 2	Lewontin's D'	R^2
SNP A	SNP B	0.9388	0.6016
SNP A	SNP C	0.9486	0.5230
SNP A	SNP D	0.9481	0.5160
SNP A	SNP E	0.9533	0.6078
SNP B	SNP C	1.0000	0.8527
SNP B	SNP D	1.0000	0.8420
SNP B	SNP E	0.8634	0.7365
SNP C	SNP D	1.0000	0.9876
SNP C	SNP E	1.0000	0.8629
SNP D	SNP E	1.0000	0.8523

Statistical analysis showed no association between the pig genetic line and genotype, nor pig line and Hp serum concentration, however there was a significant association ($p<0.01$) between genotype and Hp serum concentrations for 3 of the SNPs. Higher Hp serum concentrations were associated with allele 1 in SNPs C and D, while lower concentrations were associated with allele 2 in SNP B (Table 2).

Table 2: Association between single nucleotide polymorphisms (SNP) in the porcine Haptoglobin (Hp) gene and baseline Hp serum concentrations.

SNP	Genotype	*n*	Hp (g/L)	P-value
SNP A	11	117	0.34	0.3908
	12	96	0.31	
	22	125	0.37	
SNP B	11	99	0.41	0.0093**
	12	84	0.31	
	22	159	0.32	
SNP C	11	85	0.39	0.0051**
	12	88	0.37	
	22	170	0.31	
SNP D	11	84	0.39	0.0094**
	12	88	0.37	
	22	171	0.31	
SNP E	11	100	0.36	0.1412
	12	81	0.37	
	22	160	0.32	

**p < 0.01.

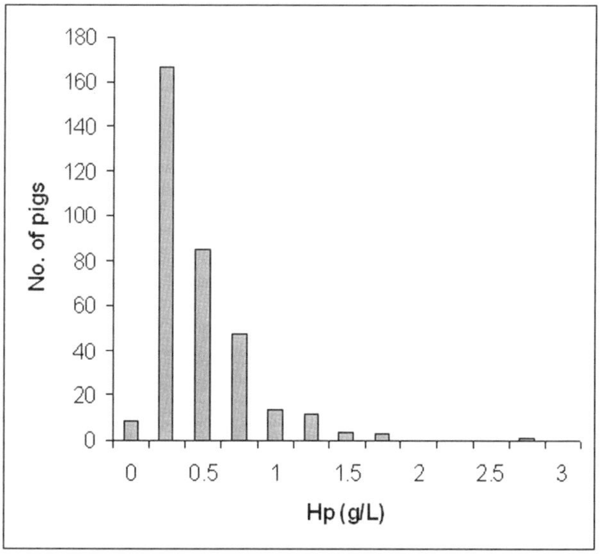

Fig. 1: Serum concentrations of Haptoglobin (Hp) across a high health boar population (n = 345).

Discussion

Serum Hp concentrations were within the accepted normal range for healthy pigs; the skewed nature seen is common in healthy populations where the concentrations are expected to be low (<1 g/L) [7].

Analysis of the individual SNPs revealed that the majority of them were not in Hardy-Weinberg distribution. This is not surprising as this study used a small pig population under selection for production traits, and it is likely that gene pooling has occurred over time. Additionally, the majority of the SNPs were found in strong LD with each other. This is likely due to the small distances between the SNPs, decreasing the chance of recombination occurring.

The association analysis showed a significant association between three of the SNPs (SNPs B, C and D) and Hp serum concentration and identified the genotypes responsible for lower Hp concentrations in SNP B (allele 2) and higher Hp concentrations in SNPs C and D (allele 1). This is the first report of an association between Hp genotype and concentration in pigs and suggests that there is a genetic component to the control of Hp baseline concentrations. It is possible that all three SNPs influence Hp concentrations, however, due to the high LD measurements and the fact that none of the SNPs are within coding regions of the gene, it is more likely that they are linked to a SNP that changes protein expression levels. Alternatively, the differences in Hp concentrations could be due to changes in gene regulation or expression caused by a non-coding SNP.

The lack of association between pig line and Hp concentration allow for the use of Hp as a marker of health across this population as line effects do not need to be considered. The association between genotype and Hp concentration would allow animals to be selected for either a high or low baseline Hp concentration and enable a comparative study of the response of the animals under disease challenge. It is likely that it is the response of an animal to infection rather than the initial baseline concentrations that will have implications for future breeding strategies, particularly in breeding for immune traits or disease resistance.

Conclusion

This is the first study to identify associations between SNPs in the porcine Hp gene and baseline Hp serum concentrations in a high health pig population. This has potential implications in breeding for resistance to disease or in the use of Hp as a marker of pig health.

Acknowledgements

Funding and support from BBSRC, Genus Plc (PIC) and Genesis-Faraday partnership are gratefully acknowledged.

References

1 Baumann H, Gauldie J: The acute phase response. Immunol Today 1994;15(2):74-80.

2 Langlois MR, Delanghe JR: Biological and clinical significance of haptoglobin polymorphism in human. Clin Chem 1996;42(10):1589-1600.

3 Eurell TE, Bane DP, Hall WF, Schaeffer DJ: Serum haptoglobin concentration as an indicator of weight gain in pigs. Can J Vet Res 1992;56(1):6-9.

4 Sorensen NS, Tegtmeier C, Andresen LO, Piñero M, Toussaint MJM, Campbell FM et al: The porcine acute phase protein response to acute clinical and subclinical experimental infection with Streptococcus suis. Vet Immunol Immunopathol 2006;113(1-2):157-168.

5 Eckersall PD, Saini PK, McComb C: The acute phase response of acid soluble glycoprotein, alpha(1)-acid glycoprotein, ceruloplasmin, haptoglobin and C-reactive protein, in the pig. Vet Immunol Immunopathol 1996;51(3-4):377-385.

6 Eckersall PD, Duthie S, Safi S, Moffat D, Horadagoda NU, Doyle S et al: An automated biochemical assay for haptoglobin: Prevention of interference from albumin. Comp Haem Int 1999;9(3):117-124.

7 Petersen HH, JP Nielsen, Heegaard PM: Application of acute phase protein measurements in veterinary clinical chemistry. Vet Res 2004;35(2):163-187.

Abigail Diack, APPH, Faculty of Veterinary Medicine, Bearsden Rd, Glasgow, G61 1QH.UK.

E-mail: abigail.diack@roslin.ed.ac.uk

Pinard M-H, Gay C, Pastoret P-P, Dodet B (eds): Animal Genomics for Animal Health. Dev Biol (Basel). Basel, Karger, 2008, vol 132, pp 261-266.

Single Nucleotide Polymorphism (SNP) in Prion Protein Gene (*PRNP*) exon-3 in Gaddi Sheep

N. Gupta, S. Choudhary, G. Malik, A. Pandey, S C Gupta

DNA Fingerprinting Unit, National Bureau of Animal Genetic Resources, Karnal, India

Abstract: Nine PCR-SSCP conformations in *PRNP* the exon-3 genomic DNA segment covering codons 136, 154 and 171 were observed. DNA sequences revealed 9 unique variants (DQ885789-DQ885796). At codon 171 of exon-3, two alleles, Q and R, had a frequency of 0.336 and 0.664, respectively. Genotypes QQ, QR and RR were observed in 25.86, 15.51 and 58.62% of animals, respectively, showing deviation from Hardy-Weinberg equilibrium. At codon 154, all genotypes were RR. At codon 136 AA, AV and VV genotypes were 86.21, 10.34 and 3.44%, respectively, and frequencies of the alanine (A) and valine (V) alleles were 0.914 and 0.086, respectively, where the VV genotype had lower frequency than expected. Polymorphisms were also observed at codons 137, 143 and 173 of exon-3.

Key words: scrapie, *PRNP* gene, SSCP, SNP, Gaddi sheep

INTRODUCTION

Scrapie is a slow progressive neurodegenerative disease of sheep and goats [1] caused by a minute infectious protease resistant isoform of sheep prion protein (PRP^{sc}). In Europe it was recognized more than 250 years ago [2], but it was seen only once in India in the sixties, when four animals were infected and died in the remote Himalayan foothills; the stock was probably imported from Britain [3]. According to Zlotnik and Katiyar [4] the animals had typical clinical symptoms of scrapie and the histological lesions in the central nervous system of affected animals were identical to those described for natural scrapie in sheep in Britain. Ten polymorphisms in the sheep prion protein gene (*PRNP*) have been reported to occur at different codons [5]. Polymorphisms within *PRNP* were associated with susceptibility/resistance to transmissible spongiform encephalopathies (TSEs) in sheep [6] and goats [7]. Codon 171 influenced the susceptibility to scrapie in Suffolk sheep under natural outbreak conditions while Cheviot sheep showed association under experimentally induced conditions [8]. Substitution of Ala (A) to Val (V) at codon 136 significantly enhanced susceptibility [9], whereas His (H) substitution to Arginine (R) at codon 154 enhanced the resistance to scrapie [10] in sheep breeds; however, clear-cut involvement of this codon has not been proven in the majority of cases. This study describes the polymerase chain reaction-single strand conformation polymorphism (PCR-SSCP) amplimers of ovine *PRNP* exon-3 covering codons 136, 154 and 171 known to influence disease susceptibility. As far as we are aware, this is the first survey of *PRNP* gene polymorphism in Indian sheep.

Materials and Methods

Blood from 58 animals at random was collected from 25 different village flocks in Himachal Pradesh. DNA was isolated in the laboratory according to Sambrook and Russell [11]. PCR-SSCP primers (forward 5'-caa ggt ggt agc cac agt ca-3' and reverse 5'-cca cca ctc gct cca tta tc-3') were designed based on GenBank accession No. DQ077504 sequences covering part of exon-3 from 344 to 711 nucleotide positions of the ovine *PRNP* gene. The primers were synthesized by Integrated DNA Technologies, Inc. Coralville, USA. PCR was performed in 25 μl volume of 10X PCR buffer, 2 mM $MgCl_2$, 200 μM each of dNTP's, 1.0 μM of each primer, 100 ng genomic DNA, and 0.75 U high fidelity Taq DNA polymerase under the following conditions: initial denaturation at 95°C for 5 min, 30 cycles of denaturation at 95°C for 30 sec, annealing at 62°C for 30 sec and extension at 72°C for 30 sec with final extension for 5 min at 72°C. The samples were denatured at 98°C for 12 min, transferred on ice for 5 min, and loaded onto a 14% neutral poly-acrylamide gels (acrylamide:bisacrylamide = 29:1). The samples were separated at 250 volts for 9 h at 4°C. The gels were stained with silver nitrate according to Sambrook and Russell [11]. PCR products were purified with shrimp alkaline phosphatase (1 U/μL) and Exonuclease III (10 U/μL). Both forward and reverse primers were used in all sequencing reactions and all reactions were analysed using an ABI 3700 automatic sequencer. All sequences were base called and analysed using DNAStar version 6.0).

Results

Nine PCR-SSCP patterns of amplimers were identified (Fig. 1). These nine haplotypes were assigned to the alleles and the nomenclature of PCR-SSCP band pattern according *PRNP* alleles, mainly based on SNPs at loci 136, 154 and 171 keeping as A/V, R/R and Q/R, respectively. SNPs at codon 137, 143, 172 and 173 were also included in the allele nomenclature. The PCR-SSCP genotype frequencies and NCBI accessions awarded for 9 novel sequences are presented in Table 1. Based on main *PRNP* loci from 136 to 173, out of 9 genotypes identified, ARR-S173/ARR-R173 was most frequent (51.70%), followed by ARRM137H143/ARQL137R143 (15.51%). The other genotypes had frequencies ranging from 3.44 to 5.17% (Table 1).

The sequence analysis of exon-3 revealed complete homology with Genbank accessions of domesticated (*Ovis aries*) and wild sheep and other species. The phylogeny (Fig. 2) developed by our sequence alignment showed that all sequences in the study were different from those previously reported. Genotypes showed that DQ885789 and DQ885791 were very close to Rupicara (AY735496). Other genotypes with accession numbers DQ885786-DQ885788, Q885790, and DQ885792-DQ885793 are closer to wild sheep species *Ovis dali* (AJ969016) rather than *Ovis aries*-MARC (DQ345757) or NCBI reference sequence (DQ077504).

Two alleles, Q and R, and three genotypes, QQ, QR and RR, were observed at codon 171 (Table 1). SSCP band analysis and further sequence analysis revealed that the R allele was most frequent (66.40%) and that the Q allele had a frequency of 33.60%. The genotypic frequencies of QQ, QR and RR were 25.86, 15.51 and

58.62% respectively. At this locus, the observed genotypic frequencies were significantly different from the expected frequencies of 11.29, 22.31 and 44.09%. Both QQ and RR type homozygote were in excess at the expense of QR heterozygotes, and the population deviated from Hardy-Weinberg equilibrium (Table 2). At codon 154, only RR genotypes were present. At codon 136, out of 58 animals screened, 52 were AA (86.21%), 4 were AV (10.34%) and 2 were VV (3.44%). Two alleles showing polymorphisms of amino acids Ala (A) and Val (V) had a frequency of 0.938 and 0.862, respectively. Taking together codons 136, 154 and 171, the most commonly occurring genotypes were AARRRR, AARRQR and AARRQQ. The AVRRRR genotype was observed in 4 animals, with AVRRQQ and VVRRQQ in 2 animals each. All of these genotypes were also detectable by SSCP band profile (Fig. 1).

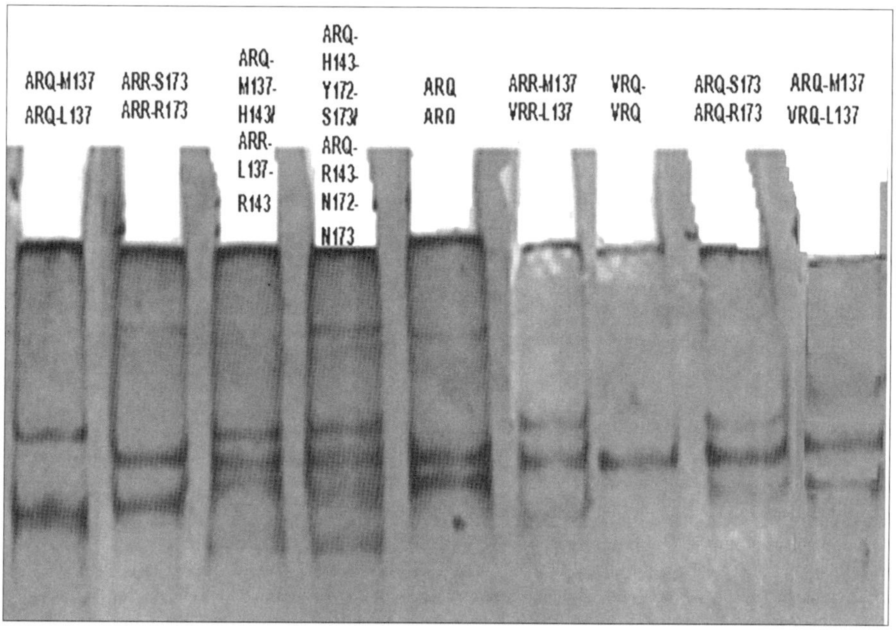

Fig. 1: PCR-SSCP allelic variants in Gaddi sheep *PRNP* gene exon-3.

At codon 137, A/N substitutions at nucleotide position (409) resulted in AA Met to Leu in 13 animals (DQ885786, DQ885788, DQ885794). At codon 143, a His to Arg mutation was observed in 7 animals. Tyr to Asn at codon 172, Ser to Arg at codon 173, was observed in 3 and 5 homozygous, animals respectively. All heterozygous genotypes at these two codons were observed in 3 and 5 animals, respectively.

Table 1: SSCP genotype and genotypic frequency in Gaddi sheep in the region(131 to 173) known for scrapie susceptibility.

SSCP Genotypes	Accession No	No	Frequency
ARQ/ARQ	DQ885793	3	5.17
VRQ/VRQ	DQ885787	2	3.44
ARR-M137/ VRR-L137	DQ885786	4	6.88
ARQ-M137/VRQ-L137	DQ885794	2	3.44
ARR-S173/ARR-R173	DQ885792	30	51.70
ARQ-M137-H143/ ARR-L137-R143	DQ885788	9	15.51
ARQ-H143-Y172-S173/ ARQ-R143-N172-N173	DQ885789	3	5.17
ARQ-S173/ARQ-R173	DQ885791	2	3.44
ARQ-M137/ARQ-L137	DQ885790	3	5.17

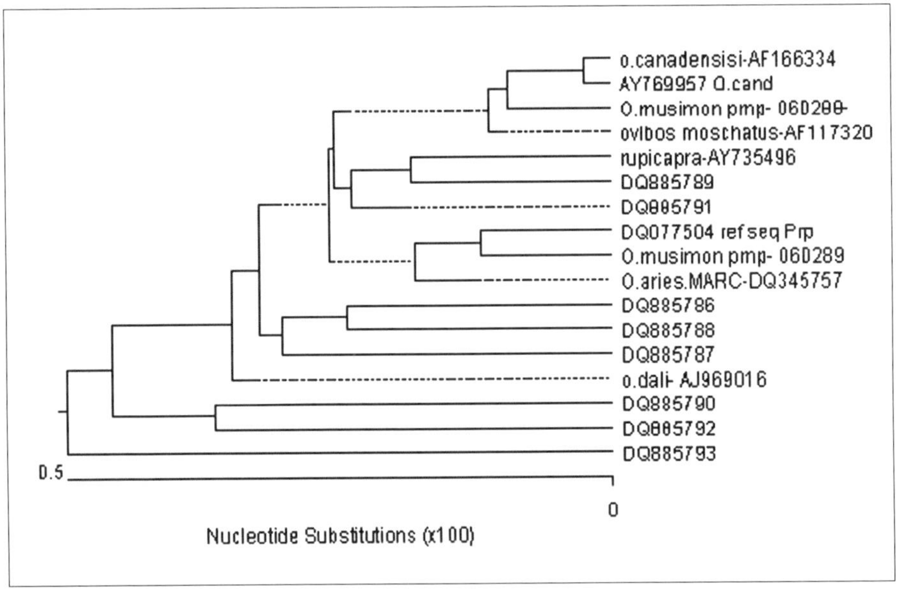

Fig. 2: Cluster diagram of nine nucleotide sequence alignment of Gaddi sheep *PRNP* gene exon-3 between other ovine and caprine species.

Table 2: Genotypic and allele frequency at major loci 136, 154 and 171 of *PRNP* gene related with susceptibility/resistance of Scrapie disease in Gaddi sheep. The expected genotypic frequencies were calculated using the formula: (D+d)2 = D2 + 2Dd + d2 where D is the wild type allele, d is mutated allele and Dd is the heterozygote.

Loci	Parameter	Genotype frequency (%)			Allele frequency (%)	
		AA	AV	VV	A	V
136	observed	86.21^{a}	10.34^{a}	3.44^{a}	91.40	8.60
	expected	83.50 ^{a}	15.70^{b}	0.74^{b}		
154		RR			R	
		100.00			100.00	
		QQ	QR	RR	Q	R
171	observed	25.86^{a}	15.51^{a}	58.62^{a}	33.60	66.40
	expected	11.29^{b}	44.62^{b}	44.09^{b}		

Superscripts ^{a} and ^{b} denote the differences between observed and expected significant frequencies ($P<0.01$) using the Chi-Square test.

Discussion

This study aimed at observing the prevalence of the polymorphic variants of codons 136, 154 and 171 in Gaddi sheep. The genotypic frequency of R alleles at codon 171, which is known to give maximum resistance to scrapie, was much higher in Gaddi sheep than reported in several breeds of Oklahoma sheep [13] and sheep breeds in USA [14]. These results of 100% frequency of RR154 are in close agreement to those reported by DeSilva et al [13] in Hampshire, Montadale and meat-type crosses. The RR 154 homozygote along with AA at 136 and QQ at 171 has been reported in sheep stock highly susceptible to scrapie [15]. The data presented here is similar to that reported in Suffolk, Hampshire, meat-type and wool-type crosses of Oklahoma sheep [13]. The prevalence of scrapie, however, has not been observed in last 45 years in Gaddi sheep. The most probable reason for the lack of scrapie may be range grazing with no supplement of concentrated feed rich in proteins, which might be preventing the protease-sensitive precursor (PRP^{c}) from being activated to PRP^{sc} .

Within this region, A/C substitution was observed in the original codon of ATG to CTG resulting in a heterozygote at codon 137 in 14 out of 58 animals analysed. Because of this substitution at this codon methionine (M) was replaced by leucine (L). At codon 143, the mutation of AA His to Arg was observed, similar to that in Oklahoma sheep [13]. The PCR-SSCP technique was found to be useful in amplifying variants at codons 136, 154 and 171 of the *PRNP* gene in single genomic DNA fragments and the sequences could be aligned perfectly with reported *PRNP* genes in other species of wild and domesticated sheep. Thus this technique can be used for *PRNP* genotyping of large number of animals at low cost in scrapie eradication programmes, through a DNA polymorphism based survey. Zhou et al [12] also observed nine different genotypes of sheep *PRNP* exon-3 using the SSCP technique. Further, the PCR-SSCP amplicons differed in one or a few base pair differences, so this technique can be considered useful for the identification of SNPs in *PRNP* alleles. Large scale genotyping of the *PRNP* gene in indigenous sheep populations where mass scale crossing has been practiced should be performed so that gene based breeding strategies can be developed for eliminating susceptible alleles from indigenous sheep.

References

1 Prusiner SO: Molecular biology and pathogenesis of prion diseases. Trends Biochem Sci 1996;21:482-487.

2 Dawson M, Hoinville LJ, Hosie BD, Hunter N: Guidance on the use of PrP genotyping as an aid to the control of clinical scrapie. Scrapie Information Group. Vet Rec 1998;142:623-625.

3 Satishchandra P, Shankar SK: Creutzfeldt-Jakob disease in India (1971-1990). Neuroepidemiology 1991;10:27-32.

4 Zlotnik I, Katiyar RD: The occurrence of scrapie disease in sheep of the remote Himalayan foothills. Vet Rec 1961;73:543-544.

5 Belt PB, Muileman IH, Schreuder BE, Bos-de RJ, Gielkens AL, Smits MA: Identification of five allelic variants of the sheep PrP gene and their association with natural scrapie. J Gen Virol 1995;76:509-517.

6 Hunter N, Goldmann W, Marshall E, O'Neill G: Sheep and goats: natural and experimental TSEs and factors influencing incidence of disease. Arch Virol Suppl 2000;16:181-188.

7 Goldmann W, Martin T, Foster J, Hughes S, Smith G, Hughes K et al: Novel polymorphisms in the caprine PrP gene: a codon 142 mutation associated with scrapie incubation period. J Gen Virol 1996; 77:2885-2891.

8 Westaway D, Zuliani V, Cooper CM, Da CM, Neuman S, Jenny AL et al: Homozygosity for prion protein alleles encoding glutamine-171 renders sheep susceptible to natural scrapie. Genes Dev 1994;8:959-969.

9 Hunter N, Goldmann W, Smith G, Hope J: The association of a codon 136 PrP gene variant with the occurrence of natural scrapie. Arch Virol 1994;137:171-177.

10 Elsen JM, Amigues Y, Schelcher F, Ducrocq V, Andreoletti O, Eychenne F et al: Genetic susceptibility and transmission factors in scrapie: detailed analysis of an epidemic in a closed flock of Romanov. Arch Virol 1999;144:431-445.

11 Sambrook J, Russell DW: 2001; In Molecular Cloning: A Laboratory Manual 3rd ed., In Cold Spring Harbour, Cold Spring Laboratory Press, NY, USA.

12 Zhou H, Hickfford JGH, Fang Q: Technical Note: Determination of alleles of the ovine *PRNP* gene using PCR-Single strand conformational polymorphism analysis. J Anim Sci 2005;83:745-749.

13 DeSilva U, Guo X, Kupfer DM, Fernando SC, Pillai AT, Najar FZ et al: Allelic variants of ovine prion protein gene (*PRNP*) in Oklahoma sheep. Cytogenet Genome Res 2003;102:89-94.

14 Guo X, Kupfer DM, Fitch GQ, Roe BA, DeSilva U: Identification of a novel lysine-171 allele in the ovine W prion protein (*PRNP*) gene. Anim Genet 2003; 34: 303-305.

15 Walsh JT: Scrapie diagnosed in QR sheep. The Shepherd 2002; 41: 1-12.

Dr. Neelam Gupta, DNA Fingerprinting Unit, National Bureau of Animal Genetic Resources, P.O. Box 129, NH-1, Karnal- 132001, India.

E-mail: neelamgnbagr@gmail.com

Pinard M-H, Gay C, Pastoret P-P, Dodet B (eds): Animal Genomics for Animal Health. Dev Biol (Basel). Basel, Karger, 2008, vol 132, pp 267-270.

Delayed-type Hypersensitivity Response to KLH in F_2 and Backcrosses of Two Immune Selected Chicken Lines: Effect of Immunisation and Selection

G. Minozzi[1], H.K. Parmentier[2], B. Bed'hom[1], F. Minvielle[1], D. Gourichon[3], M.H. Pinard-van der Laan[1]

[1] INRA /AgroParisTech, UMR1236 Génétique et Diversité Animales, Jouy en Josas, France
[2] Adaptation Physiology Group, Department of Animal Sciences, Wageningen, Institute of Animal Sciences, Wageningen University, The Netherlands
[3] UE de Génétique Avicole, INRA, Tours, France

Key words: chicken, selection, keyhole limpet hemocyanin, immune responses

Abstract: Cell-mediated immune response to keyhole limpet hemocyanin (KLH) was studied in 13 different progeny groups belonging to a second generation cross between two lines of White Leghorn previously selected for 10 generations for high antibody response to Newcastle disease vaccine (ND3) and high cell-mediated response to phytohemagglutinin (PHA) measured as wing web swelling. The cutaneous hypersensitivity response to KLH was assessed in KLH immunized and non-immunized birds. Wing thickness was measured at three time points: before the injection and 4 and 24 hours after the injection. The effect of previous immunization and genetic background was assessed. While no differences were present between progeny types or to a randomly bred control line at any of the three time points, significant differences were present between immunized and not immunized birds 24 hours after challenge. Results demonstrate that KLH immunized birds perform better and that the previous selection and crossbreeding scheme has not influenced the intensity of the skin swelling response to KLH. Apparently, neither selection for antibody responsiveness to ND3 nor selection for enhanced responsiveness to a T cell mitogen affected delayed-type hypersensitivity (DTH) to the Th 2 antigen KLH. These results may serve to facilitate further selection based on independent immune parameters.

Introduction

In today's global climate of fear of food born diseases and consumer demand to reduce the use of antibiotics in animal production, genetic selection and research on animal immunological responses could be a tool towards sustainable, efficient production. In addition, selection for general immune responses in poultry has been proposed as a sustainable alternative to selection for specific infectious diseases, since there are an extremely high number of different pathogens that an animal could face in its entire lifecycle. The objective of this study was to better understand the genetics of immune responses in poultry and to help unravel the underlying mechanisms.

Materials and Methods

Experimental model

The selection experiment started in 1994 with the selection of three chicken lines based on three different *in vivo* immune related traits: high antibody response, high phagocytic activity and high T cell-mediated activity. All birds were measured for three immune related traits, at each generation, but only one trait was selected for in each line. The traits selected for were high antibody response to Newcastle disease virus (HB1 vaccine) (ND3) three weeks after vaccination; high cell-mediated immune response, using the wing web response to phytohemagglutinin (PHA) at nine weeks of age and high phagocytic activity, measured as carbon clearance (CC). The aim was to select for one of the components of the immune system in each line, so that both innate and acquired immunity were implicated in the experiment. In addition, a control line was kept by random breeding. The results after 12 discrete generations of single-trait within-line selection were significant for all three traits, which were found to be independent [1]. Each line showed a significant increase in the immune capacity for the selected trait as evidenced in figure 1.

In addition, planned crosses between the three lines were carried out over the last two years. The selected immune trait values expressed as the difference from the contemporary control line are shown in figures 2 and 3 for the antibody (ND3) and cell-mediated repose traits (PHA). Results from the crossbred animals showed that correlations between the traits confirmed the independence of the selection criteria in the first [2] and second generation crosses.

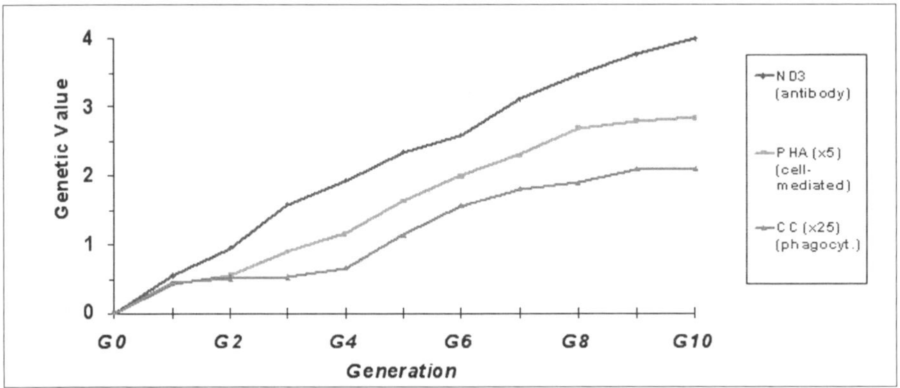

Fig. 1: Response to selection over 10 generations.

Purpose of the study and experimental approach

The objective of this study was to determine if the selection criteria applied to two of the pure lines (ND3 and PHA) changed the level of and type of cell-mediated immune response to a new antigen, keyhole limpet hemocyanin (KLH), in the second generation progeny groups. The cutaneous delayed-type hypersensitivity (DTH) response using a mitogen or T cell dependent antigens is often used as a method to measure in vivo T cell reactivity. Consequently, the cell-mediated immune response to KLH was studied in 13 different progeny groups belonging to the second generation cross between the line that had been previously selected for high antibody response to ND3 and that which had been selected for high cell-mediated response to PHA. Previous studies on the same second generation genetic groups revealed that the different genetic groups did not differ in cell-mediated response to PHA [3]. Consequently, another mitogen was tested and the cutaneous hypersensitivity response to KLH was assessed in KLH immunized and non-immunized birds to verify the effect of previous immunisation and genetic background. Wing thickness was measured at three time points: before the injection and 4 hours and 24 hours after the injection

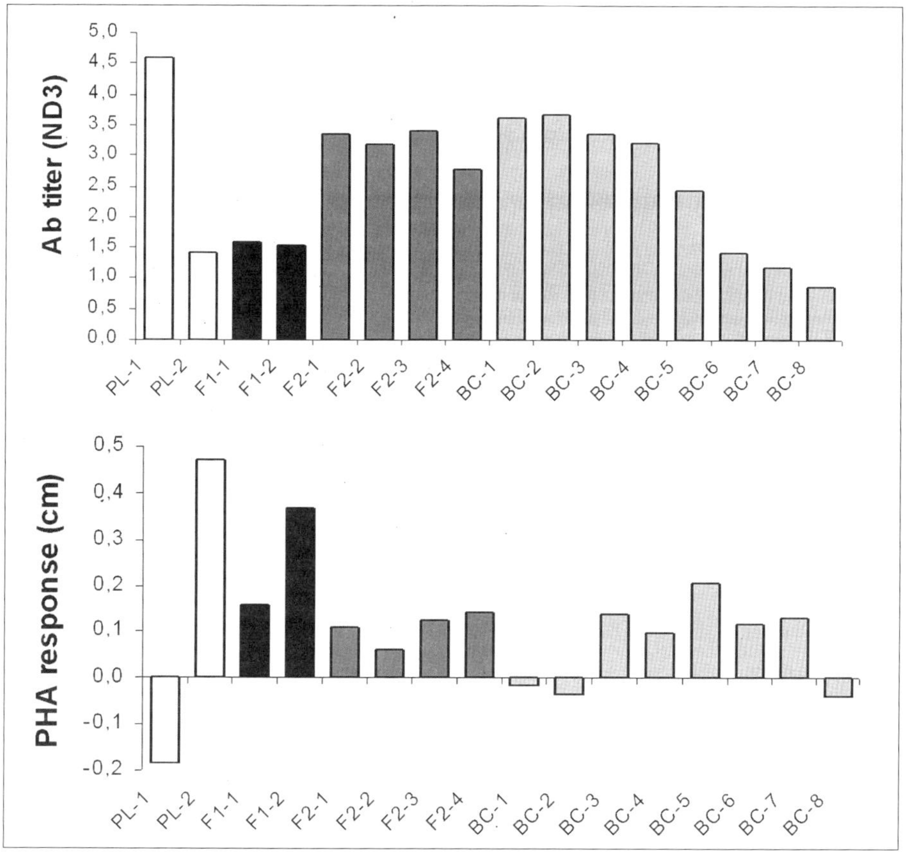

Fig. 2: Antibody response values: ND3 and *Fig. 3* T cell response: PHA; Immune values in the corresponding selected lines (PL-1 and PL-2), first and second generation crosses, expressed as the difference from their corresponding control line value.

Results and Conclusions

In this experiment, early and late measurements of the thickness of the wing web in response to subcutaneous challenge with 0.1 mg of KLH in 0.1 ml of PBS were taken in group of immunized and non-immunized birds. The first measurement, taken at 4 h consists of precipitating antibodies leading to type-III hypersensitivity responses that may result in cutaneous anaphylactic hypersensitivity (PCA) reactions, while the later measurement looks at type-IV DTH responses. Wing web responses at 4 h were not significantly affected by the immunisation protocol, confirming prior results in growing layer hens [4]. However, later measurements of DTH responses measured at 24 h evidenced the significant effect of the prior immunisation. These results (Fig. 4) demonstrate that KLH immunized birds perform better than non-immunized birds in accordance with previous results in chicken [4].

While the effect of the immunization was significant, no effect of the genetic composition of the progeny groups was found, indicating that the previous selection and crossbreeding scheme did not influence the intensity of skin swelling response to the new antigen tested (KLH).

Apparently, neither selection for antibody responsiveness to ND3 nor selection for enhanced responsiveness to a T cell mitogen (PHA) affected the DTH response to the Th 2 antigen KLH, which could be used as further selection trait of immune response selection.

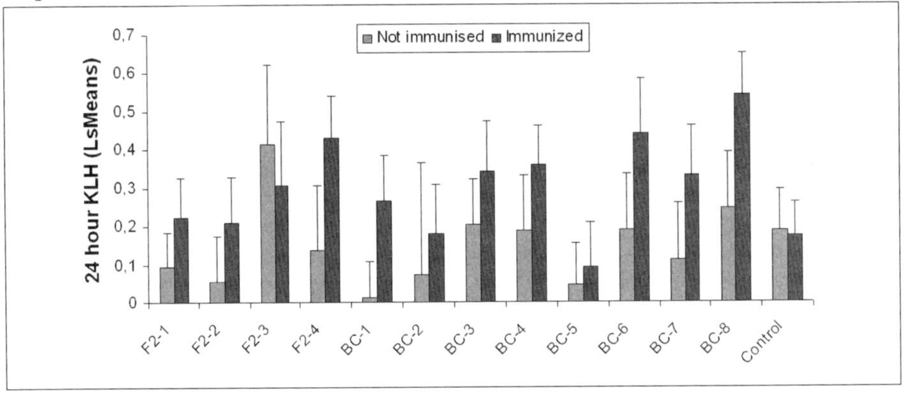

Fig. 4: LS-Means of cutaneous response to KLH in immunized (blue) and non-immunized birds (violet) measured at 24 hours by wing web swelling.

ACKNOWLEDGEMENTS

This research project is supported by a Marie Curie Early Stage Research Training Fellowship of the European Community's Sixth Framework Programme under contract number MEST-CT-2004-504854.

REFERENCES

1 Pinard-van der Laan MH: Immune modulation: the genetic approach. Vet Immunol Immunopathol 2002;87:199-205.

2 Minozzi G, Bed'hom B, Minvielle F, Monvoisin JL, Gourichon D, Pinard-van der Laan MH: Chicken lines selected for immune responses: results of a crossbreeding design. Proceedings of the 8th World Congress on Genetics Applied to Livestock Production 2006. 13-8-2006.

3 Minozzi G, Parmentier HK, Nieuwland MG, Bed'hom B, Minvielle F, Gourichon D, Pinard-van der Laan MH: Antibody responses to keyhole limpet hemocyanin, lipopolysaccharide, and newcastle disease virus vaccine in f2 and backcrosses of white leghorn lines selected for two different immune response traits. Poult. Sci.2007,86:1316-1322.

4 Sijben,JW, de,Groot H, Nieuwland MG, Schrama JW, Parmentier HK: Dietary linoleic acid divergently affects immune responsiveness of growing layer hens Poult. Sci. 2007,79:1106-1115.

Dr. Giulietta Minozzi, UMR 1236, INRA /AgroParisTech, Unité Génétique et Diversité Animale, F-78350 Jouy en Josas, France.

E-mail: giulietta.minozzi@jouy.inra.fr

Pinard M-H, Gay C, Pastoret P-P, Dodet B (eds): Animal Genomics for Animal Health. Dev Biol (Basel). Basel, Karger, 2008, vol 132, pp 271-278.

Molecular Genotype Investigation of the *Gallus gallus* Major Histocompatibility Complex

O. Chazara[1], M.H. Pinard-Van Der Laan[1], M. Tixier-Boichard[1], B. Bed'hom[1]

[1] UMR 1236 Génétique et Diversité Animales, INRA/AgroParisTech, 78352 Jouy en Josas, France

Key words: *Gallus gallus*, major histocompatibility complex, SNPs, microsatellite

Abstract: The association between genetic markers and serological major histocompatibility complex (MHC) definition was studied in experimental White Leghorn lines selected for different immune traits. Different markers, three variable number tandem repeats (VNTRs) and 10 single nucleotide polymorphisms (SNPs), distributed along the MHC sequence were genotyped. Correlation between VNTR marker alleles and serology was studied in more than 800 animals and correspondence between SNP haplotypes and allele size was defined. A complete table of correspondence between VNTR allele sizes, SNP genotypes and serology was then drawn. This research shows that integrating new molecular tools can definitely help in MHC haplotype characterization and that, since the chicken MHC locus has been associated with disease resistance, these tools are becoming more and more needed for disease resistance and immunogenomics studies.

Introduction

Chickens are the primary animal protein source of human food at the global level. Today in Europe, poultry production systems are facing a trend towards reduced use of drugs, thus strengthening the need for robust animals. Moreover, disease resistance has always been considered in selection objectives. Genetic variability for disease resistance has been shown to take place both between and within lines or breeds of chickens. However, reliable and convenient selection criteria are still under investigation. Furthermore, disease resistance is still difficult to characterize in the numerous local breeds identified worldwide for which experimental data is scarce and generally only genomic DNA is accessible. Thus, universal molecular tools are needed to characterize functional diversity regarding disease resistance across populations.

The chicken major histocompatibility complex (MHC) is a genomic region comprising genes coding for proteins mainly involved in peptide presentation to T cells (Class-I and Class-II molecules). The chicken MHC locus has been associated with disease resistance, susceptibility, and differential response to vaccine. A major concern has arisen about the fine dissection of molecular variation in this locus which is divided in two genetically independent regions (B and Y). Molecular MHC genotyping tools in the chicken would permit exploration of the diversity of more breeds at this level. Historically, the B-locus genotype has been studied using serology, and DNA studies of the B- and Y-locus polymorphisms have been based on restriction fragment length polymorphisms (RFLPs) and single-strand conformation polymorphism (SSCP). Several atypical variable number tandem repeat (VNTR) markers, including MCW0370, MCW0371, and LEI0258, are now available to replace the serological tests, but further studies are required in order to fully understand the limits of their correlation with the serology tests. They are also not polymorphic enough to describe the whole variability of coding sequences, particularly at the level of the peptide binding region sequence of the Class-I and Class-II genes. A specific single nucleotide polymorphism (SNP) strategy would help to establish and validate a correlation between classical serology and polymorphism at the gene level.

Materials and Methods

White Leghorn lines were previously selected for 12 generations at INRA for different immune traits (line 1, antibody response to Newcastle vaccine; line 2, delayed hypersensitivity measured by phytohemagglutinin; line 3, carbon clearance as an indicator of phagocytosis; line 4, control [1]). The founders were genotyped with three atypical markers, LEI0258, MCW0370 and MCW0371 as described in Fulton et al. [2] (Table 1, Fig. 1) and studied by serology, to determine the correlation between the markers and serology in those lines. Other animals from the selected lines, including one whole generation (generation 9, with 800 animals born in 2003 from 110 families) were genotyped for LEI0258 (Fig. 2). Twenty-four animals, six per selected line, from another generation of selection (generation 12, animals born in 2006) were studied for SNP using 10 SNPs predicted in the chicken genetic variation map of the International Chicken Polymorphism Map Consortium [3] (Table 1), chosen and typed by Wageningen University in the framework of an Eadgene project focusing on the genetic basis of the immune response within chickens. Eight of those SNPs are located in the B-region and two are in the Y-region (Fig. 1); they were typed using Illumina technology. PHASE 2.1 software [4,5] was used; it implements a Bayesian statistical method for reconstructing haplotypes from population SNP data. A Neighbour-Joining unrooted tree was then constructed, based on the number of differences, using MEGA 3.1 [6].

Results

LEI0258, MCW0370 and MCW0371 observed allele sizes for the founders were compared to the serology previously determined for those animals (B15, B19, B21, B34, and B124). A strong correlation was found between the MHC haplotypes determined by both techniques, and a correspondence table was made (Table 2). Genotyping selection generation 9 for LEI0258 confirmed those results, but 8.6% incompatibility was observed between the known serology and the LEI0258 alleles (261 bp, 359 bp, 422 bp, 489 bp and 544 bp). One example of a family in which such incompatibilities were found is shown in Table 3; parents are from generation 8 with known serology but no LEI0258 typing. In this family, parents' serology was consistent with that of the progeny, but three offsprings exhibited unexpected LEI0258 genotypes that were incompatible with their own serology and their parents'. This example only shows incompatibilities regarding B15 and B21 haplotypes, but similar situations of incompatibility occurred for all five observed haplotypes in the generation under study.

Table 1: SNP or marker characteristics (name, chromosome, position in bp on reversed sequence published by Shiina et al. [8], region information, known variations, 5' and 3' near 30 bp sequence for SNPs or forward and reverse primers).

Name	**Chr**	**Position**	**Region**	**Variations**	**5' near 30 bp or forward primer**	**3' near 30 bp or reverse primer**
rs16057122	16	3386	NA	[A/G]	ATGGATAATTAGACC-CATAGGTTTCACAGC	TGTGCAGAAATCC-CAACCTATGAGTACTTC
rs15026791	16	65849	BF2	[A/G]	GCTGATATGCAAGAA-CACACGGTGACCCAC	GGGACAACCAGGAGCCATGGGGACATGTGG
rs15788177	16	112368	Blec1	[A/G]	GTTGCACTG-GAAAAGCATGAAGAAAGCACC	CATGATGAGAG-GAACAGTTCATCCCACAGC
MCW0370	16	112482	Blec1	(A)n	GAAAGAGGAGAG-TAGTTCACG	TCCCACCTCATCATG-CATTCC
MCW0371	16	114542	Blec2	(A)n	TTTCATGGCATCC-TAAGATGG	CTGCTCCGAGCTG-TAATCCTG
rs15026773	16	116738	Blec2	[G/T]	ACACGGACGGAG-GCTGAGAGGCAGAAAATT	TGTCAGAATTC-CGTCAGCTCCGCCGTTTTC
LEI0258	16	124903	BG1	12-13 bp	TCGGGAAAAGATCT-GAGTCATTG	TGATTTTCA-GATCGCGTTCCTC
rs15788101	16	133829	B-BTN2	[C/T]	AGGCAGCAGC-CTCAGCTCCAACCCTTCTCA	ATTCCCGTCCTCAC-CGGGCTTTATCTCCCT
rs15788011	16	165938	TRIM27.2	[A/G]	AACGCAGCAGGGATTCTTAGGGGAATAAGG	TGTAAAAGATAAA-CACAGATGGAAGATTAA
rs15026648	16	211705	Bzfp2	[A/C]	CAGCTCCCCAAAA-GAGCAACACAACCAAAC	GTCCTTCAACCATTCC-CTTTGCAAATATGT
rs15787808	16	227307	Bzfp3	[A/C]	TGTGAATGGATGGGT-GAGTCTCCCCTCTCA	ATTAAAAAT-GTTGGGGTCTTCCTGTGAGAG
rs14096793	16	NA	Y-region	[C/T]	TGTTTCGTTCCTTCC-CTCCTCCTTTCCTTC	TTCCTTCCCTC-CTTTCTTCCTTCCTTCCTT
rs16742957	16	NA	Y-region	[C/T]	CAACTGCACTATGT-GTCTGCCCCTCCACAC	TCTGAAACCACCG-TAAGGGCCTGAAGGACA

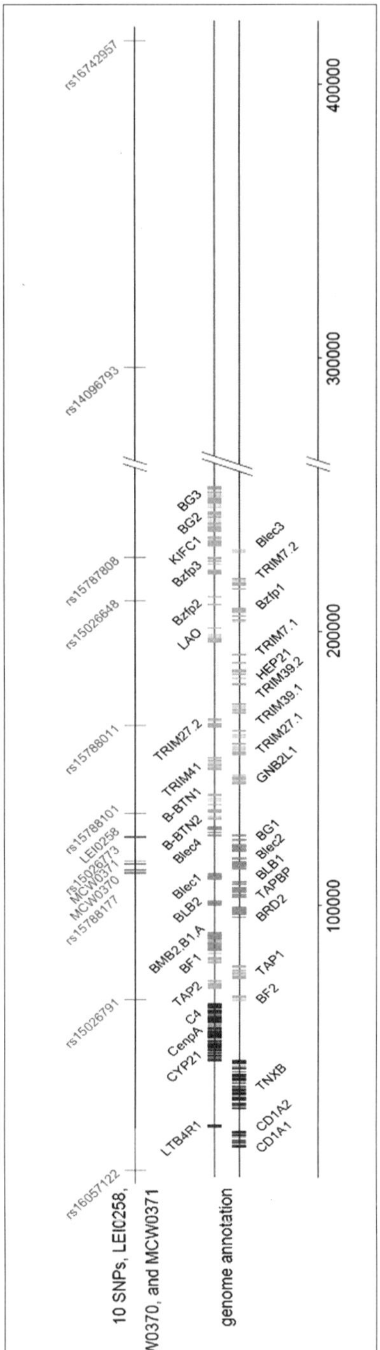

Fig 1: Schema of the 242 kb B-region showing gene content (modified from [8]) and positions of the markers (MCW0370, MCW0371, LEI0258) and the 10 SNPs.

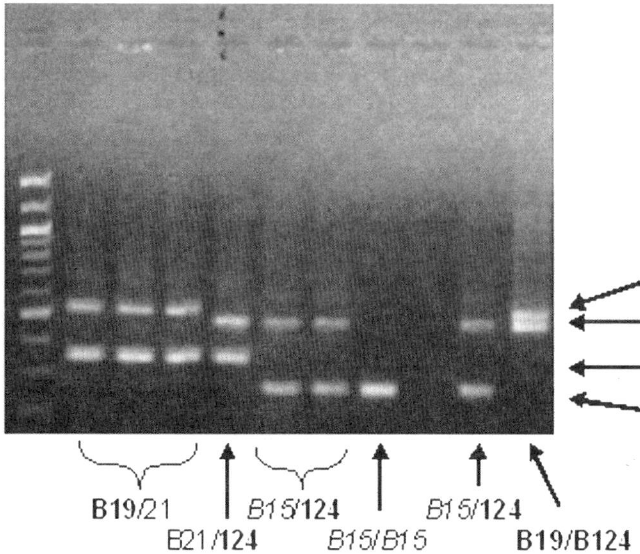

Fig 2: Marker LEI0258 PCR products separated on an agarose gel showing the correspondence between the allele sizes and the MHC B-region defined by serology.

Table 2: Table of correspondence between serology, MCW0370, MCW0371, and LEI0258 allele sizes and haplotypes defined by the eight SNPs located in the MHC B-region.

Serology	**MCW0370**	**MCW0371**	**LEI0258**	**Haplotype(s)**	
B15	*null*	*201*	*261*	*ABBBABAA*	*ABBBABAB*
B21	181	203	359	BAAAAABA	
B34	***179***	***203***	***422***	***ABABABBA***	
B19	**178**	**203**	**544**	**BAABBBBB**	**BBABBBBB**
B124	**179**	**203**	**489**	**BAABBBBB**	**BBABBBBB**

Table 3: Comparison between serology and LEI0258 allele sizes showing incompatibilities in one family from selection generations 8 and 9.

	Serology		**LEI0258**	
Male	*B15*	B21		
Female	*B15*	**B124**		
Offspring 1	*B15*	B21	359 = B21	359 = B21
Offspring 2	*B15*	B21	359 = B21	359 = B21
Offspring 3	B21	**B124**	359 = B21	**489 = B124**
Offspring 4	B21	**B124**	359 = B21	**489 = B124**
Offspring 5	*B15*	B21	*261 = B15*	359 = B21
Offspring 6	B21	**B124**	359 = B21	**489 = B124**
Offspring 7	*B15*	B21	359 = B21	359 = B21

Table 4: Comparison between LEI0258 allele sizes and haplotypes defined by 10 SNPs for 24 individuals from the four selected lines.

ind	**line**	**LEI0258**		**10 SNP haplotypes**	
1	1	359	**489**	BAAAAABA BA	**BAABBBBB BA**
2	1	359	**489**	BAAAAABA AA	**BAABBBBB AA**
3	1	359	**489**	BAAAAABA AA	**BAABBBBB BA**
4	1	**489**	**489**	**BBABBBBB AA**	**BBABBBBB AA**
11	2	**489**	**544**	**BBABBBBB BA**	**BBABBBBB AA**
23	4	**489**	**489**	**BBABBBBB BA**	**BBABBBBB AA**
5	1	***422***	**489**	***ABABABBA AA***	**BBABBBBB AA**
6	1	359	359	BAAAAABA AA	BAAAAABA AA
7	2	**489**	**544**	**BBABBBBB BA**	**BBABBBBB BA**
8	2	**489**	**544**	**BBABBBBB BA**	**BBABBBBB BA**
10	2	**489**	**489**	**BBABBBBB BA**	**BBABBBBB AA**
13	3	*261*	359	*ABBBABAB BA*	BAAAAABA AA
18	3	*261*	359	*ABBBABAB BA*	BAAAAABA AA
22	4	*261*	359	*ABBBABAB BA*	BAAAAABA AA
14	3	*261*	359	*ABBBABAB BA*	BAAAAABA BA
24	4	*261*	359	*ABBBABAB AA*	BAAAAABA AA
15	3	*261*	261	*ABBBABAA BA*	*ABBBABAB AA*
17	3	*261*	261	*ABBBABAA BA*	*ABBBABAB BA*
9	2	*261*	**544**	*ABBBABAA BA*	**BBABBBBB AA**
12	2	*261*	**544**	*ABBBABAA BA*	**BBABBBBB AA**
16	3	*261*	**544**	*ABBBABAA BA*	**BBABBBBB AA**
19	4	*261*	**489**	*ABBBABAA BA*	**BBABBBBB AA**
20	4	*261*	**489**	*ABBBABAA BA*	**BBABBBBB AA**
21	4	*261*	**489**	*ABBBABAA BA*	**BBABBBBB BA**

Genotyping of the selected lines revealed a new allele size for LEI0258, 557 bp, in one animal from the generation 11 [7]. This new allele is probably mutated from the mother's 544 bp allele, the 557 bp size would be obtained by one additional 13 bp repeat, as LEI0258 is a 12-13 bp repetition.

For the eight typed SNPs located in region B, the PHASE software predicted the most likely haplotype pairs for each individual with a probability ranging from 0.658 to 1.000. Only the information for the eight SNPs located in region B was used because one of the two SNPs located in region Y was monomorphic and the other seemed independent, so the 10 SNP genotypes could not be considered as linked. Predicted haplotypes were compared with independently typed LEI0258 alleles and with the serological typing (Table 4) and used to complete the correspondence table (Table 2). This table shows several discrepancies between sequence polymorphism on one hand and LEI0258 alleles or serology on the other: two SNP haplotypes could be found for a given LEI0258 allele (as with B15) and SNP haplotypes could not be used to discriminate B19 from B124. The haplotypes from the eight B-region SNPs were used to construct a tree (Fig. 3) which, interestingly, clustered on one hand with B15 and B34 and B21, B19 and B124 on the other hand.

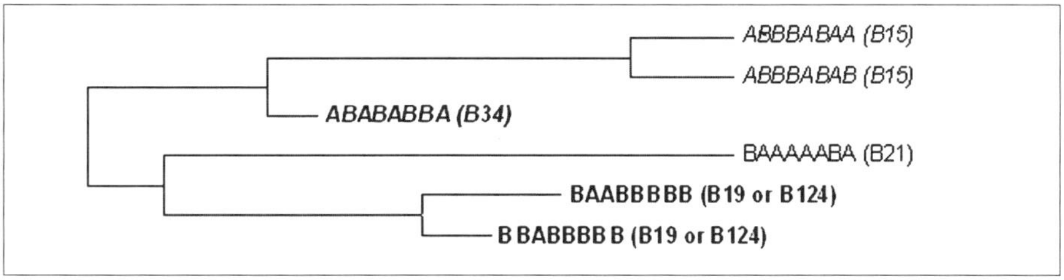

Fig. 3: Unrooted neighbour-joining tree based on the number of differences between the haplotypes constructed from the eight SNPs located in the MHC B-region.

Discussion

In this report, we describe the application of molecular approaches based on differences due to repeat length in three markers, LEI0258, MCW0370, and MCW0371, and 10 SNPs, on the characterization of the MHC region in four lines selected for different immune traits. This approach also helps to determine the extent to which integrating molecular tools can help with MHC haplotype characterization, which is becoming more and more needed for disease resistance and immunogenomics studies.

For example, in this study, we detected incompatibilities between serology and LEI0258 within families that could be explained by serology errors and parentage errors as well as by recombination within the MHC locus.

Typing SNPs yields information on other parts of the MHC B-region, indicating possible recombination spots. The 4th and 5th SNPs surrounding the LEI0258 locus are an example of this, and may indicate that B15 and B34 (BA) are the result of a recombination between B19 or B124 (BB) and B21 (AA). The 7th and 8th SNPs are an example of this as we can find the four combination : AA, AB, BA and BB.

This research focused on well-documented experimental lines that allowed for the study of allele segregation. The next step would be to examine various others breeds, including local breeds that have never been explored at this level.

REFERENCES

1 Pinard-van der Laan M-H: Immune modulation: The genetic approach. Vet Immunol Immunopathol 2002;87:199-205.

2 Fulton JE, Juul-Madsen HR, Ashwell CM, McCarron AM, Arthur JA, O'Sullivan NP, et al.: Molecular genotype identification of the *Gallus gallus* major histocompatibility complex. Immunogenetics 2006;58:407-421

3 International Chicken Polymorphism Map Consortium: A genetic variation map for chicken with 2.8 million single-nucleotide polymorphisms. Nature 2004;432:717-722.

4 Stephens M, Smith NJ, Donnelly PA: New statistical method for haplotype reconstruction from population data. Am J of Hum Genet 2001;68:978-989.

5 Stephens M, Scheet P: Accounting for decay of linkage disequilibrium in haplotype inference and missing data imputation. Am J Hum Genet 2005;76:449-462.

6 Kumar S, Tamura K, Nei M: MEGA3: Integrated software for Molecular Evolutionary Genetics Analysis and sequence alignment. Brief Bioinform 2004;5:150-163.

7 Loywyck V, Bed'Hom B, Pinard-van der Laan M-H, Pitel F, Verrier E, Bijma P: Evolution of polymorphisms at molecular markers in QTL and non-QTL regions in selected lines of chicken. Genet Select Evol 2007;in press.

8 Shiina T, Briles WE, Goto RM., Hosomichi K, Yanagiya K, Shimizu S, et al.: Extended Gene Map Reveals Tripartite Motif, C-Type Lectin, and Ig Superfamily Type Genes within a Subregion of the Chicken MHC-B Affecting Infectious Disease. J Immunol 2007;178:7162-7172.

Olympe Chazara, UMR INRA/AgroParisTech, Génétique et Diversité Animales, Domaine de Vilvert, 78352 Jouy en Josas, France.

E-mail: olympe.chazara@jouy.inra.fr

Pinard M-H, Gay C, Pastoret P-P, Dodet B (eds): Animal Genomics for Animal Health. Dev Biol (Basel). Basel, Karger, 2008, vol 132, pp 279-286.

Verification of Chromosomal Regions Affecting the Innate Immunity in Pigs Using Linkage Mapping

K. Wimmers[2], E. Jonas[1], H.-J. Schreinemachers[1], D. Tesfaye[1], S. Ponsuksili[2], E. Tholen[1], H. Juengst[1], K. Schellander[1], C. Phatsara[1]

[1] Institute of Animal Science, University of Bonn, Germany
[2] Research Institute for the Biology of Farm Animals, Dummerstorf, Germany

Key words: QTL, innate immunity, backcross, verification

Abstract: Understanding the genetic control of innate immunity in pigs would offer the opportunity to utilize natural variation and improve selective breeding programmes. As part of our porcine genome scan to identify quantitative trait loci (QTL) we examined immune response traits in a Duroc x Berlin miniature pig resource family (DUMI). Complement activity via classical (CH50) and alternative (AH50) pathways, antibody response to *Mycoplasma hyopneumoniae* (Mh), tetanus toxoid (TET) and PRRS virus (PRRSV), the complement component (C3c), and Haptoglobin serum concentration (HP) were used as phenotypes for linkage mapping. A total of 220 backcross animals were used for the QTL analysis. Blood was collected six times from each animal prior to and after vaccinations against *Mycoplasma hyopneumoniae*, tetanus toxoid and PRRS, respectively. Seventy-four microsatellites from 18 autosomes were used for QTL mapping. The analyses were performed treating the measurements of phenotypes at different time points as single traits. Forty-two significant and 24 highly significant QTL were detected, using the program QTL Express, for all immune traits using the single traits. Most QTL were detected on SSC3, SSC16, and SSC18. No significant F-value corresponded to data for SSC12 and SSC13.

INTRODUCTION

A massive number of QTL have been identified which influence economically important traits, such as growth and carcass and meat quality, on almost all chromosomes in the pig [1]. A very limited number of QTL analyses have been devoted to health, disease susceptibility or immune response traits [2]. The detection of QTL underlying experimental crosses between lines that differ in their innate and specific immune response is a general approach to dissecting the genetic background without prior hypothesis. For most infectious disease studies these methods require that the experimental populations be custom bred and challenged, in order to study genetic differences in the immune response and map genetic loci underlying these differences [3]. Antibody response has been mentioned as one of the first immune competence traits to be examined by QTL analysis [4]. Many QTL underlying immune response variations have been detected on various chromosomal regions in the mouse, chicken and human [5-9]. In this study, the focus was on traits related to the humoral immune response including innate immune mechanisms, i.e. complement activity and acute phase response, as well as acquired immunity, i.e. antibody titres.

Materials and Methods

Animals used in this study were based on the back crossing of Duroc sires and F2 Duroc x Berlin miniature dams. All animals with available immune response phenotypes were genotyped, compromising 26 full sib families (8 different sires). Information about the immune response parameters was available from 83 to 199 backcross animals. All animals were reared at the research farm in Frankenforst, the Institute of Animal Science, University of Bonn. Pigs were immunized with *Mycoplasma* (Mh), tetanus toxoid (TET) and porcine reproductive and respiratory syndrome virus (PRRSV) vaccines at 6, 9 and 15 weeks of age, respectively. Blood samples were taken at six time points: just prior to (T1, T5) and 10 days after (T2, T6) Mh and PRRSV immunizations, respectively; day 20 after Mh vaccination, just prior to TET vaccination (T3); and day 20 after TET vaccination (T4). Figure 1 shows a schematic diagram of blood sample time points.

Complement system activity was determined for both alternative and classical pathways (AH50 and CH50) at all time points (T1 - T6). Antibody response to Mh (at T1 to T3), TET (at T3 to T5), and PRRSV (at T5 and T6) was determined. Other immune parameters, including serum complement component 3 (C3c) and haptoglobin (HP) concentrations were also measured at all time points.

Seventy-four microsatellite markers were selected covering all 18 porcine autosomes. These markers were mainly selected from the USDA/MARC map. Genotyping, electrophoresis, and allele determination were performed on a LI-COR 4200 Automated Sequencer and data were analysed by OneDScan software. Data obtained from fragment size analysis were first checked for any genotyping errors using Pedcheck, version 1.1 [10]. The following multipoint analyses were carried out for female, male, and sex-average maps using CRIMAP software, version 2.4 [11]. A QTL interval mapping analysis was performed using the web-based program 'QTL express' available at http://qtl.cap.ed.ac.uk/ [12]. The QTL regression model for single analysis was:

$$y_{ijkl} = \mu + s_j + f_k + \beta\, cov_{ijk} + c_{ai}a + c_{di}d + \varepsilon_{ijkl},$$

where:

y_{ijk} = phenotype of the ith offspring;
μ = overall mean;
s_j = j^{th} fixed sex effect, j = 1, 2;
f_k = k^{th} fixed contemporary group effect;
β = regression coefficient on the covariate;
cov_{ijk} = covariate (CTM and sex);
c_{ai} = additive coefficient of the ith individual at a putative QTL in the genome;
c_{di} = dominant coefficient of the ith individual at a putative QTL in the genome;
a = additive effects of a putative QTL;
d = dominant effects of a putative QTL and;
ε_{ijk} = residual error

Significance levels as calculated using the regression analysis were obtained using 10,000 permutations.

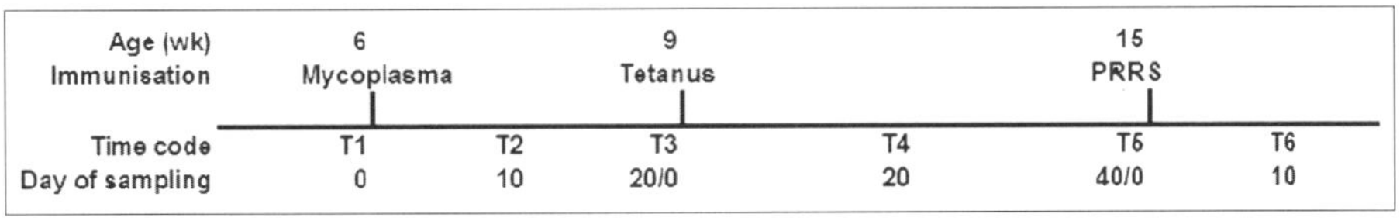

Fig. 1: Schematic diagram of the immunisation and blood sample time points in this study.

Results

Forty-two significant (p≤0.05 on a chromosome-wide level) and 24 highly significant (p≤0.01 on a chromosome-wide level) QTL were detected for all immune traits. Most QTL were detected on SSC3, SSC16, and SSC18 (nine significant F-values on each chromosome). No significant F-value was detected on SSC12 and SSC13. The most highly significant QTL was detected for antibody response to Mh (SSC 10), TET (SSC 16) and PRRSV (SSC 3), C3c (SSC 3, 11, 18) and HP (SSC 8, 14, 16, 18) concentration. For AH50 and CH50, 22 significant and 9 highly significant QTL could be detected. All results are summarized in Tables 1 and 2.

QTL for haemolytic complement activity and its component C3c: The effects on CH50 and AH50 were measured at six different time points. For CH50 at T1, significant F-values were detected on SSC3, 10 and 14. On SSC17, a highly significant F-value for the measurement at T4 and a significant F-value at T6 were detected close to the 0 cM position. For the measurement at T5, highly significant and significant F-values were detected on SSC16 and SSC5, respectively. AH50 measurements at six different time points (T1-T6) were also used for QTL detection. Significant F-values on SSC2, 3, 4, 9 and 10 were detected for only one measurement. A QTL for AH50 at T3 and T4 was found on SSC1 at 74 cM. In addition, the linked regions for similar measurements were also found on SSC18 at 100 to 130 cM. On SSC4, four significant F-values were found for the measurements at T1, T3, T5 and T6 located at approximately 50 cM. This clear picture of an interesting locus on one chromosome was not found for SSC7, where significant F-values were detected for different measurements (T3, T5 and T6) covering different regions of the chromosome. All measurements detected on SSC16 were significant, except that for T5 (Table 1). The lack of a significant QTL for the AH50 at T5 might be due to the experimental design: the largest gap without vaccination was set before T5. Additional analysis is needed for further interpretation.

For C3c, different QTL were detected in the backcross population. Only one significant F-value was detected on SSC16 and SSC17, whereas three interesting results on SSC11 for the measurements at T1, T3 and T6 were observed, indicating the region between the markers *S0071* and *S0386*. On SSC3, there were significant F-values found at 30 cM, excluding those at T5. The most interesting QTL for complement component C3c was found on SSC18, was similar for all measurements, and was located at 130 cM. All significant and highly significant F-values are shown in table 1.

QTL for antibody responses to vaccinations and HP: Four QTL affecting antibody to Mh were detected on two different chromosomes. There was a chromosome-wide significant QTL detected on SSC10 at 100 cM for all three time points (T1, T2 and T3). For antibody to TET, there was no QTL found for the measurement at T3. For the second measurement (T4), there was a significant QTL on SSC17 at 0 cM. A highly significant QTL was detected at 30 cM for the last measurement (T5).

A highly significant QTL affecting antibody to PRRS (T6) was detected on SSC3 at 14 cM. The results of detected QTL affecting antibodies in this study are shown in Table 2.

For HP concentration, significant and highly significant F-values were detected for all measurements distributed over half of the autosomes. For SSC17 only, QTL were detected for two different measurements (T4 and T5), indicating that the region next to 0 cM is interesting and comprises genes which may affect possible HP concentration. Moreover, significant F-values for one measurement on SSC3, 5, 7, 8, 11, 14, 15 and 16 were found as shown in table 2.

In summary, for most of the traits, the fraction of phenotypic variance in the backcross explained by each QTL was fairly high (20 to 30%), but high additive and dominance effects were also found for some traits (data not shown). Further studies are in progress to investigate these effects by comparing the linkage regions in studies using different populations.

Table 1: Location and estimated effects of putative QTL affecting complement activity (AH50 and CH50) and its component C3c in the backcross DUMI population.

Chromosome	Trait	Trait measurement time point (cM/F-value)					
		T1	T2	T3	T4	T5	T6
SSC1	AH50	ns	ns	74/4.94*	74/5.04*	ns	ns
SSC2	AH50	ns	ns	ns	ns	ns	58/4.25*
SSC3	AH50	ns	ns	ns	ns	ns	30/5.2*
	CH50	78/5.58*	ns	ns	ns	ns	ns
	C3c	0/5.91*	31/8.28**	30/6.38*	29/4.98*	ns	29/4.84*
SSC4	AH50	50/8.09**	ns	54/6.04*	ns	54/6.08*	62/8.31**
SSC5	AH50	66/7.13**	ns	ns	ns	ns	56/4.87*
	CH50	ns	ns	ns	ns	89/3.14*	ns
SSC7	AH50	105/5.65*	56/4.81*	ns	0/5.5*	58/5.8*	49/7.9**
	CH50	ns	ns	ns	ns	ns	66/4.41*
SSC9	AH50	29/2.12*	ns	ns	ns	ns	ns
SSC10	AH50	ns	ns	ns	0/4.33*	ns	ns
	CH50	100/5.65*	ns	ns	ns	ns	ns
SSC11	C3c	23/4.41*	ns	35/6.24**	ns	ns	42/5.48**
SSC14	CH50	120/4.21*	ns	ns	ns	ns	ns
SSC16	AH50	38/6.4*	26/9.89**	18/8.43**	21/7.93**	ns	79/4.04*
	CH50	ns	ns	ns	ns	79/9.34**	ns
	C3c	ns	ns	ns	ns	0/3.57*	ns
SSC17	CH50	ns	ns	ns	0/6.48**	ns	0/4.15*
	C3c	ns	ns	ns	0/3.45*	ns	ns
SSC18	AH50	ns	ns	130/5.78*	106/5.3*	ns	ns
	C3c	130/5.29*	130/6.35**	130/7.55**	130/6.94**	130/5.17*	130/6.55**

ns: non significant
* $p \leq 0.05$
** $p \leq 0.01$ at the chromosome-wide level

Table 2: Location and estimated effects of putative QTL affecting antibody responses to Mycoplasma hyopneumoniae, tetanus toxoid and PRRSV (Ab-Mh, Ab-TET, and Ab-PRRSV, respectively) and haptoglobin concentration (HP) in the backcross DUMI population.

Chromosome	Trait	Traits measurement time points (cM/F-value)					
		T1	T2	T3	T4	T5	T6
SSC3	Ab-PRRSV	no	no	no	ns	ns	14/6.35**
	HP	92/6.25*	ns	ns	ns	ns	ns
SSC5	HP	ns	ns	ns	ns	89/4.11*	ns
SSC6	Ab-Mh	ns	ns	16/4.53*	no	no	no
SSC7	HP	ns	127/5.44*	ns	ns	ns	ns
SSC8	HP	ns	27/7.4**	ns	ns	ns	ns
SSC10	Ab-Mh	100/4.43*	100/4.76*	100/4.18*	no	no	no
SSC11	HP	ns	ns	12/5.63*	ns	ns	ns
SSC14	HP	ns	ns	120/9.18**	ns	ns	ns
SSC15	HP	ns	ns	ns	0/4.23*	ns	ns
SSC16	Ab-TET	no	no	ns	ns	130/9.98**	no
	HP	ns	ns	ns	ns	79/14.88**	ns
SSC17	Ab-TET	no	no	ns	0/4.15*	ns	no
	HP	ns	ns	ns	0/6.05**	0/3.67*	ns
SSC18	HP	ns	ns	ns	ns	ns	114/6.46**

no: not observed
ns: non significant
* $p \leq 0.05$
** $p \leq 0.01$ at the chromosome-wide level

Discussion

A number of QTL for immune traits were detected in the backcross population in this study. Some of these QTL regions, especially those for antibody response, are chromosomal regions comparable to QTL detected in different species such as the mouse, human and chicken [5-7,13]. Only a few reports exist on QTL affecting traits related to the immune capacity of pigs. Edfors-Lilja *et al.* [4] identified four QTL with significant effects on immune function in pigs. These four QTL consist of a QTL affecting total leukocyte counts on SSC1 at 78 cM ($p \leq 0.05$), mitogen induced proliferation on SSC4 at 75 cM ($p \leq 0.05$), pre-vaccination levels of antibody response to *E. coli* Ag K88 on SSC5 at 64 cM ($p \leq 0.01$) and antibody response to the O149 antigen on SSC6 at 69 cM ($p \leq 0.01$). In addition, the other three putative QTL ($p \leq 0.10$), which affect mitogen-induced proliferation on SSC7 at 73 cM, spontaneous proliferation on SSC13 at 107 cM, and antibody response to the O149 antigen on SSC6 at 67 cM, were identified in the same study. The identification of QTL with an effect on stress induced alterations of porcine immune function was also done by Edfors-Lilja *et al.* [14], indicating several significant and suggestive QTL for a number of immune capacities. The QTL affecting hematocrit and haemoglobin levels, leukocyte numbers and in vitro leukocyte functions were also detected on

SSC1 and SSC8 in similar regions. A QTL close to the mast/stem cell growth factor receptor gene *(KIT)* was detected and candidate gene analysis showed significant effects of this gene on the immune response-related traits [15].

The results presented here partly confirm and supplement previous QTL studies in this experimental F2 population, since the same markers and similar marker distances were used. Further studies might yield better evidence, because the immunisation and sampling strategies of this study were slightly different in the F2 and the backcross population. Using an overall parameter for the immune reaction in both populations separately and combined will be one focus in future studies. Some QTL are in regions important for immunity, such as the MHC cluster on SSC7, or in regions including numerous T cell-receptor or interleukin receptor or binding factor genes (i.e. SSC3, 18 or 16).

Different genes have already been mapped on SSC3, including interleukin alpha and beta (*IL1A* and *IL1B*), immunoglobulin kappa light chain (*IGKC*) and the B cell genes, B cell CLL/lymphoma 7B (*BCL7B*) and B cell maturation factor (*BCMA*). Highly conserved regions of SSC3 include HSA16p, HSA7 and HSA2p (http://www.toulouse.inra.fr/lgc/pig/compare/compare.htm). A number of genes which are involved in the immune system were identified, e.g. immunoglobulin, variable region genes of the kappa light chain (*IGKV*), the immunoglobulin kappa constant region (*IGKC*) on HSA2p12 [16], and the T cell antigen receptor gamma subunit (*TCRG*) on HSA7p15-p14 [17].

The most interesting chromosomes for antibody related traits in this study are SSC3 and SSC10 for antibody response to PRRS and Mh vaccinations, respectively. The QTL on SSC10 for antibody response to Mh vaccination is interesting, since the same region was identified for all three measurements. In the pig, different genes involved in immune response have been mapped on SSC10q, including EST Z81203 of macrophage inflammatory protein beta 1 (*MIP1B*) at 10q11-q12, porcine EST BF713707 of inter-alpha-inhibitor heavy chain H2 at 10p17, integrin beta 1 subunit (*ITGB1*) at 10p17, porcine EST Z84039 of iron responsive element binding protein (*IREB1*) at 10q11-q12, and porcine EST of cathepsin L (AJ301219; *CTSL*) at SSC10q11-q12.

A number of genes located on HSA7, a homologous chromosomal region of SSC18, involved in the immune response were identified and proposed as candidate genes. The following genes are interesting with respect to the immune system: T cell antigen receptor, gamma subunit (*TCRG*) at 7p15-p14, mucopolysaccharidosis type VII (*GUSB*) at 7q21.11, interleukin 6 (*IL6*) at 7p21, zinc finger protein, subfamily 1A, member 1 (*ZNFN1A1*) at 7p12, Shawachman-diamond syndrome (*SDS*) at 7q11, carboxypeptidase, vitellogenic-like (*CPVL*) at 7p15-p14, and receptor activity-modifying protein 3 (*RAMP3*) at 7p13-p12.

The results of the QTL study in the backcross animals indicated at least three interesting QTL affecting the measurements for C3c on SSC3, SSC11, and SSC18. Genes already mapped on porcine chromosomes are involved in traits related to the immune system are T cell receptor beta (*TCRB*), and inhibin beta A (*INHBA*). Further studies should be done using additional population information to confirm the regions before the selection of strong candidate genes.

Conclusion

A number of QTL for immune traits were detected in the backcross population in this study. The results from this study support and supplement previous findings in terms of some of the QTL regions detected in the pig and other species. QTL for traits related to humoral innate and acquired immunity are distributed throughout the genome and are, in terms of frequency and effects, similar to QTL for other quantitative traits. A number of positional functional candidate genes, either mapped in the pig genome or in the homologous chromosome segments of the human genome were identified.

References

1 Hu ZL, Dracheva S, Jang WH, Maglott D, Bastiaansen J, Rothschild MF et al.: A QTL resource and comparison tool for pigs: PigQTLDB. Mamm Genome 2005;16:792-800.

2 Rothschild MF, Hu ZL, Jiang ZH: Advances in QTL mapping in pigs. Int J Biol Sci 2007;3:192-197.

3 de Koning DJ, Carlborg O, Haley CS: The genetic dissection of immune response using gene expression studies and genome mapping. Vet Immunol Immunopathol 2005;105:343-352.

4 Edfors-Lilja I, Wattrang E, Marklund L, Moller M, Andersson-Eklund L, Andersson L et al.: Mapping quantitative trait loci for immune capacity in the pig. J Immunol 1998;161:829-835.

5 Hall MA, Norman PJ, Thiel B, Tiwari H, Peiffer A, Vaughan RW et al.: Quantitative-trait loci on chromosomes 1, 2, 3, 4, 8, 9, 11, 12, and 18 control variation in levels of T and B lymphocyte subpopulations. Am J Hum Genet 2002;70:1172-1182.

6 Siwek M, Buitenhuis AJ, Cornelissen SJB, Nieuwland MGB, Bovenhuis H, Crooijmans R et al.: Detection of different quantitative trait loci for antibody responses to keyhole lympet hemocyanin and mycobacterium butyricum in two unrelated populations of laying hens. Poult Sci 2003;82:1845-1852.

7 Siwek M, Cornelissen SJB, Nieuwland MGB, Buitenhuis AJ, Bovenhuis H, Crooijmans R et al.: Detection of QTL for immune response to sheep red blood cells in laying hens. Anim Genet 2003;34:422-428.

8 Zhou H, Lamont SJ: Association of transforming growth factor beta genes with quantitative trait loci for antibody response kinetics in hens. Anim Genet 2003;34:275-282.

9 Zhou H, Li H, Lamont SJ: Genetic markers associated with antibody response kinetics in adult chickens. Poult Sci 2003;82:699-708.

10 O'Connell JR, Weeks DE: PedCheck: A program for identification of genotype incompatibilities in linkage analysis. Am J Hum Genet 1998;63:259-266.

11 Green P, Falls K, Crooks S: Documentation for CRIMAP, version 2.4 Washington University, School of Medicine, St Louis, MO 1990.

12 Seaton G, Haley CS, Knott SA, Kearsey M, Visscher PM: QTL Express: mapping quantitative trait loci in of simple and complex pedigrees. Bioinformatics 2002;18:339-340.

13 de Buhr MF, Mahler M, Geffers R, Hansen W, Westendorf AM, Lauber J et al.: Cd14, Gbp1, and Pla2g2a: three major candidate genes for experimental IBD identified by combining QTL and microarray analyses. Physiol. Genomics 2006;25:426-434.

14 Edfors-Lilja I, Wattrang E, Andersson L, Fossum C: Mapping quantitative trait loci for stress induced alterations in porcine leukocyte numbers and functions. Anim Genet 2000;31:186-193.

15 Wattrang E, Almqvist M, Johansson A, Fossum C, Wallgren P, Pielberg G et al.: Confirmation of QTL on porcine chromosomes 1 and 8 influencing leukocyte numbers, haematological parameters and leukocyte function. Anim Genet 2005;36:337-345.

16 Malcolm S, Barton P, Bentley DL, Fergusonsmith MA, Murphy CS, Rabbitts TH: Assignment of a IGKV locus for immunoglobulin light chains to the short arm of chromosome 2 (2p13-cen) by in situ hybridisation using a cRNA probe of H(kappa)101(lambda)Ch4A. Cytogenet Cell Genet 1982;32:296-296.

17 Bensmana M, Mattei MG, Lefranc MP: Localization of the human T-Cell receptor gamma locus (TCRG) to 7p14-p15 by *in situ* hybridization. Cytogenet Cell Genet 1991;56:31-32.

Dr. Chirawath Phatsara, Institute of Animal Science, Animal Breeding and Husbandry Group, Endenicher Allee 15, D-53115 Bonn, Germany.

E-mail: cpha@itw.uni-bonn.de

Pinard M-H, Gay C, Pastoret P-P, Dodet B (eds): Animal Genomics for Animal Health. Dev Biol (Basel). Basel, Karger, 2008, vol 132, pp 287-291.

*IGF2/Mbo*II Polymorphism in Gir and Nelore Cattle

A. Martins da Silva[1], A.F.L. Rios [1], E.S. Ramos[1,2], V.L. Cardoso[3], A.E. Vercesi Filho[4], R.B. Lôbo[1,5], M.A.R. Freitas[1]

1 Department of Genetics, Faculty of Medicine of Ribeirão Preto (FMRP), University of São Paulo (USP), Ribeirão Preto, SP, Brazil;
2 Department of Obstetrics and Gyneacology, University Hospital FMRP, USP, Ribeirão Preto, SP, Brazil.
3 Agência Paulista de Tecnologia dos Agronegócios – APTA Centro Leste, Ribeirão Preto, SP, Brazil;
4 APTA Nordeste Paulista, Mococa, SP, Brazil;
5 Associação Nacional de Criadores e Pesquisadores – ANCP, Ribeirão Preto, SP, Brazil.

Key words: insulin-like growth factor 2 (*IGF2*), imprinting, Zebu breeds

Abstract: The imprinted gene insulin-like growth factor 2 (*IGF2*) carries out important functions in the development of placental mammals, during the embryonic and fetal stages and is located on bovine chromosome 29. The main aim of this study was to verify the occurrence of a polymorphism (C/T transversion) in exon 6 of the *IGF2* gene in Gir and Nelore cattle, Zebu breeds of great economic importance in Brazil for dairy and beef production, respectively. A 193 bp fragment of bovine *IGF2* exon 6 was amplified by PCR with specific primers for this region and digested with *Mbo*II enzyme to analyze the polymorphic segment. In a total of 39 Gir animals (29 females and 10 males), the estimated frequencies of the C and T alleles were 0.42 and 0.58, respectively. The genotypic frequencies were 0.10 for CC, 0.26 for TT and 0.64 for CT. For 28 Nelore animals (17 females and 11 males), the estimated frequencies of the C and T alleles were 0.43 and 0.57, respectively. The genotypic frequencies were 0.18 for CC, 0.32 for TT, and 0.50 for CT. This is the first report of the occurrence of this polymorphism in these breeds. The polymorphic site can be a regulatory motif with functional significance to gene regulation of the *IGF2* gene. This polymorphism could be used to investigate the allele-specific expression of the *IGF2* gene, its epigenetic status, and its role in developmental, growth and reproductive traits.

Introduction

Genomic imprinting is a mechanism of gene regulation where only one member of the allelic pair is expressed, depending on the parental origin. This phenomenon is associated with differential epigenetic markers on the regions named imprinting control regions (ICRs) [1]. Insulin-like growth factor 2 (*IGF2*) is one of the first known imprinted genes and it is expressed in several tissues during mammalian development [2], influencing fetal cell division, differentiation, placental development and possibly metabolic regulation [3]. In most bovine fetal tissues, only the paternal allele of this gene is expressed, while the maternal allele is transcriptionally silent [4]. It was showed through of comparative mapping data in situ hybridization that of *IGF2* is situated in the distal end of bovine chromosome 29 [5]. A length polymorphism was identified within a poly (A) tail in the 3' UTR of the sheep (*ovis aries*) *IGF2* gene. This polymorphism was used to map *IGF2* to sheep chromosome 21 and demonstrate that *IGF2* is imprinted in these animals, being expressed from the paternal allele [6]. The main aim of this study was to verify the occurrence of a polymorphism (C/T transversion) in the exon 6 of the *IGF2* gene in Gir and Nelore cattle, Zebu breeds of great economic importance in Brazil.

Materials and Methods

A total of 39 Gir animals (29 females and 10 males) and 28 Nelore (17 females and 11 males, including three different families) were analyzed. These herds belong to the APTA (Agência Paulista de Tecnologia dos Agronegócios – Nordeste Paulista) and Breeding Program for Nelore Cattle (PMGRN), respectively. Blood samples were collected from animals and the extraction of DNA was performed by digestion with proteinase K and salt extraction as previously described [7]. For the PCR reaction, primers were designed with GeneRunner Software (3.5 version) from the sequence of bovine genome and available in GenBank (accession number AY971631). The forward and reverse primers were 5' GCCTCTCGCTGTCCTCTC 3' and 5' CAGCCCGTCCTCCCTAAAG 3', respectively (Fig. 1). The PCR reaction (25 μl) contained 50-100 ng of genomic DNA, 50 mM of $MgCl_2$, 1.5 μl of 10X PCR buffer, 5 pmol of each primer, 0.25 mM dNTPs and 0.2 U of Taq DNA polymerase and was performed in a TGradient thermocycler (Whatman-Biometra). Cycle conditions were 5 minutes at 94°C, followed by 40 cycles (30 seconds at 94°C, 30 seconds at 60°C, 30 seconds at 72°C), and a final extension for 10 minutes at 72°C. For restriction enzyme analysis of the C-T polymorphic site, the 193 bp PCR product was digested at 37°C for 2 hours in 25 μl with 0.4U of *Mbo*II restriction nuclease (BioLabs, New England, USA). The fragments obtained were visualized in a 10% polyacrylamide gel stained with 0.2% silver nitrate.

Results

The fragments obtained by enzymatic restriction (generating two single-nucleotide polymorphisms) are shown in Figure 2. A single base change or insertion/deletion or transversion in the DNA sequence is named SNP (Single Nucleotide Polymorphism); the results of this profile are summarized in Table 1. It was observed that the average of the heterozygote CT was greater when compared with data of CC and TT genotypes. Using the RegRNA web server (http://regrna.mbc.nctu.edu.tw/) it was verified that the polymorphic site is in an RNA motif to binding PTB protein, which is described as trans-acting repressor [8], Figure 3.

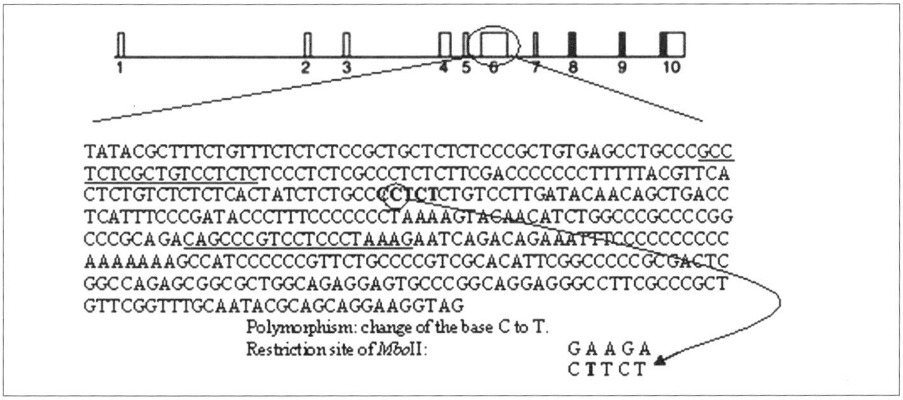

Fig. 1: Schematic representation of localization of exon 6 in the IGF2 gene locus in Bos taurus indicus (GenBank accession number AY971631). The polymorphism studied (change of one nucleotide) representing two SNPs is shown with the exons present in the IGF2 locus. Exons 8, 9 and a partial sequence of exon 10 are coding and visualized in black boxes. The remaining exons represent non-coding regions of the IGF2 gene.

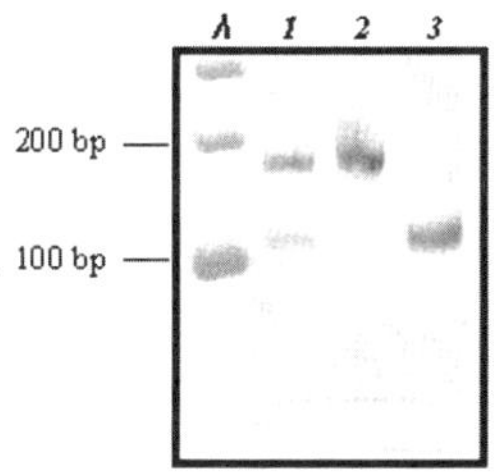

PROFILE	Breeds Gir	Nelore	AVERAGE
1 – HETEROZYGOTE (CT)	64 %	50 %	57 %
2 - HOMOZYGOTE (CC)	10 %	18 %	14 %
3 - HOMOZYGOTE (TT)	26 %	32 %	29 %

Fig. 2: Polymorphism profile of PCR products of DNA samples from Gir and Nelore breeds observed by digestion with MboII enzyme (λ: 100 bp DNA marker).

Table 1: Allelic and genotypic frequencies for polymorphism exon 6 IGF2 in Gir and Nelore animals studied.

		Allelic Frequency		**Genotypic Frequency**		
Breed	**N**	**CC**	**TT**	**CC**	**TT**	**CT**
Gir	39	0.42	0.58	0.10	0.16	0.64
Nelore	28	0.43	0.57	0.18	0.32	0.50
Total	67					

N = number of animals

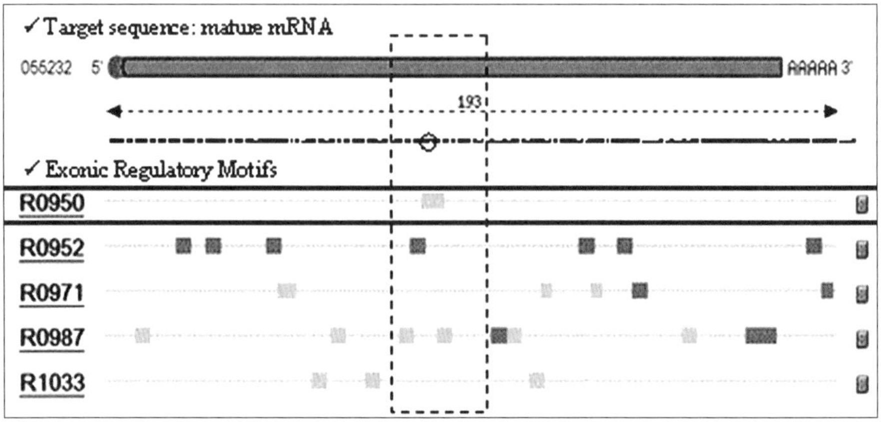

Fig. 3: Graphic representation of analysis generated by site http://regrna.mbc.nctu.edu.tw.

Discussion

The exon 6 polymorphism overlaps the 5' UTR of bovine *IGF2* (Fig.1). The sequences of mRNAs in 5' UTR can be targeted by regulatory proteins with a substantial number of mRNA structural motifs; these elements have been identified in transcriptional and post-transcription regulation of gene expression, in the transcription termination, mRNA localization and stability, mRNA alternative splicing and translation efficiency [9]. Based on our results, we hypothesized that the polymorphic site on a regulatory motif could be of functional significance to the regulation of this particular gene. Due to the biological functions of *IGF2*, the polymorphism studied here needs to be better investigated to determine possible influences on phenotypic traits of interest for production or reproduction traits in livestock. A study on the association between Birth Weight (BW), Body Mass Index (BMI) and the *IGF2/Apa*I polymorphism in adult humans was reported and although the *IGF2* genotype was not significantly associated with BW and/or BMI, the authors observed a statistically significant correlation of 0.33 (p=0.023) between BW and BMI in GG homozygotes whose BW was higher than 3.5kg [10]. The *IGF2* gene is potentially a very important marker for production traits in domestic animals, including cattle [11]. For *IGF2* genes, the future breeding model should consider the non-Mendelian inheritance of imprinted genes. This is the first study on *Mbo*II polymorphism of exon 6 in the *IGF2* gene in Brazilian Zebu breeds.

Conclusion

The results obtained suggest that the in vitro molecular protocols can be used to identify new molecular markers, such as the functional SNPs and can to contribute with additional informations for breeding strategies of livestock. Others investigations of this polymorphism should be perform to verify possible influences on traits of interest in cattle.

Acknowledgements

This work was supported by PRONEX, FAPESP, CNPq, ANCP and CAPES.

References

1 Reik W, Walter J: Genomic imprinting: parental influence on the genome. Nat Rev Genet 2001;2:21-30.

2 Hitchins MP, Moore GE: Genomic imprinting in fetal growth and development. Expert Rev Mol Med 2002;1-19.

3 Gicquel C, Le Bouc Y: Hormonal regulation of fetal growth. Horm Res 2006;65(Suppl. 3):28-33.

4 Dindot SV, Kent KC, Evers B, Loskutoff N, Womack J, Piedrahita JA: Conservation of genomic imprinting at the *XIST, IGF2*, and *GTL2* loci in the bovine. Mamm Genome 2004;15:966-974.

5 Goodall JJ, Schmutz SM: Linkage mapping of *IGF2* on cattle chromosome 29. Anim Genet 2003;34:313.

6 McLaren RJ, Montgomery GW: Genomic imprinting of the insulin-like growth factor 2 gene in sheep. Mamm Genome 1999;10:588-591.

7 Olerup O, Zetterquist H: HLA-DR typing by PCR amplification with sequence-specific primers (PCR-SSP) in 2 hours: An alternative to serological DR typing in clinical practice including donor-recipients matching in cadaveric transplantation. Tissue Antigens 1992;39:225-235.

8 Ji Y, Xu X, Stormo GD: A graph theoretical approach for predicting common RNA secondary structure motifs including pseudoknots in unaligned sequences. Bioinformatics 2004;20:1591-1602.

9 Wilkie GS, Dickson KS, Gray NK: Regulation of mRNA translation by 5' and 3' UTR binding factors. Trends in Biochemical Sciences 2003; 28:182-188.

10 Gomes MVM, Soares MR, Pasqualim-Neto A, Marcondes CR, Lôbo RB, Ramos ES: Association between birth weight, body mass index and *IGF2/ApaI* polymorphism. Growth Horm IGF Res 2005;15:360-362.

11 Flisikowski K, Adamowicz T, Switonski M, Hiendleder S, Pareek C: Nucleotide sequence and variation of IGF2 gene exon 6 in *Bos taurus* and *Bos indicus* cattle. Anim Biotechnol 2005;16:203-208.

Maria Armênia Ramalho de Freitas, Medicine College of Ribeirão Preto, University of São Paulo, Bandeirantes, Avenue 3900, Monte Alegre, Ribeirão Preto, São Paulo, Brazil- CEP 14049-900.

E-mail: mfreitas@genbov.fmrp.usp.br

Pinard M-H, Gay C, Pastoret P-P, Dodet B (eds): Animal Genomics for Animal Health. Dev Biol (Basel). Basel, Karger, 2008, vol 132, pp 293-299.

A Systematic, Data-driven Approach to the Combined Analysis of Microarray and QTL Data

C. Rennie[1,2], H. Hulme[2], P. Fisher[2], L. Hall[3], M. Agaba[4], H.A. Noyes[1], S.J. Kemp[1,4], A. Brass[2]

1. School of Biological Sciences, Biosciences Building, Liverpool, UK
2. School of Computer Science/Faculty of Life Sciences, University of Manchester, UK
3. Roslin Institute, Roslin, Midlothian, Scotland, UK
4. ILRI, Nairobi, Kenya

Keywords: automated analysis, microarray, QTL, workflow

Abstract: High-throughput technologies inevitably produce vast quantities of data. This presents challenges in terms of developing effective analysis methods, particularly where the analysis involves combining data derived from different experimental technologies. In this investigation, a systematic approach was applied to combine microarray gene expression data, quantitative trait loci (QTL) data and pathway analysis resources in order to identify functional candidate genes underlying tolerance to *Trypanosoma congolense* infection in cattle. We automated much of the analysis using Taverna workflows previously developed for the study of trypanotolerance in the mouse model.

Pathways represented by genes within the QTL regions were identified, and this list was subsequently ranked according to which pathways were over-represented in the set of genes that were differentially expressed (over time or between tolerant N'dama and susceptible Boran breeds) at various timepoints after *T. congolense* infection. The genes within the QTL that played a role in the highest ranked pathways were flagged as good targets for further investigation and experimental confirmation.

INTRODUCTION

The analysis of microarray gene expression data can present difficulties due to the vast size of the datasets. Depending on the purpose of the study, analysis may be further complicated by the need to combine data produced using different experimental techniques or by the underlying complexity of the phenotype being investigated. A systematic, data-driven, semi-automated analysis pipeline was developed for the pathway-based combined analysis of microarray and quantitative trait locus (QTL) data as part of a study investigating the genetics underlying tolerance to African bovine trypanosomiasis (nagana).

Nagana is transmitted by the tsetse fly, leading to loss of productivity and often death in infected cattle. It represents a major constraint on livestock production in Africa [1].

Some breeds of cattle, such as the Boran, are susceptible to the pathological consequences of trypanosomiasis. Others, such as the N'dama, are more resistant to these effects (trypanotolerant) [2]. However, the susceptible breeds have desirable traits, such as greater size, and may be preferred by farmers. Identification of genes that influence response to trypanosomiasis might inform new treatment approaches, or even pave the way for creating transgenic breeds that combine the desirable traits of susceptible and trypanotolerant cattle.

Trypanotolerance is a complex phenotype including several distinct components, likely to involve separate genetic control mechanisms. Features include the ability to control anaemia, control parasitaemia and maintain bodyweight. Previous studies provide evidence of the complexity of trypanotolerance. The trypanosomiasis response of haematopoietic chimeric twins bred from one Boran and one N'dama parent was studied, demonstrating that control of anaemia depends on bone marrow from a trypanotolerant background, whereas control of parasitaemia does not [3]. The mapping study that provided QTL data used in this analysis showed that the proportion of phenotypic variation explained by each QTL was between 6 and 20%, suggesting that multiple genes, or complex epistatic or environmental effects, may influence each trait [4].

A microarray gene expression time course study was carried out to investigate gene expression differences between (trypanotolerant) N'dama and (trypanosusceptible) Boran cattle infected with *T. congolense* strain IL1180. This study generated a vast dataset. Thousands of probe sets on the array generated signals that were significantly different between timepoints and/or between the two breeds (in T-tests or paired T-tests with $p \leq 0.01$).

A mapping study identified QTL for 16 phenotypic traits associated with trypanotolerance in Boran and N'dama cattle [4]. The gene underlying a QTL is not assumed to be differentially expressed. However, it is expected to connect biologically with differentially expressed genes. The known pathways that included a gene within one of the five QTL with the largest effect were identified and compared with the known pathways that included a differentially expressed gene. The rationale behind this approach was to establish the possible connections between the QTL and the differentially expressed genes.

A systematic strategy was used to enable an objective triage of the datasets, resulting in a short list of strong candidate pathways that included both a differentially expressed gene and a gene within a trypanotolerance QTL. These pathways were ranked according to the results of a Fisher exact test performed using the Database for Annotation, Visualisation and Integrated Discovery (DAVID) [5]. A literature search was carried out to determine whether the biological function of each pathway was likely to be linked to the phenotypic trait influenced by the QTL.

Large sections of the analysis were automated by adapting Taverna workflows originally developed for the study of trypanotolerance in the mouse model [6]. This allowed for the entire analysis to be repeated consistently and relatively quickly and, for example, for the incorporation of information on the bovine genome from a different EnsEMBL build. It was also possible to adapt the analysis procedure to examine a different species or a different phenotype for which QTL data was available.

Materials and Methods

Microarray gene expression data was acquired using Affymetrix Bovine Genome '100 format (Midi)' microarrays for liver samples harvested from Boran and N'dama cattle at 0, 12, 15, 18, 21, 26, 29, 32 and 35 days post-infection.

This data was analysed with dChip [7] to identify and remove outliers before normalisation using the robust multi-array (RMA) method. Principal components analysis (PCA) was used to check that hybridisations grouped as expected.

T-tests were used to compare gene expression between breeds at each time point. Paired T-tests (using data for the same individual animals at different timepoints) were used to compare gene expression for each time point with day 0. Lists of probes that showed differential gene expression ($p \leq 0.01$) between breeds or over time were compiled.

A previous study identified trypanotolerance QTL in N'dama and Boran cattle [4]. Five QTL were selected to be included in this analysis based on phenotypic trait, mapping resolution and strength of effect. Base pair positions of QTL relative to the EnsEMBL bovine genome preliminary build Btau2.0 were determined manually. Names and phenotypes for the five QTL are shown in Table 1.

Table 1: Name and phenotype for the five trypanotolerance QTL used in this analysis. For more detailed information, please refer to the original mapping study [5].

QTL	Phenotype
BTA2	Anaemia
BTA4	Parasitaemia
BTA7	Anaemia and parasitaemia
BTA16	Anaemia
BTA27	Anaemia

To combine microarray and QTL data, a Taverna workflow previously developed for the study of trypanotolerance in the mouse model was adapted [6]. The paper cited provides a full description. In brief, lists of differentially expressed genes (over time or between breeds) were associated with Kyoto Encyclopaedia of Genes and Genomes (KEGG) pathways. A separate process identified genes within QTL and associated these with KEGG pathways. A third process compared these lists to produce a list of KEGG pathways that contained both differentially expressed genes and genes from the QTL.

Some adaptations of the workflow were necessary. Rather than the mouse EnsEMBL build and IDs, the bovine EnsEMBL preliminary build (Btau2.0) was used. Bovine gene IDs were retrieved for Affymetrix probes then mapped to human homologues (using EnsEMBL data for NCBI build 36) so that human IDs could be used for the remainder of the analysis (available annotation on bovine genes is very limited). Output was in the same form as the original, comprising a list of KEGG pathways that included at least one differentially expressed gene and at least one gene from the QTL.

This list was ranked based on the p-value of each pathway in a Fisher exact test performed on the microarray data using DAVID indicating whether pathway genes showed more differential expression than expected by chance. The list was annotated to add gene symbols for pathway genes in the QTL and to indicate the breeds and timepoints in which pathway genes were differentially expressed. Further annotation was derived from gene and pathway resources including GenBank, iHOP, GenMAPP and GeneGo: MetaCore.

Figure 1 summarises the analysis protocol described above.

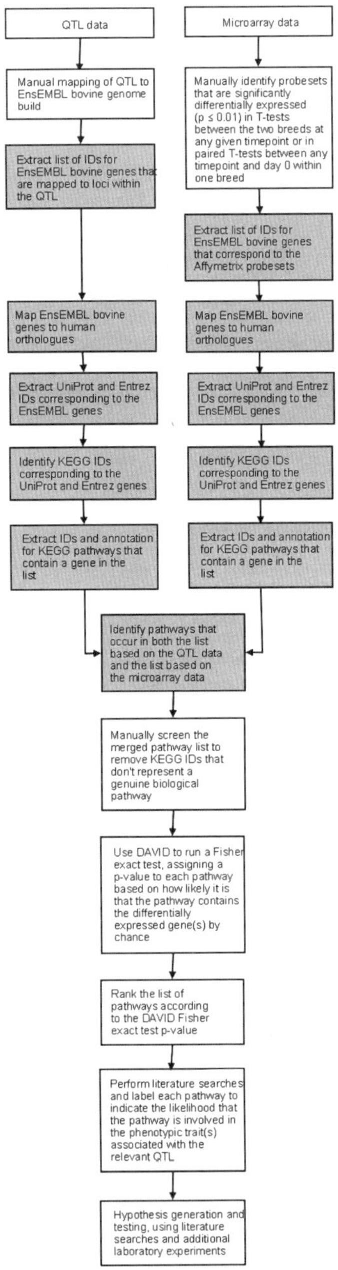

Fig. 1: Summary of the analysis procedure. Automated sections are indicated using grey shading.

Results

This analysis procedure could be re-used or adapted to examine another species or phenotype for which QTL data are available. The modular nature of the protocol and of Taverna workflows facilitates adding or altering analysis stages. Workflow scripts and supplementary data (e.g. files from intermediate stages) are available from the authors.

In the bovine trypanotolerance study, the result of the analysis procedure was to provide a short list of targets for further investigation. This result can be quantified by assessing the numbers of genes requiring further investigation based on the combined analysis results or on the original data.

Out of 24,128 probe sets on the array, 12,591 were significantly differentially expressed ($p \leq 0.01$) in T-tests or paired T-tests comparing expression between breeds or over time. Of these probe sets, 8,342 were mapped to a known gene, in total representing 7,071 unique gene symbols.

After combining the pathway lists for differentially expressed and QTL genes, pathway genes within QTL provided a list of 127 targets. Restricting the pathway list to those with a significant ($p \leq 0.05$) score in the DAVID Fisher exact test reduced this to 51 targets (it could be reduced more by checking whether expression changes are downstream of the QTL gene and whether the pathway function is related to the QTL phenotype).

The list of pathways with a significant score ($p \leq 0.05$) in the DAVID Fisher exact test is displayed in Table 2. Pathway genes lying within each QTL are also listed. Note that these data are based on an analysis using the EnsEMBL Btau2.0 preliminary build. A more recent preliminary build is available, and the analysis will be repeated, and key findings discussed, in a future publication.

Discussion

When studying complex phenotypes, analysis based on biological processes already known to be involved may be insufficient. It is possible that other key biological pathways, or complex interactions between them, could be missed. Data-driven approaches are useful to identify the biological processes showing strongest variation in the results.

The aim of a pathway-based approach to analysing microarray and QTL data is to identify biologically meaningful links between the two datasets. The gene underlying a QTL is not necessarily differentially expressed, but may influence the expression of other genes downstream in a known pathway. This approach allows such genes to be identified without detailed investigation of every gene in the QTL regions or every gene that is differentially expressed.

Automation is increasingly necessary to handle the vast quantities of data produced by high-throughput technologies, where manual analysis of the entire dataset is not feasible. Automated approaches are systematic, promoting consistency and reducing bias. Consistent replication of automated analyses is relatively simple, allowing separate studies to produce comparable results and allowing analyses to be repeated in order to incorporate new information (e.g. from updates to genome build and gene information available in public databases). Automated analysis can be an effective triage process, producing a short list of strong targets for thorough manual investigation.

Table 2: Pathways with a significant ($p \leq 0.05$) score in a Fisher exact test to determine whether the differential expression of pathway genes is higher than expected by chance. The columns on the right give the gene symbols for pathway genes within each of the QTL.

KEGG pathway name	BTA2	BTA4	BTA7	BTA16	BTA27
Leukocyte transendothelial migration			*VAV1*		*CLDN23*
Regulation of actin cytoskeleton	*FN1* *PIP5K3*	*CHRM2*	*VAV1*		*BRAF* *FGF20*
Cell cycle	*MCM6* *ORC4L*		*CDKN2D*		
Gap junction			*PRKACA* *TUBB4*	*GNAO*	
Focal adhesion	*FN1*	*ZYX*	*COL5A3* *VAV1*	*CAPN2*	*BRAF*
MAPK signalling pathway	*CASP8*	*CASP2*	*ECSIT* *PRKACA*	*DUSP10*	*BRAF* *DUSP4* *FGF20* *IKBKB*
Hematopoietic cell lineage			*EPOR* *FCER2*		
Huntington's disease	*CASP8*				
Glycerolipid metabolism	*LCT*	*DGKI*			*AGPAT6*
Axon guidance	*EFNB1*	*EPHA1* *EPHB6*			*UNC5D*
Glycerophospholipid metabolism		*DGKI*	*ARD1A*		*AGPAT6*
Adherens junction			*INSP*		*FGFR1*
Neurodegenerative disorders	*CASP8*				
T cell receptor signalling pathway	*CD28* *CTLA4* *ICOS*		*VAV1*		*IKBKB*
Long-term potentiation			*PRKACA*	*GNAQ*	*BRAF*
Apoptosis	*CASP8*		*IRAK1* *PRKACA*	*CAPN2*	*IKBKB*
Toll-like receptor signalling pathway	*CASP8*		*IRAK1* *TICAM1*		*IKBKB*
Wnt signalling pathway			*PRKACA*		*DKK4* *SFRP1*
Glutathione metabolism	*IDH1*	*GSTK1*			
Calcium signalling pathway		*CHRM2*	*PTGER1* *PRKACA*	*GNAO* *GNA14* *ITPKB*	*ADRB3* *VDAC3*

Conclusion

Systematic data-driven automated approaches offer an excellent means to triage data from high-throughput technologies, providing a shortlist of viable targets for thorough manual analysis and experimental confirmation.

Acknowledgements

This work was wholly funded by The Wellcome Trust.

References

1 Kristjanson PM, Swallow BM, Rowlands GJ, Kruska RL, de Leeuw PN: Measuring the costs of African animal trypanosomosis, the potential benefits of control and returns to research. Agric Syst 1999;59:79-98.

2 Murray M, D'Ieteren G, Teale AJ. Trypanotolerance, in Maudlin I, Holmes PH, Miles MA (eds): The Trypanosomiases. Wallingford UK, CABI Publishing, 2004, pp 461-477.

3 Naessens J, Leak SG, Kennedy D, Kemp SJ, Teale AJ: Responses of bovine chimaeras combining trypanosomosis resistant and susceptible genotypes to experimental infection with Trypanosoma congelense. Vet Parasitol 2003;111:125-142.

4 Hanotte O, Ronin Y, Agaba M, Nilsson P, Gelhaus A, Horstmann R et al: Mapping of quantitative trait loci controlling trypanotolerance in a cross of tolerant West African N'dama and susceptible East African Boran cattle. Proc Natl Acad Sci USA 2003;100(13):7443-7448.

5 Dennis GJ, Sherman BT, Hosack DA, Yang J, Gao W, Lane HC et al: DAVID: Database for Annotation, Visualisation, and Integrated Discovery. Genome Biol 2003;4(9):R60.

6 Fisher P, Hedeler C, Wolstencroft K, Hulme H, Noyes H, Kemp S et al: A systematic strategy for large-scale analysis of genotype-phenotype correlations: identification of candidate genes involved in African trypanosomiasis. Nucl Acids Res 2007;35(16):5625-5633.

7 Li C, Wong WH: Model-based analysis of oligonucleotide arrays: Expression index computation and outlier detection. Proc Natl Acad Sci USA 2001;98:31-36.

Catriona Rennie, LF8, Kilburn Building, The University of Manchester, Oxford Rd, Manchester, M13 9PL, UK.

E-mail: catriona.rennie@postgrad.manchester.ac.uk

Pinard M-H, Gay C, Pastoret P-P, Dodet B (eds): Animal Genomics for Animal Health. Dev Biol (Basel). Basel, Karger, 2008, vol 132, pp 301-306.

Evaluation of *MAFbx* Expression as a Marker for Congenital Splay Leg in Piglets

D. Boettcher[1], R. Schmidt[1], C. Rehfeldt[2], G. Thaller[3], H.H. Swalve[1], S. Maak[2]

[1] Institute of Agricultural and Nutritional Sciences, Martin-Luther-University Halle-Wittenberg, Halle, Germany

[2] Research Unit Muscle Biology and Growth, Research Institute for the Biology of Farm Animals (FBN), Dummerstorf, Germany

[3] Institute for Animal Breeding and Husbandry, Christian-Albrecht-University Kiel, Kiel, Germany

Key words: swine, splay leg, *MAFbx*, gene expression

Abstract: Splay leg is a hereditary syndrome observed in highly varying frequency in newborn piglets. Although the phenotype indicates a muscular weakness, the etiology is still poorly understood. Only recently, the gene expression of muscle atrophy F-box (*MAFbx; FBXO32*) was proposed as being of diagnostic value for splay leg in piglet. In this study, total RNA from three healthy and three affected male piglets was isolated. Samples were collected from *M. gracilis, Mm. adductores,* and *M. sartorius*. Further samples were taken for histological and biochemical analyses. Expression of *MAFbx* was analysed by real-time RT-PCR and with the GeneChip® Porcine Genome Array (Affymetrix). No significant differences ($p>0.05$) were observed in relative MAFbx expression, either between the three muscles or between splay leg and healthy piglets for each muscle. The expression of further atrophy-related genes was unchanged, indicating that splay leg is not characterized by general muscular atrophy in the affected hind limbs. This is supported by histological and biochemical data that does not demonstrate signs of atrophy in splay leg muscles. We conclude that the diagnostic value of *MAFbx* expression for congenital splay leg in piglets is doubtful and that the disease is characterized by heterogeneous alterations in skeletal muscle.

Introduction

Congenital splay leg is the most frequent hereditary disorder in pig production. ThCongenital splay leg is the most frequent hereditary disorder in pig production. The frequency of the disease varies between 1 and 2% of the piglets born alive. This comprised almost 50% of all birth defects observed in a continuous survey of 16,000 to 20,000 litters in the years 2003 to 2006 (http://www.gfs-topgenetik.de). Affected piglets suffer from impaired mobility and are often crushed by the sow [1]. Despite numerous investigations the etiology and genetic background are still unknown [2]. There are several hypotheses that are partly conflicting regarding pathogenesis: (i) incomplete maturation of the neuromuscular system at birth [3], (ii) defects in the cholinergic system [4] and (iii) incomplete myelinization of nerves and the spinal cord [5]. Furthermore, environmental factors (e.g. pen floors and nutrition of the sow during gestation) are discussed in the modulation of the phenotype of the disease [6,7]. Only recently, muscle fibre atrophy was proposed as a clinical sign of congenital splay leg based on histological and molecular analyses [8]. The authors suggested increased expression of the atrophy factor *MAFbx* (muscle atrophy F-box; *FBXO32*: F-Box only 32) in combination with a reduced expression of the gene for *P311* as diagnostic markers for the disease. Additionally, they described a significantly reduced fibre diameter combined with increased fibre densities in several muscles of splay leg piglets. The objective of this study was to evaluate this hypothesis by measurement of gene expression (real-time RT-PCR and GeneChip® Porcine Genome Array, Affymetrix) and determination of histological and biochemical traits in different skeletal muscles.

Materials and Methods

Samples were collected from *Mm. sartorius, gracilis,* and *adductores* of each of three male healthy and splay leg piglets immediately after birth. The tissue was stored in RNAlater (Qiagen) and total RNA was isolated with the Trizol reagents and purified with RNeasy (Qiagen). An aliquot of the RNA was labeled with the One-Cycle Target Labeling kit (Affymetrix) and analysed with the GeneChip® Porcine Genome Array (Affymetrix) at the Resource Center for Genome Research (RZPD Berlin). Reverse transcription was performed with random hexamer primers and TaqMan RT Reagents (Applied Biosystems). For real-time PCR, ImmoMix reagents (Bioline) were used on an ABIPrism 7000SDS (Applied Biosystems). Primer sequences are given in Table 1. The values were normalized for individual expression of the gene for *18S rRNA*. The analysis of the results was done according to the standard curve method as previously described [9]. For histological analyses, samples of *Mm. gluteus* and *longissimus* were collected and immediately frozen in liquid nitrogen. Cross sections (12-15 µm) were cut using a cryostat-microtome and stained by ATPase histochemistry (pH = 9.4) [10] without pre-incubation of the samples. Concentrations of protein and DNA and the activity of creatine phosphate kinase (CPK) in muscle homogenates were determined as described previously [11]. Different options of SAS v9.1 were used for statistical analysis of the data.

Table 1: Primers used for real-time RT-PCR.

Primer		Sequence 5'-3'
MAFbx1	forward	AAGCGCTTCCTGGATGAGAA
[8]	reverse	GGCCGCAACATCATAGTTCA
P311	forward	GAGGGAAGGCCTAAGGG
[8]	reverse	CGGTCTCGCCATCCTTCTT
MAFbx2	forward	AACTCTGCCAGTACCACTTCTCC
(this study)	reverse	GTCTCCATATTGCTCTTTCCGA
18S rRNA	forward	GACCATAAACGATGCCGACT
	reverse	GGTGCCCTTCCGTCA

Results

The relative abundance of *MAFbx* transcripts was measured with real-time RT-PCR with primers developed for this study (Fig. 1). There were no significant differences between the muscles of both groups, although there was a trend towards higher values in some splay leg muscles. Additional analysis of samples of *M. biceps femoris* from a previous unpublished study, with six healthy and six splay leg piglets per group, confirmed that *MAFbx* is not differentially expressed in splay leg muscle (left part of Figure 1). Similar results were obtained with a different amplicon of the MAFbx gene (data not shown). The analysis of the expression values for *MAFbx* on the GeneChip® Porcine Genome Array confirmed the results obtained with real-time PCR. The normalized values for the expression of *MAFbx* on the array reflect the large variation between different sib-pairs obtained by real-time PCR. The comparison of the data for *M. gracilis* is shown in Figure 2. The primers for *P311* [8] did not yield reproducible results under our real-time PCR conditions (data not shown). No significant differential expression of myosin heavy chain isoforms (*MyHC 2a, 2b* and *2x*) was observed between healthy and splay leg piglets. Furthermore, the expression of atrophy related genes like *SMN1, MUSK, AR, FOXO1, SOD1,* and *MSTN* was not affected by the health status of the piglets (array data, not shown). Cross sections of *M. gluteus* and *M. longissimus* of splay leg piglets partly revealed differences compared with their healthy siblings (Fig. 3). However, the alterations were heterogeneous, comprising reduced bundle density and a lower number of secondary fibres (splay leg piglet F3, upper panel) as well as reduced size of primary and secondary fibres (splay leg piglet F7 lower panel). The biochemical data (Table 2) of the small sample studied here underlines the heterogeneity of the histochemical observations. No significant differences were observed in the investigated criteria. There was a trend towards an increased DNA/protein ratio in *M. gluteus* of splay leg piglets ($p = 0.06$) indicating disturbed protein accretion in this muscle.

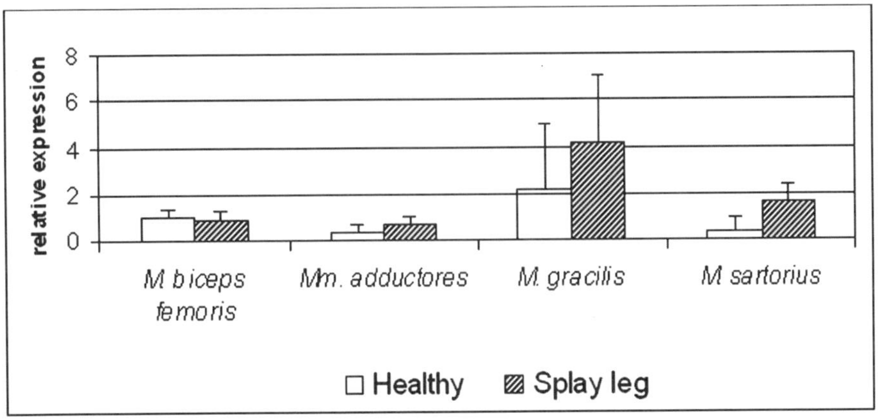

Fig. 1: Relative transcript abundance of *MAFbx* in different muscles of healthy and splay leg piglets (n = 3 vs n = 3 for *Mm. adductores, gracilis* and *sartorius*; n = 6 vs. n = 6 for *M. biceps femoris* from a previous study).

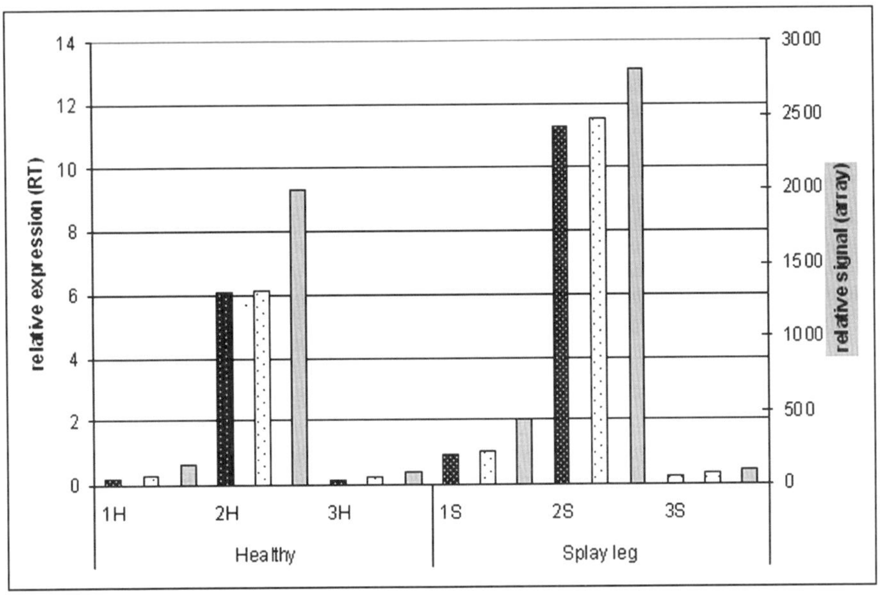

Fig. 2: Comparison of the transcript abundance of *MAFbx* measured by real-time RT-PCR with primers *MAFbx1* (dotted black bars), and *MAFbx2* (dotted white bars) and by GeneChip® Porcine Genome Array (grey). Identical numbers on the x-axis indicate sib-pairs.

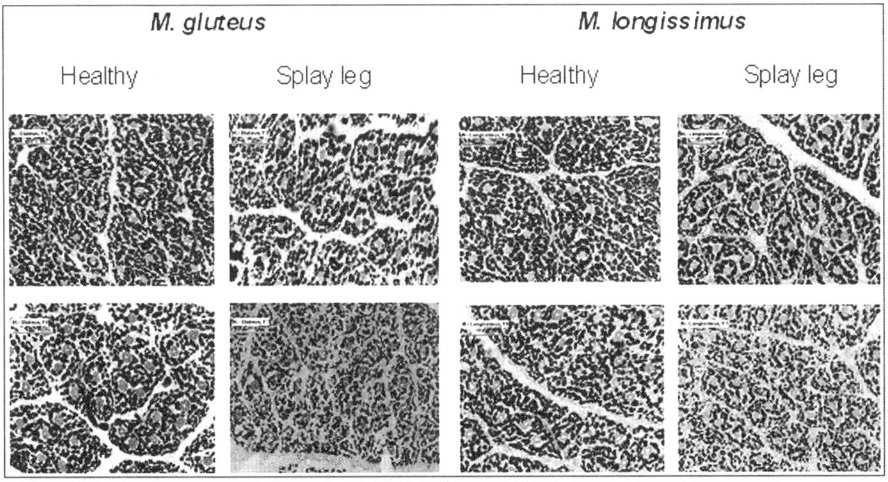

Fig. 3: Comparison of cross sections of *M. gluteus* and *M. longissimus* of healthy and splay leg piglets. (myofibrillar ATPase staining, pH 9.4). Each line represents a sib-pair.

Table 2: Comparison of biochemical traits in *M. gluteus* and *M. longissimus* of healthy (H) and splay leg (S) piglets.

Trait	n	*M. gluteus*	n	*M. longissimus*
CPK (IU/mg protein)	H (4)	6.72 ± .59	H (2)	6.80 ± .92
	S (2)	6.16 ± .84	S (2)	5.59 ± .92
DNA (µg/g)	H (4)	2150 ± 189	H (2)	2279 ± 190
	S (3)	2258 ± 218	S (3)	2541 ± 155
Protein (µg/g)	H (4)	70.29 ± 7.28	H (2)	74.40 ± 7.04
	S (3)	61.58 ± 8.41	S (3)	83.66 ± 5.75
DNA/Protein (µg/mg)	H (4)	30.82 ± 1.80	H (2)	30.67 ± 3.53
	S (3)	37.39 ± 2.08	S (3)	30.80 ± 2.88

Discussion

The results of this study do not support fibre atrophy as the cause of congenital splay leg in piglets as previously proposed [8]. Neither consistent differences in *MAFbx (FBXO32)* transcript abundance nor a change in the expression of other markers for atrophic conditions of skeletal muscle as previously described [12] were found. There, a common pattern of gene expression in atrophic muscle from different pathological conditions including a strong increase in *MAFbx* expression was demonstrated. This is in contrast to the results presented here. Quantitative trait loci (QTL) for the disease were mapped to porcine chromosomes 3, 5, and 11 [13]. *MAFbx* and *P311* are located on porcine chromosomes 4 and 2, respectively and therefore, may be excluded as positional candidates for the disease. The histological data presented here demonstrates alterations in skeletal muscles of splay leg piglets like reduced fibre density and partly decreased size and number of primary and secondary fibres. However, these changes are not consistent between litters and even splay leg samples without obvious changes were observed. The effect of the litter of the piglets observed in this study was much larger than differences between healthy and splay leg piglets. Therefore, we conclude that fibre atrophy is not a general feature of congenital splay leg in piglets and that *MAFbx* expression is of limited value as a marker for the disease.

References

1 Holl JW, Johnson RK: Incidence of splayleg pigs in Nebraska litter size selection lines. J Anim Sci 2005;83:34-40.

2 Boettcher D, Paul S, Bennewitz J, Swalve HH, Thaller G, Maak S: Exclusion of NFYB as candidate gene for congenital splay leg in piglets and radiation hybrid mapping of further five homologous porcine genes from human chromosome 12 (HSA12). Cytogenet Genome Res 2007;118:67-71.

3 Bergmann V: Electronmicroscopic findings in the skeletal muscles of newborn swine with congenital splayleg. Arch Exp Veterinarmed 1976;30:239-260.

4 Tucek S, Svoboda T, Ricny J, Bass A, Soukup T, Vitek V: The concentration of choline and the activities of cholinesterases, creatine kinase and lactate dehydrogenase in the blood plasma of piglets with the syndrome of splayleg (congenital myofibrillar hypoplasia). Zentralbl Veterinarmed A 1985;32:1-10.

5 Szalay F, Zsarnovsky A, Fekete S, Hullar I, Jancsik V, Hajos F: Retarded myelination in the lumbar spinal cord of piglets born with spread-leg syndrome. Anat Embryol 2001;203:53-59.

6 Christison GI, Lewis NJ, Bayne GR: Effects of farrowing crate floors on health and performance of piglets and sows. Vet Rec 1987;121:37-41.

7 Dobson KJ: Failure of choline and methionine to prevent splayleg in piglets. Aust Vet J 1971;47:587-590.

8 Ooi PT, da Costa N, Edgar J, Chang KC: Porcine congenital splayleg is characterised by muscle fibre atrophy associated with relative rise in MAFbx and fall in P311 expression. BMC Vet Res 2006;2:23.

9 User Bulletin #2: http://www3.appliedbiosystems.com/cms/groups/mcb_support /documents/generaldocuments/cms_040980.pdf

10 Brooke MH, Kaiser KK: Muscle fiber types: How many and what kind? Arch Neurol 1970;23:369-379

11 Rehfeldt C, Adamovic I, Kuhn G: Effects of dietary daidzein supplementation of pregnant sows on carcass and meat quality and skeletal muscle cellularity of the progeny. Meat Sci 2007;75:103-111.

12 Lecker SH, Jagoe RT, Gilbert A, Gomes M, Baracos V, Bailey J et al: Multiple types of skeletal muscle atrophy involve a common program of changes in gene expression. FASEB J 2004;18:39-51.

13 Schwarz S: Kartierung von Loci für kongenitales Ausgrätschen beim neugeborenen Ferkel. Thesis, Kiel, Germany. http://e-diss.uni-kiel.de/diss_827/ (2002).

Dr. S. Maak, Research Unit Muscle Biology and Growth, Research Institute for the Biology of Farm Animals (FBN), W.-Stahl-Allee 2, D-18196 Dummerstorf, Germany.

E-mail: maak@fbn-dummerstorf.de

Pinard M-H, Gay C, Pastoret P-P, Dodet B (eds): Animal Genomics for Animal Health. Dev Biol (Basel). Basel, Karger, 2008, vol 132, pp 307-313.

Organization of D_H-Gene Locus is Distinct in Cattle

M. Koti[1], G. Kataeva[2], A.K. Kaushik[1]

[1] Department of Molecular and Cellular Biology, University of Guelph, Guelph, Ontario, Canada.

[2] Mobix Lab, McMaster University, Hamilton, Ontario Canada

Keywords: bovine, D_H gene, bovine D gene locus, D gene organization

Abstract: Some cattle antibodies are amongst the largest known to exist, in a species where CDR3 of the heavy-chain variable region (CDR3H) is of exceptional size as it may extend up to 61 amino acids. The origin of such an exceptionally long CDR3H in cattle antibodies is not yet understood. For these reasons, we have completely characterized D_H gene locus in a Holstein cow. In contrast to other species, such as mice and humans, the D_H gene locus in cattle, extending over 68 kb, is organized in subclusters comprising two to four D_H gene segments. Analysis of D_H genes reveals the presence of characteristic repetitive GGT and TAT codons. No evidence for a D_H gene segment longer than 148 base pairs has been noted. Therefore, mechanistic factors contribute significantly to CDR3H size generation. Such comparative genomics perspectives of bovine D gene loci have revealed the complexity of evolution, across animal phyla, of functional significance in the generation of antibody diversity.

INTRODUCTION

Significant advances have been made to contemporary understanding of the genetic elements and the mechanisms involved in creating the enormous antibody diversity essential to host defence. The jawed vertebrates, from shark to humans, generate clonally diverse lymphocytes via rearrangement of segmental genetic elements that encode B cell receptors mediated by recombinase-activating proteins RAG-1 and RAG-2 [1].

Comparative genomics approaches have revealed that the basic mechanistic components, such as RAG-1/RAG-2 homologues and DNA repair factors, involved in immunoglobulin (Ig) diversification are conserved across the deuterostome split [2]. While such comparative knowledge is based on a few more than 30 extant animal phyla, additional knowledge from these and other phyla is likely to reveal complexity in the evolution of immunity and immune receptor diversification. Despite the fact that processes of $V(D_H)J$ recombination in generating immunoglobulin B cell antigen receptor are common to all vertebrates, differences in diversification processes exist across species that hinge upon factors such as germline sequence divergence and placental and/or colostral transfer of immunoglobulin (Fig. 1). Significant germline-encoded combinatorial diversity in humans and mice results from a number of divergent germline variable (V)-region genes (VH, Vκ or Vλ). In contrast, species like the chicken [3], rabbit [4], pig [5], cow [6] and sheep [7], express restricted combinatorial diversity due to limited germline sequence variation in the V-region genes. Thus, antibody diversification mechanisms differ across species, for example, gene conversion in the chicken [3] and rabbit [4], and antigen-independent somatic hyper-mutations in sheep [8]. Further, it has been suggested that select members of gut microflora drive the gut-associated lymphoid tissue (GALT) development in rabbits, apart from gene conversion and antigen-independent somatic hyper-mutations [9].

Fig.1: Antibody diversification mechanisms in vertebrates.

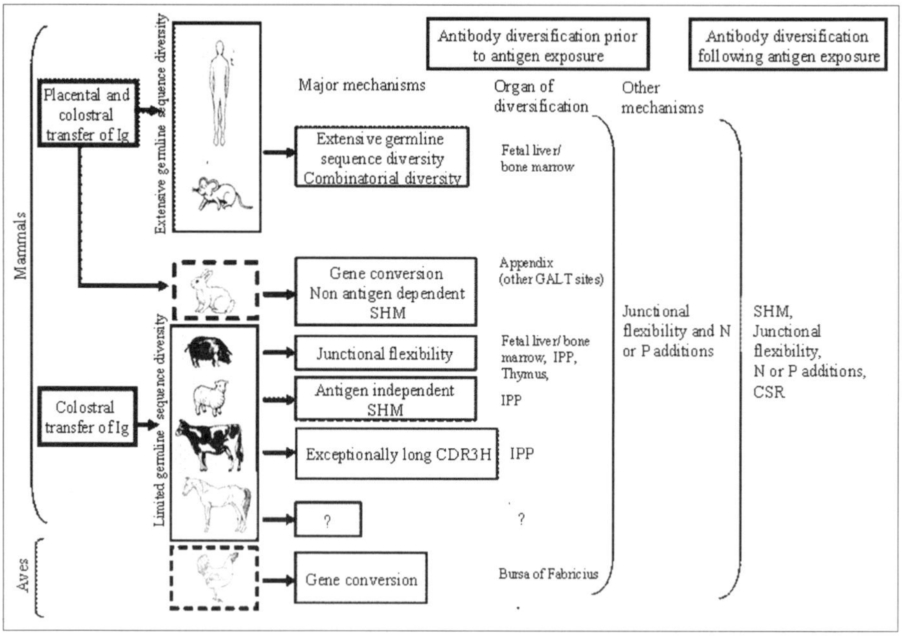

The bovine immune system develops in the absence of influences from maternal antibodies (Fig. 1) since syndesmochorial placentation prevents the trans-placental passage of maternal Ig to the developing foetus [10]. Given the limited germline sequence divergence at both variable heavy- [6] and λ light-chain [10, 11] loci, some cattle antibodies diversify, unlike other species, by generating an exceptionally long CDR3H (up to 61 amino acids) with multiple cysteine residues [10]. Such an atypical CDR3H in cattle antibodies develops early during B cell ontogenesis [12]. This provides a novel mechanism of antibody diversification in cattle where an antigen-combining site of hitherto unknown conformational diversity is created via possible inter- or intra-CDR3H disulfide bridging between even numbered cysteine residues. These bovine antibodies with exceptionally long CDR3H preferentially pair with λ-light chains, exclusively encoded by closely related Vλ1x, Vλ1d and Vλ1e genes, that seem to provide conformational support to the variable heavy-chain domain rather than directly contributing to epitope recognition [11].

The presence of an exceptionally long sized CDR3H in cattle antibodies in contrast to other vertebrate species (Table 1) raises questions about its origin. Partial characterization of a 2.3 kb genomic fragment corresponding to the bovine D_H gene locus revealed the presence of two short (14-19 codons) and an unusually long D_H gene directly capable of encoding 49 codons [13]. We have now completely characterized the D_H gene locus in a Holstein cow and demonstrated that the organization of the bovine D_H gene locus is distinct form other species, with D_H genes arranged in subclusters, and thus provides a different perspective to the evolution of D gene loci across phyla.

Table 1: CDR3H size and CDR3H diversification across species.

Species	CDR3H Size (codons)	Diversification Mechanism		
		N*	P**	D-D fusion
Human	févr-26	+	+	+
Mouse	févr-19	+	+	+
Rabbit	avr-19	+	+	?
Sheep	mars-23	+	+	?
Cattle	mars-61	?	?	?
Pig	mars-25	+	+	?
Camel	oct-24	+	+	?
Chicken	15-30	?	+	+

*N nucleotide addition; **P nucleotide addition

Materials and Methods

A bovine genomic DNA library, in lambda FIX II vector, with an insert size ranging from 9-23 kb, prepared from the liver of a 2 year-old Holstein cow (Stratagene, La Jolla, USA) was screened for detecting bovine D_H genes. Luria Bertani (LB) broth supplemented with 0.2% maltose and 10 mM $MgSO_4$ was inoculated with the XL-1 Blue MRA-P2 strain of *E. coli* (Stratagene, La Jolla, USA) and grown overnight at 30°C. The bacterial cells (200 µl) were diluted to OD 600 = 0.5 in 10 mM $MgSO_4$ and infected with one µl of serially diluted lambda phage. The mixture was incubated at 37°C for 15 min followed by the addition of 3 ml of NZY top agar (48°C) and poured on NZY agar plates. The plates were incubated at 37°C overnight and plaques were counted. The titre of the phage library was calculated to be 1.36×10^{10} plaque forming units (pfu)/ml. A total of 1×10^6 phage clones of the bovine genomic library were plated for screening with D_H gene specific probe. The plaques were transferred onto a nitrocellulose membrane (Boehringer Mannheim, Germany) for 2 min. and denatured in 1.5 M NaCl, 0.5 M NaOH for 2 min, followed by neutralization in 1.5 M NaCl, 0.5 M Tris-HCl for 5 min. The membranes were rinsed in 0.2 M Tris-HCl (pH 7.5), 2X SSC and the phage DNA was crosslinked in a UV crosslinker (Stratalinker; 12,000 µJ) for 30 s. The membranes were pre-hybridized (using 50% formamide, 5X Denhardt's solution, 5X SSPE, 0.5% SDS, and 100 µg/ml sonicated and denatured salmon sperm DNA) overnight at 42°C. The 2.3 kb D_H gene specific probe was prepared by digesting the plasmid pBGD14s [13] with *Eco*RI. The membranes were hybridized with the [α-32P]dCTP labeled 2.3 kb D_H gene specific probe overnight at 42°C. The membranes were washed and exposed to XAR-5 film (Eastman Kodak Company, Rochester, NY, USA) for 24 h at -70°C. Ten positive plaques were identified upon secondary screening and their stocks were stored at -70°C until used. The phage DNA was purified (Wizard Lambda kit, Promega, USA) and subjected to automated DNA sequencing using forward and reverse internal primers. The DNA sequences were analyzed using the DNA Star program (South Park St. Madison, USA).

Results and Discussion

Complete screening of the bovine genomic library originating from a Holstein cow upon hybridization with a DNA probe specific to bovine D_H genes (Fig. 2) led to the isolation of three distinct lambda phage clones, BDG2-4, BDG1-2 and BDG5-3. While the BDG2-4 and BDG1-2 clones showed distinct restriction patterns, BDG5-3 was identical to the 2-11-C clone (Fig. 3). All three clones, BDG5-3/2-11-C, BDG1-2, and BDG2-4, were subjected to DNA sequencing and analyzed for the presence of D_H genes. Complete sequencing of the BDG5-3/2-11-C phage clone did not reveal the presence of any additional D genes other than the cluster containing the D_{H1}, D_{H2} and D_{H3} genes (Fig. 4) identified earlier [13]. Further, DNA sequencing revealed that BDG1-2 and 2-11-C were overlapping clones resulting in the assembly of approximately 16 kb of nucleotide sequence of the bovine D_H gene locus. The BDG2-4 clone spanning ~18 kb was completely sequenced, and had another D gene subcluster comprising four D gene segments. Thus, seven DH gene segments were identified upon complete screening ($1x10^6$ phage clones representative of 99% of the genome) of the bovine genomic library originating from a single Holstein cow. Surprisingly, a single long D_{H2} gene [13], potentially capable of encoding 49 codons, is present in Holstein cattle and is responsible for contributing to large sized CDR3H in cattle antibodies. The presence of D genes in subclusters at the cattle D gene locus is distinct from other species, such as mice [14] and humans [15]. This is relevant to the origin and evolution of the D gene locus in the context of each species across animal phyla. These observations extend our previous studies involving the bovine D_H locus at chromosome 21q23-24 in which both long and short D_H genes with relatively short intervening introns were identified [13]. The experiments outlined here have led to complete characterization of the bovine D_H gene locus in a Holstein cow in which the structure and organization of bovine D_H genes is distinct from other species, since they are organized in subclusters spread over a genomic region of over 68 kb. The bovine D gene locus is highly polymorphic; here, seven DH gene segments were identified in a single Holstein cow, and genomic complexity is likely to vary across cattle breeds. The bovine DH2 represents the single longest D_H gene segment capable of directly encoding 49 codons with characteristic repetitive GGT (glycine) and TAT (tyrosine) codons. This however, does not explain the origin of an exceptionally long CDR3H in some cattle antibodies where it may extend up to 61 codons [10] and it is likely that other mechanistic factors contribute to CDR3H length. Although the organization of the bovine D_H locus in subclusters is distinct from other species, it is likely that gene duplication events contributed to its expansion and polymorphism.

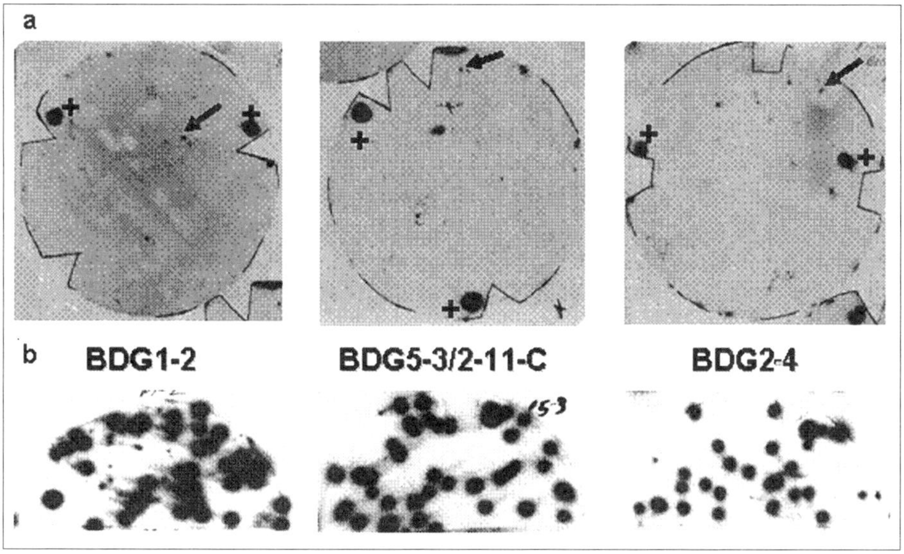

Fig. 2: Primary (a) and tertiary screening (b) of bovine genomic library screened with D-gene specific probe. Arrow indicates a positive lambda phage clone.

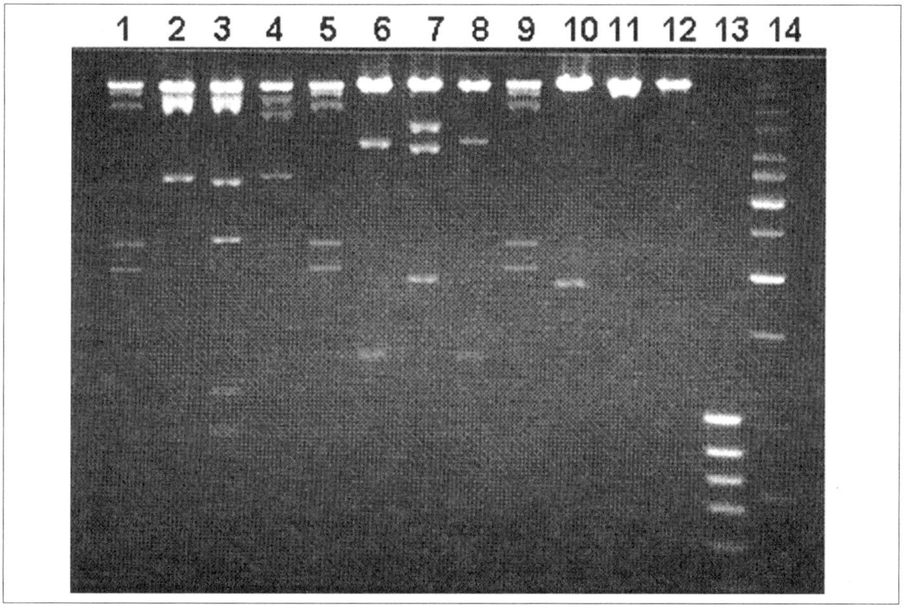

Fig. 3: Restriction digestion pattern demonstrates two distinct clones when compared to previously isolated 211C clone [13].
Lane 1, 5 and 9: Lambda DNA marker; **Lane 2:** *Bgl*II digested BDG1-2 DNA; **Lane 3:** *Bgl*II digested BDG2-4 DNA; **Lane 4:** *Bgl*II digested 211C DNA; **Lane 6:** *BamHI* digested BDG1-2 DNA; **Lane 7:** *BamHI* digested BDG2-4 DNA; **Lane 8:** *BamHI* digested 211C DNA; **Lane 10:** *Eco*RI digested BDG1-2 DNA; **Lane11**: *Eco*RI digested BDG2-4 DNA; **Lane 12:** *Eco*RI digested 211C DNA; **Lane 13:** 100bp DNA marker; **Lane 14:** 1kb DNA marker.

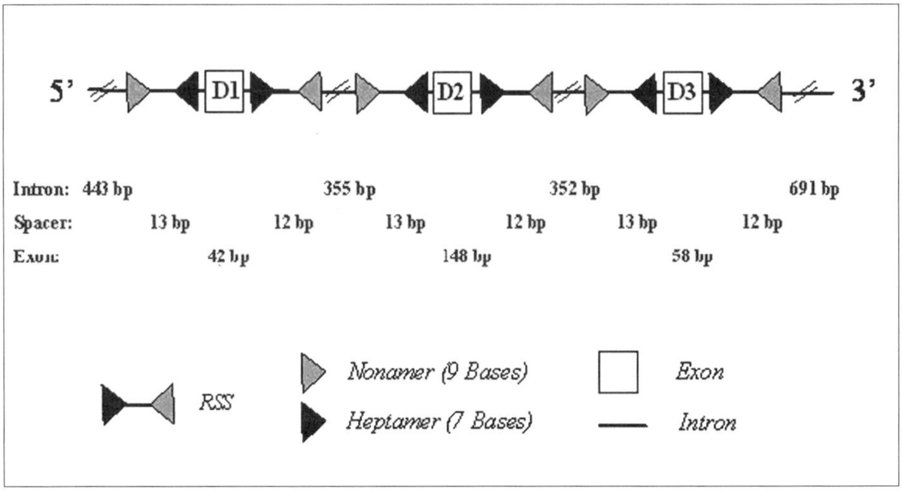

Fig. 4. Bovine D-gene subcluster showing the recombination signal sequence together with intervening introns.

Acknowledgements

The authors would like to thank Mobix Laboratory, McMaster University, Hamilton, Ontario for lambda DNA sequencing support. This research was supported by NSERC discovery and OMAF grants to Dr. Azad K. Kaushik.

References

1 Litman GW, Cannon JP, Dishaw LJ: Reconstructing immune phylogeny: new perspectives. Nat Rev Immunol 2005;5:174-180.

2 Fugmann SD, Messier C, Novack, LA, Cameron RA, Rast JP: An ancient evolutionary origin of the Ragl/2 gene locus. Proc Natl Acad Sci 2006;103:3728- 3733.

3 Reynaud CA, Dahan A, Anquez V, Weill JC: Somatic hyperconversion diversifies the single Vh gene of the chicken with a high incidence in the D region. Cell 1989;59:171-183.

4 Becker RS, Knight KL: Somatic diversification of immunoglobulin heavy chain VDJ genes: evidence for somatic gene conversion in rabbits. Cell 1990;63: 987-997.

5 Sun J, Kacskovics I, Brown WR, Butler JE: Expressed swine VH genes belong to a small VH gene family homologous to human VHIII. J Immunol 1994;153:5618-5627.

6 Saini SS, Hein WR, Kaushik A: A single predominantly expressed polymorphic immunoglobulin VH gene family, related to mammalian group, I, clan, II, is identified in cattle. Mol Immunol 1997:34:641-651.

7 Dufour V, Malinge S, Nau F: The sheep Ig variable region repertoire consists of a single VH family. J Immunol 1996;156: 2163-2170. 8. Reynaud CA, Anquez V, Weill JC: The chicken D locus and its contribution to the immunoglobulin heavy chain repertoire. Eur J Immunol 1991;21:2661-2670.

8 Reynaud CA, Anquez V, and Weill JC: The chicken D locus and its contribution to the immunoglobulin heavy chain repertoire. Eur J Immunol 1991;21:2661-2670.

9 Lanning DP, Sethupathi KJ, Rhee S, Zhai K, Knight KL: Intestinal microflora and diversification of the rabbit antibody repertoire. J Immunol 2000;165:2012-2019.

10 Saini SS, Allore B, Jacobs RM, Kaushik A: Exceptionally long CDR3H region with multiple cysteine residues in functional bovine IgM antibodies. Eur J Immunol 1999;29:2420-2426.

11 Saini SS, Farrugia W, Ramsland PA, Kaushik AK: Bovine IgM antibodies with exceptionally long complementarity-determining region 3 of the heavy chain share unique structural properties conferring restricted VH + Vlambda pairings. Int Immunol 2003;15:845-853.

12 Saini SS, Kaushik A: Extensive CDR3H length heterogeneity exists in bovine foetal VDJ rearrangements. Scand J Immunol 2002;55:140-148.

13 Shojaei F, Saini SS, Kaushik AK: Unusually long germline DH genes contribute to large sized CDR3H in bovine antibodies. Mol Immunol 2003;40:61- 67.

14 Kompfner E, Oliveira P, Montalbano A,.Feeney AJ: Unusual germline DSP2 gene accounts for all apparent V-D-D-J rearrangements in newborn, but not adult, MRL mice. J Immunol 2001;167:6933-6938.

15 Ichihara Y, Matsuoka H, Kurosawa Y: Organization of human immunoglobulin heavy chain diversity gene loci. EMBO J 1988;7:4141-4150.

Dr. Azad K. Kaushik, DVM, MVSc, DSc (Paris), Associate Professor of Immunology, Department of Molecular and Cellular Biology University of Guelph, Room SCIE3504, Guelph, Ontario, N1G 2W1 Canada.

E-mail: akaushik@uoguelph.ca

Pinard M-H, Gay C, Pastoret P-P, Dodet B (eds): Animal Genomics for Animal Health. Dev Biol (Basel). Basel, Karger, 2008, vol 132, pp 315-320.

Differential Gene Expression of High and Low Immune Responder Canadian Holstein Dairy Cows

M.I. Nino-Soto[1]=, A. Heriazón[1]=, M. Quinton[2], F. Miglior[3], K. Thompson[1], B.A. Mallard[1]

[1] Department of Pathobiology, Ontario Veterinary College and 2Department of Animal and Poultry Science, Centre for Genetic Improvement of Livestock, University of Guelph, Guelph, Ontario, Canada
[3] Canadian Dairy Network, Guelph, Ontario, Canada
= Equal contribution of the authors

Keywords: immune response, phenotype, gene expression, cDNA microarrays, Holstein cattle

Abstract: Infectious diseases are an important cause of economic loss in the agri-food business. This study investigates indicators of bovine high (HR) and low (LR) immune response and their associated patterns of gene expression. Holstein cows were immunized to induce antibody (AMIR) and cell-mediated (CMIR) immune responses. Based on the results of enzyme-linked immunosorbent assay (ELISA) and delayed-type hypersensitivity (DTH), cows were ranked as HR, LR or average (AR) immune responders. For microarray analysis, phenotypic HR and LR status in both groups was confirmed and total RNA from blood mononuclear cells (BMCs) was obtained. RNA from a pool of AR cows was used as a common reference for hybridization to an in-house cDNA microarray. Results of microarray analysis showed transcriptional differences in several immune-related genes between the HR and LR groups. Genes identified as differentially expressed include transcription factors, cytokines, MHC, and TCR-related genes. These results can aid in the establishment of selection programmes based on broad-based disease resistance, aimed at improving general health in cattle herds.

INTRODUCTION

Infectious diseases are an important cause of economic loss in the livestock business [1-4]. Because the immune system is involved in the prevention and control of infectious diseases, measurable phenotypic immune response traits have been proposed as candidates for improving overall health through genetic selection [5,6]. However, the genetic profiles associated with these phenotypes have not been fully explored. Since the immune response is controlled by many genes, it is likely that suitable candidates for disease resistance can be found using genomic approaches [7,8]. Microarray technology allows examination of the transcriptional activity of hundreds or thousands of genes simultaneously, facilitating the comparison of gene expression patterns between different phenotypes and the search for candidate genes [9]. The objectives of the present study were to streamline a method to classify the phenotypic immune response of dairy cattle and describe the patterns of gene expression between the groups using microarray technology.

Materials and Methods

A total of 182 lactating Canadian Holstein cows were immunized intramuscularly (im) with specific antigens known to induce antibody-mediated (AMIR) and cell-mediated (CMIR) immune responses, respectively. The antigen-specific IgG AMIR response was measured by ELISA using sera collected on days 0, 14 and 21. To evaluate CMIR using DTH, the cows were sensitized intradermally (id) in the neck on day 21 with specific CMIR-inducing antigens and phosphate saline solution as control. Double skin-fold thickness measurements were taken before sensitization and at 24 and 48 hours post-injection. These results were analyzed using SAS v 9.0. Cows were classified as high (HR), low (LR) or average (AR) phenotypes for each trait using the mean ± 1 standard deviation (SD) of each trait [10]. Cows above 1 SD were considered HR while cows below 1 SD were considered LR. Cows between ± 1 SD were considered AR. For microarray analysis, 20 HR and LR cows (n = 5 per HR-AMIR, LR-AMIR, HR-CMIR and L-CMIR groups) were selected and their status reassessed. Only cows with confirmed phenotypic status were analyzed further (n = 3 per group). Blood mononuclear cells (BMCs) were isolated and total RNA was extracted. Total RNA from AR cows (n = 12) was pooled and used as a common reference for microarray hybridizations. Total RNA was treated with Dnase I and amplified. Microarray hybridizations were performed using a common reference design with dye swap and four technical replicates per animal. The microarray platform was submitted to the GEO database (GPL5172). Amplified RNA was reverse transcribed and indirectly labelled. Slides were scanned and datasets were analyzed using SAS v 9.0. Analysis included array-specific normalization using the lowess algorithm, data filtering and analysis using a two-step mixed model approach [11].

Results

Cows were classified in the different groups as shown in Table 1. The majority of cows were AR for both AMIR and CMIR. As anticipated from the statistical approach, the number of cows in both HR-AMIR or CMIR and LR-AMIR or CMIR was in the range of 7 – 9% of examined animals. Results of the microarray analysis are summarized in Tables 2 and 3. The magnitude (expressed in fold change) of differential expression was larger for genes with increased transcription than for genes with lower transcriptional activity. Cows with the HR phenotype for either AMIR or CMIR showed the higher number of differentially expressed genes. In the HR groups, increased transcriptional activity was detected for genes associated with stimulation of T cells, including *CD69* and genes of the TCR/CD3 complex. On the other hand, the HR groups showed lower transcription of genes associated with inflammation. Cows classified as LR-AMIR or LR-CMIR showed only a few genes with differential transcriptional activity compared to the AR common reference.

Table 1: Percentage of cows by phenotypic immune response group.

	HR	AR	LR
AMIR	7.2%	85.0%	7.8%
CMIR	9.4%	82.5%	8.1%

Table 2: Genes with higher transcriptional activity in cows classified by phenotypic antibody (AMIR) or cell-mediated (CMIR) immune responses when compared to a common reference control of average cows.

Gene symbol	Gene name	Fold change	p-value
High AMIR			
CD69	CD69 molecule (early T-cell activation antigen)	3.9	< 0.0001
TRA@	T-cell receptor, alpha chain	3.4	< 0.0001
TRD@	T-cell receptor, delta chain	3.4	< 0.0001
CD3D	T-cell receptor CD3, delta chain	3.5	< 0.0001
CD3E	T-cell receptor CD3, epsilon chain	3.2	< 0.0001
CD3G	T-cell receptor CD3, gamma chain	3.0	< 0.0001
CD247	CD247 molecule (T-cell receptor CD3, zeta chain)	3.0	< 0.0001
SOD	Superoxide dismutase	2.9	< 0.0001
Low AMIR			
CD69	CD69 molecule (early T-cell activation antigen)	3.8	< 0.0001
High CMIR			
CD69	CD69 molecule (early T-cell activation antigen)	3.1	< 0.0001
CD3D	T-cell receptor CD3, delta chain	3.6	< 0.0001
CD3E	T-cell receptor CD3, epsilon chain	3.2	< 0.0001
CD3G	T-cell receptor CD3, gamma chain	3.2	< 0.0001
TRD@	T-cell receptor, delta chain	3.4	< 0.0001
TRB@	T-cell receptor, beta chain	3.2	< 0.0001
TRA@	T-cell receptor, alpha chain	2.9	< 0.0001
Low CMIR			
IGHG1	immunoglobulin heavy constant gamma 1	3.1	< 0.0001
IGHG2A	immunoglobulin heavy constant gamma 2A	2.9	< 0.0001
MIC	Major Histocompatibility Complex class I (BoLA)	2.8	< 0.0001

Discussion

Significant AMIR and CMIR responses were observed at all tested points of time. Phenotypic classification as HR, AR and LR was possible using the mean ± 1 SD approach [10] with the objective of identifying differential gene expression between HR and LR.

In this study, transcriptional profiles associated with the previously defined immune response phenotypes using a common reference design were explored. Both HR response groups (AMIR and CMIR) showed the largest number of genes with increased transcriptional activity compared to the common reference, with fold changes ranging between 2.8 and 3.9 (Table 2). Most genes were associated with the TCR/CD3 complex including both αβ and γδ T-cell receptors, although for the latter, only TRD@ was found to be significantly higher. In addition, *CD69*, an early activator of T cells was also found to be elevated in these groups. These genes are associated with T cell activation which is not surprising considering that these cows have been repeatedly stimulated with the specific test antigens. *CD69* is considered to be a very early T cell activator, with mRNA levels that drop within 8 h post-stimulation [12]. As most studies have been performed with cell culture and stimulation *in vitro*, it is possible that what was detected here is a different pattern of expression in vivo, likely driven by repetitive antigenic stimulation.

Cows in the low immune response groups showed very few genes with higher transcriptional activity than the AR common reference. In the LR-AMIR group, only *CD69* showed higher transcript levels, which were also inducibly expressed by immature thymocytes, B cells, natural killer cells, monocytes, neutrophils and eosinophils [12,13]. Since this incremental change in transcription was not accompanied by increased expression of any other gene associated with T cells, as in the HR groups, it is possible that these other cell types are contributing to the increase.

Very few genes were identified as having lower transcriptional activity than the common reference (Table 3). Most of these genes, such as *TLR2, CASP1* and *IL1B* are related to the inflammatory response. These results suggest that the extreme immune response phenotypes identified in this study show higher transcriptional activity of genes associated with lymphocyte activation and lower transcriptional activity of genes associated with inflammation. This may relate to the time frame post-immunization that gene expression was examined. Further characterization of transcriptional activity is required to confirm these findings. Future studies will also examine gene expression of cows identified as high or low responders based on their Estimated Breeding Value classification.

Conclusion

It was possible to identify and classify both high and low AMIR and CMIR cows using phenotypic indicators. Microarray analysis results showed differential gene expression patterns between the phenotypic immune response groups and the common AR reference mostly for genes related to T cell activation and inflammation. These findings contribute to elucidating the mechanisms that underpin high and low immune responsiveness in dairy cattle and in the future could be use in conjunction with selection programmes to improve animal health.

Table 3: Genes with lower transcriptional activity in cows classified by phenotypic antibody (AMIR) or cell-mediated (CMIR) immune responses when compared to a common reference control of average cows.

Gene symbol	Gene name	Fold change	p-value
High AMIR			
TLR2	Toll-like receptor 2	1.6	< 0.0001
High CMIR			
IL1B	Interleukin 1 beta	1.6	< 0.0001
CASP13	Caspase 13	1.6	< 0.0001
CASP1	Caspase 1	1.6	< 0.0001
STAT1	Signal Transducer and Activator of Transcription 1	1.6	< 0.0001
Low CMIR			
IL8	Interleukin 8	1.5	< 0.0001
NLRP3	NLR family, pyrin domain containing 3	1.5	< 0.0001

Acknowledgments

We want to acknowledge the assistance of the personnel at the Elora and Ponsonby Research Stations, Sophia Lim for technical support and William Sears for statistical analysis.

References

1 Collins JD, Wall PG: Food safety and animal production systems: controlling zoonoses at farm level. Rev Sci Tech 2004;23(2):685-700.

2 Rajala-Schultz PJ, Grohn YT, McCulloch C.E, Guard C: L. Effects of clinical mastitis on milk yield in dairy cows. J Dairy Sci 1999;82(6):1213-1220.

3 Wierup M: The control of microbial diseases in animals: alternatives to the use of antibiotics. Int J Antimicrob Agents 2000;14(4):315-319.

4 Zessin KH: Emerging diseases: a global and biological perspective. J Vet Med B Infect Dis Vet Public Health 2006;53 Suppl 1:7-10.

5 Pinard MH, van Arendonk J.A, Nieuwland MG, van der Zijpp AJ: Divergent selection for immune responsiveness in chickens: estimation of realized heritability with an animal model. J Anim Sci 1992;70(10):2986-2993.

6 Puel A, Mevel JC, Bouthillier Y, Feingold N, Fridman WH, Mouton D: Toward genetic dissection of high and low antibody responsiveness in Biozzi mice. Proc Natl Acad Sci USA 12-10-1996;93(25):14742-14746.

7 Abbas AK, Janeway CA, Jr. Immunology: Improving on nature in the twenty-first century. Cell 1-7-12000;100(1):129-138.

8 Hernandez A, Karrow N, Mallard BA: Evaluation of immune responses of cattle as a means to identify high or low responders and use of a human microarray to differentiate gene expression. Genet Sel Evol 2003;35 Suppl 1:S67-S81.

9 Olson NE: The Microarray Data analysis process: from raw data to biological significance. NeuroRx 2006;3(3):373-383.

10 Wagter LC, Mallard BA, Wilkie BN, Leslie KE, Boettcher PJ, Dekkers JC: A Quantitative approach to classifying Holstein cows based on antibody responsiveness and Its relationship to peripartum mastitis occurrence. J Dairy Sci 2000;83(3):488-498.

11 Madsen SA, Chang LC, Hickey MC, Rosa GJ, Coussens PM, Burton JL: Microarray analysis of gene expression in blood neutrophils of parturient cows. Physiol Genomics 1-15-2004;16(2):212-221.

12 Ahn JS, Hamilton MJ, Davis WC, Park YH: Cloning of bovine CD69. Vet Immunol Immunopathol. 9-6-2002;88(1-2):43-48.

13 Ziegler SF, Ramsdell F, Alderson MR: The activation antigen CD69. Stem Cells 1994;12(5):456-465.

Dr. B.A. Mallard, Department of Pathobiology, University of Guelph, Guelph, ON N1G 2W1.

E-mail: bmallard@uoguelph.ca

Pinard M-H, Gay C, Pastoret P-P, Dodet B (eds): Animal Genomics for Animal Health. Dev Biol (Basel). Basel, Karger, 2008, vol 132, pp 321-325.

cDNA Microarray Analysis of Gene Expression Patterns in Blood Mononuclear Cells of SLA-DRB1-Defined Yorkshire Pigs

M.I. Nino-Soto[1], R.J. Jozani[2], B. Bridle[1], B.A. Mallard[1]

1 Department of Pathobiology, Ontario Veterinary College, University of Guelph, Guelph, Ontario, Canada
2 Faculty of Veterinary Medicine, University of Tehran, Tehran, Iran

Key words: MHC class II, SLA-DRB1, cDNA microarrays, Yorkshire pig

Abstract: Three lines of commercial Yorkshire pigs with defined SLA-DRB1 alleles were developed at the University of Guelph for xenotransplantation and immune response studies. Two of the SLA-DRB1 alleles have been previously reported (SLA-DRB1*0502 and *0701), whereas the third one is a new allele. The influence of defined SLA-DRB1 alleles on transcriptional patterns of immune-related genes in blood mononuclear cells (BMCs) of pigs was explored using cDNA microarray. Microarray analysis showed significant differential expression of inflammatory genes in association with the various SLA-DRB1 alleles. A better understanding of the association between SLA genotypes and gene activity can increase the knowledge of the function of these molecules, as well as define new strategies to control animal health and optimize animal production.

INTRODUCTION

The swine leukocyte antigen (SLA) system is involved in self-nonself discrimination, constitutes a major transplantation barrier and is associated with immune response and disease resistance. The SLA is located on either side of the centromere of chromosome 7 and ~152 loci have been annotated within the region, including histocompatibility genes as well as "non-MHC genes" involved in natural immunity or of unknown function [1]. In pigs, class II DRB genes show extensive polymorphism and different SLA haplotypes have been associated with variation in immune response and disease, as well as reproduction and production traits. Therefore,

SLA-defined pigs constitute a valuable resource for the study of the immune response, disease resistance and production traits, and provide an important large animal model for biomedical research such as transplantation [2,3]. Three lines of commercial Yorkshire pigs with defined SLA-DRB1 phenotypes were developed at the University of Guelph to study xenotransplantation reactions and immune response. This study explores the association between defined MHC genotypes and transcriptional activity of immune response genes using heterologous cDNA microarray hybridization.

Materials and Methods

Six animals representing three SLA-DRB1 alleles were included. Venous blood was used to isolate BMCs using Histopaque-1077. Total RNA was extracted and treated with DNase I. A previous version of an in-house cDNA microarray [4] was updated to include more genes involved in T cell differentiation and innate host responses. Microarray datasets and platform were submitted to the GEO public repository (GSE7908). Heterologous hybridizations (n = 2 pigs per group, 3 groups) were performed in a loop design with dye-swap on the same slide for a total of 12 microarrays. Total RNA was amplified and indirectly labelled using Alexa Fluor® dyes (Invitrogen) following the manufacturers' instructions. Hybridization was performed on bovine cDNA microarrays. Slides were scanned and data analyzed using the lowess algorithm. The data was filtered and log-ratios of expression were calculated. The Student's t-test with false discovery rate (FDR) correction for multiple comparisons was used for statistical analysis.

Results

Results from the microarray hybridizations are summarized in Table 1. The allele *0502 group showed higher transcriptional activity for macrophage inflammatory protein-1β (*CCL4*) and interleukin 1 beta (*IL1B*) when compared to the two other groups. Pigs expressing the *0701 allele showed less SLA class II DQ alpha (*SLA-DQA*) transcripts when compared to the other groups. Toll-like receptor 2 (*TLR2*) and caspase 1 (*CASP1*) showed higher transcriptional activity in pigs with alleles *0502 and *0701 respectively, when compared to the new allele. All differences were statistically significant at $p \leq 0.01$.

Discussion

Due to the homology between the porcine and bovine genomes, a high level of probe and target hybridization was expected as previously described [4]. Using microarray analysis, differences in transcript amounts for several genes related to inflammation were found (see Table 1).

The genes differentially expressed included *TLR2, IL1B, CASP1* and *CCL4*. Toll-like receptors (TLRs) are involved in the innate immune system, the first line of defence against pathogens, are a crucial link between the innate and adaptive immune systems [5,6]. TLRs are type I integral membrane glycoproteins, closely related to the interleukin 1 receptor (IL1R) and expressed on various immune and non-immune cells. TLR2, together with TLR1 and TLR6, recognizes lipid-containing pathogen-associated molecular patterns and its activation results in the induction of genes involved in inflammatory responses, such as innate defence cytokines IL1B, IL6, IL12 and tumour necrosis factor (TNF) [5,6]. TLRs are also implicated in the pathogenesis of diseases such as atherosclerosis and asthma [5,6]. Monocytes and B lymphocytes express TLRs and respond to their ligands by functional activation. The significance of TLRs in these adaptive immune system cells remains to be elucidated. The expression of both rearranged adaptive and germ-line innate immune receptors can make B cells a liability in terms of non-specific activation through TLRs, leading to the secretion of antibodies that may be of harmful consequence for the host, such as in certain types of autoimmune disorders [6].

IL1B is a member of the IL-1R/TLR superfamily. It is a major mediator of inflammation and, in general, initiates and/or amplifies a wide range of effects associated with innate immunity and host responses to microbial invasion and tissue injury [7]. In addition, IL-1β has effects on the central nervous system, is involved in reproductive function and plays a role in destructive joint and bone disease among other roles. IL1B is translated as a precursor protein that requires intrinsic cleavage to become biologically active. This processing is done by the IL-1-converting enzyme (ICE or CASP1), a member of the inflammatory caspases. The biological activity of IL1B is directly associated with CASP1 activation. Therefore, regulation of CASP1 is closely related to IL1B function. [7]. For example, over-activation of CASP1 in monocytes and aberrant maturation of IL1B are associated with at least three auto-inflammatory syndromes.

CCL4 is an inflammatory chemokine of the CC subfamily that exerts its effects through CC receptor 5 (CCR5) [8,9]. CCL4 can be produced in large amounts by activated monocytes and macrophages, T and B lymphocytes, neutrophils and dendritic cells among others. Its production can be induced by various pro-inflammatory agents and cytokines including LPS, substance P, TNF and IL1B [10].

Leukocyte traffic is highly coordinated and a breakdown of the underlying control mechanisms might contribute to immune suppression and autoimmune disease and in fact, numerous inflammatory cytokines are produced in autoimmune diseases [9]. The farm personnel observed that several of these pigs, specifically in the SLA-DRB1*0502 and *0701 lines, had joint problems similar to the ones described in rheumatoid arthritis (data not shown). It is interesting that all genes shown to be differentially expressed in these two lines are related to the inflammatory response and can be linked to the origin of autoimmune processes. In humans, certain specific HLA-DR/DQ alleles have been found to be involved in the pathogenesis of autoimmune diseases such as rheumatoid arthritis, type I diabetes mellitus and multiple sclerosis among others [11].

Numerous associations have been established in swine between SLA haplotypes and physiological features. Many of these traits are not directly regulated by individual SLA genes and could rather be under the influence of non-classical MHC genes or controlled by downstream pathways yet to be described. It is tempting to treat the associations with physiological traits or disease as primary associations with classical MHC genes, however, the involvement of other closely linked genes, whose variants are in linkage disequilibrium (LD) can not be disregarded [1,12].

Although it is not possible from the results in this study to establish a direct causal relationship between a particular SLA-DRB1 allele and the differential transcription of inflammatory genes, it is undeniable that there is an association, which may be better explained by the characterization of the haplotypes associated with these alleles.

Table 1: Summary of results of cDNA microarray analysis.

Gene	Fold-change [a]	Ratio [b]	p-value [c]
Comparison A: SLA-DRB1 alleles 0502/new allele			
TLR2	2.76	1.468	< 0.0001
CCL4	2.10	1.075	0.0010
IL1B	3.21	1.683	0.0025
Comparison B: SLA-DRB1 alleles 0502/0701			
SLA-DQA	2.46	1.303	0.0038
CCL4	2.41	1.271	0.0158
IL1B	3.57	1.837	0.0079
Comparison C: SLA-DRB1 alleles 0701/new allele			
CASP1	1.79	0.840	0.0010
SLA-DQA	-2.52	-1.335	0.0035

[a] Fold change = $2^{\text{(Log-ratio of intensity)}}$ or $(-1)\ 2^{\text{(Log-ratio of intensity)}}$; [b] median of the lowess-normalized log-ratio of intensity; [c] p-value with FDR correction for multiple comparisons.

Conclusion

This study is the first to report an association between defined SLA-DRB1 alleles and the transcription rates of other genes involved in the immune response, particularly inflammation. These findings support previous studies in their association between MHC haplotypes and inflammation processes, including autoimmune disorders. Studies of gene expression patterns, such as this, provide insights into the biology of the SLA complex that will increase our understanding of physiological and immune responses processes, as well as define new strategies to control animal health.

Acknowledgments

We want to acknowledge the assistance of the personnel at Arkell Research Station, Sophia Lim for technical assistance and William Sears for statistical analysis.

References

1 Ando A, Chardon P: Gene organization and polymorphism of the swine major histocompatibility complex. Anim Sci J 2006;77:127-137.

2 Ho CS, Rochelle ES, Martens GW, Schook LB, Smith DM: Characterization of swine leukocyte antigen polymorphism by sequence-based and PCR-SSP methods in Meishan pigs. Immunogenetics 2006;58(11):873-882.

3 Lunney JK: Advances in swine biomedical model genomics. Int J Biol Sci 2007;3(3):179-184.

4 Tao W, Mallard B, Karrow N, Bridle B: Construction and application of a bovine immune-endocrine cDNA microarray. Vet Immunol Immunopathol 2004;101(1-2):1-17.

5 Mitchell JA, Paul-Clark MJ, Clarke GW, McMaster SK, Cartwright N: Critical role of Toll-like receptors and nucleotide oligomerisation domain in the regulation of health and disease. J Endocrinol 2007;193(3):323-330.

6 Pasare C, Medzhitov R: Toll-like receptors: linking innate and adaptive immunity. Microbes Infect. 2004;6(15):1382-1387.

7 Martinon F, Tschopp J: Inflammatory caspases and inflammasomes: Master switches of inflammation. Cell Death Differ 2007;14(1):10-22.

8 Zlotnik A, Yoshie O, Nomiyama H: The chemokine and chemokine receptor superfamilies and their molecular evolution. Genome Biol. 2006;7(12):243.

9 Moser B, Wolf M, Walz A, Loetscher P: Chemokines: multiple levels of leukocyte migration control. Trends Immunol. 2004;25(2):75-84.

10 Maurer M, von Stebut E: Macrophage inflammatory protein-1. Int J Biochem Cell Biol: 2004;36(10):1882-1886.

11 Bondinas GP, Moustakas AK, Papadopoulos GK: The spectrum of HLA-DQ and HLA-DR alleles, 2006: a listing correlating sequence and structure with function. immunogenetics 2007;59(7):539-553.

12 Dorak MT, Shao W, Machulla HK, Lobashevsky ES, Tang J, Park MH et al: Conserved extended haplotypes of the major histocompatibility complex: further characterization. Genes Immun 2006;7(6):450-467.

Dr. B.A. Mallard, Department of Pathobiology, University of Guelph, Guelph, ON N1G 2W1.
E-mail: bmallard@uoguelph.ca

Pinard M-H, Gay C, Pastoret P-P, Dodet B (eds): Animal Genomics for Animal Health. Dev Biol (Basel). Basel, Karger, 2008, vol 132, pp 327-330.

Opportunities for Collaborative Phenotyping for Disease Resistance Traits in a Large Beef Cattle Resource Population

R.M. Thallman, L.A. Kuehn, M.F. Allan, G.L. Bennett, M. Koohmaraie

US Meat Animal Research Center, USDA-ARS, Clay Center, NE, USA

Key Words: Disease resistance, phenotypes, beef cattle

Abstract: The Germplasm Evaluation (GPE) Project at the US Meat Animal Research Center (USMARC) is planned to produce about 3,000 calves per year in support of the following objectives: identification and validation of genetic polymorphisms related to economically relevant traits (ERT), estimation of breed and heterosis effects among 16 breeds for ERT, and estimation of genetic correlations among ERT and physiological indicator traits (PIT). Opportunities exist for collaboration in the development and collection of PIT phenotypes for disease resistance. Other areas of potential collaboration include detailed diagnosis (identification of disease causing organisms, etc.) of treated animals, collaborative development of epidemiological statistical models that would extract more information from the records of diagnoses and treatments, or pharmacogenetics. Concentrating a variety of different phenotypes and research approaches on the same population makes each component much more valuable than it would be individually.

INTRODUCTION

The USMARC is embarking on an ambitious initiative to study the genetics of disease resistance. The approach is to build a large resource population that is deeply phenotyped for as many traits as possible, with appropriate tissue samples collected at appropriate times for effective and comprehensive study of disease resistance. The general philosophy is that research in the field can proceed more rapidly if samples and phenotypes are readily available for initial studies. Then, based on the initial studies, sampling and phenotype protocols can be adjusted, as needed, to meet specific experimental requirements in subsequent years.

Scientists at USMARC will use this population to conduct research in areas of high priority to our stakeholders, subject to expertise and resources available at USMARC. Initial efforts at USMARC will be focused on bovine respiratory disease (BRD), because it is the disease with the greatest economic impact on the US beef industry [1]. However, the scope of the problem is obviously too great for the project to be conducted entirely by USMARC. Therefore, we are seeking collaboration with laboratories capable of adding value to the initiative.

The objectives of the initiative are to determine the feasibility of genetic change of disease resistance through selection, identify a suite of phenotypic measures that can be used to indicate clinical and subclincal infections, evaluate variation in these indicator measures, and characterize physiological mechanisms involved in infection and immunological response. These objectives will be assessed through a two-pronged approach of quantitative genetics and functional genomics. The quantitative genetics approach involves the generation of the populations, phenotypes, and genotypes leading to estimation of genetic parameters, QTL discovery, and ultimately marker identification. The purpose of the functional genomics approach is to understand pathogen interactions throughout the process of pathogen invasion leading to host infection followed by the cascade of host response.

Materials and Methods

The GPE Project [2] at USMARC is planned to produce about 3,000 calves per year in support of the following objectives: identification and validation of genetic polymorphisms related to economically relevant traits (ERT), estimation of breed and heterosis effects among 16 breeds for ERT, and estimation of genetic correlations among ERT and physiological indicator traits. The ERT include dystocia, survival, growth, carcass, meat quality, feed efficiency, temperament, reproductive, longevity, and maternal traits, as well as records of the diagnoses and treatments of naturally occurring diseases at USMARC. These include bovine respiratory disease, pinkeye, footrot, calf scours, failure of passive transfer of immunity, anaplasmosis, and bluetongue. Other diseases that occur at sufficient frequency at USMARC, or for which we had an occasional outbreak, will also be studied.

The USMARC Disease Resistance Population is a planned subset of the GPE population to be produced by F1 sires via natural service for several years resulting in large paternal half-sib families. It should start at about 600 progeny/yr beginning with the spring 2008 calf crop and increase to about 1,000 progeny/yr by 2010. Phenotyping of this population will be somewhat more invasive (e.g., frequent tissue sampling) than in the other populations described. Feeding of antibiotics will be restricted in this population. Disease-related phenotypes will take higher priority than phenotypes for other trait categories in this population only.

The USMARC Feed Efficiency Population is another subset of the GPE population that consists of animals measured for individual feed intake. It consists of about 2,000 animals through the spring 2007 calf crop and is expected to increase by about 600 to 900 calves per year. At least 300 of these will be measured in a new state-of-the-art feeding facility that allows measurement of feeding behaviour and water consumption in addition to individual feed intake. Additional telemetric technologies for monitoring animal health would be particularly appropriate for study in this population. Serum samples for measuring response to vaccination were collected beginning with the spring 2007 calf crop.

Other cattle populations at USMARC produce about another 3,000 calves per year, all with disease treatment records, DNA samples, and basic weight and production records. Phenotyping for other ERT varies by population. These, and the remainder of the GPE population, can be available for more extensive phenotyping for disease traits, provided the protocols are not too invasive.

About 3,600 USMARC animals, including the current USMARC Feed Efficiency Population and the sires of the USMARC Disease Resistance Population will be genotyped for about 50,000 SNP using the Illumina Infinium BovineSNP50 BeadChip. (Reference herein to any specific commercial products by trade name, trademark, manufacturer, or otherwise, does not necessarily constitute or imply its endorsement, recommendation, or favouring by the United States Government. The views and opinions of authors expressed herein do not necessarily state or reflect those of the United States Government, and shall not be used for advertising or product endorsement purposes.) The USMARC Disease Resistance Population will be extensively genotyped, using state-of-the-art technology, as funding allows.

The plan is to have a variety of different protocols for collecting samples and phenotypes, with less expensive protocols applied to large numbers of animals and the most expensive and invasive protocols applied to fewer individuals.

Disease status should be viewed as an unobservable underlying variable. The objective is that the phenotypes in the following list, when combined properly, would be reasonably effective in predicting disease status. Examples of phenotypes that could be collected at treatment include: veterinary and/or pen rider diagnosis, clinical symptoms, treatment records, temperature, respiration rate, heart rate, pathogen identification and counts from nasal swabs or other relevant tissue samples, immune cell counts and activity, cytokine levels, acute phase protein levels, and antibody titres and activity. Furthermore, lung lesion score may be collected at harvest and necropsy with culture and/or histology of affected tissues performed at death. This list likely needs to be adjusted to be effective and practical, but is provided here as a starting point for discussion. It will only be feasible to apply subsets of the list to most of the populations.

Physiological indicator traits (PIT) are those that are expected to be closely related to physiological processes that are components of disease resistance. In most cases, it should be useful to measure them in all animals in a population, whether sick or not. Ideally, it should be useful to measure them regardless of the level of natural exposure to disease. Because they are related to components, they are expected to have higher heritability than disease resistance itself. Because of higher expected heritability and greater effective numbers of observations than for disease incidence (especially when depending on natural exposure), QTL detection for PIT is likely to be considerably more successful than QTL detection for disease incidence directly. Then, QTL for PIT can be validated for effects on disease incidence in larger populations. The following list of potential PIT is presented as a starting point for discussion: antibody titres and activity, cytokine levels, immune cell counts and activity [3], or other indicators of immune response to naturally occurring diseases or vaccination (including antigens the animals would have no natural exposure to). Other possibilities include quantitative assessment of pathogen load (e.g., worm egg counts), telemetric sensing of body temperature, rumen pH, respiration rate, feeding and water behaviour and consumption, or other parameters indicative of disease status. Furthermore, gene expression, proteomics, and metabolomics are classic examples of physiological indicator traits.

Discussion

Genetic improvement of livestock is highly dependent on reliable and definitive phenotypes. However, phenotypes associated with infectious diseases have not been clearly defined. The complex nature of disease resistance and the lack of acceptable animal challenges for disease resistance have hindered previous research in this area. The number and sizes of cattle herds in research stations is decreasing rapidly. Consequently, most current research on disease resistance of cattle relies on collaborations with commercial producers. Although this can have advantages (e.g., large numbers of animals and concentrating on outbreaks) it often seriously limits the amount of information that can be recorded and samples that can be obtained. The aim of this project is to provide a cattle population in which more intensive studies of components of disease resistance can be conducted.

Conclusion

Concentrating a variety of different phenotypes and research approaches on the same population makes each individual component much more valuable than it would be individually, because it allows determination of how that component fits into the big picture. Specifically, this work should be integrated with work on thermal stress and other work on animal stress and well-being. However, this concept extends beyond animal disease and stress; the GPE population will be extensively phenotyped for many of the most economically important traits in beef production. Therefore, the economic value of genetic improvement for disease resistance could be estimated directly. Such comprehensive objectives will require the collaborative efforts of experts in animal health, quantitative genetics, and molecular genetics from multiple laboratories.

Opportunities for collaboration exist in the following areas: collection of phenotypes that allow more accurate diagnosis and description of disease states, collection of phenotypes for physiological indicator traits, testing or validation of technologies that provide earlier and/or more accurate detection of disease, and electronic monitoring of physiological parameters such as body temperature, rumen pH, respiration rate, etc. We are eager for advice to make the protocols for phenotype and sample collection more effective and practical. Phenotypes that would help to distinguish between disease resistance and disease tolerance are of particular interest. Collection of gene expression, proteomic, or metabolomic data on large numbers of animals is especially encouraged, as is study of differences in pathogen populations among animals. Studies of pharmacogenetics could be considered. It is possible to apply a limited number of treatments to some of the populations, provided sufficient novel phenotypes are collected in conjunction with the treatments. There are also opportunities for collaboration on genetic epidemiology and/or statistical genetics. Although the word "disease" is used here, similar opportunities exist for studying parasite resistance. Similar opportunities for collaboration exist in swine. USMARC does not conduct challenge experiments in which animals are infected with disease-causing agents; however, a small number of animals can be transferred to other facilities in the US for such studies.

To inquire about a specific potential collaboration, send a brief description of the proposed project to Dr. Mark Thallman (Mark.Thallman@ARS.USDA.GOV).

References

1 Fulton RW, Cook BJ, Step DL, Confer AW, Saliki JT, Payton ME, et al: Evaluation of health status of calves and the impact on feedlot performance: Assessment of a retained ownership program for preweaning calves. Can J Vet Res 2002;66:173-180.

2 Wheeler TL, Cundiff LV, Shackelford SD, Koohmaraie M: Characterization of biological types of cattle (Cycle VII): Carcass, yield, and longissimus palatability traits. J Anim Sci 2005;83(1):196-207.

3 Kelm SC, Detilleux JC, Freeman AE, Kehrli ME Jr, Dietz AB, Fox LK, et al: Genetic association between parameters of innate immunity and measures of mastitis in periparturient Holstein cattle. J Dairy Sci 1997;80:1767-1775.

Dr. R. Mark Thallman, US Meat Animal Research Center, PO Box 166, Clay Center, NE, USA 68933-0166
Email: Mark.Thallman@ars.usda.gov

Pinard M-H, Gay C, Pastoret P-P, Dodet B (eds): Animal Genomics for Animal Health. Dev Biol (Basel). Basel, Karger, 2008, vol 132, pp 331-336.

Functional Characterization of a Single Nucleotide Polymorphism in the 5' UTR Region of the Bovine Toll-like Receptor 4 Gene

B.S. Sharma, J. Mount, N.A. Karrow

Centre for the Genetic Improvement of Livestock, Department of Animal and Poultry Science, University of Guelph, Guelph, Canada

Keywords: SNP characterization, toll-like receptor 4, bovine, reporter gene assay

Abstract: Toll-like receptor 4 (TLR4) is the key receptor for the LPS component of Gram-negative bacteria. We have identified a SNP within the 5' UTR region of the bovine *TLR4* gene at position 226 (c.-226C>G) relative to start codon, which was found to be associated with health related traits in the Canadian Holstein population. *TLR4* gene expression was measured over time intervals in samples collected from 18 cows (6 cows in each genotype) challenged in vitro with LPS (20 ng/ml) to determine if this SNP alters the gene expression. *TLR4* gene expression was highest 6 h post LPS challenge, and cows with the GG genotype exhibited significantly higher expression compared to other genotypes. This was further validated by a dual luciferase reporter gene assay in which sequential progressive constructs were cloned into the pGL3 basic vector and transfected into bovine Mac-T cells. The reporter construct containing most of the 5' UTR (-456G) exhibited significantly higher luciferase activity compared to –456C, however, further inclusion of 5' upstream sequence showed a reversal phenomenon where the –913C construct exhibited significantly higher activity compared to –913G. These results indicate that the c.-226C>G SNP affects the expression of *TLR4*, suggesting that this SNP might be influencing the binding and interaction of transcription factors regulating gene expression.

INTRODUCTION

Gram-negative bacteria are responsible for approximately 40% of all clinical cases of mastitis that occur annually, and nearly 25% of those cows either die or are culled [1]. The Gram-negative bacteria are characterized by the presence of endotoxin or lipopolysaccharide (LPS) in their outer membrane. The bovine mammary gland is highly sensitive to LPS, and LPS has been implicated, in part, in the pathogenesis of Gram-negative mastitis. During intra-mammary infection, mammary epithelial cells and tissue macrophages become activated via their pattern recognition receptors (PRRs), which recognize pathogen-associated molecular patterns (PAMP). PRR–PAMP ligation induces the transcription and secretion of cytokines and chemokines that recruit the effector cells of the innate immune system for regulating inflammatory response and the subsequent immune response. The toll-like-receptors (TLRs), a multigene family, recognize a wide variety of PAMP, and therefore play a central role in the initiation of inflammatory response and subsequent adaptive immune response to microbial pathogens [2,3]. One of the best-studied TLRs is TLR4, the key receptor for the LPS component of Gram-negative bacteria and for structures of mycobacteria, fungi, and malaria parasites [4-6]. Several rodent and human studies have demonstrated the impact of *TLR4* polymorphisms on inflammation and immunologic processes [7].

We have identified a novel single nucleotide polymorphism (SNP) within the 5' UTR region of the bovine *TLR4* gene at position 226 relative to the start codon of the gene (c.-226C>G; earlier named as P-226 SNP) [8]. This SNP has been found to be associated with estimated breeding value (EBV) for lactation persistency in the Canadian Holstein bull population [8]. Further, we found that this SNP was associated with EBV for SCS in a group of clinical mastitis (CM)-resistant and CM-susceptible Canadian Holsteins cows [9]. Thus, the compelling genetic association with health related traits suggested that further investigation into the role of this SNP in gene regulation was warranted. In the present study the effect of c.-226C>G on *TLR4* regulation was demonstrated through real-time PCR analysis of *TLR4* expression and by reporter gene assay.

Materials and Methods

Study Population, Genotyping and Animals: Elora and Ponsonby Dairy Research stations of University of Guelph, Guelph, Ontario were genotyped for c.-226C>G SNP of TLR4 using the tetra-primer ARMS-PCR method [10]. Primers used for tetra-primer ARMS-PCR were as follows: outer forward, 5' CTG TTT GAC CAG TCT GAT TGC TTT CCTA 3'; outer reverse, 5' CTC TGT GAA AGC ATG ACC CTC TGT CT 3'; allele C specific, 5' TTC TTC TTC TTC CTC TAA CTT CCC CACC 3'; and allele G specific, 5' TCT CTG ATA TTT ATC TCC TCT GCC AGCC 3'. PCR was performed at 61°C annealing temperature initially keeping a touch-down (1°C in each cycle) profile for the first six cycles at 67°C. A total of 18 cows were selected with six cows for CC, CG and GG genotypes.

LPS Challenge and Blood Mononuclear Cell Separation: Blood was collected from 18 cows and challenged with 20 ng of LPS per ml of blood. LPS challenged blood samples were incubated at 37°C for 3, 6, 9 and 24 hours. After each challenge interval, blood mononuclear cells (BMC) were separated by centrifugation at 800 x g for 20 minutes at 4°C, and washed twice with RBC lysis buffer (5 mM $MgCl_2$, 10 mM NaCl, 10 mM Tris-HCl, pH 7) by centrifugation at 400 x g for 5 minutes.

RNA Preparation and Reverse Transcription: Total RNA was isolated from BMC using Trizol (Invitrogen Life Technologies, Carlsbad, CA, USA) according to the manufacturer's protocol. The quality of total RNA was assessed using the Agilent 2100 Bioanalyzer and the integrity of the 28S and 18S rRNA was determined visually and by determination of the 28S:18S ratio. Reverse transcription of total RNA was performed with 1.5 μg of total RNA using oligo(dT) primer and SuperScript II Reverse Transcriptase (Invitrogen). This reaction was performed at 42°C for 90 minutes followed by 65°C for 15 minutes to inactivate the reverse transcriptase enzyme. The obtained cDNA was quantified by PicoGreen (Molecular Probes, Oregon, USA) using a standard curve method. A cDNA pool was created by preparing 40 ng aliquots of cDNA from each sample; this was used to create a standard curve for the subsequent real-time PCR analysis.

Quantitative Real-Time Polymerase Chain Reaction: Quantitative real-time PCR (qRT-PCR) analyses were performed on an ABI PRISM 7000 Sequence Detection System (PE Applied Biosystems, Foster City, CA, USA) by monitoring the increase in fluorescence of the SYBR Green dye using Platinum SYBR Green qPCR SuperMix-UDG with ROX kit (Invitrogen). Along with *TLR4* cDNA detection, the detection of cDNA was analyzed in each sample as an internal control house-keeping gene. The primer sequences were derived with Primer Express Software version 2.0 (Applied Biosystems) for both the genes. The sequences of primers used for the *RPLP0* cDNA were as follows: forward, 5' TGC TTG ACA TCA CAG AGG AAA CTC 3' and reverse, 5' CGC TTG TACC CAT TGA TGA TAG AA 3'; and for *TLR4* cDNA were forward, 5' GTT GGG ACC CTT GCG TAC AG 3' and reverse, 5' GGC TGC CTA AAT GTC TCA GGT AGT 3'.

The cDNA was amplified by PCR in a total volume of 25 μl that contained 4 ng cDNA, 13 μl of SYBR Green, and 1 μM of each of sense and antisense primers. Thermal cycling conditions for both *RPLP0* and *TLR4* using the ABI7000 were as follows: 50°C for 2 min followed by 95°C for 3 min and 42 cycles at 95°C for 15 sec, 59°C for 15 sec and 72°C for 30 sec, followed by a dissociation stage to verify the specificity of primer sequences. Relative *TLR4* mRNA expression was then calculated in comparison to *RPLP0* mRNA expression using the standard curve method and Pfaffl's formula [11]. Standardized data was analyzed by ANOVA and compared among three genotypes for all time points (hour 0, 3, 6, 9, and 24).

Reporter Gene Constructs: Progressive constructs of the bovine *TLR4* 5' UTR and putative promoter region were prepared and the DNA fragments were cloned into pGL3 basic vector (Promega, Wisconsin, USA), which contains a *Photinus* luciferase gene without a promoter and enhancer. Various constructs were generated by PCR using different sense primers with a *Kpn*I linker and common antisense primer with a *Nhe*I linker so that the DNA fragments had a common 3' end and a different 5' end (Table 1). The numbers indicate the length of each fragment relative to the ATG start codon. These DNA fragments containing various 5' UTR and promoter sequences of the bovine *TLR4* gene harbouring C and G alleles at c.-226C>G were ligated into the pGL3 basic vector in order to construct pGL3-series plasmids. The cloning sites of all reporter plasmids were verified by DNA sequencing.

Table 1: List of primers used to construct reporter gene plasmids.

Name	Sequence (5' primer 3')
RT4AS+68	GGCATGCTAGCGTTCTCAGGCAGGAGAGGAT
RT4S-99	GGACAGGTACCGGGGAGAGACGACACTACAG
RT4S-199	CGGACGGTACCGAAACGAAAGTCCAGAAAGC
RT4S-283	GGACAGGTACCTTGTTACTTCAGCCCCTGGTC
RT4S-456	GGACAGGTACCGGCTGGAACATAGGAAATGA
RT4S-631	GGACAGGTACCCCAGCACATGCAGAGAGAAG
RT4S-732	GGACAGGTACCCACGATTAAGCGACTGAACTGA
RT4S-819	CGGACGGTACCTGGACATGAGTTTGGGTAGA
RT4S-913	GGACAGGTACCCTCACTGGAACAGACCCTGAT
RT4S-1032	GGACAGGTACCGGACTGCAAGGAGATCCAAA

All underlined nucleotides represent restriction enzyme sites.

Transient Transfection and Luciferase Activation Assay: Bovine Mac-T cells, an immortalized epithelial cell line, were grown in Dulbecco's Modified Eagles Medium (DMEM, Invitrogen) supplemented with 10% heat inactivated foetal calf serum (Gibco-BRL, Gaithersburg, MD, USA), 1 mM sodium pyruvate (Gibco), 40 U/ml pen/step (Invitrogen), 25 μg/ml insulin (Sigma), and 25 mM HEPES buffer (Gibco). Cells were incubated at 37°C in a 5% CO_2 environment. For cell culture passage, adherent cells were released by using 0.05% trypsin (Gibco).

Mac-T cells were seeded onto 24-well plates at 10^5 cells/well 24 h before transfection. The cells were transfected with a constant amount of DNA that included 720 ng of the *Photinus* luciferase reporter plasmid and 80 ng of pRL-TK plasmid (Promega) using 2 μl of lipofectamin 2000 (Invitrogen) in Opti-MEM (Gibco). The pRL-TK plasmid containing a *Renilla* luciferase gene under the regulation of a thymidine kinase (TK) promoter served as an internal control for normalizing the expression of the *Photinus* luciferase of reporter plasmids.

After incubation for ~ 40 h, the transfected cells in 24-well plates were washed twice with PBS and harvested in 100 µl of Passive Lysis Buffer (Promega). Luciferase activity of the cell extracts was determined using the Dual-Luciferase Reporter Assay System (Promega). Relative luciferase activity was calculated as observed relative light units from *Photinus* luciferase normalized to *Renilla* luciferase values, and expressed as a ratio to its promoterless pGL3-basic control vector. A total of six independent experiments was carried out and the average relative luciferase activity was obtained from triplicate readings of each experiment. The data therefore represent the means of six individual experiments. A pair-wise Student's t-test was applied to compare luciferase activities between allele C and G in the same size construct.

Results and Discussion

Quantitative Real Time-PCR Analysis:

The *TLR4* expression level was highest at 6 h post-challenge with LPS in all three genotypes. Individuals with GG exhibited higher expression compared to CC (6 h P<0.01, 9 h P<0.07, Fig. 1) and CG (6 h P<0.07, Fig. 1).

Fig. 1: Standardized relative expression of *TLR4* at various time points following LPS challenge.

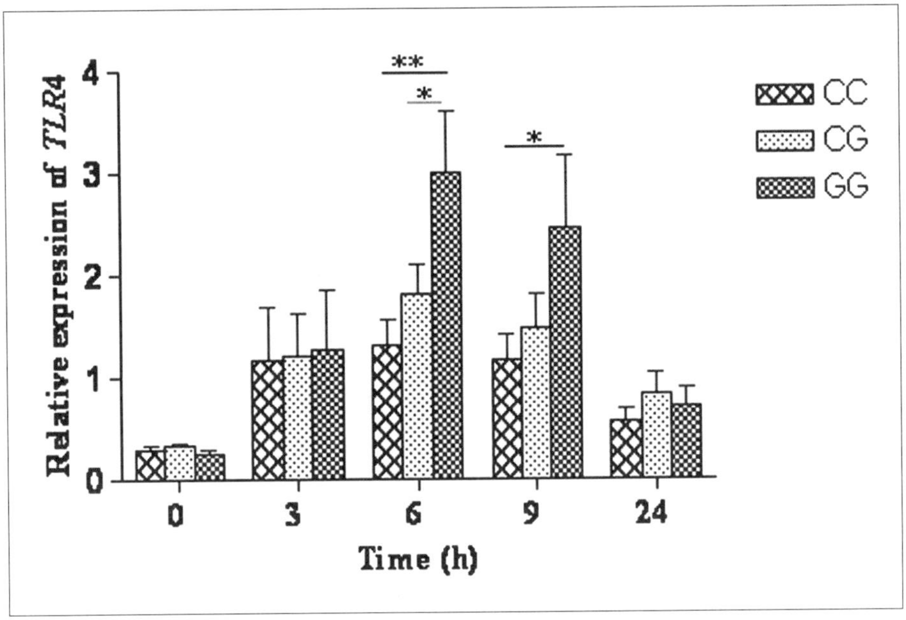

* P<0.07 ** P<0.01

Reporter Gene Construct Analyses:

To examine the functional relevance of this SNP, progressive constructs (~ up to 1kb) were created relative to the ATG start codon of the *TLR4* gene. A total of 15 progressive constructs starting in the 5' UTR region (-99, -199, -283G, -456C/G, -631C/G, -732C/G, -819C/G, -913C/G, and -1032C/G bp ~ +68 bp) were amplified using *Kpn*I and *Nhe*I linkers (Table 1). These sequential length 5' UTR-/+ promoter regions were fused to the luciferase reporter gene in the pGL3 basic vector and transfected into the bovine Mac-T cell line for the measurement of luciferase activity.

As shown in Figure 2, the relative luciferase activity in the pGL3 constructs of the 5' upstream region of *TLR4*, depending on the inserted region, was up to 20-fold higher than that of the pGL3 basic vector. The luciferase activity was significantly increased from –283 to –456 bp suggesting that this region of the TLR4 5' UTR contains potential positive regulatory element(s). Reporter gene plasmid –456G exhibited significantly higher luciferase activity compared to –456C. A similar trend was observed in the –631 bp ~ +68 bp plasmid. Further inclusion of 5' upstream region resulted in the repression of luciferase activity, suggesting the presence of negative regulatory element(s). However, further inclusion of the 5' upstream region (up to –913 bp), revealed a reverse phenomenon whereby the plasmid with allele C exhibited significantly higher activity than the plasmid carrying allele G. These results indicate that the c.-226C>G SNP affects the expression of *TLR4*, and it may influence the binding and the interaction of transcription factors regulating *TLR4* expression. More investigation into the impact of this altered gene expression on host disease resistance is warranted.

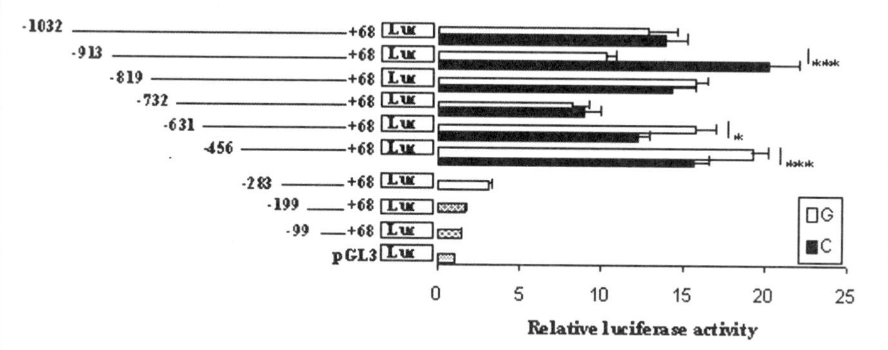

Fig. 2: Transient expression analyses of the TLR4 5' upstream regulatory region (UTR). Progressive constructs containing different lengths of the TLR4 5' UTR and promoter were subcloned upstream of the luciferase reporter gene and transiently transfected into Mac-T cells. Relative to the ATG start codon, positions of the 5' and 3' ends in each construct are marked. The relative luciferase assay results of each construct were expressed as a fold of those of the pGL3-basic promoterless vector.

* P<0.05 *** P<0.001

References

1 Eberhart RJ: Coliform mastitis. Vet Clin North Am Large Anim Pract 1984;6:287-300.

2 Sabroe I, Read RC, Whyte MK, Dockrell DH, Vogel SN, Dower SK: Toll-like receptors in health and disease: Complex questions remain. J Immunol 2003;171:1630–1635.

3 Takeda K, Kaisho T, Akira S: Toll-like receptors. Annu Rev Immunol 2003; 21:335–376.

4 Mockenhaupt FP, Cramer JP, Hamann L, Stegemann MS, Eckert J, et al: Toll-like receptor (TLR) polymorphisms in African children: Common TLR-4 variants predispose to severe malaria. Proc Natl Acad Sci USA 2006;103:177–182.

5 Netea MG, van der Graaf C, Vonk A, Verschueren I, van der Meer JWM, Kullberg BJ: The role of Toll-like receptors in the defense against disseminated candidiasis. J Infect Dis 2002;185:1483–1489.

6 Lien E, Ingalls RR: Toll-like receptors. Crit Care Med 2002;30(Suppl.):S1–11.

7 Roger T, Miconnet I, Schiesser AL, Kai H, Miyake K, Calandra T: Critical role for Ets, AP-1 and GATA-like transcription factors in regulating mouse Toll-like receptor 4 (*Tlr4*) gene expression. Biochem J 2005;387:355–365.

8 Sharma BS, Leyva I, Schenkel F, Karrow NA: Association of Toll-like receptor 4 polymorphisms with somatic cell score and lactation persistency in Holstein bulls. J Dairy Sci 2006; 89:3626–3635.

9 Sharma BS, Leyva I, Schenkel F, Karrow NA: Polymorphisms in the bovine *TLR4* gene and associations with milk somatic cell score and other health related traits in Canadian Holstein cows. Proceedings of 8th World Congress on Genetics Applied to Livestock Production, Belo Horizonte-MG, Brazil, August 13-18, 2006.

10 Ye S, Dhillon S, Ke X, Collins AR, Day INM: An efficient procedure for genotyping single nucleotide polymorphisms. Nucleic Acids Res 2001;29:e88.

11 Pfaffl MW: A new mathematical model for relative quantification in real-time RT-PCR. Nucleic Acids Res 2001;29:2002-2007.

Dr. Niel A. Karrow, Associate Professor, Department of Animal and Poultry Science, University of Guelph, Guelph, Ontario, Canada N1G 2W1.

E-mail: nkarrow@uoguelph.ca

Pinard M-H, Gay C, Pastoret P-P, Dodet B (eds): Animal Genomics for Animal Health. Dev Biol (Basel). Basel, Karger, 2008, vol 132, pp 337-342.

SNP Discovery in Genes Affecting Leg Health Traits in Pigs

S.K. Onteru, B. Fan, B. Mote, T. Serenius, M. Nikkilae, K.J. Stalder, M.F. Rothschild

Department of Animal Science and Center for Integrated Animal Genomics, Iowa State University, Ames, IA, USA.

Keywords: pig, SNP, leg and body conformation traits, association analysis

Abstract: Sow reproductive life is influenced by lameness issues. It has been reported that up to 44 percent of sows have locomotive problems. To date, few genome scans or association studies have been conducted to look at genes controlling lameness and other gait traits. In addition to health problems associated with leg and bone disorders, the pig has been suggested to be a good model for human bone disorders. Hence, the present study examined 134 porcine genes affecting skeletal development, mineral metabolism and other candidate genes for single nucleotide polymorphism (SNP) discovery. A total of over 370 SNPs have been identified to date and are being mapped. These SNPs are also being investigated for their associations with gait and locomotion problems in approximately 2,000 commercial pigs scored for various leg and locomotion traits. The association analysis of 22 genes revealed that the genes *CALCR, HDBP, CALCA, MTHFR, OXTR, IHH, ANKH, LRCH1* and *OPN* were significantly associated with leg and body conformation traits which affect the health and productivity of pigs.

INTRODUCTION

Unsoundness of feet and legs influences the reproductive performance and health of sows and boars [1]. Lameness has been identified as one of the major causes of sow mortality [2]. Structural unsoundness is mainly determined by the skeletal design of feet and legs including the hip. The skeletal design is chiefly determined by the pattern of skeletal development during prenatal and postnatal development [3]. Obviously, this development is under the influence of many genes and their regulatory pathways. Therefore, insight on genes and their variants that influence skeletal design is an essential prerequisite to better understand leg soundness in pigs. This knowledge may also be useful in further understanding of human bone disorders and their intricacies as pig is the best model for human osteoporosis due to its histologically similar bone remodelling cycle [4]. Although several researchers have studied the genetic control of structural soundness in pigs, studies on large numbers of candidate genes, or single nucleotide polymorphisms (SNPs) and their association with lameness are scarce. Hence the present study was conducted to find SNPs for genes affecting skeletal pattern development (e.g., *IHH* (Indian hedgehog)), osteoblast and osteoclast differentiation (e.g., *HDBP* (Vitamin D binding protein)), bone matrix biosynthesis (e.g., *OPN* (osteopontin)) and phosphorus metabolism (e.g., calcitonin). This initial analysis of 22 genes identified nine genes which were significantly associated with leg and body conformation traits in the pig.

Materials and Methods

Population: Two populations, a commercial population of gilts (n=2.066)from a National Pork Board (NPB) funded project and the ISU Berkshire × Yorkshire (BY) resource family (n = 572), were phenotyped for leg and body conformation traits and gene mapped, respectively.

Phenotype determination: The phenotypes were determined by scoring a total of 17 leg and body conformation traits based on the standards of the National Swine Improvement Federation (NSIF) scoring standards and modified Finnish scoring systems.

Genotype determination: The DNA was isolated from ear tissue using the DNeasy 96 Blood & Tissue Kit (QIAGEN). The genes were selected based on human genes which have been shown to play a role in skeletal pattern development, osteoblast and osteoclast differentiation, bone matrix biosynthesis, phosphorus metabolism and bone related signaling pathways. A total of 134 genes were selected and analyzed for SNPs. At present, the SNPs from 22 genes were genotyped for all animals in the populations. In order to discover SNPs, the PCR products of three animals with different leg scores were pooled and a total of 8 pooled PCR samples were sequenced (DNA facility, Iowa State University, Ames, IA, USA). The sequence was analyzed by Sequencher 4.6 software (Gene Codes, Ann Arbor, MI, USA). Once the SNP was determined, the SNP genotyping for the entire population was done with the Sequenom mass array system **iPLEX™ platform.**

Mapping of genes and association analysis: The genotyping data was used to map the genes in the BY population by using CRIMAP software and association analysis was done in the NPB population using SAS software with the following model:

$$Y_{ijklmn} = \mu + \text{Animal ID}_i + \text{Gilt line}_j + \text{Batch}_k + \text{Scorer}_l + \text{Genotype}_m + \text{Body weight}_n + e_{ijklmn}.$$

In this model gilt line, scoring data (batch), scorer and marker genotype are fixed effects. Animal ID is a random effect and body weight is a linear covariate.

Results

There were 370 SNPs found in 134 genes. To date, 22 genes have been analyzed and 9 genes among them were significantly ($P < 0.05$) associated with different leg and body conformation traits. The SNPs in these genes are A/C in intron 12 of *CALCR* (calcitonin receptor), G/A in exon 3 of *IHH*, A/G in the promoter of *OPN*, C/T in intron 4 of *HDBP*, A/G in the 5'UTR of *MTHFR* (methyelene terahydrofolate reductase), C/T in intron 11 of *ANKH* (ankylosis homolog), C/T in intron 5 of *LRCH1* (luecine rich caponin homolog 1), G/T in exon 3 of *OXTR* (oxytocin receptor) and C/G in intron 2 of *CALCA* (calcitonin related polypeptide alpha). The genotype effects of six significant genes on leg and body conformation traits are shown as an example in Figure 1. The favourable alleles associated with better phenotype scores are presented in Table 1. The genes were mapped (chromosomes) as follows: *CALCA* (2), *MTHFR* (6), *HDBP* (8), *OPN* (8), *CALCR* (9), *LRCH1*(11), *OXTR* (13), *IHH* (15), *ANKH* (16), *IL6* (9), Matrillin3 (3), *ADAM12* (14), *COL9A2* (6), *GNAS* (17), *TNF1* (7), *NPP1* (1), *HSP90* (7), *TRAF6* (2) and *NRP2* (15).

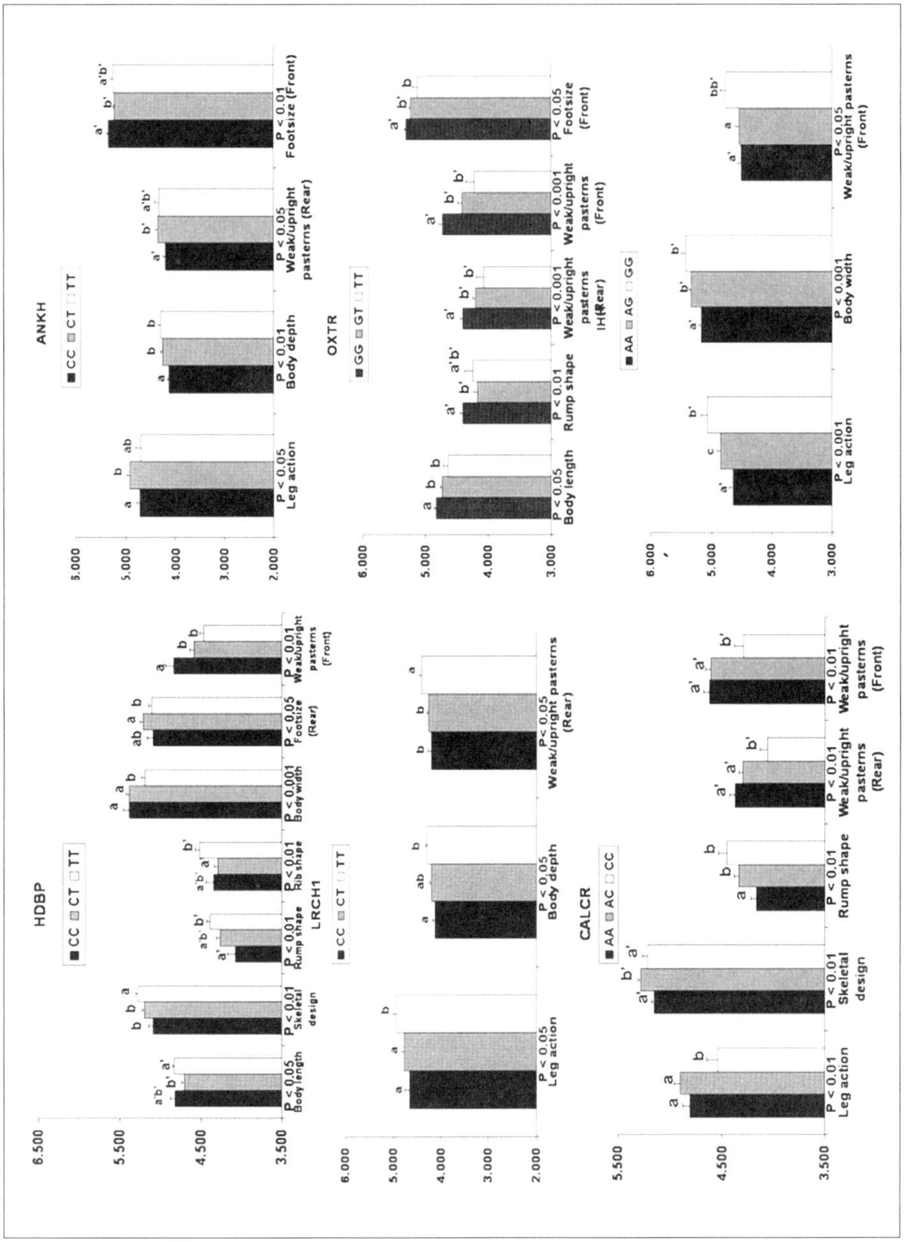

Fig. 1. Genotype effects of candidate genes on leg and body conformation traits. The X axis has leg and body conformation traits and the Y axis indicates phenotype scores. The P values mentioned above the trait indicates that overall the gene is significantly associated with the particular trait. The values with different letters (a, b, and c) for each individual genotypes differ significantly with P < 0.05 within the associated trait. The values with different letters (a', b', and c') for each individual genotypes differ significantly with P < 0.01 with in the associated trait.

Table 1: Favourable alleles for better phenotypic scores.

Trait	Favourable alleles (gene)
Leg action	C (*CALCR*), A (*IHH*), G (*MTHFR*), C (*ANKH*), C (*LRCH1*), C (*CALCA*)
Weak upright pasterns (front)	G (*IHH*), A (*CALCR*), C (*HDBP*), G (*OXTR*)
Weak upright pasterns (rear)	A (*CALCR*), T (*LRCH1*), G (*OXTR*)
Foot size (front)	T (*ANKH*), G (*OXTR*)
Small inner calws (rear)	A (*OPN*)
Body length	G (*OXTR*)
Body width	G (*IHH*), C (*HDBP*)
Body depth	T (*LRCH1*), C (*ANKH*)
Rump shape	A (*CALCR*), C (*HDBP*)
Skeletal design	C (*HDBP*)

Discussion

In the present ongoing study, to date the genes *CALCR, HDBP, MTHFR, ANKH, LRCH1, CALCA, IHH, OPN* and *OXTR* were associated with different leg and body conformation traits. From the published quantitative trait loci (QTL) mapping results for the pig and the comparative QTL mapping from dairy and human populations, swine chromosomes 1, 5, 8, 9, 13, 15 and 16 were the most common regions anchoring QTL pertaining to leg and foot scores, physical scores and osteochondrosis traits [5, 6]. The genes *HDBP, CALCR, OXTR, IHH*, and *ANKH* were assigned to swine chromosome (SSC) 8, 9, 13, 15 and 16, respectively. On SSC8, QTL relevant to physical scores and gait were mapped previously. The *HDBP* on SSC8 was also found to be associated with several body conformation traits and feet pastern set. On SSC9, QTL for front leg scores were confirmed by two different studies. *CALCR* was associated with front and rear feet pastern set and leg action traits from the present study. On SSC15, rear leg scores and osteochondrosis QTL were detected previously and our studies indicated that *IHH* was significantly related to leg action. These results appear to indicate that these genes were possibly causative QTL or they were closely linked to the causative QTL.

Calcitonin receptor mediates the action of calcitonin which is not only a hypocalcaemic hormone, but also plays a role in phosphorus reabsorption. Among leg traits in this population, *CALCR* was associated with leg action, and rear and front pastern set ($P < 0.01$). There was a correlation between leg action and pastern effects as *CALCR* was associated with both of these traits. Overall, the present results are in accordance with previous reports in humans that *CALCR* has different splicing variants which affect bone mineral density and fracture risk [7]. The SNP in *LRCH1* was also related to leg action and rear pastern set. LRCH 1 is involved in chondrocyte responses to pressure through interaction with the cytoskeleton. It has been reported that a SNP in intron1 was associated with osteoarthritis in human [8].

Like calcitonin, CALCA is a hypocalcaemic peptide. Among leg traits *CALCA* was associated with weak or upright pasterns and buck kneed front legs. Similarly, *OXTR* was also associated with three traits including rear and front pasterns, and foot size. OXTR plays a role in bone remodelling by decreasing bone resorption and up-regulating nitric oxide synthase isoforms. However, the SNP in *OXTR* did not change amino acid, hence it may not be a direct cause for different phenotypic conformation types. The *IHH* gene controls chondrocyte differentiation in the growth plate by regulating PTH related peptide expression. It was observed that *IHH* was associated with leg action and front pastern set in this study where as *HDBP* was associated with front pastern set and foot size. HDBP enhances osteoclast differentiation and increases bone resorption by inhibiting extracellular Ca^{+2} sensing by osteoclasts [9]. The *ANKH* and *OPN* genes were also associated with leg traits. For instance *ANKH* was associated with leg action, rear pastern set and foot size. *OPN* was associated with small inner claws of rear legs. ANKH modulates pyrophosphate transport in chondrocytes [10] and OPN is an extracellular structural component of bone. Among genes studied in the present population, the *CALCR, ANKH* and *OXTR* are the most important because of their apparent association with a larger number of leg traits. Among body conformation traits, *HDBP* was associated with skeletal design, rump shape, rib shape, body length and body width ($P < 0.01$). *CALCR* was associated with skeletal design and rump shape. Similarly, *OXTR* was associated with rump shape, rib shape and body length. *IHH* was associated with body width. *MTHFR, CALCA* and *LRCH1* were associated with body depth.

In conclusion, the genes in the present analysis mapped to suggested QTL regions for leg and body conformation traits. This is the first attempt using molecular markers for leg and body conformation traits in pigs. As the SNPs in these genes are not likely to be causative mutations, they could be in linkage disequilibrium with causative mutations. Hence functional analysis and association studies in other populations are warranted. These candidate genes should be strongly considered for marker assisted selection to improve leg and body conformation traits in commercial sow populations.

Acknowledgments

The authors appreciate the support of National Pork Board, Newsham Genetics, Hatch funding and the State of Iowa and the College of Agriculture for their financial support.

References

1 Wood CM: Don't Ignore Feet and Leg Soundness in Pigs. Livestock Update June 2001.

2 Kirk RK, Svensmark B, Ellegaard LP, Jensen HE: Locomotive disorders associated with sow mortality in Danish pig herds. J Vet Med A 2005;52:423–428.

3 Mossaro EJ, Rogers JM: The skeleton biochemical, genetic and molecular interactionsin development and homeostasis. Humana Press, 2004, pp v-x.

4 Hittmeier LJ, Grapes L, Lensing RL, Rothschild MF, Stahl CH: Genetic background influences metabolic response to dietary phosphorus restriction. J Nutr Biochem 2006;17:385-395.

5 Lee G J, Archibald AL, Garth GB, Law AS, Nicholson D, Barr A et al: Detection of quantitative traits loci for locomotion and osterochondrosis-related traits in Large White × Meishan pigs. Anim Sci 2003;76:155-165.

6 Andersson-Eklund, Uhlhorn H, Lundeheim N, Dalin G, Andersson L: Mapping quantitative trait loci for principal components of bone measurements and osteochondrosis scores in a wild boar Large White intercross. Genet Res Camb 2000;75:223-230.

7 Taboulet J, Frenkian M, Frendo J L, Feingold N, Jullienne A, de Vernejoul MC: Calcitonin receptor polymorphism is associated with a decreased fracture risk in post menopausal women. Hum Mol Genet 1998;7:2129-2133.

8 Spector TD, Reneland RH, Mah S, Valdes AM, Hart DJ, Kammerer S et al: Association between a variation in LRCH1 and knee osteoarthritis. A genome wide single nucleotide polymorphism association study using DNA pooling. Arthritis Rheum 2006;54 (2):524-532.

9 Adebanjo OA, Moonga BS, Haddad JG, Christopher L, Huang H, Zaid M. A possible new role for vitamin D binding protein in osteoclast control; Inhibition of extracellular Ca2 sensing at low physiological concentrations. Biochem Bioph Res Commu 1998;249:668-671.

10 Pendleton A, Johnson MD, Hughes A, Gurley KA, Ho AM, Doherty M et al: Mutations in *ANKH* Cause Chondrocalcinosis. Am J Hum Genet 2002;71:933–994.

Dr. Max F. Rothschild, 2255 Kildee Hall, Department of Animal Science, Iowa State University, Ames, IA 50011 USA.

E-mail: mfrothsc@iastate.edu

Pinard M-H, Gay C, Pastoret P-P, Dodet B (eds): Animal Genomics for Animal Health. Dev Biol (Basel). Basel, Karger, 2008, vol 132, pp 343-346.

Mapping MHC Genes in River Buffalo

E.A. Rodrigues Filho[1], N.B. Stafuzza[1], A.R. Caetano[2], C.A. Gill[3], P.K. Riggs[3], J.E. Womack[4], M.E.J. Amaral[1,4]

1 UNESP, São Paulo State University, IBILCE, Depto. Biologia, São Jose Rio Preto, Brazil
2 Embrapa Recursos Genéticos e Biotecnologia, Brasília, DF, Brazil
3 Department of Animal Science, Texas A&M University, College Station, TX, USA
4 Department of Veterinary Pathobiology, Texas A&M University, College Station, TX, USA

Key words: MHC genes, RH mapping, BBU2p, river buffalo

Abstract: The major histocompatibility complex (MHC) contains a set of genes necessary for antigen presentation in the immune system. This gene dense and polymorphic region of the mammalian genome is of considerable interest due to the role of MHC genes in immune function and animal health. Previous cytogenetic studies have indicated that the MHC in river buffalo resides on the short arm of chromosome 2 (BBU2). A 5000-rad radiation hybrid mapping panel was recently generated to enable construction of a whole genome map of river buffalo. To this end, the aims of this project were to elucidate the general organization of the MHC on BBU2, and to compare gene order within this region to the MHC in cattle. PCR primers were selected from the bovine gene map and used with the $BBURH_{5000}$ panel to map a set of ten MHC class II genes in river buffalo. Analysis indicates that these genes fall into two linkage groups, consistent with organization of the MHC in cattle. This comparison of buffalo and bovine MHC gene order provides the first insight into the organization of the MHC on river buffalo chromosome 2.

INTRODUCTION

The MHC is considered to be the most gene dense and polymorphic region of the mammalian genome and plays an important role in disease resistance [1]. Previous studies have indicated that all river buffalo chromosomes arms show homology with single bovine acrocentric chromosomes [2]. Buffalo chromosome 2 is a biarmed chromosome with banding homology to cattle chromosomes 2 and 23 (BBU 2p and 2p, respectively). In cattle, the MHC is found on chromosome 23. Likewise, *in situ* hybridization has previously confirmed that the buffalo MHC resides on BBU 2p.

To elucidate the general organization of the MHC in river buffalo and to compare gene order within this region to the MHC in cattle, a study was initiated to map a set of ten MHC class II genes in river buffalo using cattle-derived PCR primers following a radiation hybrid panel mapping strategy.

Materials and Methods

Published cattle-derived PCR primers were used to type 90 selected radiation hybrid cell lines of the river buffalo-hamster $BBURH_{5,000}$ panel [3]. PCR products were scored as 1 for presence, 0 for absent, or 2 for ambiguous amplification. All primer sets were typed twice on the RH panel and scored independently to increase the accuracy of the results. The primer pairs that showed ambiguous results were typed a third time.

The PCR reactions were performed in an MJ Research PTC-200 (Peltier Thermal Cycler) thermocycler with thermal gradient software. Each PCR reaction contained 50ng of DNA, 10 mM Tri-HC l, 1.5 mM $MgCl_2$, 50 mM KCl, pH 8.3, 10 mM dNTPs, 0.2 mM of each primer, and 0.5 unit of Taq polymerase (AmpliTaq Gold™, Applied Biosystems, Foster City, CA) in a total volume of 10 µl. The appropriate annealing temperature for each set of primers was established in preliminary experiments to yield specific amplification of buffalo PCR product in a hamster background. An initial hot start of 10 min at 95°C was followed by 35 cycles for 30 s at 94°C, 30 s at the established annealing temperature, and 45 s at 72°C, followed by a final elongation step at 72°C for 6 min. PCR products were analyzed by electrophoresis on 2% agarose gels in 1.0X TBE. CARTHAGENE software was used to generate the BBU MHC region map [4].

Results an Discussion

A total of ten cattle-derived PCR primers for MHC class II genes, were tested with buffalo DNA and all ten primer sets produced reliable scores with buffalo DNA, showing that the cattle genome is an important source of markers for buffalo gene mapping. Table 1 shows additional information about the markers presented on the buffalo MHC RH map. The retention frequency (RF) of the markers ranged from 25.5% (*BoLA-DRB2*) to 51.5% (*BoLA-DYB* and *PSMB9*).

The markers incorporated into the buffalo MHC region RH map (Fig. 1) were distributed in two linkage groups, consistent with organization of the MHC region in cattle. LG1 spanning 20.1 cR with four genes (*BoLA-DYA, BoLA-DYB, BoLA-DMA* and *PSMB9*) and LG2, spanning 72.4 cR, with six genes (*BoLA-DRA, BoLA-DRB, BoLA-DRB2, BoLA-DRB3, BoLA-DQA* and *BoLA-DQA2*). One marker from LG1 (*BoLA-DYB*) and another pair of markers (*BoLA-DQA2* and *BoLA-DQA*) from LG2 on the BBU MHC map can be placed on bovine sequence assembly build 3.1, but their order is shuffled between the two maps. These rearrangements may represent rearrangement differentiating BBU2p and BTA23, or the map and/or the assembly may be incorrect. Adding more markers is our current goal. It is also possible that as more markers are placed on the BBU MHC RH map, relative positions assigned to the markers may change. Once we have a consistent map from the region, additional techniques like fluorescent in situ hybridization may be used if the final map shows discrepancies with the cattle map.

Conclusion

Ten new genes were mapped on the MHC region from river buffalo. The comparison of MHC gene order between buffalo and bovine provided the first insight into the organization of the MHC on river buffalo chromosome 2.

Table 1: Cattle-derived primers from MHC class II mapped on river buffalo chromosome 2.

Gene	Primer Sequence (5′- 3′)	Annealing Temperature (ºC)	Reference
BoLA-DRA	F: CCCCCTTTCTTGTCTTTTCAGAG	65	Sena et al, 2003
	R: CAATTCCCAAGTCTAGGAGGACTG		
BoLA-DRB	F: GGACACGTTCTGCAGATACAACTAC	65	BOVMAP
	R: GAACTCTCCTTAAGCATACTTGCTC		
BoLA-DRB2	F: TCCAACACTCACCTGGACG	65	Muggli-Cockett et al, 1989
	R: AGGCAGCGCCGAGGTGAGCGA		
BoLA-DRB3	F: ATCCTCTCTCTGCAGCACATTTCCT	60	Groenen et al, 1990
	R: CGCGCTCACCTCGCCGCTG		
BoLA-DYA	F: CTCGAATTCGGCACTTACGGCACAAATG	65	van der Poel et al, 1990
	R: CAGGGCGCTCTGAACTTCAAG		
BoLA-DYB	F: CACGGGCTCTCCCGCAGAGAATT	65	GenBank (UniSTS:254144)
	R: CCCACAAGGCTCTGTGTGAAGGCAT		
BoLA-DQA	F: CGACTCAGATGACCACATTG	65	Hess et al, 1999
	R: GGGGACACATACTGTTGGTAG		
BoLA-DQA2	F: CACTTATCAGCTGACCAC	50	GenBank (UniSTS:254124)
	R: GGTGGACACTTACCATTG		
BoLA-DMA	F: TTTGTCTCCGCGGTTGATGG	65	Hess et al, 1999
	R: CCAAGCCAACAATGATGCCC		
PSMB9	F: TGCTGCCAACGTGGTGAG	65	Hess et al, 1999
	R: GCTGCATCCACGTAACCA		

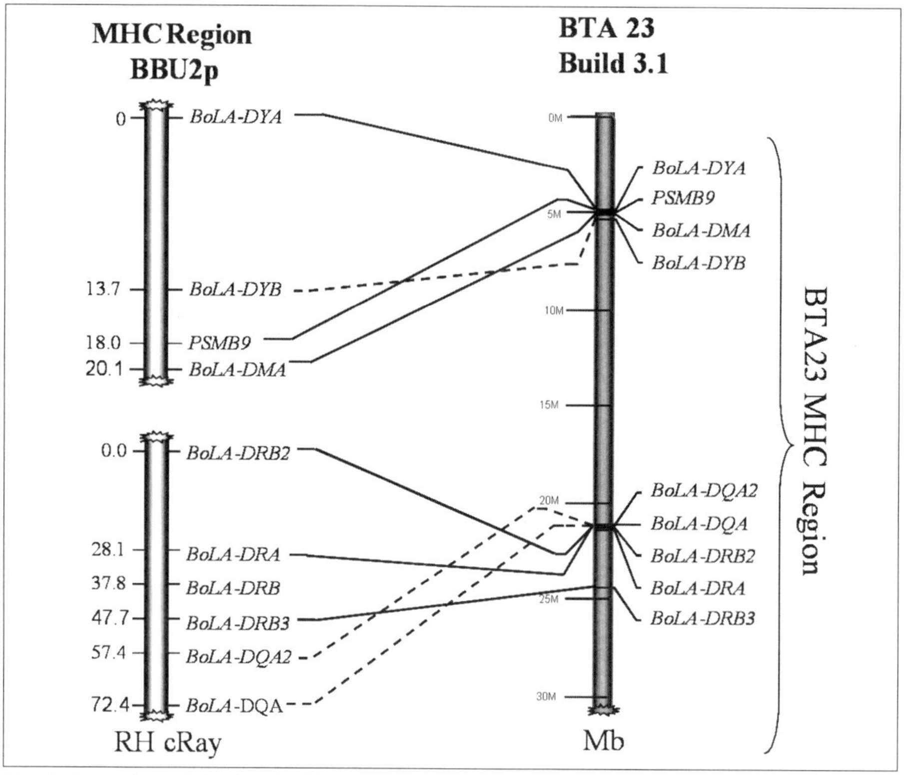

Fig. 1: Comparison of the river buffalo MHC region RH map (left) with the bovine genome build 3.1 for MHC region map (BTA23). Markers common to both BBU and BTA maps are joined by a black solid line. Black dotted lines are showing markers with different order between the two species. Positions in the BBU2p are in cR5000, while positions in the BTA23 are in Mb.

REFERENCES

1 Kelley J, Walter L, Trowsdale J: Comparative genomics of major histocompatibility complexes. Immunogenetics 2005;56:683-695.

2 Iannuzzi L, Di Meo GP, Perucatti A, Schibler L, Incarnato D, Gallagher D et al: The river buffalo (*Bubalus bubalis*, 2n = 50) cytogenetic map: assignment of 64 loci by fluorescence in situ hybridization and R-banding. Cytogenet Genome Res 2003;102:65-75.

3 Amaral MEJ, Owens KE, Elliott JS, Fickey C, Schäffer AA, Agarwala R et al: Construction of a river buffalo (*Bubalus bubalis*) whole genome radiation hybrid panel and preliminary RH mapping of chromosomes 3 and 10. Anim Genet 2007;38:311-314.

4. De Givry S, Bouchez M, Chabrier P, Milan D, Schiex T: CARHTA GENE: multipopulation integrated genetic and radiation hybrid mapping. Bioinformatics 2005;21:1703-1704.

M. Elisabete J. Amaral, Department of Veterinary Pathobiology, Texas A&M University, Vet. Med. Res., BLDG. # 1197, Raymond Stotzer Parkway, College Station, TX 77843-4467.

E-mail: eamaral@cvm.tamu.edu

Pinard M-H, Gay C, Pastoret P-P, Dodet B (eds): Animal Genomics for Animal Health. Dev Biol (Basel). Basel, Karger, 2008, vol 132, pp 347-351.

Single Nucleotide Polymorphisms of Interleukin-1 beta Related Genes and their Associations with Infection in the Horse

P. Horin[1], J. Osickova[1], M. Necesankova[1], J. Matiasovic[1], P. Musilova[2], S. Kubickova[2], D. Hubertova[1], M. Vyskocil[1], J. Rubes[2]

1 Institute of Animal Genetics, Faculty of Veterinary Medicine, Brno, Czech Republic

2 Dept. of Genetics and Reproduction, Veterinary Research Institute, Brno, Czech Republic

Key words: horse, interleukin 1 beta, interleukin 1 receptor antagonist, interleukin 1 beta converting enzyme, polymorphism, infection

Abstract: In previous work, we found significant associations of horse chromosome 15 (ECA15) microsatellite markers HMS01 and HTG06 with two horse infections, *Rhodococcus equi* and *Lawsonia intracellularis*, respectively. Interleukin-1 beta subunit and interleukin-1 receptor antagonist encoding genes (*IL1B* and *IL1RN*) could be considered as candidate genes underlying the associations reported. Therefore, we identified single nucleotide polymorphisms (SNPs) within three interleukin-1 beta functionally related genes: *IL1B, IL1RN* and *Casp1* (interleukin-1 beta converting enzyme/caspase1 encoding gene). Using appropriate restriction fragment length polymorphism (PCR-RFLP) and/or single strand conformation polymorphism (PCR-SSCP) markers, their associations with the two infections by genotyping foals from the original study were tested. In addition, the physical localization of one of the two closely located genes, *IL1RN*, was re-assessed by fluorescence in-situ hybridization (FISH). A statistically significant association between an intronic SNP of the *Casp1* gene with *R. equi* infection was found. The *IL1RN* gene was localized to 15q13-q14 in agreement with its originally reported physical position.

INTRODUCTION

In previous work, we found significant associations of horse chromosome 15 (ECA15) microsatellite markers HMS01 and HTG06 with two horse infections,

Rhodococcus equi and *Lawsonia intracellularis*, respectively [1]. Due to poor resolution of the mapping information available and to complex relationships between linkage and physical mapping data, no positional candidate genes could be identified directly. The microsatellite HMS01 associated with the presence of high numbers of *R. equi* in transtracheal aspirates is located next to the LEX46 and LEX51 loci encompassing the interleukin-1 beta subunit encoding locus (*IL1B*) on the horse RH map [2]. The *IL1B* and IL1RN (interleukin-1 receptor antagonist) genes map physically to ECA15q13-14 [3,4]. The microsatellite HTG06 was associated with the presence of *Lawsonia intracellularis* in faeces and maps physically to ECA 15q26-q27 [2]. However, a very strong population correlation (0.93) between HTG06 and *IL1B* has been reported [3]. To resolve this apparent contradiction and to test *IL1B* and *IL1RN* as possible candidate genes, single nucleotide polymorphisms (SNPs) were identified within three functionally related genes, *IL1B, IL1RN* and *Casp1*, encoding the interleukin 1 beta-converting enzyme, caspase 1. Appropriate markers were established and their associations with the two infections tested by genotyping foals from the original study. In addition, the physical localization of one of the closely located genes, *IL1RN*, was re-assessed by fluorescence in-situ hybridization (FISH).

Materials and Methods

For SNP identification, horse forward/reverse primer pairs located in exons 6 and 7 of the *IL1B* gene, in exons 2 and 3 of the *IL1RN* gene, and in exon 5 and intron 7 of the *Casp1* gene were designed based on the horse GenBank sequences U92481, U92482, and AF090119, respectively [5,6]. Exon/intron boundaries within the *Casp1* gene were determined based on the horse full genome sequence [http://www.ncbi.nlm.nih.gov/genome/guide/horse/].

The primers were as follows: for *IL1B*, forward 5' GGCCTCAAGGAAAAGAACCT 3' and reverse 5' TTGCCCTTGATTTCCATCTT 3'; for *IL1RN*, forward 5' GAAGATAGATGTGGTGCCCATT 3' and reverse 5' CTTGTTCTTGCTCAGGTCAGTG 3'; and for *Casp1*, forward 5´ GAGGATGATGCCATTAAGAAAG 3´ and reverse 5´GAAGGTGTAATGAGGGGAAGA3´.

The PCR total reaction volume was 12.5 μl with 100 ng of genomic DNA. For *IL1B*, the PCR mixture comprised 6.25 μl of HotStarTaq Master Mix (Qiagen, Carlsbad, CA, USA) and 25 pmol of each primer. The PCR protocol consisted of initial denaturation at 95°C for 15 m, 40 cycles at 94°C for 20 s, 56°C for 30 s, and 68°C for 90 s followed by final extension at 68 °C for 10 min. For *IL1RN*, the PCR mixture contained 10x NEB PCR buffer, 200 nM of DNTP's, 1% of final concentration of DMSO, 1.25 U Taq polymerase NEB (New England BioLabs, Beverly, MA, USA) and 12.5 pmol of each primer. The PCR protocol consisted of initial denaturation at 95 °C for 2 m, 35 cycles at 94°C for 20 s, 56°C for 30 s, and 68 °C for 120 s followed by final extension at 68°C for 10 min. For *Casp1*, the PCR mixture comprised 6.25 μl of HotStarTaq Master Mix (Qiagen, Carlsbad, CA, USA) and 25 pmol of each primer. The PCR protocol consisted of initial denaturation at 95°C for 15 m, 35 cycles at 94°C for 40 s, 55°C for 30 s, and 72°C for 90 s followed by final extension at 72°C for 10 min. The size of the PCR product used for PCR-SSCP was 264 bp.

All the PCR products were directly sequenced with an ABI PRISM 310 Automated Sequencer (Applied Biosystems). Two SNPs (G to A substitution at position 105 and T to A substitution at position 385) found within intron 6 of *IL1B* were identified and resolved by using BseRI and Tsp45I restriction enzymes, respectively. Due to the absence of additional restriction sites for both enzymes, haplotyping of double heterozygous genotypes by consecutive digestions was possible. PCR product, 5 μl and/or 7.5 μl, was digested overnight in the appropriate buffer with 4 U of Tsp45I at 65°C or with 4 U BseR I at 37°C, respectively. The resulting fragments were resolved on 2% agarose. For haplotyping the composed heterozygotes, Tsp45I-digested PCR products precipitated with isopropanol were subsequently cleaved with BseRI. The resulting fragments were resolved on 6% PAGE 37.5:1. One SNP (C to T substitution at position 983) was found within intron 3 of *IL1RN*. It could be identified with Alu I. PCR product, 12.5 μl, was precipitated with isopropanol and subsequently cleaved overnight with Alu I in the appropriate buffer at 37°C. The resulting fragments were resolved on 6% PAGE 37.5:1. A-T to G substitution at position 174 within intron 7 was found in *Casp1* (GenBank accession number EF492868). It could be genotyped by PCR-SSCP as follows. A 201 bp long PCR product mixed with loading dye (95% formamide, 0.05% xylene cyanol) was heated to 95°C for 10 min, then placed on ice until loaded. Gels containing 13% w/v acryl/bis-acrylamide (37.5:1) and 1x TBE were run in 1x TBE at 10 W for 19 h at 10°C. The SSCP patterns were visualized by silver staining

Variations in *L. intracellularis* and *R. equi* shedding status were described previously [1]. Briefly, foals were classified as PCR positive and/or negative for presence of *L. intracellularis* in faeces or as *R. equi* negative and/or harbouring high numbers of *R. equi* in transtracheal aspirates (Tables 1 and 2). Fisher´s exact test was used for assessing associations between the polymorphisms detected and the variation observed.

The probe for FISH mapping of *IL1RN* was prepared based on the horse sequence U92482 [7] by PCR with a horse primer pair 5' AACCAGAAGACCTTCTACATGAGG3'/5'CTTGTTCTTGCTCAGGTCAGTG 3'. The total PCR volume was 50 µl with 5 µl 10x Taq buffer, 0.2 mM dNTPs, 80 pmol primers, 2,5U Taq polymerase (Top Bio, Prague, Czech Republic), 2% of final concentration of DMSO and 100 ng DNA. The PCR protocol consisted of an initial denaturation at 95°C for 2 minutes followed by 35 cycles of 10 s at 95°C, 59°C for 30 s, and extension at 72°C for 7 min followed by a final extension at 68°C for 15 min. The PCR products obtained were cloned using the QIAGEN PCR Cloning Kit (QIAGEN, Valencia, CA, USA). Identity of the probe obtained was confirmed by partial sequencing of an approximately 2,500 bp long segment. FISH mapping was performed as described previously [8].

Table 1: Horse Casp1, IL1B and IL1RN polymorphisms and Rhodococcus equi burden in transtracheal aspirates.

Rhodococcus equi infection status	>5,000 colonies/ml	Negative foals	Total
Genotypes			
Casp1 174G carriers	6	21	27
Casp1 174G negative	0	23	23
Total			50*
IL1B 105A, 385T/*IL1RN* 983T carriers	3/3	3/3	6/6
IL1B 105A, 385T/*IL1RN* 983T negative	24/20	13/23	37/43
Total	27/23	16/26	43**/49**

* F= 0.033
** F_{corr}>0.05

Table 2: Horse *Casp1*, *IL1B* and *IL1RN* polymorphisms and *Lawsonia intracellularis* faecal shedding.

Lawsonia intracellularis infection status	PCR positive foals	PCR negative foals	Total
Genotypes			
Casp1 174G carriers	12	17	29
Casp1 174G negative	4	17	21
Total	16	44	50*
IL1B 105A, 385T/*IL1RN* 983T carriers	4/7	7/5	11/12
IL1B 105A, 385T/*IL1RN* 983T negative	23/19	9/17	32/36
Total	27/26	16/22	43*/48*

* $F_{corr.}$>0.05

RESULTS

Only two out of the three possible *Casp1* genotypes were observed in the group analyzed - the 607GG homozygote was not detected. A significant association ($p<0.03$) with the presence of *Rh. equi* in the tracheal lavage was found (Table 1). No association was observed with the *Lawsonia intracellularis* shedding status (Table 2).

Only three out of four possible *IL1B* haplotypes were observed in the group of 43 thoroughbreds, where associations with infections were originally observed. The haplotypes identified and their frequencies within the group were *105G, 385T* (0.47), *105A, 385T* (0.34), *105G, 385A* (0.19). Both *IL1RN* alleles were found within the group analyzed: *983C* (0.74) and *983T* (0.26). No statistically significant associations of these SNPs with *R. equi* and/or *L. intracellularis* shedding status were found. The *IL1RN* gene was localized to 15q13-q14 in agreement with the original report [4].

DISCUSSION

The data suggest that an immune mechanism involving interleukin-1 beta related pathways could underlie the differences in the shedding status observed. All three genes have been shown to be associated with diseases primarily of epithelial and endothelial cell origin in humans [9]. The biological plausibility of the statistical associations observed needs to be further investigated. Some of factors that could cause false interpretations, like the number of horses and/or stratification of the group analyzed were discussed in our original report [1]. It seems that the SNP in the *Casp1* gene is not neutral. We have not been able to detect the *174GG* homozygote in more than 200 horses of 8 breeds (data not shown). As the SNP is located in an intron, it could be in strong linkage disequilibrium with a polymorphism under selection pressure. On the other hand, intronic polymorphisms were also found to have functional impact [e.g. 10]. Despite the population association between *IL1B* and *HTG06* reported previously [3], a large physical distance between *IL1B* and *IL1RN* and the microsatellite locus was confirmed by FISH.

ACKNOWLEDGEMENTS

The technical assistance of Mrs. E. Hantakova, Mrs. V. Novakova, and Mrs. J. Dufkova is acknowledged. The work was supported by the Grant Agency of the Czech Republic, projects 523/02/1526 and 523/06/1402, by the Ministry of Education of the Czech Republic, project MSM 161700001, and by the Grant Agency of the Ministry of Agriculture of the Czech Republic, project MZE 0002716201.

REFERENCES

1 Horin P, Smola J, Matiasovic J, Vyskocil M, Lukeszova L et al: Polymorphisms in equine immune response genes and their associations with infections. Mam Genome 2004;15:843-850.

2 Chowdhary BP, Raudsepp T, Kata SR, Goh G, Millon LV, Allan V et al: The first-generation whole-genome radiation hybrid map in the horse identifies conserved segments in human and mouse genomes. Genome Res 2003;13:742-751.

3 Caetano AR, Shiue YL, Lyons LA, O´Brien SJ, Lauglin TF, Bowling AT et al: A comparative gene map of the horse (*Equus caballus*). Genome Res 1999;9:1239-1249.

4 Lindgren G, Breen M, Godard S, Bowling A, Murray J, Scavone M et al: Mapping of 13 horse genes by fluorescence *in-situ* hybridization (FISH) and somatic cell hybrid analysis. Chromosome Res 2001;9:53-59.

5 Howard RD, McIlwraith CW, Trotter GW, Nyborg JK: Cloning of equine interleukin 1 alpha and equine interleukin 1 beta and determination of their full-length cDNA sequences. Amer J Vet Res 1998;59:704-711.

6 Wardlow S, Penha-Goncalves MN, Argyle DJ, Onions DE, Nicolson L: Nucleotide sequence of equine caspase-1 cDNA. DNA sequencing 1999;10:133-137.

7 Howard RD, McIlwraith CW, Trotter GW, Nyborg JK: Cloning of equine interleukin 1 receptor antagonist and determination of its full-length cDNA sequence. Amer J Vet Res 1998;59:712-716.

8 Musilova P, Kubickova S, Vychodilova-Krenkova L, Kralik P, Matiasovic J et al: Cytogenetic mapping of immunity-related genes in the domestic horse. Animal Genet 2005;36:507-510

9 Arend WP: The balance between IL-1 and IL-1Ra in disease. Cytokine Growth Factor Rev 2002;13:323-340.

10 Tsukada S, Tanaka Y, Maegawa H, Kashiwagi A, Kawamori R, Maeda S: Intronic polymorphisms within TFAP2B regulate transcriptional activity and affect adipocytokine gene expression in differentiated adipocytes. Mol Endocrinol 2006;20:1104-1111.

Dr. Petr Horin, Institute of Animal Genetics, Faculty of Veterinary Medicine, 612 42 Brno, Czech Republic.
E-mail: horin@dior.ics.muni.cz

Pinard M-H, Gay C, Pastoret P-P, Dodet B (eds): Animal Genomics for Animal Health. Dev Biol (Basel). Basel, Karger, 2008, vol 132, pp 353-357.

An Integrated Approach of Genetic Resistance to *Salmonella* Carrier State in Fowls: from Genetics to Genomics and Modelling

C. Beaumont[1], F. Lecerf[2], J. Protais[3], F. Calenge[1], K. Prevost[1,4], A.C. Lalmanach[5], H. Chapuis[6], F. Pitel[2], T. Burlot[7], N. Sellier[1], P. Fravalo[3], A. Vignal[2] , P. Velge[8]

1 INRA, Department of Animal Genetics, URA, Centre de Tours, France
2 INRA, Department of Animal Genetics, LGC, Centre de Toulouse, France
3 AFSSA, Unit of Hygiene and Food Protection, Ploufragan, France
4 Faculté des sciences et techniques, Université du Havre, Rouen, France
5 INRA, Department of Animal Health, IASP, Centre de Tours, France
6 Syndicat des Sélectionneurs Avicoles et Aquacoles Français, 37380 Centre de Tours, France
7 Institut de Sélection Animale, 22000 Saint Brieuc, France
8 INRA, Department of Microbiology of Food Chain, IASP, Centre de Tours, France

Key words: genetic resistance, *Salmonella*, genomics, marker genes, fowl

Abstract: Increasing resistance to acute Salmonellosis (that is, contamination level shortly after infection) is not sufficient to reduce the risk for consumers to be contaminated by *Salmonella*. Indeed, animals may remain contaminated at a low level for weeks or months. Increased resistance to the *Salmonella* carrier state, *i.e.* animals' ability to clear bacteria, is needed; it involves measuring bacterial contamination several weeks after inoculation with a low dose. To study such resistance traits, three convergent approaches were used. A quantitative trait loci (QTL) study was performed, taking advantage of inbred lines differing in resistance. Several QTLs controlling resistance at a younger age were identified and are currently being confirmed in a new cross before finer mapping, using advanced intercross lines. These inbred lines are also presently being compared using functional genomics. In parallel, a selection experiment for increased or decreased resistance at a younger and a later age was undertaken. Besides providing genetic models differing in their levels of resistance, it underlined the importance of the choice of selection criterion, whether marker assisted or not. Indeed, genes controlling resistance are strongly dependant on age; selecting for resistance at a younger age might result in increased susceptibility at an older age. Finally, the results of this experiment were used in a model of the intra-flock propagation of *Salmonella*. It showed that introducing a proportion of resistant animals within a flock of susceptible hens could dramatically change the evolution of contamination. Moreover, it demonstrated the magnitude of synergy between selection and vaccination, which should enhance the interest of increased resistance. The results show that selection for increased resistance to the *Salmonella* carrier state may be efficient, providing that the appropriate criteria of selection are used.

Introduction

While food safety is an increasing request of consumers, *Salmonella* remains one of the major sources of toxic infections, most often as a result of egg consumption [1]. This point is of particular importance since the European Commission has just implemented a new regulation aiming at reducing *Salmonella* prevalence in poultry flocks to less than 2% while the average value for European countries is about 30%, with large variations from values close to zero to more than 80%. To control the incidence of *Salmonella* in poultry flocks, many prophylactic means have been investigated, including genetic resistance. A pioneer study [2] evidenced differences in resistance to mortality due to *Salmonella* between poultry lines. Increasing resistance to acute Salmonellosis, however, is not sufficient to reduce the risk for consumers to be contaminated by Salmonella. Indeed, animals may remain contaminated at a low level for weeks or months without showing any symptoms. Increased resistance to the *Salmonella* carrier state, i.e. animals' ability to clear bacteria, is needed. It involves taking advantage of protocols of infection mimicking the carrier state and resulting in long-lasting asymptotic contaminations. Such protocols have been developed in chicks [3] and hens [4]. The role of genetics in the control of bacterial contamination several weeks after inoculation with a low dose has been shown by estimating the heritability of measures of resistance to the carrier state at 0.20 in chicks [5] and at 0.35 in hens [6]. This divergent selection was therefore initiated, which resulted in very valuable estimations of genetic parameters and genetic models for further comparisons. Another important spin-off of this study was to make it possible to test the effect, in a population close to commercial hens, of genes assumed to be involved in the control of bacterial contamination, as was done for two candidate genes, *SLC11A1* and *TLR4* [7]. Finally, the data was used to fit parameters of mathematical models and investigate the most appropriate strategy for genetic improvement.

The goal of this report is to give the most important results derived from this selection experiment, *i.e.* feasibility, the estimation of genetic parameters, and the results from marker gene tests and modelling.

Materials and Methods

Selection experiment

The L2 line, which had previously shown a high susceptibility to the *Salmonella* carrier state in chicks [8] and hens [4], was used for selection. Within this base population, four lines were selected for decreased or increased resistance, the first two at a young age and the other two at an adult age. Since the Salmonella carrier state may only be assessed upon necropsy, sib-selection was performed: in each line and at each generation, chickens were produced in two hatches issued from the same parents. One hatch was kept to produce the breeders of the next generation and the other was measured for contamination level. In lines selected for contamination level at a younger age, chicks were orally contaminated at one week of age, as previously described [3], and level of caecal contamination assessed and used as the selection criteria. In the "adult lines," hens were orally contaminated at the peak of lay, as described preciously [4], and the caeca, spleen, liver and ovary were tested for bacteria four weeks later. Selection was based on a trait coded "1" if the caeca, spleen or liver was found positive and "0" for all other cases, i.e. presence of the bacteria in at least one of these three organs *versus* absence of bacteria in any of them. Breeders of the next generation were chosen among the siblings of the animals using estimated breeding values predicted by Gibbs sampling and Thrgibbs90 software [9]. Since divergent selection was achieved, animals with a higher genetic risk of *Salmonella* contamination were selected in the line with a higher rate of *Salmonella* carrier state (computed as the mean of estimated values at each age) and *vice versa*. From the third generation, the "young" and "adult" lines were separated and selection was based on increased contamination at a young (Saly+) or adult (Sala+) age or a decreased level in the Saly- and Sala- lines. This was continued for five generations and resulted in a very large set of measurements of pedigreed animals (2,165 chicks and 2,097 hens), which was used to estimate genetic parameters by Gibbs sampling.

Candidate region testing

The interest of some of the genomic regions identified for their effect on level of intestinal contamination [10] in a cross between inbred lines was investigated by genotyping the animals that had been measured for *Salmonella* contamination either at a young or an older age with polymorphic markers (microsatellites or SNPs). One marker gene was chosen in the vicinity of each QTL region (with the exception of the QTL on chromosome 11 which was considered as uncertain). Data for which it was not possible to infer the allele transmitted by the sire was discarded. The effect of the level of caecal contamination in chicks was tested using analysis of variance (ANOVA) fitting the effects of sire, dam within the sire, cage-generation and allele transmitted by the sire (nested within sire). For the rate of adult contamination, an analysis dedicated to all-or-none trait was used and the model only took into account the effects of the sire and allele within the sire.

Modelling

Models of *Salmonella* dissemination within a laying hens' flock [11;12] were used. They distinguished three steps of infection: digestive contamination, systemic infection (when systemic organs such as the liver or spleen were contaminated after translocation of the bacterium through the digestive barrier), and bacterial clearance leading to recovery. Observations from the aforementioned selection experiment were used to compute the model parameters and investigate the interest of mixing animals with a lower or higher propensity to carry *Salmonella*.

Results and Discussion

Selection experiment

Selection is efficient but not very rapid. As a matter of fact, in the fourth generation, there was 40.7% of infected hens in the Sala- line and 60.25% in the Sala+ line, while the level of caecal contamination between the Saly+ and Saly- lines differed by 0.37 log(cfu). A larger response could be obtained by increasing the number of generations. Selection was more efficient in adult lines. Indeed, selection pressure was higher, especially during the first three generations, where breeders were common to both sets of lines. Moreover, adult contamination is more heritable. Heritability was estimated at 0.15 for the level of caecal contamination in chicks and 0.18 for the number of contaminated organs in adults. Since both 95% confidence intervals are strictly positive, these results definitely show the role of genetics in the control of these traits. Moreover, results obtained per organ were heritable with values ranging from 0.11 for the rate of ovary contamination to 0.23 for caecal contamination. Genetic correlations between these rates and the global measurements are higher than 0.75: selecting for the global response should efficiently modify results in each organ. On the contrary, selection based on chick contamination should have a small, but detrimental, effect on adult resistance since a major and unexpected result of the study was the estimated value of the genetic correlation between chicken caecal contamination and adult global contamination, which was found to be -0.40.

Candidate region testing

Two of the QTLs found previously [10] had significant (or close to significant (P<0.07) effects on the level of caecal contamination, confirming the interest of these QTLs. On the contrary, fewer effects were found on adult contamination. The role of the genome region carrying *SLC11A1* was nearly significant (P=0.07) in adults, as previously reported [7], as well as in chicks, which confirmed that this gene is likely involved in both resistance to mortality due to *Salmonella* [13] and *Salmonella* contamination. Differences in gene expression such as those observed for gallinacin 1 and 2 were also confirmed [14;15]. Functional genomics is currently being used to identify other differentially expressed genes. These results strongly suggest that a set of marker genes (presumably SNPs) could be derived in a rather short term and make large scale selection for increased resistance possible. After a few generations of selection, the resulting lines could be introduced into commercial crosses.

Modelling

Rearing both genotypes in equal proportions resulted in a reduction by half of the maximal percentage of contaminated animals but it did not accelerate the extinction of the disease. These results show the interest of the introduction, even at a rather low percentage, of animals selected for a reduced rate of *Salmonella* carrier state within commercial flocks. This could be achieved by using one or more selected lines in commercial crosses. Vaccination and selection should be synergic, since prior contamination reduces the maximal prevalence by 45 and 71% in Sala+ or Sala- flocks, respectively. These results must be confirmed experimentally while the mathematical model could be extended with minor modifications to other animal species or pathogenic species. It may be used to estimate what might be expected from various prophylaxis strategies.

Acknowledgements

The grants from the Network of Excellence EADGENE for modelling, from the French Ministries of Agriculture and Fisheries and Research, Aliment Demain and Aliment Qualité Sécurité and the region Centre for the selection experiment are gratefully acknowledged. The authors thank all those who contributed to this work, especially the experimental units "Plate-Forme d'Infectiologie Experimentale" and "Unité Expérimentale Avicole" from the Tours Research Center. We are grateful to Florence Vignoles, Katia Fève, Axel Boucard and Nicole Millet for their excellent technical assistance. Genotypings were performed in Labogena and in the CRGS platform of the Genopole Toulouse Midi-Pyrenees.

References

1 Velge P, Clockaert A, Barrow P: Emergence of *Salmonella* epidemics: the problems related to *Salmonella enterica* serotype Enteritidis and multiple antibiotic resistance in other major serotypes. Vet Res 2005;36(3):267-288.

2 Bumstead N, Barrow PA: Genetics of resistance to *Salmonella typhimurium* in newly hatched chicks. British Poultry Science 1988;29:521-529.

3 Duchet-Suchaux M, Lechopier P, Marly J, Bernardet P, Delaunay, Pardon P: Quantification of experimental *Salmonella enteritidis* carrier state in B13 Leghorn chicks. Avian Dis 1995;39:796-803.

4 Protais J, Colin P, Beaumont C, Guillot JF, Lantier F, Pardon P, Bennejean G: Line differences in resistance to *Salmonella enteriditis* PT4 infection. Br Poult Science 1996;37:329-339.

5 Berthelot F, Beaumont C, Mompart F, Girard-Santosuosso O, Pardon P, Duchet-Suchaux M: Estimated heritability of the resistance to cecal carrier state of Salmonella enteritidis in chickens. Poult Science 1998;77:797-801.

6 Beaumont C, Protais J, Guillot JF, Colin P, Proux K, Millet et al : Genetic resistance to mortality of day-old chicks and carrier-state of hens after inoculation with Salmonella enteritidis. Avian Pathol 1999;28:131-135.

7 Beaumont C, Protais J, Pitel F, Leveque G, Malo D, Lantier F, et al: Effect of two candidate genes on the *Salmonella* carrier state in fowl. Poult Science, 2003;82:721-726.

8 Duchet-Suchaux M, Mompart F, Berthelot F, Beaumont C, Léchopier P, Pardon P: Differences in frequency, level, and duration of cecal carriage between four outbred chicken lines infected orally with Salmonella enteritidis. Avian Dis 1997;41:559-567.

9 Misztal I, Tsuruta S, Strabel T, Auvray B, Druet T, Lee DH: BLUPF90 and related programs (BGF90). In: Proceedings of the seventh world congress on genetics applied to livestock production, Montpellier, France. CD-ROM communication 28:07. http://www.nce.ads.uga.edu/~ignacy/programs.html

10 Tilquin P, Barrow PA, Marly J, Pitel F, Plisson-Petit F, Velge P, et al: A genome scan for quantitative trait loci affecting the Salmonella carrier-state in the chicken. Genet Sel Evol 2005;37:539-561.

11 Prévost K, Beaumont C, Magal P: A model of Salmonella infection within an industrial hen house. J Theor Biol 2006;242:755-763.

12 Prévost K, Beaumont C, Magal P: Asymptotic behaviour in a *Salmonella* infection model. Math Model Nat Phenom 2006;2:1-22

13 Hu J, Bumstead N, Barrow P, Sebastiani G, Olien L, Morgan K, Malo D: Resistance to salmonellosis in the chicken is linked to NRAMP1 and TNC. Genome Res 1997;7:693-704.

14 Sadeyen, JR Trotereau J, Velge P, Marly J, Beaumont C, Barrow PA, Lalmanach AC: *Salmonella* carrier state in chicken: comparison of expression of immune response genes between susceptible and resistant animals, Microbes Infect 2004;6:1278-1286.

15 Sadeyen JR, Trotereau J, Protais J, Beaumont C, Sellier N, Salvat P, et al: *Salmonella* carrier-state in hens: study of host resistance by a gene expression approach. Microbes Infect 2006;8:1308-1314.

Dr. Catherine Beaumont, INRA-Unité de Recherches Avicoles, Centre de Tours, 37380 Nouzilly.
E-mail : beaumont@tours.inra.fr

Pinard M-H, Gay C, Pastoret P-P, Dodet B (eds): Animal Genomics for Animal Health. Dev Biol (Basel). Basel, Karger, 2008, vol 132, pp 359-361.

Mapping the BDA20 Gene to BTAX

N.B. Stafuzza[1], S. Kata[2], J.E. Womack[2], M.E.J. Amaral[1,2]

1 UNESP, São Paulo State University, IBILCE, Depto. Biologia, São Jose Rio Preto, Brazil.

2 Department of Veterinary Pathobiology, Texas A&M University, College Station, TX, USA.

Key words: BTAX, major cow dander allergen, X chromosome, BDA20

Abstract: The lipocalin family is a large group of proteins that exhibits great structural and functional variation both within and among species, including a significant number of animal-derived aeroallergens, such as the bovine BDA20 (major cow dander allergen). This protein is classified as an occupational allergen causing asthma and other work-related allergic disorders among dairy farmers. Using a somatic cell panel the *BDA20* gene was assigned to the bovine X chromosome (BTAX) with a significant concordant value of 97% to the previously mapped reference marker *MAF45*. A radiation hybrid (RH) mapping approach confirmed the assignment of *BDA20* to BTAX. Two-point LOD scores showed that *BDA20* is linked to *XBM451* with a LOD score of 22.1 for a θ value of 0.03.

INTRODUCTION

The lipocalin protein family is found in a wide variety of tissues in vertebrate and invertebrate animals, plants and bacteria and is characterized by three molecular recognition properties: the binding of small hydrophobic substances, the binding to cell-surface receptors and the formation of complexes with other soluble macromolecules [1].

The bovine BDA20 protein, also know as Bos d 2, is considered an occupational allergen among dairy farmers [2]. Abundant information about structural, biochemical and functional aspects of this protein are available in the literature. However, no assignment of this gene to the bovine genome has been reported. Thus, the aim of this study was to assign the *BDA20* gene to a bovine chromosome using a somatic and a radiation hybrid panels.

Materials and Methods

The PCR primer was designed from a Bos *taurus* mRNA sequence (GenBank accession number L42867) using the software program Primer3 (http://www.genome.wi.mit.edu/cgi-bin/primer/primer3_www.cgi).

The PCR reactions were performed in a PTC-200 MJ ResearchTM thermal cycler with thermal gradient software. Each PCR reaction contained 50 ng of template DNA, 10 mM Tris-HCl, 1.5 mM $MgCl^2$, 50 mM KCl, 100 mM of each dNTP, 0.2 mM of each primer, and 0.5 U of Taq DNA polymerase (AmpliTaq™ Gold, Applied Biosystems, Foster City, CA, USA) in a total volume of 10 ml. PCR conditions included 94°C for 10 min; 35 cycles at 94°C for 30s; 54°C for 30s; with a final extension cycle of 72°C for 6 min. PCR products were visualized on 2% agarose gels in 1.0X TBE buffer and stained with ethidium bromide.

Initial bovine chromosomal assignment for *BDA20* was obtained by PCR analysis using a somatic hybrid cell panel of 31 hamster/bovine and mouse/bovine hybrid somatic cell lines described previously [3]. A whole-genome radiation hybrid ($WGRH_{5000}$) panel [4] composed of 90 selected radiation hybrid cell lines was used to determine the specific chromosomal location for *BDA20*.

The primer pair was optimized to maximize cattle-specific product, typed twice against and scored independently to increase the accuracy of the results. PCR results were scored as present (1), absent (0), or ambiguous (2). RHMAPPER 1.22 software [5] was used in assignment of the PCR product to a chromosome and identification of the nearest marker in the bovine $WGRH_{5000}$ map. Minimum LOD score accepted for linkage was 10.

Results and Discussion

PCR primers were used to amplify a region of about 450 bp of the bovine BDA20 gene. The primer sequences used were BDA20-F: 5'-TCTCCTCTTTG-GACTGGTTTG-3' and BDA20-R: 5'-TCCTTCTTGTCTCTTTGCCAC-3'.

The bovine-specific band was detected in 13 of the 31 hybrid cell lines for the BDA20 gene. The result obtained with the somatic cell was: 0100011011111000000010011110000 (1 = present in the cell line; 0 = not present in the cell line). The gene was assigned on BTAX with a significant concordant value of 97% to the previously mapped reference marker MAF45.

RH mapping confirmed the assignment of the BDA20 to BTAX, resulting in the following vector:
0010100000000000010010010000100000010101000011101001001000001110
1000110000100000011100001011.

A retention frequency (RF) of 30% was calculated based on 28 bovine-specific PCR bands in the RH panel with 90 hybrid cell irradiate lines. A two-point LOD score showed that *BDA20* is linked to the XBM451 microsatellite with a LOD score of 22.1 for a θ value of 0.03.

Conclusion

We have integrated a new marker with the previously reported markers in BTAX to provide a valuable map for trait mapping in cattle.

References

1 Akerstrom B, Flower DR, Salier JP: Lipocalins: unity in diversity. Biochim Biophys Acta 2000;1482:1-8.

2 Saarelainen S, Zeiler T, Rautiainen J, Närvänen A, Rytkönen-Nissinen M, Mäntyjärvi R et al: Lipocalin allergen Bos d 2 is a weak immunogen. Int Immunol 2002;14:401-409.

3 Womack JE, Mool YD: Gene map of the cow: conservation of linkage with mouse and man. J Hered 1986;72:2-7.

4 Womack JE, Johnson JS, Owens EK, Rexroad CE III, Schläpfer J, Yang Y: A whole genome radiation hybrid panel for bovine gene mapping. Mamm Genome 1997;8:854-856.

5 Slonin D, Kruglyak L, Stein L, Lander E: Building human genome maps with radiation hybrids. J Comput Biol 1997;4:487-504.

Dr. M.E.J. Amaral Department of Veterinary Pathobiology, Texas A&M University, Vet. Med. Res. BLDG. # 1197, Raymond Stotzer Parkway, College Station, TX 77843-4467.

E-mail: eamaral@cvm.tamu.edu

Pinard M-H, Gay C, Pastoret P-P, Dodet B (eds): Animal Genomics for Animal Health. Dev Biol (Basel). Basel, Karger, 2008, vol 132, p 363.

SESSION IV

Genomics in the Development of New Tools to Prevent and Control Animal Diseases

Pinard M-H, Gay C, Pastoret P-P, Dodet B (eds): Animal Genomics for Animal Health. Dev Biol (Basel). Basel, Karger, 2008, vol 132, pp 365-372.

Using Integrative Genomics to Elucidate Genetic Resistance to Marek's Disease in Chickens

H. Cheng[1], M. Niikura[1,2], T. Kim[1,2], W. Mao[1,2], K.S. MacLea[1], H. Hunt[1], J. Dodgson[2], J. Burnside[3], R. Morgan[3], M. Ouyang[4,†], S. Lamont[5], J. Dekkers[5], J. Fulton[6], M. Soller[7], W. Muir[8]

1 USDA, ARS, Avian Disease and Oncology Laboratory, East Lansing, MI, USA
2 Department of Microbiology and Molecular Genetics, Michigan State University, East Lansing, MI, USA
3 Department of Animal and Food Sciences, University of Delaware, Newark, DE, USA
4 Informatics Institute, University of Medicine and Dentistry of New Jersey, Piscataway, NJ, USA
5 Department of Animal Science and Center for Integrated Animal Genomics, Iowa State University, Ames, IA, USA
6 Hy-Line International, Dallas Center, IA, USA
7 Department of Genetics, The Hebrew University of Jerusalem, Jerusalem 91904, Israel
8 Department of Animal Sciences, Purdue University, West Lafayette, IN, USA

Key words: genomics, QTL, DNA microarray, protein interactions, disease, virus, chicken

Abstract: While rearing birds in confinement and at high density are very successful practices for producing poultry meat and eggs, these conditions may promote the spread of infectious diseases. Consequently, the poultry industry places great emphasis on disease control measures, primarily at the animal husbandry level. The field of genomics offers great promise to complement these current control measures by providing information on the molecular basis for disease, disease resistance, and vaccinal immunity. This briefly summarizes some of our efforts to apply several genomic and functional genomics approaches to identify genes and pathways that confer genetic resistance to Marek's disease (MD), a herpesvirus-induced T cell lymphoma of chickens. By utilizing the "top-down" approach of QTL to identify genomics regions, and integrating it with "bottom-up" approaches of transcript profiling and Marek's disease virus (MDV)-chicken protein-protein interactions, three genes that confer resistance to MD are revealed, plus a number of other positional candidate genes of high confidence. These genes can be further evaluated in poultry breeding programmes to determine if they confer genetic resistance to MD. This integrative genomics strategy can be applied to other infectious diseases. The impact of the genome sequence and other technological advancements are also discussed.

Introduction

Poultry is the third largest agricultural commodity and the primary meat consumed in the US and much of the world. Due to advanced breeding programmes and efficient animal husbandry, tremendous progress in production traits has been made to meet the growing demands of consumers. Several major issues confront the poultry industry today. Infectious diseases are certainly at or near the top of the list, e.g., avian influenza, *Salmonella*, and others, which harm the poultry industry through loss of birds, reduced public confidence, and lost market accessibility via trade restrictions. Disease outbreaks are enhanced by more concentrated chicken rearing, reduced genetic diversity from industry consolidation, and breed uniformity to meet consumer preferences. Changes in animal husbandry and new vaccines have alleviated some of the problems. However, improved or alternative control measures are still needed to address current diseases and counter emerging threats.

Genomics offers one of the more exciting avenues for addressing these issues. The ultimate goal for genomics is to relate genetic variation at the DNA sequence level to variation at the phenotypic level. By identifying quantitative trait loci (QTL) and genes that control heritable traits of agricultural importance, it is possible to select for birds with superior agricultural traits via marker-assisted selection (MAS). Other positive attributes commonly cited for MAS include greater speed and accuracy compared to traditional breeding. For infectious diseases, MAS would reduce or eliminate the expensive and environmentally hazardous pathogen-exposure trials needed to collect accurate disease-resistance phenotypes. The recent release of the chicken genome sequence [1] has increased the power of genomics. This review shows how integration of genomic and functional genomic approaches can lead to the identification of genes and pathways that confer resistance to Marek's disease (MD), the most serious chronic disease problem facing the poultry industry.

Marek's Disease

Marek's disease (MD) is a lymphoproliferative disease of poultry caused by the Marek's disease virus (MDV). The pathological manifestations of MD include immunosuppression, nerve enlargement, and lymphomas that metastasize to various visceral organs. During the 1960s as the industry converted to high-intensity rearing, MD generated tremendous economic losses. One of the great success stories was the introduction of vaccines that effectively controlled MD in the 1970s. Since then, the condemnation rate attributed to MD has decreased dramatically. However, although vaccination prevents the formation of lymphoma and other MD symptoms, it does not prevent MDV infection, replication, or horizontal spread. Moreover, increasingly frequent unpredictable and spontaneous vaccine breaks due to highly virulent strains of the MDV combined with the incomplete immunity that is elicited by vaccination [2] cause devastating damage to poultry farms. Consequently, MD remains one of the most serious chronic infectious disease threats to the poultry industry.

MDV is a herpesvirus that has lymphotropic properties similar to those of γ-herpesviruses, such as Epstein–Barr virus (EBV) in humans, where the host range is restricted and latency is often evident in lymphoid tissue. Viruses of this group are capable of transforming cells in natural hosts. MDV was re-classified, however, because its molecular structure and genomic organization more closely resembles that of α-herpesviruses, such as herpes simplex virus (HSV) and Varicella-Zoster virus (VZV). The complete sequence of several MDV strains has been determined [2].

Genetic resistance

The best documented case for involvement of genetic resistance to MD involves the MHC or, as it is known in the chicken, the *B* complex. By measuring the frequency of specific blood groups or using *B* congenic chickens, it has been observed that certain *B* alleles can be associated with resistance or susceptibility. The *B*-haplotype also influences vaccinal immunity as some haplotypes develop better protection with vaccines of one serotype than of a different serotype. As shown by several studies, the *B* haplotype effect is dependent on genetic background. Other studies show that genes, other than the *B* complex contribute to MD resistance. In particular, in 1939, the Agricultural Research Service (ARS) laboratory in East Lansing, MI, initiated the development of inbred lines to study what was then known as the avian leukosis complex. Of the fifteen lines initially developed, lines 6 and 7 proved to be the most interesting with respect to MD. These lines share the same B^2 haplotype yet differ greatly with respect to resistance to MD. When line 6 chicks are inoculated with the JM strain of MDV at 1 day of age, typically less than 5% of the chicks will succumb to the disease. In contrast, similar inoculations into line 7 chicks will result in greater than 90% mortality. These lines are still maintained and are over 99% inbred. The level of disease resistance as measured by mortality among F1 siblings of a cross between lines 6 and 7 is intermediate to that of the parents (~60%), while levels of resistance observed in the F_2 population derived from the above cross encompass a wide, continuous spectrum indicating that there is more than one gene for resistance involved. Thus, analysis of these lines and their crosses indicate that MD resistance is a typical polygenic trait, controlled by multiple genes.

Integrating Genomics

The presence of genetic variation in MD resistance raises the possibility of selection for genetic resistance as a complementary control strategy to augment MD vaccinal protection. Also, as noted above, MD resistant birds have the added advantage that, in general, they respond better to MD vaccines. Ideally, selection for MD resistance would be based on genetic markers within disease resistance genes or tightly linked to them. The method is known as marker-assisted selection (MAS), and is expected to accelerate genetic progress by increasing the accuracy of selection and allowing for gender-independent selection at an earlier age. Furthermore, the use of MAS for developing disease resistance avoids the necessity of challenging breeding stock with hazardous pathogens.

With this goal in mind, we have been implementing and integrating three genomic approaches to identify the genes that are involved in genetic variation in MD resistance: genome wide QTL scans, gene profiling, and virus-host protein-protein interaction screens. The rationale for using more than one approach is that the strengths of each system can be combined to yield results of higher confidence. Another justification is that given the large volume of data produced by genomics, each method provides an additional screen to limit the number of targets to verify and characterize in further experiments. The approaches are briefly described below.

Genome-wide QTL scans

With the development of a molecular genetic map of the chicken genome [3], it became possible through linkage mapping to identify QTLs (i.e., chromosomal regions containing genes or other genomic elements) controlling complex traits, including disease resistance. Our efforts have utilized resource populations based on experimental inbred lines [4,5] and commercial strains [6]. The use of the experimental lines allows for more precisely controlled environmental conditions and trait measurements while the use of commercial strains is more agriculturally relevant and permits many more birds to be produced and evaluated. In all populations, many QTLs of small to moderate effect were identified, which agrees well with earlier conclusions as to the multigenic nature of MD resistance. Some of the identified QTL locations are common across experimental and commercial populations, providing additional confidence in these QTLs and suggests that results from experimental populations can be transferred to commercial lines [6]. It should be noted, however, that all experiments to date have been implemented in White Leghorn lines, which have a known narrow genetic base and consequently, a high degree of interrelatedness. It is not known to what extent these QTLs are also responsible for genetic variation in MD resistance in broiler lines.

Genome-wide scans for QTLs conferring MD resistance have been conducted on crosses between ADOL lines 6 (MD resistant) and 7 (MD susceptible). Specifically, in a 6 x 7 F2 population genotyped with 127 genetic markers, 14 QTLs (7 significant and 7 suggestive) were identified that explain one or more MD-associated trait. The QTLs were of small to moderate effect explaining 2 to 10% of the phenotypic variance, with standardized additive gene substitution effects from 0.01 to 1.05 phenotypic standard deviations. Collectively, the QTLs explained up to 75% of the genetic variance. Interestingly, 10 of the 14 QTLs displayed non-additive gene action: 3 were over-dominant and 7 were recessive. Theoretically, MAS would be useful in exploiting non-additive QTLs. Three of the QTLs were associated almost exclusively with viremia levels while the remaining QTLs could account for disease incidence, survival, tumours, nerve enlargement, and other disease associated traits. This suggests that disease resistance occurs at least at two levels: initial viral replication and cellular transformation, which occurs later. It also highlights the added value of measuring several components as this may functionally separate a complex trait into more highly heritable physiological components, as well as provide clues for identifying positional candidate genes.

More recently, with the advent of the genome sequence, millions of high-confidence SNPs [7], and high-throughput genotyping systems, it became possible to genotype large resource populations very quickly and economically. Using the Illumina platform, we were able to re-evaluate the 6 x 7 F2 MD line resource population with 578 additional genetic markers [8]. Most interestingly, partly due to the higher accuracy of calling genotypes with SNPs, a search for two-locus epistatic interactions was conducted. A large number of highly significant two-way interactions were found that could account for viremia; no highly significant interactions were revealed for MD or survival. A total of 239 highly marker-to-marker significant interactions (LRS >57.8; genome-wide P <0.001) were identified [8]. The location of loci and their interacting partners appears to be distributed throughout the genome. On the other hand, loci in specific regions on chromosomes 1 and 4 are frequently involved; a single region on chromosome 1 accounted for 166 of the 239 highly significant interactions. Interestingly, most if not all of the interacting loci are not in the QTL regions previously identified.

These results imply that (1) interactions between loci can make a substantial contribution to variation in complex traits, (2) because of the growing awareness of biological complexity, genome-wide QTL scans should attempt to incorporate resource populations for which the genotyping and analytical capabilities available today or in the near future can be used, and (3) the challenge remains to identify the underlying genes and genetic variation for QTL as the confidence intervals are fairly large and encompass many genes.

Gene profiling

The population sizes and structures that have been used for mapping QTLs affecting MD resistance can only map the causative gene or genetic element to rather broad chromosomal regions extending for tens of centimorgans and containing many hundreds of genes. For this reason, it is clear that it will be extremely difficult to identify positional candidate genes for MD resistance using approaches based on general functional annotation and positional genetics only. Gene expression profiling using microarray hybridization technology is a powerful tool for gene function studies. Our hope is that DNA microarrays will identify genes and pathways specifically involved in MD resistance, which combined with linkage mapping and functional annotation of the genome can narrow the field of positional candidate genes [9]. In other words, positional candidate genes are those that have a genetic association, are functionally relevant, and are identified as being relevant to MD resistance through gene expression analyses.

Gene profiling has been conducted to identify differentially expressed genes between lines 6 and 7 after MDV challenge [9] among B (MHC) congenic lines of chickens following inoculation with different MD vaccines, and in chicken embryo fibroblasts (CEF) infected with MDV. Analyses of these experiments have identified a number of genes and pathways that are consistently associated with either MD resistance or MDV infection. Remarkably, the results suggest that chickens with immune systems that are more stimulated by MDV infection are more susceptible. Initially, this seems counterintuitive but upon further reflection, MDV is thought to only infect activated lymphocytes. Thus, chickens with immune systems that are more responsive may present more targets for MDV to infect and later transform.

Although DNA microarray experiments can be very powerful and the results enlightening, our experiences also suggest that it is critical to study the response of cells in a limited or local environment. For example, MHC class I expression was increased in MDV-infected CEF [10]. However, upon closer examination of individual cells, virus-infected cells actually expressed reduced levels of MHC on their cell surfaces but neighbouring uninfected cells had elevated levels of MHC class I. To partially circumvent the problem of heterogeneous response of individual components of samples, laser capture microdissection (LCM) and cell sorting has been used to isolate the cells of interest.

Virus-host protein-protein interaction screens

The third component of this approach to identifying the genes associated with MD resistance is to systematically examine protein interactions associated with MD resistance. Specifically, a two-hybrid screen has been used to detect MDV-chicken protein-protein interactions. Our hypothesis was that some chicken proteins that interact with MDV proteins are involved in the immune response and genetic resistance to MD. Thus, the two-hybrid system could be used to quickly identify interacting proteins which, when combined with genetic mapping, would identify positional candidate genes for MD resistance.

The MDV *SORF2* gene was initially chosen as bait due to its potential role in the attenuation of the MDV RM1 strain [11]. Using the yeast two-hybrid system and a splenic cDNA library, growth hormone (GH) was found to specifically interact with SORF2 [11]. To corroborate the detected interaction, an *in vitro* protein binding assay using GST-fusion proteins was performed to confirm direct binding of GH to SORF2. The results show that, while SORF2 protein was not retained by GST protein alone, SORF2 can be retained by GST-GH fusion protein, presumably due to the presence of GH. This result is in agreement with the result of the yeast two-hybrid system assay and indicates that the interaction between SORF2 and GH is a direct and specific protein-protein interaction.

Based on these results, the GH gene (*GH1*) was treated as a candidate gene for MD resistance. To determine if *GH1* contributed to the genetic basis of MD resistance, an association study involving an SNP in GH1 was conducted in an MD resource population derived from commercial White Leghorn lines. Indeed, it was found that *GH1* allelic variation was significantly associated ($P < 0.01$) with a number of MD-associated traits in MHC B^2/B^{15} chicks [11]. Furthermore, providing functional information support, DNA microarray results indicate that GH is differentially expressed between MD resistant (line 6) and susceptible (line 7) chicks after MDV challenge [9].

Thus, the combined results of a specific MDV-chicken protein interaction, differential expression of GH between MD resistant and susceptible chickens, and association of *GH1* allelic variation with MD-related traits and selected lines for MD resistance, all strongly suggest that *GH1* is a MD resistance gene. This conclusion is supported by reports demonstrating that GH modulates the immune system in many species, and frequency of *GH1* alleles differs in the expected direction in chicken strains in response to selection for MD resistance. Most importantly, it exemplifies the power of combining genetic and molecular approaches to identify positional candidate genes for QTL.

With the complete sequence of the MDV genome, it became feasible to conduct a systematic screen of the relevant MDV genes for interacting chicken partners. All the MDV genes that are considered unique to serotype I (virulent) strains were screened, and all potential MDV-host protein interactions were tested by an *in vitro* binding assay to confirm the initial two-hybrid results. As a result, 8 new MDV-chicken protein interactions were identified [12]. More importantly, genetic mapping and association analyses of the encoding chicken genes revealed that *LY6E* [lymphocyte complex 6, locus E; also known as stem cell antigen 2 (*SCA2*) and thymic shared antigen 1 (*TSA1*)] is another MD resistance gene [13] and further suggest that *BLB*, the gene for MHC class II β chain, is a strong positional candidate gene [12].

Higher order protein-protein interactions can be revealed with gentle cell lysis and immunoprecipitation of a protein complex using antibodies directed against one member of the complex. The identity of the other interacting proteins can be quickly revealed through mass spectrometry, which is now facilitated by the availability of the whole genome sequence. To extend the MDV-chicken protein interactions, we are employing tandem affinity purification (TAP) tag to several chicken-MDV protein complexes. Rather than expressing the viral gene product alone, we plan to take advantage of our infectious MDV BAC clones [14] and the ability to make defined recombinant viruses. The use of these viruses expressing TAP-tagged MDV proteins during infection allows for the use of a system closer to normal viral infection, which will hopefully yield useful information on the biology of viral infection and the role of host-virus interactions in that context.

Conclusion

It has been over 30 years since MDV was identified as the causative agent of MD, and effective vaccines produced. In spite of our ability to detect and protect against MD, there is a surprising lack of information on the components of the chicken immune system that confer disease resistance and contribute to vaccinal immunity. With the advent of new technologies, we believe that our ability to tangibly improve upon MD control mechanisms is at hand. The key to making rapid gains will be the ability to integrate various methods and information. If this occurs, then we should be able to accelerate the transition from serendipity to rational, mechanism-based control of disease and food production in the poultry industry.

Acknowledgements

We would like to thank current and past lab members, and our many collaborators within academia and the poultry breeding industry. Financial support was provided in part from USDA awards 00-03473, 2001-52100, 2002-03407, 2003-05414, and 2004-05434.

References

1 Hillier LW, Miller W, Birney E, Warren W, Hardison RC, Ponting CP, et al: Sequence and comparative analysis of the chicken genome provide unique perspectives on vertebrate evolution. Nature 2004;432:695-716.

2 Osterrieder N, Kamil JP, Schumacher D, Tischer BK, Trapp S: Marek's disease virus: from miasma to model. Nature Rev Micro 2006;4:283-294.

3 Groenen MAM, Cheng HH, Bumstead N, Benkel BF, Briles WE, Burke T, et al: A consensus linkage map of the chicken genome. Genome Res 2000;10:137-147.

4 Vallejo RL, Bacon LD, Liu HC, Witter RL, Groenen MAM, Hillel J, Cheng HH: Genetic mapping of quantitative trait loci affecting susceptibility to Marek's disease virus induced tumors in F2 intercross chickens. Genetics 1998;148:349-360.

5 Yonash N, Bacon LD, Witter RL, Cheng HH: High resolution mapping and identification of new quantitative trait loci (QTL) affecting susceptibility to Marek's disease. Anim Genet 1999;30:126-135.

6 Heifetz EM, Fulton JE, O'Sullivan NP, Arthur JA, Wang J, Dekkers JCM Soller M: Mapping QTL affecting susceptibility to Marek's disease in a backcross population of layer chickens. Genetics (in press).

7 Wong GK, Liu B, Wang J, Zhang Y, Yang X, Zhang Z, et al: A genetic variation map for chicken with 2.8 million single-nucleotide polymorphisms. Nature 2004; 432:717-722.

8 Cheng HH, Zhang Y, Muir WM: Evidence for widespread epistatic interactions influencing Marek's disease virus viremia levels in chicken. Cytogen Genome Res 2007;117:313-318.

9 Liu HC, Cheng HH, Sofer L, Burnside J: A strategy to identify positional candidate genes conferring Marek's disease resistant by integrating DNA microarrays and genetic mapping. Anim Genet 2001;32:351-359.

10 Morgan RW, Sofer L, Anderson AS, Bernberg EL, Cui J, Burnside J: Induction of host gene expression following infection of chicken embryo fibroblasts with oncogenic Marek's disease virus. J Virol 2001;75:533-539.

11 Liu HC, Kung HJ, Fulton JE, Morgan RW, Cheng HH: Growth hormone interacts with the Marek's disease virus SORF2 protein and is associated with disease resistance in chicken. Proc Natl Acad Sci USA 2001;98:9203-9208.

12 Niikura M, Liu HC, Dodgson JB, Cheng HH: A comprehensive screen for chicken proteins that interact with proteins unique to virulent strains of Marek's disease virus. Poultry Sci 2004;83:1117-1123.

13 Liu HC, Niikura M, Fulton J, Cheng HH: Identification of chicken stem lymphocyte antigen 6 complex, locus E (*LY6E*, alias *SCA2*) as a putative Marek's disease resistance gene via a virus-host protein interaction screen. Cytogen Genome Res 2003;102:304-308.

14 Niikura M, Dodgson JB, Cheng HH: Direct evidence of host genome acquisition by the alphaherpesvirus Marek's disease virus. Arch Virol 2006;151:537-549.

Dr. Hans H. Cheng, USDA, ARS, Avian Disease and Oncology Laboratory, East Lansing, MI 48823, USA.
E-mail: Hans.Cheng@ars.usda.gov

Pinard M-H, Gay C, Pastoret P-P, Dodet B (eds): Animal Genomics for Animal Health. Dev Biol (Basel). Basel, Karger, 2008, vol 132, pp 373-376.

Machine Learning Classification Procedure for Selecting SNPs in Genomic Selection: Application to Early Mortality in Broilers

N. Long[1], D. Gianola[1, 2], G.J.M. Rosa[2], K.A. Weigel[2], S. Avendano[3]

[1] Department of Animal Sciences, University of Wisconsin, Madison, WI, USA
[2] Department of Dairy Science, University of Wisconsin, Madison, WI, USA
[3] Aviagen Ltd., Newbridge, Midlothian, UK

Key words: machine learning, filter-wrapper feature selection, mortality, SNP

Abstract: In genome-wide association studies using single nucleotide polymorphisms (SNPs), typically thousands of SNPs are genotyped, whereas the number of phenotypes for which there is genomic information may be smaller. A two-step SNP (feature) selection method was developed, which consisted of filtering (using information gain), and wrapping (using naïve Bayesian classification). This was based on discretization of the continuous phenotypic values. The method was applied to chick early mortality rates (0-14 days of age) on progeny from 201 sires in a commercial broiler line, with the goal of identifying SNPs (over 5,000) related to progeny mortality. Sires were clustered into two groups, low and high, according to two arbitrarily chosen mortality rate thresholds. By varying these thresholds, 11 different "case-control" samples were formed, and the SNP selection procedure was applied to each sample. To compare the 11 sets of chosen SNPs, predicted residual sum of squares (PRESS) from a linear model was used. Naïve Bayesian classification accuracy was improved over the case without feature selection (from 50% to 90%). Seventeen SNPs in the best case-control group (with smallest PRESS) accounted for 31% of the variance among sire family mortality rates.

Introduction

A massive number of genetic markers, known as single nucleotide polymorphisms (SNPs), is now available. Because of their low mutation rate, SNPs are useful for association studies. A focus may be identifying patterns of SNPs that vary concomitantly with phenotypic values or disease status [1]. There is usually a small number of phenotypic observations (n) and a large number of SNPs (p) typed (the "large p, small n" problem). This creates difficulties in genome-wide association studies. The objective of this study was to explore machine learning methods for identifying SNPs associated with chick mortality in broilers. The machine learning approach consisted of discretizing continuous mortality rates of sire families into two classes, to coin the problem in a classification context. SNP discovery had two-steps. The first, or "filter" step, used information gain and reduced the thousands of SNPs to a small number. The second, or "wrapper" step, optimized the performance of the filtered SNPs via a naïve Bayesian classifier, with classification accuracy used as a gauge.

Materials and Methods

Data was from the Genomics Initiative Project at Aviagen, Ltd. (Newbridge, Scotland, UK) and consisted of mortality rates and SNPs in an elite broiler chicken line raised in a low hygiene environment. Mortality records were for early age (0-14 days) in chicks which were the progeny of 231 sires, 5,523 SNPs were available only for sires.

Different partitions of sire samples into two classes were generated by moving the thresholds c1 and c2, where c1 is the α ($0 \leq \alpha \leq 1$) quantile from the distribution of adjusted mortality rates over all sires and c2 is the 1-α quantile. Eleven different sets of values for (c1, c2) pairs corresponding to 11 different α values were adopted. These values were 0.025, 0.05, 0.10, 0.15, 0.20, 0.25, 0.30, 0.35, 0.40, 0.45 and 0.50. Sires with mortality rates $\leq c_1$ were referred to as "low-progeny-mortality" (LPM) sires; sires with mortality rates $\geq$ c2 were referred to as "high-progeny-mortality" (HPM) sires. Only LPM and HPM sires were adopted as training samples. The purpose of a feature (SNP) selection algorithm is to find relevant SNPs that can be used in the classification of items; this would allow for the identification of SNPs associated with variation in mortality rates. The filter and wrapper are illustrated in Figure 1. An information gain [2] based filter was used as the first step to reduce the feature space from thousands of SNPs to a small arbitrary number (e.g., 50), to eliminate irrelevant features. In the second step, the wrapper optimized the performance of the feature subset by adopting a naïve Bayesian classifier [3] to score the subset. The filter was coded in Java, and the wrapper procedure was run on the Weka platform [4].

The two-step SNP selection was applied to each of the 11 case-control groups of sires in exactly the same way. Different training samples result in different selected subsets of SNPs. In order to choose among the 11 subsets, a fixed-effects linear model was used in which explanatory variables were the selected SNPs from each case-control group, and the response variable was the (weighted) raw mortality rate for each sire family. Focus was on the *p*-value of the model and on the PRESS (predicted residual sum of squares) criterion [5]. A subset of SNPs was considered "best" if it produced the smallest PRESS among the 11 competing subsets.

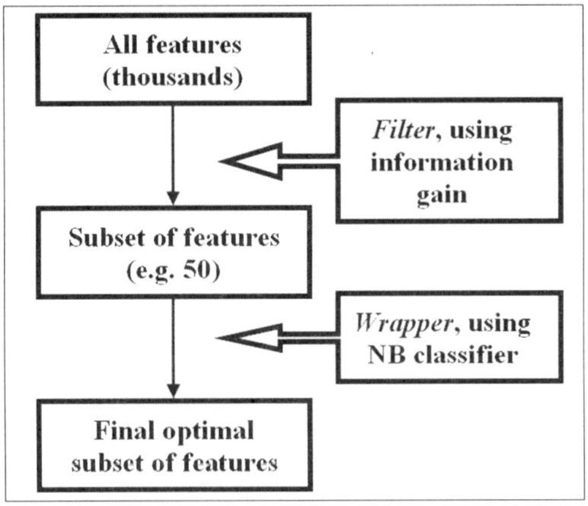

Fig. 1: The filter + wrapper feature selection approach. The subset of features obtained from the filter step, having a size that is reduced from the original input set, is fed into a wrapper step, which uses a naïve Bayesian (NB) classifier.

Results

Information gain of the 50 top-ranked SNPs was higher in Groups 1-3 than in other groups, meaning that individual SNPs reduced uncertainty considerably when sires were extreme, but to a much lesser extent when mean divergence between groups was low. In Groups 5-11 the distribution of IG was narrow, indicating that individual SNPs did not differ by much in their ability to reduce uncertainty about the mortality rate distribution.

Figure 2 depicts prediction accuracy achieved from using either all 50 SNPs or the reduced subset. For all 11 sample groups, feature selection increased classification accuracy noticeably.

Linear model with SNPs from Groups 1-3 were not significant (p-value > 0.05). Among the other groups, the best predictive performance was achieved with the partition of sires corresponding to Group 4 (63 sires in the training sample, 17 SNPs in the model). This corresponding main effects model had a PRESS statistic of 0.0027 and accounted for 31% of the variance among sire family mortality rates. Thus, the Group 4-generated SNP subset was regarded as optimal among those contrasted, all starting from the 50 top SNPs, based on IG.

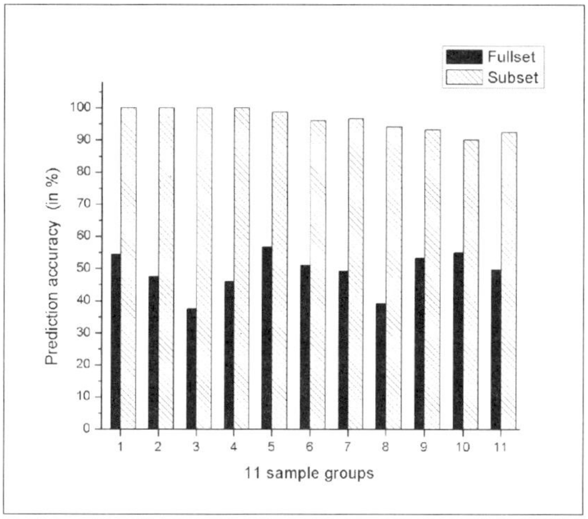

Fig. 2: Prediction accuracy of the top 50 SNPs (full set) and of those in the subset.

Discussion

The great improvement of naïve Bayesian classifier's classification accuracy after feature selection implies that, in the development of genomic-based predictive models, predictive performance of a model may be hampered by using all available markers which may be redundant with each other.

The main-effect linear model used for evaluating a set of SNPs did not account for interaction effects, but it gave a gauge for comparing different subsets of SNPs. Even with only 17 SNPs, a model accounting for some or all possible interactions may lack parsimony and may have a poorer predictive ability. More advanced curve fitting methods such as semiparametric regression [6] should be explored as candidates for capturing interaction effects.

CONCLUSION

This study focused on SNP subset identification, and on use of a filter-wrapper hybrid feature selection. An approach for using continuous phenotypic values in a classification framework was proposed, and preliminary results showed its feasibility and potential. Results suggested that the feature selection method, coupled with sample partition and subset evaluation procedures, provides a useful tool for finding important SNPs relevant to a trait of interest. This set could eventually be incorporated into a genetic evaluation model, more refined from a statistical point of view.

REFERENCES

1 Balding DJ: A tutorial on statistical methods for population association studies. Nature Rev Genet 2006;7:781-791.

2 Mitchell TM: Machine Learning. Hightstown, NJ, USA, McGraw-Hill, 1997.

3 Elkan C: Boosting and naïve Bayesian learning. Technical Report CS97-557 1997, University of California, San Diego.

4 Witten IH, Frank E: Data mining: practical machine learning tools and techniques. San Francisco, CA, USA, Morgan Kaufmann, 2005.

5 Ruppert D, Wand MP, Carroll RJ: Semiparametric regression. New York, NY, USA, Cambridge University Press, 2003.

6 Gianola D: Genomic-assisted prediction of genetic value with semiparametric procedures. Genetics 2006;173:1761-1776.

Nanye Long, Department of Animal Sciences, University of Wisconsin – Madison, 454 Animal Sciences Building, 1675 Observatory Dr., Madison, WI 53706, USA.

E-mail: nlong@wisc.edu

Pinard M-H, Gay C, Pastoret P-P, Dodet B (eds): Animal Genomics for Animal Health. Dev Biol (Basel). Basel, Karger, 2008, vol 132, pp 377-382.

The Use of RNAi and Transgenics to Develop Viral Disease Resistant Livestock

T.G. Wise, D.S. Schafer, J.W. Lowenthal, T.J. Doran

CSIRO Livestock Industries, Australian Animal Health Laboratory, Geelong, Australia

Key words: RNAi, transgenic, influenza, disease resistance

Abstract: The possibility of genetically engineering poultry to make them resistant to avian influenza is attracting attention and has now become a real possibility with improved methods for genetic modification and the emergence of RNAi as an antiviral strategy. In order to test this possibility, we have generated transgenic mice that express RNAi molecules targeting a conserved region of the influenza A NP gene and are testing these mice for resistance to influenza infection. Transgenes were initially developed that express short hairpin RNAs (shRNAs) targeting multiple influenza A viral genes. The shRNAs were tested for inhibition of H1N1 PR8 virus in vitro. Two potent shRNAs that target the NP and PA genes were chosen for lentiviral mediated generation of transgenic mice. Transgenic founders for the NP shRNA construct and also a negative control shRNA targeting EGFP were generated. The constitutive expression of the shRNA molecules in a range of tissue types including lung, was confirmed and so far stable transmission of the RNAi transgenes from the F0 to F3 generation has been observed. Resistance to influenza infection in these transgenic mice is now being confirmed.

INTRODUCTION

Clark and Whitelaw [1] proposed in their review "A future for transgenic livestock" that the advent of new methods such as RNA interference (RNAi) for modifying genomes will underpin a resurgence of research using transgenic livestock. They suggest this may be an important alternative to traditional breeding and could lead to the generation of farm animals that are more resistant to infectious disease such as foot-and-mouth disease or influenza. For example, the possibility of genetically engineering poultry to make them resistant to avian influenza is attracting attention and has now become a real possibility with improved methods for genetic modification and the emergence of RNAi as an antiviral strategy. In order to test this possibility, we have generated transgenic mice that express RNAi molecules targeting a conserved region of the influenza A NP gene and are testing these mice for resistance to influenza infection.

Materials and Methods

Construction of shRNA expression plasmids

shRNA expression plasmids (pshNP-1496, pshPA-2087 and pshEGFP) were made by Polymorphic (USA). Briefly, shRNA expression cassettes were synthesized to contain the mouse U6 promoter from pSilencer plasmid (Ambion) and the hairpin sequence containing the sense and antisense sequence of the influenza siRNAs, plus the loop and terminator sequences. The influenza siRNA sequences NP-1496 and PA-2087 [2] and the EGFP siRNA sequence [3] are as previously described. The hairpin loop and terminator sequences are also as previously described [4]. The synthesized shRNA genes were inserted into the Polymorphic cloning vector (Fig. 1) and the sequence was then confirmed.

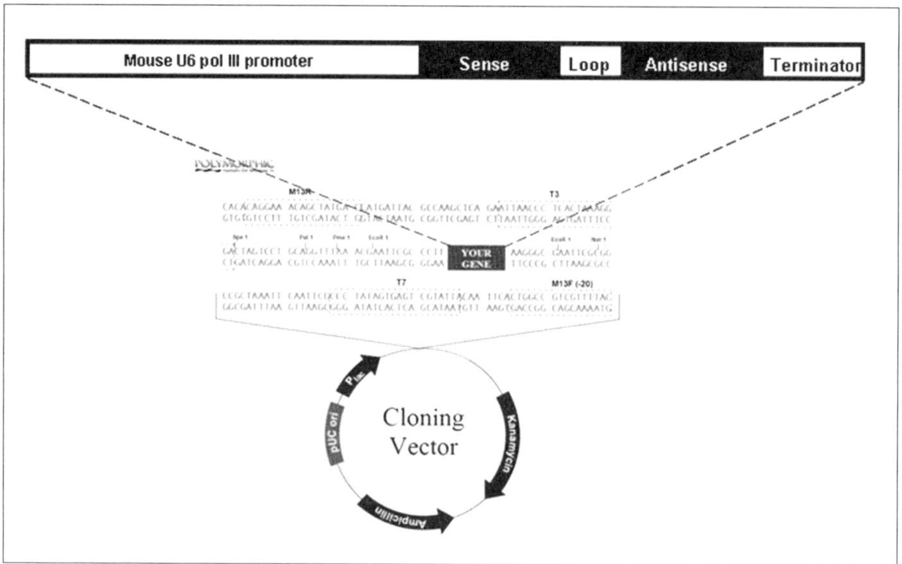

Fig. 1: Construction of shRNA expression plasmids.
shRNA expression cassettes containing the mouse U6 promoter, and hairpin RNA sequence were synthesized and ligated into the polymorphic cloning vector.

In vitro silencing of PR8 genes using shRNA plasmids

MDCK cells (MDCK-ATCC CCL-34) were cultured in Earls Modified Eagle's Medium containing 10% Foetal Calf Serum (FCS), 10 mM Hepes, 2 mM glutamine, supplemented with penicillin (100 U/ml) and streptomycin (100 µg/ml) at 37°C in a humidified atmosphere containing 5% CO_2. Logarithmic phase cells were trypsinized and 1.5 million cells used for each electroporation. Cells were resuspended in 100 µl of electroporation solution T (Amaxa), mixed with 2.5 µg of shRNA plasmid DNA and electroporated using program T20 of the Amaxa Nucleofector. Cells were plated out in a 24 well plate at 250,000 cells per well in duplicate for each infection. Cell cultures were infected with influenza A/PR/8/34 (PR8) virus (stock of allantoic fluid from ten day-old embyronated chicken eggs) 29 hours post transfection at a multiplicity of infection (MOI) of 0.01 and incubated at 37°C for 68 hours. Supernatants from infected cultures were assayed for the presence of virus in a heamagglutination (HA) assay. Also, cells from infected cultures were processed for total RNA isolation using a Qiagen RNeasy kit and viral mRNA was quantified by real time PCR with specific primers for influenza A M gene [5] using an ABI 7700 sequence detection system (Applied Biosystems).

Generation of transgenic mice

Transgenic mice (C57BL/6) were generated at Ozgene (Bentley, Australia). Briefly, the shPA-2087, shNP-1496 and shEGFP transgene cassettes were cloned into a lentiviral gene transfer vector. Transgenic constructs were then packaged into lentiviral particles and the lentiviral titres were determined. Lentiviral particles were injected into the perivitelline space of early stage embryos which were then implanted into pseudo-pregnant females. Resulting pups were then screened for presence of the transgene by PCR and Southern blot. Transgenic founders were only generated for the shNP-1496 and shEGFP transgenes. All experimental procedures involving mice were carried out according to protocols approved by the Australian Animal Health Laboratory Animal Ethics Committee and the Animal Resources Center/Ozgene Animal Ethics Committee.

Confirmation of shRNA expression in F1 transgenic mice

Confirmation of expression of the shNP-1496 and shEGFP transgene in F1 transgenic mice was carried out using an RNAse protection assay (RPA) on extracts enriched for small RNAs isolated from mouse tail snips and selected organs using the mirVana microRNA (miRNA) Isolation Kit (Ambion). RNA oligonucleotides complementary to EGFP-126 and NP-1496 were end labelled with [γ-^{32}P] ATP using the mirVana Probe & Marker Kit (Ambion) and hybridized to the enriched small RNAs using the mirVana miRNA Detection Kit (Ambion). RNAse protected fragments were then separated by electrophoresis on a 15% denaturing polyacrylamide/8 M urea gel and detected by autoradiography. Expression of shEGFP and shNP-1496 was confirmed with strong signals of expected size for each shRNA.

Results

Recent studies have shown dramatic inhibition of influenza A viruses using chemically synthesized small interfering RNAs (siRNAs) targeting highly conserved sequences of influenza A genes [2]. We have taken two of these siRNAs targeting PA and NP genes, and an irrelevant siRNA targeting EGFP [3], and expressed them as shRNAs from a plasmid system utilising the mouse U6 promoter in pSilencer plasmid (Ambion).

The shRNA vectors were generated to ensure placement of the hairpin sequence in the +1 position relative to the native mouse U6 transcript. The EGFP siRNA sequence was obtained from Kim and Rossi [3], the PA-2087 and NP-1496 siRNA sequences are from Ge et al. [2]. All three shRNA plasmids were sequenced to confirm that components of the shRNA expression cassettes were intact. The shRNA plasmids were then transfected into MDCK cells which were then challenged with influenza A PR8 at a multiplicity of infection (MOI) of 0.01.

Supernatants from the infected cultures were assayed for the presence of virus in an haemagglutination (HA) assay. Also, cells from infected cultures were processed for total RNA isolation and viral mRNA was quantified by real time PCR with specific primers for influenza A M gene [5]. The results of the HA assay showed that both shPA-2087 and shNP-1498 greatly inhibited viral HA activity at the tested MOI, with the more potent shNP-1496 giving total inhibition (Fig. 2). These results also correlated with observed CPE of transfected MDCKs. Cells transfected with the two PR8 shRNA constructs showed much reduced or no CPE compared with the shEGFP control plasmid, where total destruction of the cell monolayer was observed (data not shown). Real time PCR measurements of viral RNA confirmed HA titres and CPE observations, showing a potent and significant reduction in M gene mRNA of >98% for shNP-1496 and > 80% for shPA-2087 (Fig. 2).

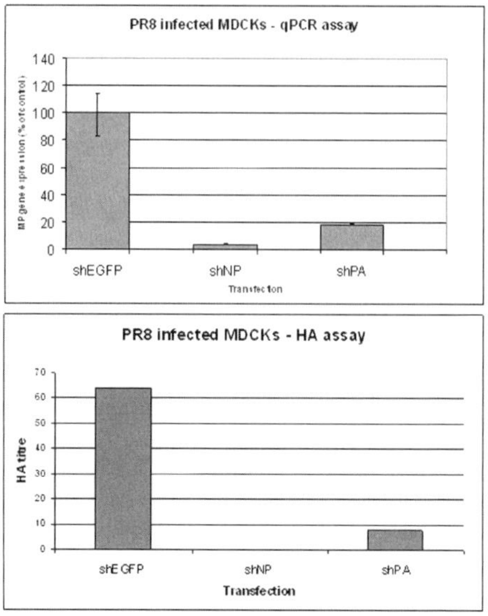

Fig. 2: Inhibition of influenza A virus replication in MDCK cells.
Cells were transfected with either pshNP-1496, pshPA-2087 or pshEGFP and then infected with PR8 virus. Inhibition of virus replication as a result of RNAi gene silencing was then measured using a haemagglutination (HA) assay or quantitative PCR (qPCR) assay.

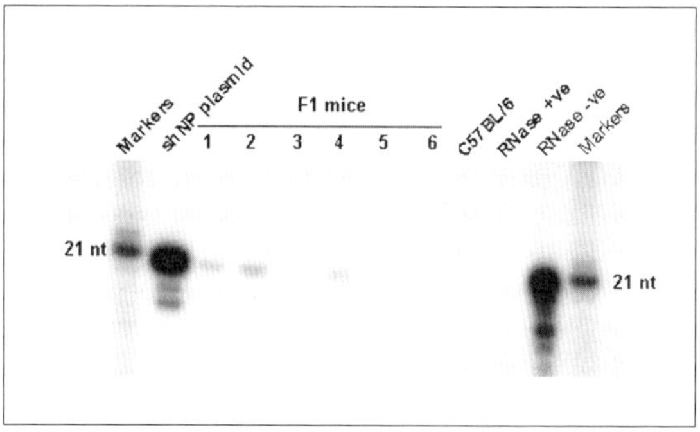

Fig. 3: Expression of shNP-1496 in transgenic mice.
Confirmation of expression of the shNP-1496 transgene in F1 transgenic mice was carried out using an RNAse protection assay (RPA) on extracts enriched for small RNAs isolated from mouse tail snips and selected organs. RNA oligonucleotides complementary to NP-1496 were end labelled with [γ-^{32}P] ATP and hybridized to the enriched small RNAs. RNAse protected fragments were then separated by electrophoresis on a 15% denaturing polyacrylamide/8 M urea gel and detected by autoradiography. Expression of shNP-1496 was confirmed with strong signals of expected size in the expected lanes.

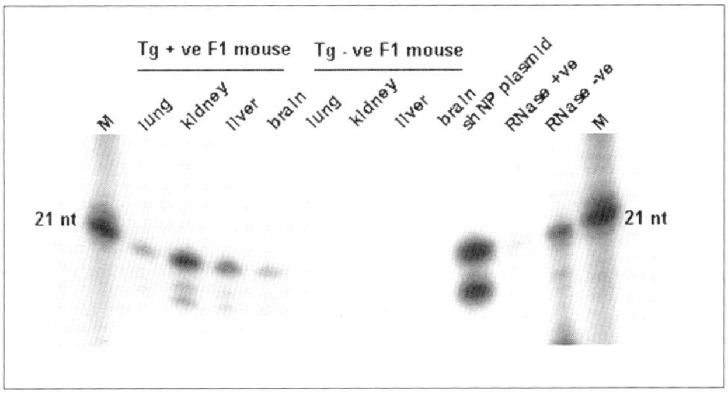

Fig. 3

Based on the in vitro silencing results, the shRNA genes were used to generate transgenic mice. Only transgenic founder mice containing the shNP-1496 and control shEGFP transgenes could be generated. The transgenic mice are currently being bred to establish homozygous lines and will then be tested for resistance to influenza infection in challenge experiments. Thus far, stable transmission of the RNAi transgenes from the F0 to F3 generation has been observed. Confirmation of expression of both shRNAs in tail snips and selected organs from transgenic mice has been done using an RNAse protection assay (RPA) on extracts enriched for small RNAs isolated from the mice. Expression of shEGFP, and shNP-1496 was confirmed with strong signals of expected size for each shRNA (Fig. 3).

Discussion

We have developed and described transgenes that express short hairpin RNAs (shRNAs) targeting several influenza A viral genes [2]. The shRNAs were tested for inhibition of influenza A PR8 virus in vitro. Two potent shRNAs that target the NP and PA genes were chosen for lentiviral mediated generation of transgenic mice. Only transgenic founders for the NP shRNA construct and a negative control shRNA targeting EGFP could be generated. The constitutive expression of the shRNA molecules in a range of tissue types including lung, has been confirmed and so far stable transmission of the RNAi transgenes from the F0 to F3 generation has been observed. Resistance to influenza infection in these transgenic mice is now being confirmed.

Overall, this study confirms previous work indicating that NP-1496 and PA-2087 are potent inhibitors of influenza A viruses. The development of mice that can express these molecules from an integrated transgene will provide a significant test of the efficacy and value of using RNAi transgenes as a transformational approach to controlling viral disease in livestock species. For example, if successful, the development of transgenic poultry that are resistant to avian influenza would certainly be of benefit to the poultry industry and would also help to reduce the risk of a human influenza pandemic by reducing the potential for virus re-assortment within intensively reared birds.

Conclusion

Transgenes were initially developed that express short hairpin RNAs (shRNAs) targeting several influenza A viral genes [2]. The shRNAs were tested for inhibition of influenza A PR8 virus in vitro. Two potent shRNAs that target the NP and PA genes were chosen for lentiviral mediated generation of transgenic mice. Only transgenic founders for the NP shRNA construct and a negative control shRNA targeting EGFP could be generated. The constitutive expression of the shRNA molecules in a range of tissue types including lung, has been confirmed and so far stable transmission of the RNAi transgenes from the F0 to F3 generation has been observed. Resistance to influenza infection in these transgenic mice is now being confirmed.

References

1 Clark J, Whitelaw B: A Future for Transgenic Livestock. Nat Rev Genet 2003;4:825-833.

2 Ge Q, McManus MT, Nguyen T, Shen CH, Sharp PA, Eisen HN, Chen J: RNA interference of influenza virus production by directly targeting mRNA for degradation and indirectly inhibiting all viral RNA transcription. Proc Natl Acad Sci USA 2003;100:2718-2723.

3 Kim DH, Rossi JJ: Coupling of RNAi-mediated target downregulation with gene replacement. Antisense Nucleic Acid Drug Dev 2003;13:151-155.

4 Wise TG, Schafer DJ, Lambeth LS, Tyack SG, Bruce MP, Moore RJ, Doran TJ: Characterisation and comparison of chicken U6 promoters for the expression of short hairpin RNAs. Anim Biotechnol 2007;18:153-162.

5 Heine HG, Trinidad L, Selleck P, Lowther S: Rapid detection of highly pathogenic avian influenza H5N1 virus by TaqMan reverse transcriptase -polymerase chain reaction. Avian Dis 2007;51(Suppl 1):370-372.

Dr. Timothy Doran, CSIRO Livestock Industries, Private Bag 24, Geelong VIC 3220, Australia.
E-mail: timothy.doran@csiro.au

Pinard M-H, Gay C, Pastoret P-P, Dodet B (eds): Animal Genomics for Animal Health. Dev Biol (Basel). Basel, Karger, 2008, vol 132, pp 383-389.

The Use of High Density Genotyping in Animal Health

M.E. Goddard

School of Agriculture and Food Systems, University of Melbourne, Melbourne Australia and Department of Primary Industries, Victoria, Australia

Key words: genomic selection, genome-wide association study, single nucleotide polymorphism, complex traits, causative mutation

Abstract: The sequencing of genomes, such as that of the cow, has led to the discovery of thousands of single nucleotide polymorphisms (SNPs). By combining this knowledge with new methods that can genotype thousands of SNPs efficiently, it has become possible to carry out genome-wide association studies in domestic animals to map genes for complex traits, including disease resistance, using the linkage disequilibrium between the SNPs and the unknown genes affecting the trait of interest. Although experiments using 10,000 SNPs and 384 animals have found many significant associations, power calculations suggest that we need >50,000 SNPs and >1000 animals to map genes explaining most of the genetic variance for complex traits. Such experiments are now underway and the results will have two applications. Firstly, they will lead to panels of SNPs that can be used to accurately select animals with high breeding value for desired traits leading to a great increase in the rate of genetic improvement. Secondly, they will form the first step in identifying the genes and mutations that cause variation in complex traits. A collaborative approach to achieving this second goal is proposed.

INTRODUCTION

Genetic differences between animals are an important cause of differences in susceptibility to disease. In classical Mendelian abnormalities, a mutation in a single gene causes the disease. Most of these disorders are rare because natural and artificial selection decrease the frequency of the abnormal allele. However, for most common diseases (e.g., mastitis in dairy cows) many genes and environmental factors determine susceptibility. In the human disease literature these are referred to as 'complex diseases.'

For simple Mendelian disorders, linkage analysis has been quite successful in finding the gene and mutation responsible for the disease. Linkage analysis has also been used to map genes for complex diseases such as somatic cell score. However, because the effect of individual genes is small, the confidence interval with which they are mapped is large, commonly up to 50 cM. Consequently it has proven very difficult to identify the gene responsible or to find genetic markers that can be used for marker assisted selection (MAS).

Linkage mapping uses genetic markers that are linked to a gene with an effect on the trait of interest. Because they affect quantitative traits, these genes are called quantitative trait loci (QTL). In theory, genetic markers that are linked to a QTL but in linkage equilibrium with it can be used for MAS [1] and this has been demonstrated in practice in dairy cattle [2]. It requires that the linkage phase between the marker and QTL be established in every family in which MAS is to be practiced and in most cases this is not feasible or too expensive.

An alternative to linkage mapping is to use markers in linkage disequilibrium (LD) with the QTL in an association study [3,4]. Linkage disequilibrium means that one allele of the marker occurs on the same chromosome as the disease causing allele more often than expected by chance. This causes an association between the marker allele and the incidence of disease. LD generally occurs only between genes that are close together on the chromosome. Therefore, if we wish to be able to find QTL located anywhere in the genome, we need a dense coverage of the whole genome with markers. This has become feasible in the last two years due to genome sequencing projects that have discovered thousands of single nucleotide polymorphisms (SNPs) and to the development of genotyping platforms that can genotype thousands of SNPs efficiently. These developments have lead to genome-wide association studies (GWAS) using 10,000, 25,000 or 50,000 SNPs in cattle and up to 1,000,000 SNPs in humans.

The results from GWAS are useful for two purposes in livestock. Firstly, the markers that are associated with a trait of interest can be used for MAS. GWAS, which uses markers covering the whole genome, has been called genomic selection [5] to distinguish it from older forms of MAS that used only a few markers. Secondly, the markers that are associated with the trait can be used to find the genes that have an effect on the trait and, ideally, the mutations in the genes that have the effect. This information will hopefully lead to a better understanding of the biology controlling susceptibility to the disease and hence to improved control strategies.

This paper reviews the results from GWAS, the application of genomic selection, and methods for finding the causal mutations, and put forth a proposal for a large project to find hundreds of causal mutations for disease susceptibility in livestock.

Genome-wide Association Studies

Number of genes affecting complex traits

We have performed GWAS in dairy and beef cattle using assay of approximately 10,000 SNPs commercialised by the company Parallele (the 10k assay). When so many markers are tested, some significant results are expected by chance alone. This implies that very stringent significance levels are necessary to be confident that a particular association between an SNP and a trait is not due to chance. Calculation of false discovery rates (FDR) is a useful method for deciding on the significance threshold that should be used. However, this depends on the purpose of the analysis. If the focus is on a particular SNP, one might want a low FDR before investing in further, expensive research on that SNP. However, we are interested in the total number of genes affecting traits of interest and can estimate this using the logic of FDR. Table 1 lists the number of SNPs from the 10k assay that were significant for four traits in dairy cattle. For all four traits, there are many more significant SNPs than expected by chance. For instance, out of approximately 8000 SNPs assayed, 654 were significant at $p<0.03$ for milk protein yield, whereas only 240 would be expected by chance. From this we can calculate that approximately 414 of the SNPs have true associations with milk protein yield. This does not imply that 414 genes affect protein yield because several SNP associations might be due to the same QTL. In Hayes et al [6], we allowed for this and estimated the number of genes for protein yield in dairy cattle and net feed intake in beef cattle at 150 - 200. Table 1 implies that a large number of genes also affect the variation in other traits.

A number of GWAS have been reported in humans for diseases such as diabetes [7]. These reports find that for markers validated in independent samples, each copy of an allele that predisposes to a disease typically increases the risk of that disease by a factor of 1.1 to 1.2. This means that individual genes have a small effect on disease susceptibility and that hundreds or thousands of genes must have an effect on susceptibility to each disease.

Thus, both human and cattle experiments are finding that complex traits are typically controlled by hundreds of genes, most of which have a small effect.

Table 1: Number of significant SNPs in GWAS.

Trait	**Number of significant SNPs**	
	($p<0.03$)	($p<0.001$)
Protein yield	654	66
Somatic Cell Score	362	12
Longevity	669	66
Fertility	545	41
Number expected by chance	240	8

Power of GWAS

Despite these encouraging results, we have concluded that the 10k assay does not find all the genes affecting a complex trait. Macleod et al [8] used the 10k SNP genotypes but simulated phenotypes by assuming that one of the SNPs was in fact a QTL with an effect on the trait. Even when the QTL explained 20% of the phenotypic variance, another SNP within 5 Mb had a significant association with the trait in only 30% of cases. This is because the 10,000 unevenly spaced SNPs do not cover the whole genome densely enough to detect all QTL.

A simple 'back of the envelope' calculation can be done to indicate the size of experiment necessary to have the power to detect most QTL. For each SNP, we can calculate the correlation (r) between the trait and the SNP where the SNP is scored as 0 or 1 or 2 for homozygote, heterozygote, and other homozygote. Then, under the null hypothesis that the true correlation is zero, Nr^2 is distributed as an F statistic with degrees of freedom (1, N-2) where N is the number of animals included in the correlation. The expected value of r^2 can be calculated as follows:

$$r^2(\text{marker, trait}) = r^2(\text{marker, QTL}) * r^2(\text{QTL, trait})$$

r^2(QTL, trait) is the proportion of variance in the trait explained by this QTL.

If there are 100 genes plus environmental factors affecting the trait, the average QTL must explain less than 0.01 of the phenotypic variance. However, not all genes have equal effects and at least some will explain 0.02 of the variance. r^2 (marker, QTL) is the traditional statistic to measure LD between the marker and the QTL. We do not know how high this is but we could estimate it by looking at the LD between two markers. For SNPs 300 kb apart (the average distance with 10,000 SNPs in cattle), the average r^2 is about 0.1 - 0.2. Therefore, optimistically, the expected r^2 (marker, trait) = 0.02 x 0.2 = 0.004. If we use a significance level of $p<0.001$, F must be >10 and so N must be at least 2,500, so that the expected value of F = 2,500 x 0.004 = 10.

However with 50,000 markers, the average distance between markers is 60 kb and the average LD between markers 60 kb apart is about $r^2 = 0.4$. Consequently to get an expected F = 10, we require N = 1250.

The conclusion from these power calculations is that GWAS need large numbers of markers (e.g., 50k) and large numbers of animals (>1000).

If markers are to be used for genomic selection, it would be of great practical benefit if the same markers could be used in all breeds. We have investigated this question by examining whether the LD between two markers is consistent across breeds in the expectation that the result will also be applicable to the LD between a QTL and a marker. Goddard et al [9] found that, if markers were only 10 kb apart, the LD was similar in Angus and Holsteins, but if the markers were 300 kb apart, the LD was quite different in the two breeds. Thus, we also need dense markers if we hope to find markers that can be used for selection in multiple breeds.

The extent of LD affects the ease with which markers in LD with a QTL can be found and it also affects the accuracy with which the QTL can be mapped. If LD exists at large distances, less markers are needed to detect the QTL but they cannot be mapped to a precise position because markers over a wide distance are associated with the trait. The precision of mapping can be improved by using multiple breeds. If a marker shows a consistent association with the trait across several breeds then it must be close to the QTL.

Genomic Selection

The obvious practical benefit of discovering markers associated with a trait is that they can be used to select animals for that trait. However, until recently the practical benefits from MAS have been small [10]. To obtain a large benefit, most of the genetic variance must be tracked by the markers and the markers must be in strong LD with the QTL [11]. These two objectives can be achieved by using dense markers covering the whole genome. It may not be necessary to use all markers when genomic selection is put into practice but it is at least necessary to use dense markers in order to discover the subset of markers that are needed to predict the breeding value of selection candidates.

Meuwissen et al [5] used simulation to show that genomic selection was feasible. They found that breeding value could be predicted with an accuracy of 0.8 even in animals with no phenotypic record and no offspring. That is, breeding value could be predicted at birth with an accuracy comparable to a progeny test. This would be useful for all traits and is expected to speed up the rate of genetic improvement considerably [12]. It would be especially useful for disease resistance because it is often difficult to select for this by conventional means. Wray et al [13] considered prediction of genetic risk to disease in humans and concluded that high accuracy of prediction was possible provided the prediction equation was derived from large GWAS (e.g. 1,000 to 10,000 patients and controls).

Identifying the Causal Gene and Mutation

The markers used in GWAS are usually random polymorphisms and are not expected to have a functional effect on any trait. The association they have with a trait is due to LD between the marker and the QTL which has a functional effect on the trait. We would like to identify the QTL, i.e., to know the gene and the polymorphism in that gene, that actually cause an effect on the trait. It is very difficult to prove that a particular change in DNA sequence is the cause of the effect on a trait. Generally, circumstantial evidence is gradually built up for a particular hypothesis. First, the QTL can be mapped as accurately as possible. Using dense markers is beneficial, but there is a limit to the precision achievable due to the extent of LD. Using multiple breeds should lead to greater precision because LD is consistent across breeds only for genes closely linked. The problem of mapping a QTL can be converted to the problem of mapping a single gene if the genotype of animals at the QTL is established, for instance, by progeny testing sires. Genome sequencing within the confidence interval of the QTL can then be used to find all the polymorphisms in the region. Hopefully, only one polymorphism is consistent with the known QTL genotypes. If there are more than one possible polymorphism, we must rely on functional data. A polymorphism in a gene with a known role in the physiology of the trait is more likely to be the causal mutation than a polymorphism in another gene. A polymorphism that changes amino acid sequence or affects the regulation of the gene is more likely to cause an effect on the trait than one that does neither of these things. The hypothesis concerning a particular polymorphism can be supported by demonstrating a chain of functional effects such as differences in enzyme activity or differences in circulating hormone concentration. In theory, the hypothesis can be tested by making a transgenic animal that contains just the change in DNA sequence that is thought to cause the effect, but this is seldom possible in livestock.

A Proposal

Many breeds of livestock have a small effective population size of 50-100. This means that most chromosome segments found in animals of the current generation trace back to one of <100 key ancestors within a few generations. This short coalescence time means that the chromosome segments are large and could be recognised by their haplotype at a group of markers. Consequently, we can carry out *'in silico'* genotyping as follows. The key ancestors are genotyped for a dense set of markers.

Current generation animals are genotyped for just enough markers to enable chromosome segments to be matched to the segments carried by the key ancestors. Once they are matched, all the markers that have been genotyped on the key ancestor can be imputed to the matching chromosome segment in the current animal. In this way, we could genotype a large number of current animals for a moderate number of markers but obtain *in silico* genotypes for a large number of markers. As the cost of genome (re)sequencing drops, this *in silico* genotyping can be extended to *in silico* sequencing. That is, the key ancestors are sequenced and then the full genome sequence is imputed for chromosome segments in current animals that match the chromosome segment in key ancestors.

This *in silico* genotyping and sequencing is similar to the use of tagged SNPs. However, an important difference is that this *in silico* method makes use of the known relationships between current animals and key ancestors only a few generations ago. This greatly reduces the number of markers that must be genotyped for the current animals.

This method could be used to greatly reduce the cost of genomic selection by reducing the number of markers needed to be typed for selection candidates. The method could also be used to identify the causal polymorphisms underlying QTL. As set out above, genome sequencing is an important step in identifying a QTL. If each QTL is attacked separately, the genome sequencing is targeted to a particular region. However, since there are so many QTL affecting many traits of interest, this is a very inefficient approach. It would be much more efficient to do complete genome (re)sequencing and search for many QTL simultaneously. By using *in silico* sequencing, a large experiment with thousands of fully genome sequenced animals will be possible within five years, if sequencing costs drop as expected. The proposal would involve genotyping for a moderate number of markers a large sample of animals that had been recorded for many traits of interest (e.g., disease traits), sequencing the genomes of the key ancestors, tracing the chromosome segments of the current animals back to their ancestors, imputing the full genome sequence on each animal, and carrying out a GWAS based on full genome sequence.

This proposed experiment would identify hundreds of mutations affecting disease susceptibility or other traits. It is an ideal project for international collaboration because it replaces many small experiments with one large experiment with enormous savings in cost. Once the key ancestors have been sequenced, any researchers could use this information to impute genome sequence on their own group of experimental animals.

Conclusion

Most complex traits are controlled by over 100 genes. GWAS are a powerful tool for mapping these QTL but >50,000 markers and >1,000 animals are needed to have the power to detect genes explaining most of the genetic variance. Genomic selection has the potential to greatly increase the rate of genetic improvement especially for disease resistance. A proposed, large international project based on *in silico* genome sequencing of thousands of animals has the potential to discover many of the genes causing variation in disease resistance and other traits.

References

1 Fernando RL, Grossman M: Marker assisted selection using best linear unbiased prediction. Genet Sel Evol 1989;21:467-477.

2 Boichard D. Fritz S, Rossignol MN, Guillaume F, Colleau JJ, Druet T: Implementation of marker assisted selection: practical lessons from dairy cattle. Proc 8th WCGALP, 2006; 22-11.

3 Goddard ME: Mapping genes for quantitative traits using linkage disequilibrium. Genet Sel Evol 1991;23(suppl. 1):1315-1345.

4 Meuwissen THE, Karlsen A, Lien S, Olsaker I, Goddard ME: Fine mapping of a quantitative trait locus for twinning rate using combined linkage and linkage disequilibrium mapping. Genetics 2002;161:373-379.

5 Meuwissen THE, Hayes BJ, Goddard ME: Prediction of total genetic value using genome wide dense marker maps. Genetics 2001;157:1819-1829.

6 Hayes BJ, Chamberlain A, Goddard ME: Use of linkage markers in linkage disequilibrium with QTL in breeding programs. Proc. 8th WCGALP, 2006; 30- 06.

7 Frayling T: Genome-wide association studies provide new insight into type 2 diabetes aetiology. Nat Rev Genet 2007; 8:657-662.

8 MacLeod IM, Hayes BJ, Goddard ME: Efficiency of dense bovine single-nucleotide polymorphisms to detect and position quantitative trait loci. Proc. 8th WCGALP, 2006; 20-04.

9 Goddard ME, Hayes B, McPartlan H, Chamberlain AJ: Can the same genetic markers be used in multiple breeds? Proc. 8th WCGALP, 2006; 8:22-16.

10 Dekkers JC: Commercial application of marker- and gene-assisted selection in livestock: strategies and lessons. J Anim Sci 2004;82:E313-328.

11 Goddard ME, Hayes BJ: Optimisation of response using molecular data. Proc. 7th WCGALP 2002;33:3-10.

12 Schaeffer LR: Strategy for applying genome-wide selection in dairy cattle. J Anim Breed Genet 2006;123:218-223.

13 Wray NR, Goddard ME, Visscher PM: Prediction of individual risk to disease from genome-wide association studies. Genome Res 2007;17:1520-1528.

Dr Mike Goddard, School of Agriculture and Food Systems, University of Melbourne, Parkville, Victoria, Australia.

E-mail: Mike.Goddard@dpi.vic.gov.au

Pinard M-H, Gay C, Pastoret P-P, Dodet B (eds): Animal Genomics for Animal Health. Dev Biol (Basel). Basel, Karger, 2008, vol 132, pp 391-398.

Epigenetic Regulation of Genomes: Nutrient-specific Modulation of Genetic Networks in Bovine Cells

C.-J. Li, T.H. Elsasser, R.W. Li

USDA-ARS, Bovine Functional Genomics Laboratory, Beltsville, MD, USA

Keywords: butyrate, cell cycle, epigenetics, genetic network, ruminant

Abstract: The modern version of epigenetics includes the molecular mechanisms that influence the phenotypic outcome of a gene or genome, in absence of changes to the underlying DNA sequence. A host of genomic interrelationships with the diet evidently exist. The broad topic of nutrigenomics is defined as the interaction between nutrition and an individual's genome. Ruminant species have evolved to metabolize the short-chain volatile fatty acids (VFAs, acetate, propionate, and butyrate) to fulfill up to 70% of their nutrient energy requirements. The potential biological roles of VFAs were investigated using the established Madin-Darby bovine kidney epithelial cell line. Butyrate induces cell cycle arrest and apoptosis in bovine cells. Gene expression profiling indicated that butyrate induces many significant changes in the expression of genes associated with regulatory pathways that are critical to cell growth, immune response and signal transduction. Functional category and pathway analyses of the microarray data revealed that several canonical pathways (the cell cycle G2/M DNA damage checkpoint and G1/S checkpoint regulation; pyrimidine metabolism; and purine metabolism insulin-like growth factor axis components) were significantly affected.

INTRODUCTION

The modern version of epigenetics includes the molecular mechanisms that influence the phenotypic outcome of a gene or genome, in the absence of changes to the underlying DNA sequence. As this area continues to unfold, it is becoming increasingly evident that a host of genomic interrelationships with the diet exist which encompasses the broad topic of nutrigenomics, defined as the interaction between nutrition and an individual's genome.

It is now well established that microbial fermentation in the gastrointestinal tract contributes to the energy balance of all mammalian species [1,2]. Short-chain fatty acids (acetate, propionate, and butyrate) are formed during microbial fermentation of dietary fibre in the gastrointestinal tract and then are directly absorbed at the site of production. Ruminant species have evolved to metabolize the short-chain volatile fatty acids (VFAs, acetate, propionate, and butyrate) to fulfill up to 70% of their nutrient energy requirements [2]. While acetate and propionate hold a prominent position in providing energy to ruminant metabolism, butyrate, although low in its relative concentrations, appears to be involved in aspects of metabolism beyond its role as a nutrient. The presented study, using genomics and molecular techniques, bioinformatics and computation, as well as knowledge integration, reveals that the short-chain volatile fatty acids (VFA, acetate, propionate, and butyrate), especially butyrate, are involved in metabolism not only as nutrient, but also as genomic regulator.

Materials and Methods

Madin-Darby bovine kidney epithelial cells (MDBK) were cultured in Eagle's minimal essential medium and supplemented with 5% foetal bovine serum At approximately 50% confluence (during the exponential phase), the cells were treated with 10mM of sodium butyrate for 24h. Cell populations (cell cycle progression and DNA synthesis) were monitored using flow cytometric analysis. The bovine microarray platform used was described previously [3]. The data generated from microarray studies was explored using Ingenuity Pathways Analysis (Ingenuity® Systems, www.ingenuity.com), a web-delivered application that enables the discovery, visualization and exploration of molecular interaction networks in gene expression data. Canonical pathway analysis, functional analysis, pathways analysis and network generation and functional analysis of a Network were performed.

Results

Butyrate treatment of MDBK cells arrests cells in G1 phase and induces apoptosis. Average flow cytometric cell cycle distribution from three experiments using butyrate treatment showed an arrest of cells in the G1 phase with the majority of cells in G0/G1 (76 ± 2.9%), while 8.9 ± 1% and 15 ± 2.4% of the cells were in S and G2/M phases, respectively (Figure 1).

To understand the mechanism of cell cycle arrest induced by butyrate, a systematic search for butyrate-responsive genes in MDBK cells was conducted using a high-density oligonucleotide microarray, followed by detailed canonical pathway analysis and regulatory network identification. Using Significance Analysis of Microarray (SAM), 450 genes significantly regulated by sodium butyrate at a very stringent false discovery rate (FDR) of 0% were identified. The largest category of genes regulated by butyrate was cell cycle control related. Fourteen genes related to apoptosis and extracellular matrix (ECM) were also significantly regulated by butyrate.

The functional category and pathway analysis of differentially expressed genes in cells treated with butyrate were explored using the Ingenuity Pathway Knowledge base. Four canonical pathways (the cell cycle G2/M DNA damage checkpoint and G1/S checkpoint regulation; pyrimidine metabolism; and purine metabolism) were significantly ($P<0.05$) regulated. Cell cycle checkpoint pathways (the G2/M DNA damage checkpoint and G1/S checkpoint regulation) were impaired due to treatment with butyrate; for the first time, two pathways critical for the regulation of purine and pyrimidine metabolism were linked to the induced biological effects of butyrate (Table 1). These findings underlie mechanisms of butyrate induced cell arrest in the progression of the cell cycle. Global functional analysis also indicated that the levels of many genes involved in the cell cycle, cellular movement, DNA replication, recombination and repair, cellular growth and proliferation are significantly and consistently dysregulated. The top 10 dysregulated functions induced by butyrate are listed in Table 2.

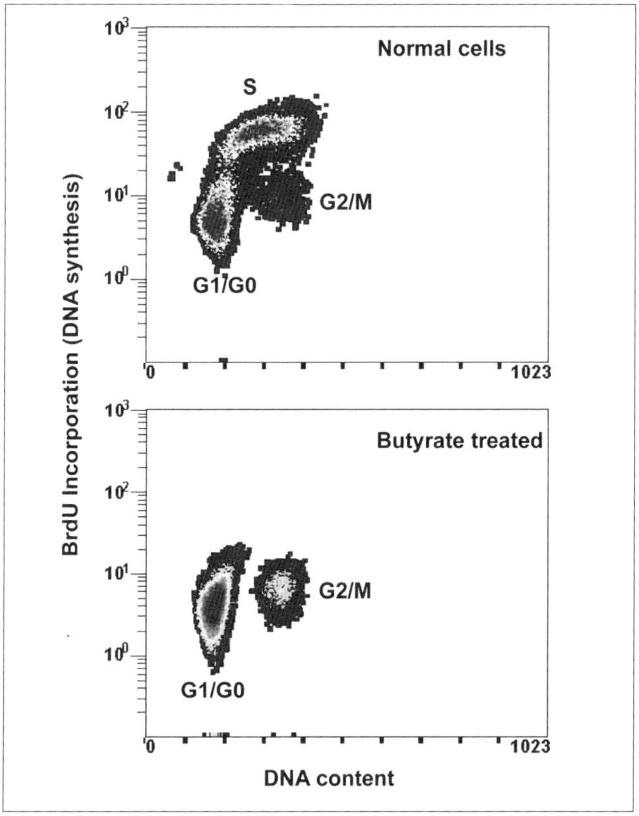

Fig. 1: Butyrate treatment arrested cells in G1/G0 phases.
For flow cytometric analysis, exponential growing cells (normal cells) and butyrate treated cells were first pulse labeled with BrdU for 30 min. Collected cells were stained with diluted fluorescent (Fluorescent isothiocyanate, FITC) anti-BrdU antibody and then stained with DNA marker (7-ADD). The fluorescent signal generated by FITC was acquired in a logarithmic mode (DNA synthesis), and fluorescent signal from the DNA-content marker 7-ADD was normally acquired in the linear signal amplification mode. Two colour staining separated cells into three populations, G1/G0, S and G2/M phases. Butyrate treatment arrested cells in G1/G0 and almost eliminated the S cell population.

Table 1: Canonical pathways regulated by butyrate in MDBK cells.

Pathway	-log(p-value)	Differential expressed genes
Cell cycle: G2/M DNA damage checkpoint regulation	3.60E+00	*TP53, CDC25B, CDC25C, CCNB1, CCNB2, CDC2, CHEK1*
Pyrimidine metabolism	3.32E+00	*NME5, ITPA, ENTPD2, DKC1, RRM2, POLA2, REV3L, DTYMK, NXN, TK1, RFC3, NME6*
Cell cycle: G1/S checkpoint regulation	2.88E+00	*TP53, E2F4, PA2G4, TGFB1, CDK4, CDC25A, HDAC, E2F, EBP1*
Purine metabolism	2.83E+00	*NME5, PKLR, RRM2, REV3L, NME6, ADSS, IMPDH2, ITPA, ENTPD2, IMPDH1, MPHOSPH1, ADA, POLA2, PDE5A, GMPR, RFC3*

Table 2: Major cellular functions targeted by butyrate.

Function	-log(p-value)	Number of genes targeted
Cell cycle	1.9E-11-3.56E-02	82
Cellular movement	2.84E-07-3.64E-02	44
DNA replication, recombination, and repair	4.57E-06-3.52E-02	75
Cellular assembly and organization	1.75E-05-3.1E-02	48
Cellular growth and proliferation	8.51E-05-3.35E-02	97
Cellular function and maintenance	3.2E-04-3.27E-02	9
Cell death	6.15E-04-3.27E-02	94
Cell-to-cell signaling and interaction	9.14E-04-3.61E-02	27
Nucleic acid metabolism	9.14E-04-1.36E-03	5
Cellular development	1.04E-03-3.56E-02	23

To determine the biologically relevant networks and pathways of the differentially expressed genes, pathway analysis was done on the integrated total data sets using the Ingenuity Pathways Knowledge Base. A total of 26 regulatory networks were linked to butyrate. These networks had an Ingenuity Pathway Analysis (IPA) score ≥ 10, indicating a less than 10^{-10} chance that the genes in the network are associated together solely due to random events. The networks describe functional relationships between gene products based on known interactions reported in the literature. The top five networks that are targeted by butyrate are listed in Table 3. The first network was centred on upregulated TGFB1. In addition, the most significant characterization of this network was the downregulated expression of the genes such as *IGFBP6*, *E2F4*, *CDC2*, *CKS*, *CCNA2*, *CCNB1* and *CHEK1*, which are all related to cell cycle progression. The significant changes to this network may indicate the important role of *TGFB1* in the cell cycle regulation induced by butyrate. The second network associated with up- and downregulated genes is also related to the cell cycle as well as cellular movement. Genes associated with this network were mostly downregulated genes such as *TP53*, *CDCA8*, *CCNG1*, *CHD3*, *CDC20*, and *MAD2L1*. Only three upregulated genes (*BIRC5*, *THBS2* and *CTSF*) were involved in this network. The third network identified contains upregulated *IGF2*, *IGFBP3*, *INHBA*, *MMP1*, *MMP13*, etc., as well as downregulated genes such as *IGFBP4*, *MCM2*, *MCM3*, *MCM4*, *MCM5*, *MCM6*, and *ORC1L*. Major cellular functions or diseases associated with this network are cancer, cellular movement and cell death.

In a graphical representation of the molecular relationships between genes/gene products (Network Pathway), genes or gene products are represented as nodes, and the biological relationship between two nodes is represented as an edge (line). All edges are supported by at least one reference from the literature, from a textbook, or from canonical information stored in the Ingenuity Pathways Knowledge Base. Figure 2 shows the third network listed in Table 3 as an example of graphical representations. This network contains upregulated *IGF2*, *IGFBP3*, *INHBA*, *MMP1*, *MMP13*, etc., as well as downregulated genes such as *IGFBP4*, *MCM2*, *MCM3*, *MCM4*, *MCM5*, *MCM6*, and *ORC1L*.

Table 3: Major genetic networks targeted by butyrate regulation.

Network	Molecules in network	Score	Focus genes	Top functions
1	***↓ACTN1, ↓ASGR2, ↓CCNA2, ↓CCNB1, ↓CDC2, ↓CDC25A, ↓CDC25B, ↓CHEK1, ↓CKS1B, ↑CLDN4, ↓CYC1, ↓E2F4, ↓EEF1A1, ↓ELAVL1, ↓FEN1, ↓HNRPK, ↑HTRA1, ↓IGFBP6, ↓ITPA, ↓LOXL1, ↑LTC4S, ↑MXI1, ↓PPIA, ↓PPIG, ↑PRODH, ↓RAD51, ↓RAD51AP1, ↓RAD54L, ↓RBM3, ↓SERPINH1, ↓SFRS3, ↑TGFB1, ↑TNFAIP2 (includes EG:7127), TXNIP, ↓UBE2C***	54	35	Cell cycle, cancer, reproductive system disease
2	***↓ANLN, ↓AURKA, ↓iAURKAIP1, ↓AURKB, ↑BIRC5, ↓BUB1, ↓BUB1B, ↓CCNB2, ↓CCNG1, ↓CDC20, ↓CDCA8, ↓CDK4, ↓iCENPA, ↓CHD3, ↑CTSF, ↓DLG7, ↓DNMT1, ↓FUBP1, ↓GNL3, ↓GTSE1, ↓HSP90AA1, ↓HSPA9B (includes EG:3313), ↓INCENP, ↓KIF14, ↓KIF23, ↓MAD2L1, ↓NASP, ↓NUSAP1, ↓PRC1, ↓REV3L, ↑THBS2, ↓TK1, ↓TP53, ↓TUBB4, ↓VRK1***	54	35	Cell cycle, cellular movement, cancer
3	***↑ANKRD1, ↓C1QA, ↓CDC45L, ↓CENPF, ↑CTGF, ↑CTSD, ↑DUSP1, ↓FST, G1P2, ↓HELLS, ↓HGF, ↓HSPB2, ↑IGF2, ↑IGFBP3, ↓IGFBP4, ↑INHBA, ↓KIF22, ↓MCM2, ↓MCM3, ↓MCM4, ↓MCM5, ↓MCM6, ↑MMP1, ↑MMP7, ↑MMP13, ↓ORC1L, ↓PFDN6, ↓PGAM1, ↑PLAT, ↓PLAUR, ↑PSAP, ↓RRM2, ↓TFPI2, ↑TIMP3, ↓UHRF1***	54	35	Cancer, cellular movement, cell death
4	*ANXA2, ARG1,* ***↓BCAT1,*** *↓CCNA2,* ***↓CDC2L1,*** *CEBPA,* ***↓CKS2, ↑CREG1, ↑CTSB,*** *DDX21,* ***↓E2F4, ↓ESPL1,*** *FBL, GADD45A,* ***↓HNRPH1, ↓HNRPK, ↑LGMN, ↓MTHFD1,*** *MYC,* ***↓NCL,*** *NFYA, NFYC,* ***↓NOL5A, ↓PABPC4, ↓PCBP2, ↑PLAT,*** *PTTG1,* ***↓RPL14, ↓RPS12, ↓RPS4X, ↑SERPINI1,*** *SNRP70, TNC, TUBB2A,* ***↓UBE2C***	25	22	Gene expression, cell cycle, cardiovascular disease
5	***↓ANXA3,*** *ARID4A,* ***↓BDP1,*** *BRF1, BRMS1, CBFA2T3, CBX1,* ***↓CBX5,*** *CCNA1,* ***↓CDC25A,*** *CDC25C,* ***↓CDK4,*** *CDKN2D,* ***↓CHAF1A, ↓CHEK1, ↓DNMT1, ↑GJA1, ↓GJA7, ↓GTF3C2, ↓HDAC8,*** *HIST1H1B, HIST2H3C, HRAS,* ***↓PHF5A,*** *PLOD1, ↓RAD51, RB1, RBBP4, RUNX1T1,* ***↓SAP30, ↓SDCCAG10, ↓SHFM1,*** *SUV39H1,* ***↓TLK1,*** *TOP2A*	18	18	Cancer, cell cycle, reproductive system disease

Differentially expressed genes are in bold (↑ up or ↓ down regulated)

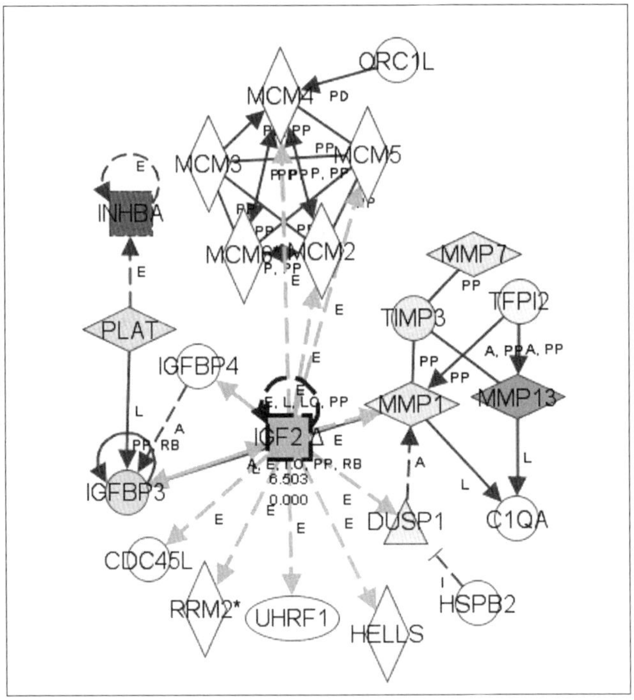

Fig. 2: An integrated network with both up- and downregulated genes in cells treated with butyrate for 24 h. Data set was analyzed by the Ingenuity Pathways Analysis software (Ingenuity® Systems, www.ingenuity.com). The graphical pathway network showed is the third network from Table 3. This network contains upregulated *IGF2, IGFBP3, INHBA, MMP1, MMP13*, etc., as well as downregulated genes such as *IGFBP4, MCM2, MCM3, MCM4, MCM5, MCM6*, and *ORC1L*. Notes and edges are displayed with various shapes and labels that present the functional class of genes and the nature of the relationship between the notes, respectively.

Discussion

During the last two decades, we have come to realize that lipids function as both energy stores and signaling molecules. They play pivotal roles in regulating a wide variety of cellular processes. As a particular group of lipids, short-chain fatty acids have been studied intensively [1,4]. Short-chain fatty acids are a major energy source in ruminants and contribute up to 70% of their energy requirements [2,5,6]. Of these short-chain fatty acids, butyrate is the major source of energy for colonocytes [7,8]. Short-chain fatty acids are organic acids produced by anaerobic fermentation of dietary fibre in the gastrointestinal tract of mammalian species and then directly absorbed at the site of production. Short-chain fatty acids, especially butyrate, are important nutrients and function as signaling molecules. The molecular mechanisms involved in biochemical and molecular actions of these nutrients remain largely unknown. Few definitive studies in cattle, if any, have addressed the capabilities of short-chain fatty acids as signaling molecules to modulate gene expression and the proteometric outcomes at cellular and molecular levels.

Since the short-chain volatile fatty acids (VFAs, acetate, propionate, and butyrate) fulfill up to 70% of their nutrient energy requirements, cells derived from ruminant species are presumed to be equipped to respond to VFA with increased sensitivity compared to other species [2]. Our studies revealed, however, that butyrate induced cell cycle arrest and apoptosis in an established bovine MDBK cell line [9]. In this study, the detailed mechanisms by which butyrate induces cell growth arrest and apoptosis were analyzed using global gene expression profiles with high-density oligonucleotide microarrays and the Ingenuity Pathways Knowledge Base. Gene expression profiling indicated that butyrate induces many significant changes in gene expression involving many regulatory pathways that are critical to cell growth, immune response and signal transduction. The functional category and pathway analyses of the microarray data revealed that four canonical pathways (the cell cycle G2/M DNA damage checkpoint and G1/S checkpoint regulation; pyrimidine metabolism; and purine metabolism) were significantly perturbed.

The significance calculated for each canonical pathway is a measurement of the likelihood that the pathway is associated with the dysregulated genes by random chance. Interestingly, four pathways (the cell cycle G2/M DNA damage checkpoint and G1/S checkpoint regulation; pyrimidine metabolism; and purine metabolism) showed significant regulation by butyrate. Cell cycle checkpoint pathways (G2/M DNA damage checkpoint and G1/S checkpoint regulation) are impaired due to the treatment with butyrate. For the first time, two pathways critical for the regulation of purine and pyrimidine metabolism were linked to the induced biological effects of butyrate. These findings underlie mechanisms of butyrate induced cell arrest in the progression of the cell cycle.

Insulin-like growth factor binding proteins (IGFBPs) modulate IGF action and regulate cell growth and apoptosis by preventing IGF from interacting with their own receptors. In this study, insulin-like growth factor (IGF2) was upregulated by butyrate, which is consistent with other published data [10]. Microarray and real-time RT-PCR results confirmed that *IGFBP6* was downregulated by butyrate. IGFBP-6 has a 100-fold higher affinity for IGF2 than IGF1 [11]. It seems paradoxical that while *IGF2* is upregulated significantly by butyrate, its highest affinity binding protein is downregulated. These results may suggest different functions for various IGFBP members in regulating apoptosis and cell cycle progression. It would be intriguing to see how IGFBPs exert their actions in cell growth and apoptosis via an IGF-independent fashion. In addition, extensive repression of cyclin-dependent kinases as well as cell cycle related genes, such as *CDC2/CDK1*, *CDC20*, *CDC25A*, *CCNG1*, *CCNB1*, *CCNB2*, *CCNA2*, *CCNG1*, and *PCNA*, may be closely associated with the cell growth arrest induced by butyrate. They are consistent with the growing body of evidence suggesting that disruption of the coordinate between regulation of DNA synthesis and cyclin-dependent kinase activity is an important feature of apoptosis.

The results also showed that *TGFB1* was significantly upregulated. Studies have shown that TGFB1 regulates anterior pituitary cell proliferation and hormone secretion. There is also a report that indicates that butyrate and TGFB1 inhibit pituitary cell proliferation and regulate the expression of 7B2, PC1, and PC2 in a cell culture model of pituitary tumours [12]. However, to our best knowledge, there is no previously reported relation between butyrate and the expression of TGFB1. TGFB is a multifunctional peptide that controls proliferation, differentiation, and other functions in many cell types. TGFB acts synergistically with TGFA in inducing transformation. It also acts as a negative autocrine growth factor. We speculate that the upregulated expression of both IGF2 and TGFB1 may have a synergistic effect in inducing cell apoptosis.

Conclusion

The present findings provide an example of the epigenetic regulation of the genome at work and a basis for understanding the full range of the biological roles and molecular mechanisms that butyrate may play in human and animal cell growth, proliferation, and energy metabolism. The results illustrate the potential for the nutritional manipulation of gene activity for the purpose of enhancing human health, such as in cancer prevention. This study also generated comprehensive information on the experimental system, which can be used in many functional genomic studies in bovines and provided a basis for understanding the full range of the biological roles and molecular mechanisms of butyrate in animal cell growth, proliferation, and energy metabolism. The results illustrate the potential for exploiting the nutritional manipulation of gene activity to enhance animal production efficiency in a drug residue-free format.

References

1 Andoh A, Tsujikawa T, Fujiyama Y: Role of dietary fiber and short-chain fatty acids in the colon. Curr Pharm Des 2003;9(4):347-358.

2 Bergman EN: Energy contributions of volatile fatty acids from the gastrointestinal tract in various species. Physiol Rev 1990;70(2):567-590.

3 Li RW MM, Van Tassell CP, Sonstegard TS, Connor EE, Van Amburgh ME,Boisclair YR, Capuco AV: Identification of estrogen-responsive genes in the parenchyma and fat pad of the bovine mammary gland by microarray analysis. Physiol Genomics 2006;27:42-453.

4 Hague A, Butt AJ, Paraskeva C: The role of butyrate in human colonic epithelial cells: an energy source or inducer of differentiation and apoptosis? Proc Nutr Soc 1996;55(3):937-943.

5 Siciliano-Jones J, Murphy MR: Production of volatile fatty acids in the rumen and cecum-colon of steers as affected by forage:concentrate and forage physical form. J Dairy Sci 1989;72(2):485-492.

6 Sharp WM, Johnson RR, Owens FN: Ruminal VFA production with steers fed whole or ground corn grain. J Anim Sci 1982;55(6):1505-1514.

7 Bugaut M: Occurrence, absorption and metabolism of short chain fatty acids in the digestive tract of mammals. Comp Biochem Physiol B 1987;86(3):439-472.

8 Roediger WE: Role of anaerobic bacteria in the metabolic welfare of the colonic mucosa in man. Gut 1980;21(9):793-798.

9 Li CJ, Elsasser TH: Butyrate-induced apoptosis and cell cycle arrest in bovine kidney epithelial cells: involvement of caspase and proteasome pathways. J Anim Sci 2005;83(1):89-97.

10 Chiba T, Yokosuka O, Fukai K, Kojima H, Tada M, Arai M, et al: Cell growth inhibition and gene expression induced by the histone deacetylase inhibitor, trichostatin A, on human hepatoma cells. Oncology 2004;66(6):481-491.

11 Baxter RC, Saunders H: Radioimmunoassay of insulin-like growth factor-binding protein-6 in human serum and other body fluids. J Endocrinol 1992;134(1):133-139.

12 Kobayashi I, Jin L, Ruebel KH, Bayliss JM, Hidehiro O, Lloyd RV: Regulation of cell growth and expression of 7B2, PC2, and PC1/3 by TGFbeta 1 and sodium butyrate in a human pituitary cell line (HP75). Endocrine 2003;22(3):285-292.

Dr. Congjun Li, Bovine Functional Genomics Laboratory, Animal and Natural Resources Institute, ARS, USDA. 10300 Baltimore Ave. BARC East, Building 200, Rm 209, Beltsville, MD 20705, USA.

E-mail: congjun.li@ars.usda.gov

Pinard M-H, Gay C, Pastoret P-P, Dodet B (eds): Animal Genomics for Animal Health. Dev Biol (Basel). Basel, Karger, 2008, vol 132, pp 399-403.

Diagnostic Markers for Diseases: SELDI-TOF Profiling of Pig Sera for PRRS

S. Genini[1], M. Cantu[2,3], S. Botti[1], R. Malinverni[1], A. Costa[2], D. Marras[2,4], E. Giuffra[1]

[1] Parco Tecnologico Padano – CERSA, Lodi, Italy
[2] BIOTRACK s.r.l., Lodi, Italy
[3] Current address: LaP – Proteomic Core Facility and Research Group, CIRB–CRBA–St. Orsola Malpighi University Hospital, Bologna, Italy
[4] Current address: BioLab s.p.a., Vimodrone, Italy

Key words: porcine reproductive and respiratory syndrome (PRRS), SELDI-TOF MS analysis, proteomic fingerprint profiling

Abstract: Porcine reproductive and respiratory syndrome (PRRS) is a highly infectious viral disease causing severe losses to the pig industry. Most weaning piglets are likely to be exposed to the infection and show at least asymptomatic PRRS viremia strongly related to productive performance. The aims of this study were to set up experimental conditions for pig sera proteomic profiling and to identify biomarkers that differentiate weaning asymptomatic piglets positive to PRRS viremia from negative controls (PCR tested) with potential predictive value for the subsequent occurrence of clinical PRRS. Protein profiles were generated by SELDI-TOF MS using the Bio-Rad Chips WCX, IMAC30 and H50. The discovery phase revealed that a consistent number of highly significant protein peaks can be detected by the WCX and IMAC30 surfaces; however none of these peaks were statistically confirmed by the subsequent validation phase, highlighting that serum concentration of the contaminant and most abundant proteins is a crucial parameter for SELDI-TOF MS studies. Current protocols are being further optimized and adapted to pig sera to reduce the unfavourable effects of the most abundant proteins and to increase the number of potential detectable biomarkers. Furthermore, proteomic fingerprint profiling has been shown to be a promising diagnostic tool that, in the future, may be useful to provide also insights into the mechanisms of early viral infection in vivo.

INTRODUCTION

Worldwide, PRRS is one of the most important infectious disease of swine and has high economic and ethical implications in food production and animal health [1,2].

The responsible agent is a single stranded RNA virus (PRRSV), which belongs to the family of the Arteriviridae [3], and causes huge economic losses for the pig industry by increasing mortality of nursery pigs, reproductive failure in sows, respiratory distress in pigs of all ages and influenza-like symptoms in grow-finish swine [1,4]. PRRSV also reduces the daily growth rates and nutritional efficiency [4] of affected swine and triggers high medication costs and marketing problems, as well as considerable animal health costs needed to control secondary pathogens [5].

Several recent studies demonstrated the usefulness of the surface enhanced laser desorption ionization, time of flight mass spectrometry (SELDI-TOF MS) technology for biomarker discovery [6,7]. This powerful tool rapidly generates high-throughput protein profiles from a large number of samples and permits the representation of, by mass spectral analysis, the proteins of a specific extract through the interaction of the protein components with chromatographic surfaces with a defined chemistry.

The overall goal of this study is to identify discriminating markers for an early, possibly predictive diagnosis of PRRS in weaning piglets. Here, a SELDI-TOF MS method for protein profiling of pig sera was set up and tested on two groups of pigs without clinical symptoms of the disease. The pigs were sampled from a reference population (MISAGEN project) [8] and differed for being positive or negative to PRRS viremia.

Materials and Methods

Collection of serum samples

A total of 48 swine whole blood samples were obtained from Large White piglets at weaning (45-50 days of age). The piglets were selected from a large repository database [8] presently containing more than 3,000 well characterized purebred pig samples from 12 different farms of the Lombardia region, Italy. Eight negative and 8 positive piglets to PRRS (verified as described below) were analysed during the discovery phase of the study; 16 negative and 16 positive piglets were compared in the subsequent validation phase. Two mL of whole blood were collected in tubes without anticoagulants, stored at room temperature for 4 h, and then centrifuged at 3,500 rpm for 4 min for serum separation. The sera were then distributed into aliquots and stored frozen at -80°C prior to further experiments.

Verification of the presence/absence of PRRSV

Viral RNA extraction from pig sera was performed with standard Roche protocols (High Pure Viral RNA Kit, Roche Diagnostics GmbH, Germany) and the presence or absence of the virus was determined by standard PCR at the Istituto Zooprofilattico Sperimentale (IZS), Lodi, Italy.

Veterinary inspection of the overall clinical status at the day of serum collection evidenced that the piglets did not show clinical symptoms of PRRS.

SELDI-TOF MS analysis

A total of 30 µl of full serum were processed following the instructions and the optimized experimental conditions, provided by the supplier, for human samples using SELDI-TOF MS (ProteinChip Applications Guide, Bio-Rad Laboratories, Hercules, CA, USA). Three arrays with different surface chemistries, WCX (weak cation exchange), IMAC30-CU (immobilized metal affinity capture-copper) and H50 (reverse phase), were placed in a ProteinChip System Series 4000 Enterprise mass spectrometer reader (Bio-Rad Laboratories). The detected peaks were processed by Ciphergen Express software with baseline correction, normalization (Total Ion Current), mass accuracy calibration and clustering. Proteomic peak features that differed significantly between the two groups were identified by calculating the p-value, ROC/AUC (Receiver Operating Characteristic/Area Under Curve) and PCA (Principal Component Analysis).

RESULTS

In the initial discovery phase of the study, 18 statistically highly significant ($p < 0.01$) and 34 significant ($0.01 < p < 0.05$) unique peaks were found to allow for discrimination between PRRSV positive and negative piglets. Table 1 shows that the overall highest number of discriminating peaks was found with the surface IMAC30; the surface WCX allowed for the identification of a lower number of peaks, most of which had highly significant p-values, while only one peak with p-value below 0.01 was found by H50. As IMAC30 and WCX provided the major number of potential biomarkers, H50 was not used further in the validation phase of the study.

In the validation phase, performed on a more consistent number of piglets, none of the 18 highly significant peaks with $p < 0.01$ previously identified during the discovery phase could be confirmed, because the p-values were not statistically significant ($p > 0.05$) (Table 2).

Table 1: Number of discriminatory protein peaks identified in the discovery phase and associated with PRRS infection. The peaks are divided by the three different array surfaces analysed and by the p-value ranges.

Array surface	p-value ranges		
	<0.01	**0.01< p <0.05**	**Total (<0.05)**
IMAC30	10	19	29
WCX	7	4	11
H50	1	11	12

Table 2: Proteomic features of the 18 highly significant peaks identified by SELDI-TOF MS in the discovery phase and associated with PRRS infection.
P-values are reported for both phases.

Array surface	Molecular weight (Da)	p-value (discovery phase)	p-value (validation phase)
WCX	75511	0.000003	>0.05
WCX	75999	0.000034	>0.05
WCX	37726	0.000044	>0.05
WCX	8933	0.000089	>0.05
WCX	8922	0.000148	>0.05
WCX	8320	0.000574	>0.05
IMAC30	39306	0.001629	>0.05
WCX	8322	0.002985	>0.05
IMAC30	69588	0.003275	>0.05
H50	12663	0.004574	not performed
IMAC30	10007	0.004574	>0.05
IMAC30	8926	0.006322	>0.05
IMAC30	14302	0.006322	>0.05
IMAC30	17849	0.006322	>0.05
IMAC30	54021	0.008651	>0.05
IMAC30	69180	0.008651	>0.05
IMAC30	9140	0.008651	>0.05
IMAC30	138079	0.008651	>0.05

Discussion

Because the protein concentrations in pig and human sera are very similar, the experimental conditions of the present profiling experiment were largely based on those previously reported in human studies on severe acute respiratory syndrome (SARS). Indeed, SARS shows very similar pathological symptoms to PRRS, and recent studies demonstrated the efficacy of the SELDI-TOF MS for the characterization and identification of diagnostic biomarkers associated with viral infection and for improving the knowledge on this severe pathology [9-11]. The preliminary results presented here revealed that the parameters and conditions already optimized for human samples could be successfully transferred to swine sera. Furthermore, they highlighted IMAC30 and WCX as being the most representative surfaces in terms of potential biomarkers.

The discovery phase identified 18 highly significant unique protein peaks that differentiated 8 piglets with and 8 without PRRSV. However, none of these potential biomarkers could be validated in the subsequent phase in terms of p-value and ROC/AUC. A likely explanation is that this may have been caused by differences in the concentrations of the proteins in the serum, as the serum protein concentration is a crucial parameter for SELDI-TOF MS in order to avoid saturation of the chromatographic surface and to maintain specificity of the analysis. Moreover, the piglets did not show symptoms of the disease, and PRRSV-related early diagnostic biomarkers may be in such a low concentration that they are hidden by the most abundant proteins. The latter, by competing with the other serum components for the binding site of the chromatographic surfaces and for the protein ionization, cause signal suppression of the medium-low abundant proteins and, for instance, the reduction of the number of potential biomarkers.

Conclusion

This in vivo study was possible only because we had access to a large repository database [8] in which different groups of piglets of the same breed and age, and grown under the same conditions, can be selected based on well defined phenotypes. Current efforts are being oriented i) to increase the sample size of the discovery and validation phases, and ii) to develop new sample processing protocols (e.g. serum fractionation) aimed at reducing the effects of the most abundant proteins (e.g. albumin) and contaminant proteins (e.g. haemoglobin), and to increase the number of potential biomarkers by evidencing the medium-low abundant proteins.

Acknowledgements

The authors are very thankful to Maria Cecere, Roberto Grande, Elisa Filippi and Silvia Rossini for technical assistance and helpful comments. This project was supported by a grant of the Italian Ministry of Research (MIUR project, art.10 D.M. 593/00).

REFERENCES

1 Rossow KD: Porcine reproductive and respiratory syndrome. Vet Pathol 1998;35:1-20.

2 Cho JG, Dee SA: Porcine reproductive and respiratory syndrome virus. Theriogenology 2006;66:655-662.

3 Meulenberg JJ, Hulst MM, de Meijer EJ, Moonen PL, den Besten A, de Kluyver EP, et al: Lelystad virus belongs to a new virus family, comprising lactate dehydrogenase-elevating virus, equine arteritis virus, and simian hemorrhagic fever virus. Arch Virol Suppl 1994;9:441-448.

4 Nodelijk G: Porcine reproductive and respiratory syndrome (PRRS) with special reference to clinical aspects and diagnosis. A review. Vet Q 2002;24:95-100.

5 Dee SA, Joo HS, Polson DD, Park BK, Pijoan C, Molitor TW, et al: Evaluation of the effects of nursery depopulation on the persistence of porcine reproductive and respiratory syndrome virus and the productivity of 34 farms. Vet Rec 1997;140:247-248.

6 Au JS, Cho WC, Yip TT, Yip C, Zhu H, Leung WW, et al: Deep proteome profiling of sera from never-smoked lung cancer patients. Biomed Pharmacother 2007;61:570-577.

7 Ward DG, Cheng Y, N'Kontchou G, Thar TT, Barget N, Wei W, et al: Changes in the serum proteome associated with the development of hepatocellular carcinoma in hepatitis C-related cirrhosis. Br J Cancer 2006;94:287–292.

8 Botti S, Caprera A, Gaita L, Mondin P, Ossani N, Palermo S, et al: The MISAGEN project: toward the genetic improvement of disease resistance of pig commercial populations. 8th World Congress on Genetics Applied to Livestock Production (WCGALP). 13-18 August 2006, Belo Horizonte, Brazil.

9 Pang RT, Poon TC, Chan KC, Lee NL, Chiu RW, Tong YK, et al: Serum proteomic fingerprints of adult patients with severe acute respiratory syndrome. Clin Chem 2006;52:421-429.

10 Poon TC, Chan KC, Ng PC, Chiu RW, Ang IL, Tong YK, et al: Serial analysis of plasma proteomic signatures in pediatric patients with severe acute respiratory syndrome and correlation with viral load. Clin Chem 2004;50:1452-1455.

11 Kang X, Xu Y, Wu X, Liang Y, Wang C, Guo J, et al: Proteomic fingerprints for potential application to early diagnosis of severe acute respiratory syndrome. Clin Chem 2005;51:56-64.

Dr. Sem Genini, Parco Tecnologico Padano (PTP) - CERSA, Livestock Genomics I, Loc. Cascina, Codazza, Via A. Einstein, 26900 Lodi, Italy.

E-mail: sem.genini@tecnoparco.org

Pinard M-H, Gay C, Pastoret P-P, Dodet B (eds): Animal Genomics for Animal Health. Dev Biol (Basel). Basel, Karger, 2008, vol 132, p 405.

SESSION VI

Advancing Animal Genomics in Animal Health

Pinard M-H, Gay C, Pastoret P-P, Dodet B (eds): Animal Genomics for Animal Health. Dev Biol (Basel). Basel, Karger, 2008, vol 132, pp 407-424.

Animal Genomics for Animal Health Report: Critical Needs, Problems to be Solved, Potential Solutions, and a Roadmap for Moving Forward

A. Archibald[1], J.-C. Audonnet[2], L. Babiuk[3], S.C. Bishop[1], C.G. Gay[4], J. McKay[5], B. Mallard[6], G. Plastow[3], M.-H. Pinard van der Laan[7], M. Torremorell[8]

1 Roslin Institute and R(D)SVS, University of Edinburgh, Roslin Biocentre, Midlothian, Scotland, United Kingdom
2 Merial Inc., Laboratoire Lyon Gerland, Lyon, France
3 University of Alberta, Alberta, Canada
4 Agricultural Research Service, Beltsville, Maryland, USA
5 EW Group, Newbridge, Midlothian, Scotland, United Kingdom
6 University of Guelph, Guelph, Ontario, Canada
7 Institut National de Recherche Agronomique, Jouy-en-Josas, France
8 Pig Improvement Company, Hendersonville, Tennessee, USA

Key words: animal genomics, animal genomes, animal health, animal selection, genome mapping, bioinformatics, comparative genomics, functional genomics, genetic selection, disease resistance, disease susceptibility, systems biology.

Abstract: The first International Symposium on Animal Genomics for Animal Health, held at the World Organisation for Animal Health (OIE) Headquarter, 23-25 October, 2007, Paris, France, assembled more than 250 participants representing research organizations from 26 countries. The symposium included a roundtable discussion on critical needs, challenges and opportunities, and a forward look at the potential applications of animal genomics in animal health research. The aim of the roundtable discussion was to foster a dialogue between scientists working at the cutting edge of animal genomics research and animal health scientists. In an effort to broaden the perspective of the roundtable discussion, the organizers set out four priority areas to advance the use of genome-enabled technologies in animal health research. Contributions were obtained through open discussions and a questionnaire distributed at the start of the symposium. This symposium report provides detailed summaries of the outcome of the roundtable discussion for each of the four priority areas. For each priority, the problems needing to be solved, according to the views of the participants, are identified, including potential solutions, recommendations, and lastly, concrete steps that could be taken to address these problems. This report serves as a roadmap to steer research priorities in animal genomics research.

Introduction

One of the primary aims of the International Symposium on Animal Genomics for Animal Health (AGAH) was to identify opportunities for integrating genome-enabled technologies in animal health research. To achieve this aim, a roundtable discussion of leading experts in the field of genomics and animal health was organized. With contributions from the symposium participants, open discussions and a questionnaire distributed at the start of the symposium, four priorities were identified that are likely to have the greatest impact on animal genomics research:

1 Quantitative genetic studies to identify markers of health traits
2 Functional genomics studies to understand host-pathogen interactions
3. Translating genomic information to tools for controlling diseases
4. Integrating stakeholder support to advance animal genomics in animal health

This report captures the problems to be solved within the four priority areas and provides potential solutions, recommendations, and the next steps for moving forward.

Four Priorities for Animal Genomics Research

Quantitative genetic studies to identify markers of health traits

Genetic studies which aim to quantify between-host variation in health-related traits or intend to identify genetic markers associated with health have the potential to impact many areas of disease control. However, the interpretation of genetic studies has often highlighted the different perspectives held by the veterinary and animal genetics communities. For example, many in the veterinary medical field have viewed quantitative genetics as a means to emulate the success of the biomedical research community in identifying 'candidate genes' responsible for specific diseases. However, successes with livestock and poultry diseases have been relatively rare. Conversely, livestock and poultry production scientists have viewed quantitative genetics as a means of identifying diseases for which control strategies exploiting host genetics are feasible, possibly through marker-assisted selection of animals with desirable health traits. However, quantitative genetic studies to identify heritable health traits have notoriously been difficult and often remain unrealized; although successes have been achieved for some endemic diseases, genetically improving animal health remains a challenge.

One of the outcomes of the AGAH symposium is the realization that there are no generic solutions for solving the problems associated with quantitative genetic studies. However, the rapid and sustained development of genome-enabled technologies of the last few years provides extraordinary opportunities for advancing this field of research. One of the first and most exciting options to explore is the possibility of using dense single nucleotide polymorphism (SNP) maps coupled with haplotype information. Still, participants overwhelmingly agreed that the key challenge remains the lack of validated animal phenotypes. Critical needs include standardized protocols for collecting tissues and phenotypes for disease resistance traits in both experimental resource and relevant commercial animal populations. Several participants recommended taking a 'big picture' view of the problems and suggested that it may be more productive to study many diseases in the same population than to conduct many smaller, 'more focused' disease-specific studies, as very large populations are required to obtain sufficient information for meaningful genetic analysis. .

The main issues:

- Availability and acquisition of appropriate and reliable health records
- Availability and acquisition of animal pedigrees with defined phenotypes
- Development and availability of genome-associated research tools
- Refining quantitative population genetic studies to usable markers

Availability and acquisition of appropriate and reliable health records

a. Problem: The rate limiting step in studies aiming to detect genetic markers for health is invariably the collection of phenotypes, rather than genotypes. This is a general problem in many genomic studies, including studies on non-disease traits, and it is known generically as the 'phenotype gap.' However, the situation is exacerbated for disease traits as phenotypes of disease susceptibility/resistance to infectious diseases are often more difficult and/or expensive to measure than those for performance traits. Their measurement usually requires the involvement of infection (either deliberate or through clinical history of naturally infected animals).

There are many issues to be addressed in the definition and recording of health traits, and these issues should be the subject of debate between specialists with different areas of expertise. Each infectious disease will have its own unique solution, and whilst attempts should be made to capture as much information as possible in resource populations, practical considerations will more likely lead to the selection of animals with targeted disease or health-related traits.

Some diseases are multifactorial. For example, mastitis may be caused by many pathogens (e.g. *Escherichia coli, Staphylococcus aureus, Mycoplasma bovis, Streptococcus uberis, Streptococcus agalactiae,* etc), and the pathogenesis resulting from these specific infections may differ dramatically. Therefore, an important issue is whether with these different forms of the disease it is possible to have generic solutions to the disease or, conversely, pathogen-specific solutions are required – invoking considerable complexity in data recording.

In many cases, disease outcomes may be the result of infection coinciding with stress induced immune suppression in livestock. An example is the respiratory diseases of livestock and poultry in intensive animal production systems caused by several different pathogens (viral and/or bacterial) in association with transport, co-mingling of pooled animals, dietary changes and administration of various veterinary treatments. Mastitis is another example of a disease where clinical cases often have an underlying issue of immune suppression in early lactation. In such cases, addressing animals' ability to cope with stress may be as effective as concentrating on animals' resistance to infection.

b. Potential Solutions: As highlighted by many contributors, solutions and recommendations invariably involve large scale, multidisciplinary, sometimes multi-national, projects which 'think big', i.e., maximize the use of animal resources and human expertise. In particular, it is critical that dialogue is established or strengthened between contributors with complementary expertise in animal health, animal genetics/genomics and other relevant disciplines.

Animal health experts should also be aware that opportunities to record phenotypes may occur in populations that are not standard commercial populations or well-characterized experimental lines. For example, livestock exposed to extremes of environment or disease may be a rich reservoir of traits and genes related to health issues.

In general terms, the solutions must be tailored to the problem, and avoid wasted effort if host genetic studies are not appropriate. Several different situations can be envisaged. Consider the following scenarios:

- For predictable widespread diseases such as nematodiasis, mastitis, PRRS, PMWS and PCVAD – simple standardized measurement protocols are required that describe relative levels of infection or impact on the animals.
- For emerging diseases such as bovine tuberculosis, bluetongue, aquaculture viral diseases – data capture from herds/flocks where disease outbreak has occurred is required. Recording of cases and unaffected (or uninfected) 'controls', along with DNA capture will enable genome scans to be performed efficiently to address the question of host genetic control
- Since data from disease outbreaks may be influenced by environmental factors, another approach to consider is the collection of data from challenge studies in a controlled laboratory setting to increase the power of detecting genetic factors involved in disease resistance phenotypes, provided that data on sufficient animals can be measured. It is advantageous to use the phenotypes of disease resistance to discover markers which are defined at the molecular level – these can then be used to enable selection to be performed on non-infected animals.
- For general health issues, e.g., when performance in commercial herds is lower than expected, or when general health issues are observed such as high incidence of respiratory diseases, optimal solutions have not yet been defined. Possibilities include detailed recording of specific health disorders, recording breeding-animal performance in environments that replicate commercial conditions, the use of immune measures to assess immunoresponsiveness or diagnose infection. From a purely genetics point of view, general health issues are more important than disease resistance to a specific disease because pathogens can evolve much faster than animal populations; therefore, genetic improvement in general health may be more sustainable.
- The option of selecting livestock with superior resistance to the immunosuppressive effects of stress should be explored. This may have applications for many livestock and poultry species, and potentially impact on common diseases such as respiratory diseases (e.g., the bovine respiratory disease complex), reproductive diseases (e.g., retained placenta & metritis) and enteric diseases.

c. Recommendations: The goal should be to move toward more 'proactive' sample collection, with specific recommendations on optimum timing and handling of blood (or other tissue) sample collection for a wide array of immunological and other disease-related studies. Development of high-throughput technologies to evaluate the disease status of individuals in large populations would also be advantageous.

The following scenarios can be envisaged:

- For predictable widespread diseases, international agreement is required for measurement protocols for predictable diseases. Extensive data needs to be

collected and stored on accessible databases.
- For emerging diseases, dialogue with governmental and other animal health policy makers is needed to ensure timely capture of epidemic outcome data, for use in population-level quantitative genetic or epidemiological analyses.
- For general health issues, arguably, the best practice has not yet been defined. Large-scale research into the best means of describing host genetic effects is required. Support research to define heath traits at the molecular level are needed. Narrowly defined phenotypes can significantly increase the power of detecting genetic factors involved in disease susceptibility and disease resistance.
- Include the collection of data on health traits in animal disease studies conducted in controlled laboratory settings to avoid the complication of associated environmental factors.
- Research into the identification of animals with superior ability to withstand the immunosuppressive effects of stress may open new opportunities.

d. Next Steps: International-level dialogue is needed between geneticists, animal health experts and government policy makers, aiming to:

1. Define critical disease and health issues at the sector level;
2. Address technical problems of data capture;
3. Address funding of basic research to understand health traits at the molecular level;
4. Support collaborations between research centres with large resource populations and laboratories with capabilities to collect health-related phenotypes from appropriate samples;
5. Establish livestock DNA banks for use by the International Community to enable future opportunities. Wherever possible, utilizable samples (e.g. semen) should be retained from all animals in structured breeding programmes, irrespective of their genotype, for traits currently under selection.

Availability of animal pedigrees (on animals with detailed health records)

a. Problem: Quantitative genetic studies require knowledge of animal pedigrees relating to animals upon which selection is performed.

b. Potential Solutions: Vertical integration seen with pigs, poultry and aquaculture, reduces the unit cost of pedigree recording; however, in many situations, large-scale vertical integration is not feasible (e.g. dairy cattle, and most beef and sheep enterprises). Marker panels, particularly dense SNP chips, offer a possibility to reconstruct pedigree relationships in cases where they are unknown, and also the possibility to improve pedigree descriptions (i.e., to get actual rather than average genetic relationships) in cases where pedigree is recorded. Another attractive solution that eliminates the need for animal pedigrees is to conduct genome-wide association studies that use high density SNP chips, in which predictions of animals' genetic merit can be made independent of pedigrees.

c. Recommendations:

- Where possible, encourage the recording of animal pedigrees, particularly in industry-wide schemes targeted at recording animal health attributes.

- In all disease recording schemes retain tissue, a blood sample or a blood spot for future use - resultant DNA can be used for pedigree reconstruction as well as marker detection.
- Develop inter-operable databases for large-scale storage of pedigree information, phenotypes, and associated environmental factors.

d. Next Steps:
1. Encourage dialogue between geneticists, animal breeding companies, animal health experts and policy makers, and epidemiologists to ensure maximum pedigree recording, or collection of samples from which DNA can be extracted.
2. Develop user-friendly tools to reconstruct pedigrees from marker data.
3. Support meetings and workshops to enhance communication and facilitate collaborations between quantitative geneticists and experts in animal health.

Development and availability of genome-associated research tools

a. Problem: Dense SNP chips (tens or hundreds of thousands of SNPs) are the key to effective and portable genetic marker studies. Currently these are costly and only truly available in species with a genome sequence (e.g., man, mouse, chicken). Another significant issue with animal genome sequences is the lack of annotation, which is mostly based on homology to genes in other animal species and not actual experimental testing in the species of interest.

b. Potential Solution: Genome sequences are needed for all major livestock species and discovery of hundreds of thousands of validated SNPs. Also consider other high throughput technologies that measure RNA (DNA chips), protein (protein arrays or mass spectrometry), and metabolites.

c. Recommendations: Continue and increase efforts in genome sequencing and SNP discovery. The formation of large international consortia is required to reduce the per-unit cost of using SNP chips. Also, support other high throughput technologies such as DNA chips and protein arrays.

d. Next Steps:
1. Increase international effort in sequencing the remaining agricultural animal genomes and in SNP discovery;
2. Develop formation of international consortia, enabling cost-effective use of SNP chips (irrespective of the IP situation of individual partners);
3. Include the development of other high throughput technologies.

Refining quantitative population genetic studies to usable markers

a. Problem: The step from discovery of significant marker associations to utilizable markers for animal health is long, difficult and expensive.

b. Potential Solutions: There are at least three potential scenarios:
- In some situations, where you have a readily measurable phenotype, it is not necessary to resolve the genetic marker to a causative mutation or tightly

linked marker - in this case 'genome-wide selection' can be used. The sum of all SNP effects across the genome can be used to predict next generation performance/ resistance/health status.

- Where phenotypes are difficult to capture, or few/single genes dominate host genetic variation, a usable marker must be found. An SNP-based genome scan alone is unlikely to be sufficient, and information must be used from other sources ('screening filters'), including comparative genomics, functional genomic studies and verification in independent animals.
- Once a region on the genome is found, you should (1) refine the position using many more markers, (2) confirm the association, and (3) when possible, sequence as many animals as possible for novel genetic variants.

c. Recommendations:

- The development and demonstration of tools for genome wide selection is needed, as well as the capture of data (e.g., from endemic diseases) and DNA for use in such studies.
- Large-scale coordinated research programmes aimed at elucidating relevant aspects of host-pathogen interactions for high profile disease (e.g. FMD, CSF, PRRS, bovine tuberculosis, mastitis, coccidiosis, salmonella, Marek's disease, etc) are needed.
- Develop and integrate bioinformatics infrastructures in animal health research institutions.
- Develop statistical models and software to assess marker association studies.

d. Next Steps:

1. Build capacity by establishing international cooperations in host genomic studies, host-pathogen interactions and development of analytical/bioinformatic approaches and tools.
2. Develop an international working group of experts to assess and develop statistical models and software.

Functional genomics to understand host-pathogen interactions

New research strategies employing high-throughput gene expression analysis are providing novel platforms for more comprehensive understanding of host–pathogen interactions. In particular, functional genomics is rapidly revolutionizing the analysis of whole genome responses of the host to environmental, nutritional, or disease factors that affect the health of animals. The symposium included several presentations on gene expression and functional genomics studies that had the aim of looking for genes that were both differentially expressed and located within a QTL region. Several of the comments reflected the importance of basic research and the inclusion of genome-enabled technologies. Many of the questions and comments from the audience pointed out that the causative polymorphisms may reside on completely different chromosomes from the genes that are differentially expressed and the importance of considering epigenetic effects.

The main issues:

1. Availability of validated animal disease models
2. Animals with defined disease susceptible/disease resistant phenotypes

3. Bioinformatics tools
4. Identification of genes that are involved in the activation/repression of key regulatory pathways
5. Development and availability of reagents to study gene function
6. Discovery of biomarkers.

Availability of validated animal disease models

a. Problem: There is a critical need to develop animal disease models that mimic field infections using statistically significant numbers of animals per group and standardized challenge inocula. Also critical is the need to share challenge inoculums from a single source to reduce experimental variations.

b. Potential Solutions: Define validated disease challenge models with respect to challenge dose, challenge inoculum that are similar to the field situation with respect to virulence and pathogen genomic sequence, and designate resource centres specialized in given disease models to share information through public websites.

c. Recommendations:
- Define the acceptable variability in the models and share expertise and know-how in the conduct of challenges.
- Identify research centres that maintain genetically well-defined animal breeds with different disease phenotypes as a resource for developing animal disease models.

d. Next Steps: Develop an international working group with disease challenge expertise for each disease model that can interact closely and share information and resources.

Animals with defined disease susceptible/disease resistant phenotypes

a. Problem: There is a lack of livestock and poultry breeds with well characterized genotypes and disease phenotypes whose variations can be correlated with innate and acquired immune status of the host.

b. Potential Solutions:
- Identify disease phenotypes important to industry and determine how host and pathogen genetic polymorphisms such as SNPs influence the outcome of host-pathogen interactions and disease resistance phenotypes.
- Increase the genetic analysis of animals with characterized animal phenotypes in challenge experiments to build SNP maps that define the influence of genetic polymorphisms on host immune response to pathogens.

c. Recommendations:
- Conduct genetic analyses to investigate how genetic variations of host and pathogens influence disease phenotypes in infection and vaccination studies.
- Identify breeds that vary in phenotypic expression of health traits and open these resources to all scientists conducting research in animal genomics

d. Next Steps:

1. Carry out studies that can lead to the identification of potential candidate genes that control host-pathogen interactions.
2. Determine QTL and SNPs of host and pathogens that correlate with disease phenotypes, and carry out host whole genome expression analysis using tissue-specific microarray to identify genes located near QTL that influence important disease parameters.
3. Integrate functional genomics research results in host-pathogen studies to identify potential candidate markers for priority diseases and validate these markers in large population studies using validated animal disease challenge models.

Bioinformatics tools

a. Problem: Animal health research laboratories lack bioinformatics infrastructure: computer hardware, software, and trained scientists in computational biology. There is also a lack of animal and pathogen species-specific genomics and proteomic tools with well-annotated genome databases to predict key immunogenic epitopes in the context of different target species to understand the triggering mechanisms of the protective immune response and to evaluate vaccine-induced immunity for the animal health research community.

b. Potential Solutions:

- Train animal health scientists in computational biology.
- Build bioinformatics infrastructure within animal health research centres.
- Build international animal genome databases.
- Utilize available genomics resources developed in mouse and human systems, and build comparison maps for immune system genes in the target animal species.
- Align international consortia with the aim of building animal genome databases as well as genome databases for important classes of animal pathogens.

c. Recommendations:

- Build specific databases for each target animal species and pathogen type and establish a systematic programme for training and sharing predictive tools.
- Use computational biology and statistical bioinformatics tools to analyze differences between pathogenic and non-pathogenic strains, and differential host response to immunogenic epitopes using genetically disparate inbred lines with well-characterized disease phenotypes.
- Initiate projects for comparative pathogenomics of similar microorganisms in different animal species.
- Integrate genetic, functional genomic, and bioinformatics data in a systems biology approach to understand complex diseases.

d. Next Steps:

1. Establish bioinformatics infrastructure with state-of-the-art genomic tools and resources that can integrate genotype, phenotype and functional genomics data for each animal species and priority diseases.

2. Establish core centres with trained computational biologists for each animal species and establish a user-friendly bioinformatics database that other scientists can access for analysis of functional genomics data.
3. Establish core centres for classes of priority animal pathogens and build user friendly annotated genome databases. Whenever possible, align animal pathogen databases with human pathogens databases or the same class; e.g., apicomplexa parasites.

Identification of genes that are involved in the activation/repression of key regulatory pathways

a. Problem: There is a need to identify molecular pathways of host and pathogens that are involved in transmission, survival, virulence, innate and adaptive immune responses, and protective immune response to natural infection and vaccination. The ultimate goal is the need to identify the 'operational limits' in the genomic 'imbalance' associated with current genetic selection.

b. Potential Solutions:

- We need a more comprehensive approach for using functional genomics to lead us to polymorphisms that will be useful for predicting genetic merit for animal health traits. One potential approach is to use QTL studies to identify small sets of animals that are expected to differ only in whether they have favorable or unfavorable genotypes at the QTL. Use highly parallel gene expression technology (microarrays or SAGE) to determine differences in expression in relevant tissue samples that are due to the QTL. Develop RT-PCR or other similar assays for the transcripts that are most highly related to QTL genotype in the small sample. Run these RT-PCR assays on the entire QTL population and redo the QTL scan. The expression data are expected to be much higher in heritability, and therefore, a much better data set for QTL detection, than direct disease data. This QTL scan should provide much better information for locating the position of the QTL and establishing relationships between haplotypes and disease resistance that should hold up across multiple breeds.
- Develop genome-wide high throughput transcriptional profiling systems to study host-pathogen interactions and identify key candidate host and microbial genes that are associated with infection, pathogenicity, antigenic variation and host range specificity.
- Invest more resources and time into the interactomics and metabolomics area.

c. Recommendations:

- A comprehensive approach to using functional genomics to discover polymorphisms for predicting genetic merit will require significant resources and could be greatly enhanced by effective collaborations between institutions that have expertise in functional genomics, those that have appropriate populations for QTL detection and validation, and those with resources and expertise in phenotyping animal health traits.
- Build key regulatory pathways relevant to each disease trait using well annotated genome and protein databases for each species.

- Identify host and pathogen genes critical for host-pathogen interactions; e.g., protective immunity.
- Validate the role of key host and pathogen genes important in disease resistance using gene knock-out and transgenic animal models.

d. Next Steps:

1. Funding agencies should consider funding longer-term projects to support the infrastructure needed to develop animal populations with desirable health traits, the extensive phenotyping of these populations, and genome-wide genotyping of these populations.
2. Identify centres with a bioinformatics capability for each animal species and pathogen and establish genomics databases for key immune and metabolic pathways related to host-pathogen interaction.
3. Develop and support proteomics tools to identify key host molecules important in protective immune mechanisms.
4. Integrate functional genomics databases with host-pathogen studies to identify signature regulatory pathways for target diseases.

Development and availability of reagents to study gene function

a. Problem: There is a need to develop a full set of specific research reagents for the key gene products involved in immune functions.

b. Potential solutions:

- Identify research centres with the capability to develop immune reagents and specific knock-out tools (e.g., siRNAs).
- Expand existing research projects on the development of immunological reagents for different animal species.

c. Recommendations:

- Establish an international consortium to advance and enable the development of reagents to study gene function of animal species.
- Build and support formal working groups such as the US Veterinary Immune Reagent Network and align with other working groups for priority animal species that can interact closely to identify, develop and commercialize a complete list of immune reagents (and other tools) identified and validated by the scientific community.
- Increase and foster communication among scientists to identify new technologies and enhance collaborations to maximize the impact of available resources in animal genomics.

d. Next Steps:

1. Using national and international immunology meetings, organize a regular workshop to update the progress of veterinary reagent development efforts and establish a publicly accessible database to list all available immune reagents.
2. Designate commercial partners who can organize marketing of recombinant proteins and antibodies for immunology research and establish national repositories to store DNA, recombinant proteins and antibodies for easy access by the animal health research community.

3. Increase funding opportunities for basic immunology and reagent development research by governmental granting agencies, e.g. the USDA, DHS, NIH, NSF, BBSRC, and the European Commission.
4. Consider profiling the microbial environment, which is now possible with high throughput sequencing (e.g., metagenomics). In the best case scenario, this would identify pathogenic organisms or interactions that promote disease.

Discovery of biomarkers

a. Problem: It is paramount that the animal health research community has access to measurable biomarkers linked to disease susceptibility, disease resistance, and mechanisms of protective immunity.

b. Potential Solutions: Conduct biomarker discovery studies with 'standardized' validated animal models that have well defined phenotypes.

c. Recommendations: Increase the use of proteomics and microarray technologies in biomarker discovery studies and implement the rational use of proteomics data to develop specific biochips/protein chips for genomics research.

d. Next Steps: Set up multidisciplinary (clinicians, pathologists, biochemists, computational biologist/biomathematicians…) research centres for clinical models and proteomics analyses.

Translating genomic information to tools for controlling diseases

Genomic tools are providing new opportunities for understanding genetic diversity, the genetic variation associated with diseases processes, and host-pathogen interactions. The scientific information generated from the use of these tools in animal health research will enable the rational design of highly effective diagnostics, vaccines, drugs, biotherapeutics, and animals resistant to diseases.

The main issues:

1. Using animal genomes to understand genetic diversity and inform best management practices to improve animal health and animal production systems worldwide;
2. Easy to use diagnostic platforms for genetic tests, forensic veterinary medicine, and traceability;
3. Selection of animals with disease susceptible and resistant traits;
4. Application of genome-enabled technologies in drug and vaccine discovery, including transgenic animals that are resistant to disease and the selection of animals that respond effectively to drugs and vaccines.

Using animal genomes to understand genetic diversity and inform best management practices to improve animal health and animal production systems

a. Problem: There is a critical need to preserve animal genetic diversity and to understand the role of genetic variations on host responses to diseases, drug treatments, and vaccination.

b. Potential Solutions:

- Use genome wide genetic markers and novel methods to actively manage animal biodiversity.
- Study the dynamics of specific and general response to disease challenges, especially innate immunity.
- Evaluate animal disease models that try and replicate 'real world' disease complexes, such as the polymicrobial infections associated with the bovine respiratory disease complex.
- Establish general guidelines for using genomic tools in drug and vaccine clinical trials to identify the genotype of study animals and their response to the challenge and/or test treatment.

c. Recommendations:

- Support initiatives that use animal genomic innovations to assess and manage domestic and wildlife animal genetic diversity.
- Support studies that will investigate the use of genome-wide markers to monitor and prevent the loss of important non-selected components of animal genomes.
- Integrate genetic profiling in the development of disease models for drug and vaccine discovery research.
- Integrate genomics tools in drug and vaccine clinical trials.

d. Next Steps:

1. Make preserving animal genetic diversity a priority.
2. Support comparative genomics studies, within and across animal species, which integrate host and pathogen genetic variations.

Easy to use diagnostic platforms for genetic tests, forensic veterinary medicine and traceability

a. Problem: There is a lack of rapid, robust, accurate and cost effective genetic tests to use in analyzing disease outbreaks for both the disease agent and the host.

b. Potential Solutions:

- Develop molecular diagnostics for genotyping hosts and pathogens.
- Develop diagnostic platforms to predict new and emerging high-virulence pathogenic strains and the expected susceptible animal populations.
- Develop molecular diagnostics to facilitate the traceability of animals with desired or undesired health traits.

c. Recommendations:

- Develop better description and characterization of genetic variations in target host populations and the pathogens that pose the greatest threats to these populations.
- Support the development of diagnostics to detect gene expression signatures that can predict the emergence of new pathogens and the identification of susceptible animal populations.
- Support research that will lead to the discovery of rapid and user-friendly diagnostics for the traceability of animals.

d. Next Steps:

1. Support more cooperation in describing all the genetic resources of agricultural animals and their pathogens.
2. Support cooperation with medical groups involved in the description of disease processes and their effects on intervention strategies.

Selection of animals with disease susceptible and resistant traits

a. Problem: There is a need to integrate studies on disease resistance with selection programmes. This will require strategic collaborations with commercial breeding companies.

b. Potential Solutions:

- Establish collaborations from an early stage using animals in well structured populations with good records and already well characterized phenotypes.
- Animal challenge models must be commercially realistic and include polymicrobial /combined disease challenges.
- Establish correlated responses to selection for any marker.

c. Recommendations:

- Support research programmes that include collaborations between public research institutions and commercial breeding companies.
- Integrate disease resistance traits with existing multi-trait selection programmes.
- Use data from large scale disease exposures, especially in sib-test situations.
- Use case control studies in commercial environments.

d. Next Steps:

1. Improve the evaluation of marker gene effects in multi-trait selection programmes.
2. Assess and monitor the impact of genomic-based marker selection.
3. Improve statistical analyses of case control studies.

Application of genome-enabled technologies in drug and vaccine discovery, including transgenic animals that are resistant to disease and the selection of animals that respond effectively to drugs and vaccines.

a. Problem: The application of animal genome-enabled tools will demand fundamental changes in the approaches used to discover and develop animal drugs and vaccines, including their delivery and application in disease control programmes. One of the main challenges is the total absence of regulatory guidelines for the applications of these new technologies, including transgenic animals. The ability to design drugs and vaccines for genetically defined animal populations will require additional planning. For instance, in the area of vaccines, there is a lack of established 'vaccine profiles' to identify specifically what a vaccine is expected to achieve (e.g., efficacy with one dose, mass vaccination capability, prevent disease transmission, etc.) for specific diseases in individual animals and targeted commercial populations. Vaccines must not only be designed to achieve the desired vaccine profile but must also anticipate the genetic variations of the host and pathogens to enable predicted outcomes in vaccination campaigns and the selection of good responders to vaccination.

b. Potential Solutions:

- Support vaccine discovery research programmes that target vaccines specifically designed for control and eradication.
- Support vaccine studies that assess new pathogenic strains, especially 'escapes' from vaccine protection.
- Integrate genomic tools in drug and vaccine studies to assess the effects of genetic variation on efficacy and the control of disease in defined animal populations.
- Support vaccine studies that take into account age-dependent effects of vaccines and adjuvants with maturation of the immune system.
- Support research programmes that integrate genomic and transgenic technologies to achieve disease resistance.
- Include regulatory agencies in the early phases of research programmes that integrate genome-enabled tools in study designs.
- Establish strategic research collaborations with commercial breeding companies and pharmaceutical companies.

c. Recommendations:

- Establish a more integrated approach to the development of drugs and vaccines that includes genetic variations in host responses.
- Support drug and vaccine research projects in commercially relevant populations.
- Support transgenic animal research programmes that target disease resistance.
- Support international efforts that provide science-driven regulatory guidelines for integrating genome-enabled tools in animal health research.

d. Next Steps:

1. Implement vaccine development plans in commercially relevant populations.
2. Integrate drug and vaccine development with population genetics of target populations.
3. Enable information on genetic variation and population structure to be used in the design of drugs and vaccines.
4. Implement computational biology and genomics in drug and vaccine design and evaluation.
5. Support OIE efforts to prepare guidelines for the application of animal genomics in animal health.

Integrating stakeholder support to advance animal genomics in animal health

The promise of animal genomics will not be achieved without stakeholder support. The burden of communicating the significant opportunities that genome-enabled technologies offer for improving animal health lies squarely on the shoulders of the animal health research community.

The main issues:

1. Collaboration between the public and private sectors;
2. Intellectual property and technology transfer to industry;
3. Impact on veterinary education and animal and poultry science;
4. Contributions of veterinary practitioners, animal health professionals, and animal agriculture allied industries;
5. Support from regulatory agencies to enable these new technologies.

Collaborations between the public and private sectors

a. Problem: Animal health is a very broad area with different diseases affecting different sectors, species, and regions. There is a lack of understanding of the opportunities for new approaches and in particular what is required for progress. There is a critical dearth of funding for animal health research and bias against non-hypothesis led approaches.

b. Potential Solutions:

- Establish priorities for research – specific diseases, general 'robustness.'
- Establish funding mechanisms that enable researchers to work with producers and provide access to large numbers of relevant animals.
- Plan to exploit disease problems at the commercial level including 'SWAT teams' to collect phenotypic information and appropriate samples (including environmental data).

c. Recommendations:

- Establish a cost/benefit analysis (the business case) of research approaches in order to support priorities and build funding initiatives.
- Develop integrated approaches that include the development of sample repositories and databases that support research networks
- Support the funding of innovations in animal genomics research specifically designed to solve problems in animal health

d. Next Steps: Build dialogue between researchers, industry, and funding agencies.

Intellectual property and technology transfer to industry

a. Problem: Too many groups are involved in 'protecting' IP before any is developed, inhibiting cooperation or blocking progress. Adversarial approaches to protection predominate. Possible competition between genetic and 'pharma' approaches may reduce investment.

b. Potential Solutions: Seek to develop approaches based on pre-competitive research.

Seek an 'open' shared approach to generating 'next generation' IP. Create 'incubator' opportunities to take results to application.

c. Recommendations: Seek to develop a culture that seeks faster application and focuses on protection of 'solutions' rather than concepts.

d. Next Steps: Continue to organize and provide incentives for public/private partnerships.

Impact on veterinary education and animal and poultry science

a. Problem: There is little dialogue between veterinarians and animal and poultry scientists particularly in relation to the genetics of animal health. The focus tends to be on disease challenge or in vitro approaches with limited numbers of animals or genetic variation.

b. Potential Solutions:

- Seek to stimulate multidisciplinary teaching across schools.
- Seek to stimulate joint research across disciplines.
- Seek to stimulate 'host-centered' approaches.

c. Recommendations: Develop more 'epidemiological' approaches to the genetics of animal health and integrate them into veterinary and animal science educational programmes. Bring experts on infectious diseases together with genomics/genetics groups addressing animal health.

d. Next Steps: Foster cross-fertilization between veterinary and animal science university programmes.

Contributions of veterinary practitioners, animal health professionals, and animal agriculture allied industries

a. Problem: There is a lack of a unified vision on the opportunity of genomics for animal health, a lack of understanding of actions required to investigate the genetics of animal health, and poor definition of new tools required to undertake the required research.

b. Potential Solutions:

- Develop a strategic vision for animal genomics for animal health setting out the roles of each group of industry professionals.
- Create a communication programme to stimulate the interest of, and dialogue between, these groups.
- Focus on solutions that work at the population level.

c. Recommendations:

- Establish a mechanism to assess the contribution of different genomic approaches (e.g., marker discovery, gene profiling, genetic variations associated with vaccine responses) to improving animal health.
- Establish demonstration projects to deliver examples of the value of genomics approaches.

d. Next Steps: Organize strategic workshops to seek stakeholder input and support.

Support from regulatory agencies to enable these new technologies

a. Problem: The regulatory environment for 'gene-based' technologies, e.g., genetic modification, cloning, RNA interference, and transgenics, is unclear and inhibits research (as well as commercialization).

b. Potential Solutions:

- Develop a science-based approach to 'technical' aspects of regulation.
- Utilize 'participatory' approaches to stimulating public debate and inform the regulatory process.

- Develop a regulatory system to deal with issues of public acceptability in parallel with the technical aspects.
- Benchmark other industries in relation to 'licensing' of new and/or advanced technology.

c. Recommendations: Establish a regulatory framework that enables the potential benefits of genome-enabled technologies to be established and provides a clear path for registering products derived from biotechnology.

d. Next Steps: OIE to seek to establish leadership in harmonizing approaches to the regulation of gene-based technology applied to animal health.

Conclusion

The availability of new animal genomics tools provides extraordinary opportunities for solving some of the most pressing and difficult problems in animal health research. Animal genomics also provides new innovative approaches for controlling diseases affecting domestic animals and wildlife. Quantitative genetics, functional genomics, translating genomics information into tools for controlling diseases, and integrating stakeholder support were identified as four priority areas for animal health research. Investing in basic research was also identified as critical for the sustainability and success of animal genomics research in animal health. The application of new genome-enabled technologies in animal health research has the potential of significantly improving the health and welfare of animals, resulting in improved efficiency, sustainability, biosecurity, and the protection of animals worldwide. This report provides a roadmap for moving forward by providing recommendations and concrete steps to achieve the impact that animal genomics offers animal health research.

Pinard M-H, Gay C, Pastoret P-P, Dodet B (eds): Animal Genomics for Animal Health. Dev Biol (Basel). Basel, Karger, 2008, vol 132, pp 425-427.

Key Recommendations

Quantitative Genetic Studies to Identify Markers of Health Traits

Problem areas:
- Availability and acquisition of appropriate and reliable health records;
- Availability and acquisition of animal pedigrees with defined phenotypes;
- Development and availability of genome associated research tools; and
- Refining quantitative population genetic studies to usable markers.

Recommendations:
1. Support research to define health traits at the molecular level with the aim of detecting genetic factors involved in disease susceptibility and disease resistance.
2. Support initiatives at every level (national, regional, private, and research project) that aim to proactively record defined health traits and collect samples from either commercial populations, well-characterized experimental animal lines, or where livestock and poultry are exposed to environmental or epizootic extremes.
3. Support the development of inter-operable databases for large-scale storage of pedigree information, phenotypes, and associated environmental factors.
4. Support national and international efforts that aim to sequence animal genomes, develop genome-enabled technologies, and enable their cost-effective use in research.
5. Support the building and integration of bioinformatics infrastructures in animal health research institutions, including the development of statistical models and software for animal genomics research.
6. Support research programmes that focus on the the selection of animals with superior resistance to the immunosuppressive effects of stress.
7. Support large-scale coordinated research programmes aimed at elucidating relevant aspects of host-pathogen interactions for high priority diseases.
8. Support meetings and workshops to enhance communication and facilitate collaborations between quantitative geneticists and experts in animal health.

Functional Genomics to Understand Host-Pathogen Interactions

Problem areas:
- Availability of validated animal disease models;
- Animals with defined disease susceptible/disease resistant phenotypes;

- Bioinformatics tools;
- Identification of genes that are involved in the activation/repression of key regulatory pathways;
- Development and availability of reagents to study gene function; and
- Discovery of biomarkers.

Recommendations:

1. Support the development of experimental animal challenge models that provide measurable phenotypes relevant to actual diseases under field conditions.
2. Support long-term projects that will build the infrastructure needed to conduct the extensive phenotyping and genotyping of animal populations relevant to animal health priorities.
3. Support research programmes that aim to identify genetic markers that correlate with disease phenotypes, and carry out whole genome expression analyses using tissue-specific microarrays to identify genome sequences that influence disease parameters.
4. Support functional genomics research programmes that aim to identify genetic markers that are linked to regulatory pathways that modulate host-pathogen interactions.
5. Support research programmes in relevant animal hosts that aim to confirm the association of genetic markers with host responses in large animal populations.
6. Support the establishment of core bioinformatics infrastructures with state-of-the-art genomic tools and resources that can integrate genotype, phenotype and functional genomics data for agricultural animal species and their respective priority diseases.
7. Support genome database initiatives for priority animal species and pathogens.
8. Develop proteomics tools to support animal health research projects that have the aim of identifying host molecules and biomarkers that are associated with or have a role in modulating host-pathogen interactions.
9. Establish an international consortium to advance and enable the development of reagents to study gene regulation, gene function, and disease pathways in priority animal species.
10. Support initiatives to integrate genetic, functional genomic, and bioinformatics data in a systems biology approach to understand complex diseases.

Translating Genomic Information to Tools for Controlling Diseases

Problem areas:

- Using animal genomes to understand genetic diversity and inform best management practices to improve animal health and animal production systems worldwide;
- Easy to use diagnostic platforms for genetic tests, forensic veterinary medicine, and traceability;
- Selection of animals with disease susceptible and resistant traits; and
- Application of genome-enabled technologies in drug and vaccine discovery, including transgenic animals that are resistant to disease and the selection of animals that respond effectively to drugs and vaccines.

Recommendations:

1. Support initiatives that will use animal genomic innovations to assess and manage domestic and wildlife animal genetic diversity.
2. Support studies that will lead to the development of rapid, accurate and cost-effective molecular genetic tests for host and pathogens to analyze field disease outbreaks and to develop predictive diagnostics.
3. Integrate disease resistance traits with existing multi-trait selection programmes.
4. Integrate genetic profiling in the development of disease models for drug and vaccine discovery research.
5. Support initiatives that integrate genomic approaches to vaccine development, including genetic and immune variations in target animal populations.
6. Support transgenic animal research programmes that target disease resistance.
7. Support research programmes that include collaborations between animal health research institutions, commercial breeding companies, and pharmaceutical companies.

Integrating Stakeholder Support to Advance Animal Genomics in Animal Health

Problem areas:

- Collaborations between the public and private sectors;
- Intellectual property and technology transfer to industry;
- Impact on veterinary education and animal and poultry science;
- Contributions of veterinary practitioners, animal health professionals, and animal agriculture allied industries;
- Support from regulatory agencies to enable these new technologies; and
- A lack of funding targeting the application of genomics in animal health research.

Recommendations:

1. Support the funding of innovations in animal genomics research specifically designed to solve problems in animal health.
2. Develop integrated approaches that include the development of sample repositories and databases that support research networks.
3. Develop more "epidemiological" approaches to the genetics of animal health and integrate them into veterinary and animal science educational programmes.
4. Establish and support demonstration projects to deliver concrete examples of the value of integrating genome-enabled technologies in animal health research programmes.
5. Establish a progressive international regulatory framework that is science-based and provides a clear path for registering animal disease control tools derived from genome-enabled technologies.
6. Support meetings to align veterinarians, animal scientists, and genomics/genetics experts to foster a dialogue and educate on the importance of multidisciplinary approaches addressing the application of genome-enabled technologies to animal health.

Pinard M-H, Gay C, Pastoret P-P, Dodet B (eds): Animal Genomics for Animal Health. Dev Biol (Basel). Basel, Karger, 2008, vol 132, pp 429-432.

Closing Remarks

B. Vallat

Director General, World Organisation for Animal Health (OIE), Paris, France

On behalf of the OIE Members ,I thank you for the honour of allowing me to draw the conclusions of your symposium.

The core mandate of the OIE, the World Organisation for Animal Health, is "to improve animal health worldwide." We were particularly proud to host this symposium which has provided the opportunity for leaders in the field of animal genomics and animal health to come together and discuss new technologies that are changing the way we approach animal health research. Recent advances in biotechnology and genomics present unique opportunities to address global animal health challenges through multidisciplinary scientific collaborations.

These collaborations will pave the way for integrating core competencies by connecting those who are currently engaged in cutting edge genomics research with animal health scientists who are disease experts and understand the real world challenges facing animal health.

The World Organisation for Animal Heath has always had a particular interest in biotechnology related to animal health and has established an *ad hoc* group of international experts in biotechnology in charge of advising the organisation. Some of them attended this symposium.

Biotechnology has already had a significant impact on animal health, mainly by improving available diagnostic tools and vaccines. One of the best examples is the use of a recombinant vaccinia-rabies vaccine in Europe and the United-States to eliminate terrestrial wildlife rabies with great success. Last May our organisation hosted an international conference jointly organised with the World Health Organisation, the International Association for Biological Standardisation and the European Union entitled "Towards the Elimination of Rabies in Eurasia," the ultimate goal being to eradicate terrestrial rabies worldwide.

The OIE expected a lot from your Symposium and it has not been disappointed. The recommendations emerging from your presentations and your discussions will be extremely useful and will help us to accomplish our mandates. We are particularly interested in research into the interaction between the genomes of pathogens and their hosts in order to dramatically improve our ability to control and eliminate diseases through the discovery of highly effective diagnostics, vaccines and biotherapeutics, and the promise of being able to select livestock with desirable health traits. This is essential for the correct implementation of our core mandate.

Vaccination, when available, is undoubtedly the most cost-effective means of preventing and controlling, and even eradicating infectious diseases, as exemplified by the foreseen eradication of rinderpest. In recent years vaccination has also been used for other purposes in animal production and welfare. In fact, the impact of vaccination goes far beyond the control of infectious diseases. Acting through natural mechanisms, vaccination of animals serves many different purposes, e.g. controlling animal infections and infestations, thus improving animal health and animal welfare and safeguarding our food supply; controlling anthropozoonoses and food pathogens, thereby protecting public health; solving problems associated with antibiotic and anthelmintic resistance; helping to leave food-producing animals free of chemical residues; controlling the reproduction of certain animal species; protecting the environment and biodiversity; and ensuring animal farming sustainability, thereby helping to alleviate poverty.

Vaccination will help to reach many of the objectives of the United Nations "Millennium Development Goals" especially in the light of the foreseen livestock revolution. We expect an increase of 50% in the world demand for meat in the next 15 years. One billion people will shift from poverty to the middle class, becoming meat consumers.

Last but not least, vaccination is presently the only available means to control viral vector-borne emerging diseases such as Bluetongue associated with serotype 8 in Northern Europe.

In light of this we were very interested in the presentation on the "Genetics of the Response to Vaccination." The OIE recently received a request from Codex Alimentarius to assess the safety of animal products derived from vaccinated animals, including with recombinant vaccines. We will soon organise an expert group that will work with our expert group on biotechnology. We hope that the recommendations will allow the development of new fields in vaccinology.

One approach could lie in the selection of animals that respond well to vaccination. An analysis of the immune response in different dog breeds to rabies vaccination, in the framework of the "pet scheme" for the introduction of properly vaccinated animals into the United Kingdom, showed considerable variation between breeds. Animal response to vaccination is thus subject to great variability, a factor which could be used in the selection, with or without markers, of good vaccination responders. It should, however, be noted that there is a risk of encountering the same sort of unexpected results as when selecting for resistance to specific diseases.

In fact, there is a growing trend to orient livestock selection not only toward production, but also toward animal health objectives by selecting animals resistant to certain diseases. However, this approach may be limited as it is difficult to imagine a return to the production of more "robust" animals while losing the economic and social advantages of previous selection. Genetic resistance must also be well defined: is it resistance to the infection or to disease transmission, or to the disease itself? For instance, the selection of sheep resistant to scrapie remains controversial because one may argue that the resistance is not complete but only partial and could hide the pathogen. Despite these potential drawbacks we have many things to learn from your research and we are eagerly looking forward to their applications. Like all other biological technologies, the selection of animals against some specific diseases will find its own niche whenever the benefit exceeds the cost.

Keeping this in mind, it is becoming urgent to maintain the biodiversity of domestic animals and the OIE fully supports the initiative of its sister organisation, the Food and Agriculture Organisation of the United Nations to promote and preserve the diversity of domestic breeds as recommended by the recent Interlaken Conference entitled "From the State of the World's Animal Genetic Resources for Food and Agriculture to the Global Plan of Action for Animal Genetic Resources". Some endangered breeds are our main resource for mining resistance genes. In this regard, there should be public financial support for keeping endangered breeds alive both in developing and developed countries.

We must also provide the research community with phenotypically well characterized clinical cases to support their genomics work and accurate clinical data resulting from either field or experimental studies. As far as vaccination or chemotherapy is concerned, there should be good collaboration between the public and the private sectors to share the necessary and available information. We learned about the importance of epigenetic control of gene expression, which is also of great interest and opens a new, fascinating, and particularly complex area. We also learned that there are two individuals, James Watson and Craig Venter who had their genome entirely sequenced in 2007. The original human genome project took 13 years and cost 3 billion dollars, but they were able to do it in just two months for a mere 1 million dollars for each individual. This exemplifies the speed with which new technologies can impact your research area.

Our organisation is also interested in the impact of your work on animal welfare. First of all, whenever a disease is prevented, it has a direct positive impact on animal welfare, animal health being a key component of animal welfare. Moreover there may be other applications of animal genomics in animal welfare. For instance in the framework of the application of the three Rs rules (Replacement, Reduction, Refinement) in animal experimentation, an American group has recently reported the development of painless transgenic strains of mice by inserting a mutated human gene. This is not without ethical implications. Pain has always been a necessary physiological mechanism under strong evolutionary pressure.

To conclude, I want to thank all the members of the Steering Committee of this symposium, the scientific Committee and all the speakers and chairpersons. We look forward to the publication of the symposium proceedings, and especially the results of the questionnaire, which will help map the way forward for animal genomics and animal health. We would also recommend that you consider a follow-up meeting in a couple of years to continue to pave the way forward.

We are at a turning point. Further improvements in animal health and welfare will notably depend on your multidisciplinary approach. I wish you all great success in your endeavours.

Index of Authors

Developments in Biological Standardization

Vol. 36	Washington DC, 1976	Freeze-Drying of Biological Products
Vol. 37	Geneva, 1976	Standardization of Cell Substrates for the Production of Virus Vaccines
Vol. 38	London, 1977	Biological Preparations in the Treatment of Cancer
Vol. 39	Geneva, 1977	Influenza Immunization
Vol. 40	Marburg, 1977	Standardization of Rabies Vaccines for Human Use Produced in Tissue Culture (Rabies III)
Vol. 41	Guadeloupe, 1978	Vaccines in the Developing Countries
Vol. 42	Paris, 1978	Second General Meeting of ESACT
Vol. 43	Brussels, 1978	Immunization: Benefit Versus Risk Factors
Vol. 44	Villars, 1979	Test methods for the Quality Control of Plasma Proteins
Vol. 45	San Antonio, 1979	The Standardization of Animals to Improve Biomedical Research, Production and Control
Vol. 46	Oxford, 1979	Third General Meeting of ESACT
Vol. 47	Bilthoven, 1980	Reassessment of Inactivated Poliomyelitis Vaccine
Vol. 48	Geneva, 1980	Standardization of Albumin, Plasma Substitutes and Plasmapheresis
Vol. 49	Leetown, 1981	Fish Biologics: Serodiagnostics and Vaccines
Vol. 50	Heidelberg, 1981	The Use of Heteroploid and Other Cell Substrates for the Production of Biologicals
Vol. 51	Lyons, 1981	Immunization of Adult Birds with Inactivated Oil Adjuvant Vaccines
Vol. 52	Lyons, 1981	Herpes Virus of Man and Animal: Standardization of Immunological Procedures
Vol. 53	Dublin, 1982	Enteric Infections in Man and Animals: Standardization of Immunological Procedures
Vol. 54	Athens, 1982	Viral Hepatitis: Standardization in Immunoprophylaxis of Infections by Hepatitis Viruses
Vol. 55	Copenhagen, 1982	Fifth General Meeting of ESCAT
Vol. 56	Algiers, 1983	Third International Symposium on Brucellosis
Vol. 57	Paris, 1983	Monoclonal Antibodies: Standardization of their Characterization and Use
Vol. 58	Budapest, 1983	International Symposium on BCG Vaccines and Tuberculins
Vol. 59	Geneva, 1983	Standardization and Control of Biologicals Produced by Recombinant DNA Technology
Vol. 60	Gardone Riviera, 1984	Production and Exploitation of Existing and New Animal Cell Substrates
Vol. 61	Geneva, 1984	Proceedings of the Fourth International Symposium on Pertussis
Vol. 62	Stockholm, 1985	Diagnostics and Vaccines for Parasitic Diseases
Vol. 63	San Francisco, 1985	Use and Standardization of Chemically Defined Antigens
Vol. 64	London, 1985	Reduction of Animal Usage in the Development and Control of Biological Products
Vol. 65	Amsterdam, 1985	Use and Standardization of Combined Vaccines

Vol. 66	Vienna, 1985	Advances in Animal Cell Technology: Cell Engineering, Evaluation and Exploitation
Vol. 67	Melbourne, 1986	Standardization in Blood Fractionation Including Coagulation Factors
Vol. 68	Geneva, 1986	Cells - Products - Safety
Vol. 69	London, 1987	Cytokines Laboratory and Clinical Evaluation
Vol. 70	Arlington, VA, 1988	Continuous Cell Lines as Substrates for Biologicals
Vol. 71	Utrecht, 1989	Monoclonal Antibodies for Therapy, Prevention and In vivo Diagnosis of Human Disease
Vol. 72	Annecy, 1989	Progress in Animal Retroviruses
Vol. 73	Shizuoka, 1990	Pertussis: Evaluation and Research on Acellular Pertussis Vaccines
Vol. 74	Bethesda, MD, 1990	Biological Product Freeze-Drying and Formulation
Vol. 75	London, 1990	Virological Aspects of the Safety of Biological Products
Vol. 76	Bethesda, MD, 1991	Continuous Cell Lines - An International Workshop on Current Issues
Vol. 77	Annecy, 1991	Standardization of the Immunopharmacology of Natural and Synthetic Immunomodulators
Vol. 78	Bethesda, MD, 1991	Poliovirus Attenuation: Molecular Mechanisms and Practical Aspects
Vol. 79	Ploufragan, 1992	The First Steps Towards an International Harmonization of Veterinary Biologicals: 1993 and Free Circulation of Vaccines within the EEC
Vol. 80	Heidelberg, 1992	Transmissible Spongiform Encephalopathies - Impact on Animal and Human Health
Vol. 81	Cannes, 1992	Virological Safety Aspects of Plasma Derivatives
Vol. 82	Albany NY, 1993	Recombinant Vectors in Vaccine Development
Vol. 83	Annecy, 1993	Genetic Stability and Recombinant Product Consistency
Vol. 84	Langen, 1993	Non-Target Effects of Live Vaccines
Vol. 85	Santa Fé, NM, 1994	Genetics of Streptococci, Enterococci and Lactococci
Vol. 86	Langen, 1994	Replacement, Reduction and Refinement of Animal Experiments in the Development and Control of Biological Products
Vol. 87	Geneva, 1995	New Approaches to Stabilisation of Vaccines Potency
Vol. 88	Bethesda, MD, 1995	Viral Safety and Evaluation of Viral Clearance from Biopharmaceutical Products
Vol. 89	Rome, 1995	Pertussis Vaccine Trials
Vol. 90	Oslo, 1996	Fish Vaccinology
Vol. 91	San Francisco, CA, 1996	Development of Specifications for Biotechnology Pharmaceuticals Products
Vol. 92	Bergen, 1996	Modulation of the Immune Response to Vaccine Antigens
Vol. 93	Annecy, 1996	Safety of Biological Products Prepared from Mammalian Cell Culture
Vol. 94	Bethesda, MD, 1997	Simian Virus 40 (SV40): A Possible Human Polyomavirus
Vol. 95	Paris, 1997	Preclinical and Clinical Development of New Vaccines
Vol. 96	Washington, DC, 1995	Characterization of Biotechnology Pharmaceutical Products
Vol. 97	London, 1997	Biological Characterization and Assay of Cytokines and Growth Factors
Vol. 98	London, 1997	Inactivated Influenza Vaccines Prepared in Cell Culture
Vol. 99	Strasbourg, 1998	Animal Sera, Animal Sera Derivatives and Substitutes Used in the Manufacture of Pharmaceuticals: Viral Safety and Regulatory Aspects
Vol. 100	Geneva, 1998	A Celebration of 50 Years of Progress in Biological Standardization and Control at WHO
Vol. 101	London, 1998	Alternatives to Animals in the Development and Control of Biological Products for Human and Veterinary Use

Developments in Biologicals

Vol. 102	San Francisco, CA, 1999	Advances in Transfusion Safety
Vol. 103	Annecy, 1999	Physico-Chemical Procedures for the Characterization of Vaccines
Vol. 104	Langen, 1999	Development and Clinical Progress of DNA Vaccines
Vol. 105	Paris, 2000	Progress in Polio Eradication: Vaccine Strategies for the End Game
Vol. 106	Rockville, MD, 2000	Evolving Scientific and Regulatory Perspectives on Cell Substrates for Vaccine Development
Vol. 107	London, 2000	The Design and Analysis of Potency Assays for Biotechnology Products
Vol. 108	Langen, 2001	Advances in Transfusion Safety
Vol. 109	Washington, DC, 2000	Biologics 2000 – Comparability of Biotechnology Products
Vol. 110	Palm Cove, 2001	Orphan Vaccines – Bridging the Gap
Vol. 111	Utrecht, 2001	Advancing Science and Elimination of the Use of Laboratory Animals for Development and Control of Vaccines and Hormones
Vol. 112	Bethesda, MD, 2001	Immunogenicity of Therapeutic Biological Products
Vol. 113	Berlin, 2001	Process Validation for Manufacturing of Biologics and Biotechnology Products
Vol. 114	Ames, IA, 2002	Vaccines for OIE List A and Emerging Animal Diseases
Vol. 115	Bergen, 2002	Laboratory Correlates of Immunity to Influenza – A Reassessment,
Vol. 116	Los Angeles, CA, 2003	Development of Therapeutic Cancer Vaccines
Vol. 117	Langen, 2003	Consideration of Alternative Licensing Procedures for Vaccines for Minor Species, Minor Indications and Autogenous/Autologous Products
Vol. 118	Langen, 2003	PDA/EMEA European Virus Safety Forum
Vol. 119	Buenos Aires, 2004	Control of Infectious Animal Diseases by Vaccination
Vol. 120	Bethesda, MD, 2003	Advances in Transfusion Safety
Vol. 121	Bergen, 2003	Progress in Fish Vaccinology
Vol. 122	Bethesda, MD, 2003	State of the Art Analytical Methods for Characterization of Biological Products and Assessment of Comparability
Vol. 123	Rockville, MD, 2004	Vaccine Cell Substrates 2004
Vol. 124	Paris, France, 2005	OIE/FAO International Scientific Conference on Avian Influenza
Vol. 125	Kiev, Ukraine, 2005	First International Conference on Rabies in Europe
Vol. 126	Saint-Malo, France, 2005	New Diagnostic Technology: Applications in Animal Health and Biologics Controls
Vol. 127	Sydney, Australia, 2005	Advances in Transfusion Safety – Volume IV
Vol. 128	Florianopolis, Brazil, 2006	First International Conference of the OIE Reference Laboratories and Collaborating Centres
Vol. 129	Bergen, Norway, 2006	The OIE Global Conference on Aquatic Animal Health
Vol. 130	Verona, Italy, 2007	Vaccination: A Tool for the Control of Avian Influenza
Vol. 131	Paris, France, 2007	Towards the Elimination of Rabies in Eurasia